Fundamentals of Electromagnetics for Engineering

Nannapaneni Narayana Rao

Edward C. Jordan Professor Emeritus of Electrical and Computer Engineering
University of Illinois at Urbana–Champaign, USA

Distinguished Amrita Professor of Engineering
Amrita Vishwa Vidyapeetham (Amrita University), India

The author and publisher of this book have used their best efforts in preparing this book. These e. include the development, research, and testing of the theories and programs to determine their effectiveness. The author and publisher make no warranty of any kind, expressed or implied, with regard to these programs or the documentation contained in this book. The author and publisher shall not be liable in any event for incidental or consequential damages in connection with, or arising out of, the furnishing performance, or use of these programs.

About the Front Cover Image: Figure 7.14 (right), application of smith Chart in transmission line analysis and its mirror image (left), inside the circle of figure 9.5 (c), the field intensity radiation pattern of the Hertzian dipole antenna, superimposed on a historic depiction of the two hemispheres of the globe, reflecting the spirit of the dedication.

ISBN 978-81-317-2415-6

First Impression, 2009
Fourth Impression, 2010
Fifth Impression, 2011

This edition is manufactured in India and is authorized for sale only in India, Bangladesh, Bhutan, Pakistan, Nepal, Sri Lanka and the Maldives. Circulation of this edition outside of these territories is UNAUTHORIZED.

Published by Dorling Kindersley (India) Pvt. Ltd., licensees of Pearson Education in South Asia.

Head Office: 7th Floor, Knowledge Boulevard, A-8 (A), Sector-62, Noida 201309, UP, India.
Registered Office: 11 Community Centre, Panchsheel Park, New Delhi 110 017, India.

Printed in India by Chennai Micro Print.

"Fill your heart with love
and express it in everything you do."
—Amma Mata Amritanandamayi Devi,
Chancellor, Amrita Vishwa Vidyapeetham

To students all over the world,
I offer to you this book on Electromagnetics,
the "Mother of Electrical and Computer Engineering,"
with the spirit of the above message from Amma,
the "Mother of Compassion!"

Contents

Preface

"... I am talking about the areas of science and learning that have been at the heart of what we know and what we do, that which has supported and guided us and which is fundamental to our thinking. It is electromagnetism in all its many forms that has been so basic, that haunts us and guides us. . . ."

> —Nick Holonyak, Jr., the John Bardeen Endowed Chair Professor of Electrical and Computer Engineering and Physics at the University of Illinois at Urbana–Champaign, and the inventor of the semiconductor visible LED, laser, and quantum-well laser

"The electromagnetic theory, as we know it, is surely one of the supreme accomplishments of the human intellect, reason enough to study it. But its usefulness in science and engineering makes it an indispensable tool in virtually any area of technology or physical research."

> —George W. Swenson, Jr., Professor Emeritus of Electrical and Computer Engineering, University of Illinois at Urbana–Champaign

The above quotes from two of my distinguished colleagues at the University of Illinois underscore the fact that electromagnetics is all around us. In simple terms, every time we turn on a switch for electrical power or for electronic equipment, every time we press a key on our computer keyboard or on our cell phone, or every time we perform a similar action involving an everyday electrical device, electromagnetics comes into play. It is the foundation for the technologies of electrical and computer engineering, spanning the entire electromagnetic spectrum, from d.c. to light. As such, in the context of engineering education, it is fundamental to the study of electrical and computer engineering. While the fundamentals of electromagnetic fields remain the same, the manner in which they are taught may change with the passing of time owing to the requirements of the curricula and shifting emphasis of treatment of the fundamental concepts with the evolution of the technologies of electrical and computer engineering.

Three decades ago, I wrote a one-semester textbook, the first edition of *Elements of Engineering Electromagnetics*, dictated solely by the reduction in the curricular requirement in electromagnetics at the University of Illinois from a three-semester required sequence to a one-semester course, owing to the pressure of increasing areas of interest and fewer required courses. The approach used for the one-semester book was to deviate from the historical treatment and base it upon dynamic fields and their engineering applications, in view of the student's earlier exposure in engineering physics to the traditional approach of static fields and culminating in Maxwell's equations. Less than ten years after that, a relaxation of the curricular requirements coupled with the advent of the PC resulted in an expanded second edition of the book for two-semester usage. Subsequent editions have essentially followed the second edition.

Interestingly, the approach that broke with the tradition with the first edition has become increasingly relevant from a different context, because with the evolution of the technologies of electrical and computer engineering over time, the understanding of the fundamental concepts in electromagnetics based on dynamic fields has become increasingly important. Another feature of the first edition of *Elements of Engineering Electromagnetics* was the treatment of the bulk of the material through the use of the Cartesian coordinate system. This was relaxed in the subsequent editions, primarily because of the availability of space for including examples involving the geometries of cylindrical and spherical coordinate systems, although the inclusion of these examples is not essential to the understanding of the fundamental concepts.

This book, which is a one-semester textbook, combines the features of the first edition of *Elements of Engineering Electromagnetics* with the treatment of the fundamental concepts in keeping with the evolution of technologies of electrical and computer engineering. Specifically, the approach of beginning with Maxwell's equations to introduce the fundamental concepts is combined with the treatment of the different categories of fields as solutions to Maxwell's equations and using the thread of statics-quasistatics-waves to bring out the frequency behavior of physical structures. Thus, some of the salient features of the first nine chapters of the book consist of the following:

1. Using the Cartesian coordinate system for the bulk of the material to keep the geometry simple and yet sufficient to learn the physical concepts and mathematical tools, while employing other coordinate systems where necessary
2. Introducing Maxwell's equations for time-varying fields first in integral form and then in differential form early in the book
3. Introducing uniform plane wave propagation by obtaining the field solution to the infinite plane current sheet of uniform sinusoidally time-varying density
4. Introducing material media by considering their interaction with uniform plane wave fields
5. Using the thread of statics-quasistatics-waves to bring out the frequency behavior of physical structures, leading to the development of the transmission line and the distributed circuit concept
6. Covering the essentials of transmission-line analysis both in frequency domain and time domain in one chapter
7. Introducing metallic waveguides by considering the superposition of obliquely propagating uniform plane waves and dielectric waveguides following the discussion of reflection and refraction of plane waves
8. Obtaining the complete solution to the Hertzian dipole fields through a successive extension of the quasistatic field solution so as to satisfy simultaneously the two Maxwell's equations, and then developing the basic concepts of antennas

The final chapter is devoted to six supplementary topics, each based on one or more of the previous six chapters. It is intended that the instructor will choose one or more of these topics for discussion following the corresponding previous chapter(s). Material on cylindrical and spherical coordinate systems is presented in appendices so that it can be studied either immediately following the discussion of the corresponding material on the Cartesian coordinate system or only when necessary.

From considerations of varying degrees of background preparation at different schools, a greater amount of material than can be covered in an average class of three semester-hour credits is included in the book. Worked-out examples are distributed throughout the text, and in some cases, extend the various concepts. Summary of the material and a number of review questions are included for each chapter to facilitate review of the chapters.

I wish to express my gratitude to the numerous colleagues at the University of Illinois at Urbana–Champaign (UIUC) who have taught from my books over a period of 35 years, beginning with my first book in 1972, and to the numerous users of my books worldwide. Technological advances in which electromagnetics continues to play a major role have brought changes in this span of time beginning with the introduction of the computer engineering curriculum in my department at UIUC in 1972, followed by the name change of the department from electrical engineering to electrical and computer engineering in 1984, to transforming the way of life in the present-day world from "local" to "global."

The title of this book is a recognition of the continuing importance of a core course in electromagnetics in both electrical engineering and computer engineering curricula, in this high-speed era. My joint affiliation with UIUC, my "home" institution in the United States in the West, and Amrita Vishwa Vidyapeetham in my "homeland" of India in the East is a gratifying happening owing to the state of the world that, with the transformation from "local" to "global," East is no longer just East, and West is no longer just West, and the twain have met!

N. NARAYANA RAO

About the Author

Nannapaneni Narayana Rao was born in Kakumanu, Guntur District, Andhra Pradesh, India. Prior to coming to the United States in 1958, he attended high schools in Pedanandipadu and Nidubrolu; the Presidency College, Madras (now known as Chennai); and the Madras Institute of Technology, Chromepet. He completed high school in Nidubrolu in 1947, and received the B.Sc. degree in Physics from the University of Madras in 1952 and the Diploma in Electronics from the Madras Institute of Technology in 1955. In the United States, he attended the University of Washington, receiving the M.S. and Ph.D. degrees in Electrical Engineering in 1960 and 1965, respectively. In 1965, he joined the faculty of the Department of Electrical Engineering, now the Department of Electrical and Computer Engineering, at the University of Illinois at Urbana–Champaign (UIUC), Urbana, Illinois, and served on the faculty of that department until 2007.

Professor Rao retired from UIUC in 2007 as Edward C. Jordan Professor of Electrical and Computer Engineering, to which he was named to be the first recipient in 2003. The professorship was created to honor the memory of Professor Jordan, who served as department head for 25 years, and to be held by a "member of the faculty of the department who has demonstrated the qualities of Professor Jordan and whose work would best honor the legacy of Professor Jordan." During the 42 tears of tenure at the University of Illinois, Professor Rao was engaged in research, teaching, administration, and international activities.

Professor Rao's research focused on ionospheric propagation. In his teaching, he taught a wide variety of courses in electrical engineering. He developed courses in electromagnetic fields and wave propagation, and has published undergraduate textbooks: *Basic Electromagnetics with Applications* (Prentice-Hall, 1972), six editions of *Elements of Engineering Electromagnetics* (Prentice-Hall, 1977, 1987, 1991, 1994, 2000, and 2004), and a special Indian Edition of the sixth edition of *Elements of Engineering Electromagnetics* (Pearson Education, 2006). In administration, he served as Associate Head of the Department for Instructional and Graduate Affairs for 19 years, from 1987 to 2006.

Professor Rao has received numerous awards and honors for his teaching and curricular activities. These include the first Award in Engineering in 1983 from the Telugu Association of North America (TANA), an association of Telugu-speaking people of origin in the State of Andhra Pradesh, India, with the citation, "Dedicated teacher and outstanding contributor to electromagnetics"; a plaque of highest appreciation from the

Faculty of Technology, University of Indonesia, Jakarta, Indonesia, for curriculum development in 1985–1986; the Campus Undergraduate Instructional Awards in 1982 and 1988, the Everitt Award for Teaching Excellence from the College of Engineering in 1987, the Campus Award for Teaching Excellence and the first Oakley Award for Innovation in Instruction in 1989, and the Halliburton Award for Engineering Education Leadership from the College of Engineering in 1991, all at the University of Illinois at Urbana–Champaign; election to Fellow of the IEEE (Institute of Electrical and Electronics Engineers) in 1989 for contributions to electrical engineering education and ionospheric propagation; the AT&T Foundation Award for Excellence in Instruction of Engineering Students from the Illinois–Indiana Section of the ASEE (American Society for Engineering Education) in 1991; the ASEE Centennial Certificate in 1993 for exceptional contribution to the ASEE and the profession of engineering; the IEEE Technical Field Award in Undergraduate Teaching in 1994 with the citation, "For inspirational teaching of undergraduate students and the development of innovative instructional materials for teaching courses in electromagnetics"; and the Excellence in Education Award from TANA in 1999. He is a Life Fellow of the IEEE and a Life Member of the ASEE.

Professor Rao has been active internationally in engineering education. He was involved in institutional development at the University of Indonesia in Jakarta during 1985–1986. In summer 2006, he offered the first course on the EDUSAT satellite network from the Amrita Vishwa Vidyapeetham (Amrita University) in Ettimadai, Coimbatore, Tamil Nadu, India, under the Indo-U.S. Interuniversity Collaborative Initiative in Higher Education and Research. In October 2006, Amrita University named Professor Rao as its first Distinguished Amrita Professor.

Professor Rao will be continuing his academic activities, as Edward C. Jordan Professor Emeritus of Electrical and Computer Engineering at the University of Illinois and Distinguished Amrita Professor of Engineering at Amrita University.

Gratitude and "Grattitude"

I came to the United States 50 years ago in 1958 with $50, a passport from my mother-land, India, and undergraduate education in my then-technical field of electronics from the Madras Institute of Technology in India. I received my Ph.D. in electrical engineering from the University of Washington and joined what is now the Department of Electrical and Computer Engineering (ECE) at the University of Illinois at Urbana–Champaign (UIUC) in 1965, attracted by the then-department head, Edward C. Jordan, who brought the department to national and international fame as its head for 25 years from 1954 to 1979. After 42 years of tenure in this department, I retired, effective June 1, 2007, as the Edward C. Jordan Professor Emeritus of Electrical and Computer Engineering.

In recent years, I have been engaged in engineering education in India. In December 2005, I got connected to the "Hugging Saint," and "Mother of Compassion," the humanitarian and spiritual leader Amma Mata Amritanandamayi Devi, Chancellor of Amrita Vishwa Vidyapeetham (Amrita University), popularly known as "Amma," meaning "Mother," all over the world. Since then, I have been involved with Amrita University, where I now have the position of Distinguished Amrita Professor of Engineering, offered to me in October 2006. My involvement with Amrita began in a special way, as the first faculty member from the United States teaching from the Amrita campus in Ettimadai, Coimbatore, Tamil Nadu, to students at remote locations on the interactive satellite E-learning Network, under the Indo-U.S. Inter-University Collaborative Initiative in Higher Education and Research, in summer 2006.

I am grateful to many individuals, beginning with my late parents, and for many things. I came with the solid foundation laid at my alma mater in India and acquired more education at my alma mater in the United States and prospered in my profession at Illinois. For all of this, I am grateful to my two Lands, the land of my birth, India, for the foundation, and the land of my work, America, for the prosperity. I am grateful to Amma Mata Amritanandamayi Devi for attracting me to Amrita University, thereby giving me the opportunity for "serving the needs of students of various parts of the world," in the words of former President of India, Bharat Ratna, Dr. A. P. J. Abdul Kalam, with this book, bearing my joint affiliation with Illinois and Amrita.

In the words of the late Gurudeva Sivaya Subramuniyaswami of the Kauai Aadheenam, Kauai, Hawaii: "Gratitude and appreciation are the key virtues for a better life. They are the spell that is cast to dissolve hatred, hurt and sadness, the medicine which heals the subjective states of mind, restoring self-respect, confidence, and security." I am grateful that I am the author of this book and its predecessor books, over the span of more than 35 years, for introducing electromagnetic theory, commonly known as electromagnetics (EM), to students all over the world. Here, I would like to reconstruct the trail of this gratitude beginning in the 1950s.

One day during the academic year 1957–1958, I had the pleasure of having afternoon refreshments with William L. Everitt in the dining hall of the Madras Institute of Technology (MIT), Chromepet, along with some others in the electronics faculty of MIT. William L. Everitt was then the dean of the College of Engineering at the University of Illinois, Urbana, as it was then known. Dean Everitt was visiting India because the University of Illinois was assisting with the development of IIT (Indian Institute of Technology), Kharagpur, the first of the IITs. Dean Everitt came to Madras (presently Chennai) at the invitation of William Ryland Hill, who was the visiting head of the electronics faculty of MIT during that one year, on leave from the University of Washington in Seattle, Washington.

I happened to be on the staff of the electronics faculty then, having completed my diploma in electronics after three years of study during 1952–1955 and six months of practical training, following my B.Sc. (Physics) from the University of Madras, having attended the Presidency College. One of the subjects I studied at MIT was electromagnetic theory, from the book *Electromagnetic Waves and Radiating Systems*, by Edward C. Jordan, who was then the head of the Department of Electrical Engineering at the University of Illinois. I can only say that my learning of electromagnetic theory at that time was hazy at best, no reflection on Jordan's book.

While I was a student at MIT, one of our great lecturers, by the name of S. D. Mani, was leaving to take a new job in Delhi, for which we gave him a send-off party. After the send-off party, we all went to the Chromepet Railway Station adjacent to the Institute to bid a final goodbye to him on the platform. While on the platform waiting for the electric train to arrive from the neighboring station, Tambaram, he specifically called to me and said, "Narayana Rao, someday you will become the president of a company!"

Contrary to what S. D. Mani said, with his great characteristic style, I did not go on to even work in a company. Instead, William Ryland Hill "took" me to the EE Department at the University of Washington in 1958, then chaired by Austin V. Eastman, a contemporary of Edward Jordan. There, I pursued my graduate study in electrical engineering and received my Ph.D. in 1965, with Howard Myron Swarm as my advisor, in the area of ionospheric physics and propagation, and taking courses from Akira Ishimaru, among others. Eastman gave me the opportunity of teaching courses just like a faculty member, as an instructor, because of my teaching experience at MIT, and the good word of Ryland Hill. That was when I fell in love with the teaching of "transmission lines," from the electromagnetics aspect, which then extended beyond transmission lines and later led to the writing of my books.

Never did I envision during those years that in 1965, after completing my Ph.D. at the University of Washington, I would become a faculty member and be writing my books in the Jordan-built Department of Electrical and Computer Engineering (as it is now called) in the Everitt-built College of Engineering at the University of Illinois at Urbana-Champaign, as it is now known. Never did I envision that I would spend my entire professional career since 1965 in the hallowed halls of the William L. Everitt Laboratory of Electrical and Computer Engineering, which I call the "Temple of Electrical and Computer Engineering," along with personalities such as distinguished colleagues Nick Holonyak, Jr., and George W. Swenson, Jr. Never did I envision that not only would I be writing books for teaching electromagnetics, following the tradition of Jordan, but also would be holding a professorship, and now an emeritus professorship, bearing his name.

I believe that gratitude is something you can neither express adequately in words nor demonstrate adequately in deeds. Nevertheless, I have tried on certain occasions to express it in words, and demonstrate it in deeds, which I would like to share with you here:

To my alma mater, the Madras Institute of Technology, on the occasion of the Institute Day on February 26, 2004, in the presence of the then-Governor of Tamil Nadu, Sri P. S. Ramamohan Rao, a classmate of mine while in Presidency College, for presenting the sixth edition of my book, *Elements of Engineering Electromagnetics:*

> *So, Madras Institute of Technology, my dear alma mater*
> *Where I went to school fifty years ago this year*
> *Today I present to you this historic volume*
> *The product of the work of my lifetime*
> *For which fifty years ago you laid the foundation*
> *That I cherished all these years with much appreciation*
> *Please accept this book as a token of my utmost gratitude*
> *Which I offer to you in the spirit of "Revere the preceptor as God"*
> *Hopefully I will be back with Edition No. 7*
> *To express my gratitude to you again in 2007!*

And I did go back to my alma mater in January 2007, not to present Edition No. 7, but rather a special Indian Edition of Edition No. 6, which could be considered as Edition No. 7!

At the conclusion of the response speech on the occasion of my investiture as the Edward C. Jordan Professor of Electrical and Computer Engineering, on April 14, 2004:

> *To Edward C. Jordan, the "father" of my department*
> *Fifty years ago, I may have studied EM from your book with much bewilderment*
> *But today, I offer to you this book on EM which I wrote with much excitement*
> *In appreciation of your profound influence on my professional advancement.*

To my alma mater, the EE Department at the University of Washington, giving the keynote speech and presenting the sixth edition of *Elements of Engineering*

Electromagnetics, at the kick-off event for the Centennial Celebration of the Department on April 28, 2006:

> *To the EE Department at the University of Washington*
> *From this grateful alumnus who received from you his graduate education*
> *Not just graduate education but seven years of solid academic foundation*
> *For my successful career at the University of Illinois at Urbana–Champaign*
> *During which I have written six editions of this book on electromagnetics*
> *Besides engaging in the variety of all the other academic activities*
> *I present to you this book with utmost appreciation*
> *On the occasion of your centennial celebration!*

And when you are grateful in life, things continue to happen to you to allow you to be even more grateful. Even as late as November 2005, I did not envision that I would become connected to Amrita University of Amma Mata Amritanandamayi Devi. The opportunity came about as a consequence of the signing of a memorandum of understanding (MOU) in December 2005 between a number of U.S. Universities, including UIUC and the University of Washington, and Amrita University in partnership with the Indian Space Research Organization (ISRO) and the Department of Science and Technology of the Government of India. The MOU had to do with an initiative, known as the Indo-U.S. Inter-University Collaborative Initiative in Higher Education and Research, and allowed for faculty from the United States to offer courses for e-learning on the ISRO's EDUSAT Satellite Network and to pursue collaborative research with India. The Initiative was launched by the then President of India, Bharat Ratna, A. P. J. Abdul Kalam, from New Delhi on the EDUSAT Satellite Network on December 8, 2005.

A delegation from the United States went to India on this occasion, and following the launching ceremony at Ettimadai, Coimbatore, Tamil Nadu, where the main Amrita campus is located, the delegation went to Amritapuri in the state of Kerala to meet with Amma on December 9. That was when I got connected to Amma, and things began to happen. Within the next year, I became the first professor to offer a course on the EDUSAT Satellite Network—a 5-week course in summer 2006, entitled "Electromagnetics for Electrical and Computer Engineering," in memory of Edward C. Jordan, using as the textbook a special Indian Edition of *Elements of Engineering Electromagnetics, Sixth Edition*, published in this connection by Pearson Education and containing a message by former President Abdul Kalam, forewords by UIUC Chancellor Richard Herman, UIUC Provost Linda Katehi, and ECE Professor Nick Holonyak, Jr., and an introductory chapter called "Why Study Electromagnetics?" offering 18 very thoughtful responses to that question, most of them provided by UIUC ECE faculty members.

So, I did not become the "president" of a company, as S. D. Mani proclaimed on the platform of the Chromepet Railway Station. Instead, I went on to become a "resident" of the William L. Everitt Laboratory of Electrical and Computer Engineering, the "Temple of Electrical and Computer Engineering,"—the crown jewel of the campus that provided education to numerous presidents of companies—located at the northeast corner of the intersection of Wright and Green Streets in Urbana, Illinois, on the Campus of the University of Illinois at Urbana–Champaign!

And from the "Temple of Electrical and Computer Engineering" in Urbana, shown above, my gratitude took me to my motherland, halfway around the world, as an "IndiAmerican," a word that I coined implying that the "Indian" and the "American" are inseparable, and which inspired former President Abdul Kalam. There, I reached the destination in my journey at Amma Mata Amritanandamayi Devi's Amrita Vishwa Vidyapeetham, where I got connected to the "young minds" of my motherland, shown in the picture below, along with some staff and my wife and our daughter, taken on August 11, 2006, the last day of the class in front of the beautiful main building of the campus.

I have read somewhere that destination is a journey and not a success in itself. And therefore, the journey began at Amrita and is continuing! As though for this purpose and owing to a combination of circumstances, I became the first Distinguished Amrita Professor of Engineering in October 2006, at which time I decided to write this book, and hence began working on it while at Amrita in Ettimadai. Subsequently, I retired from UIUC effective June 1, 2007, becoming the Edward C. Jordan Professor Emeritus of Electrical and Computer Engineering, so that my journey is now continuing as Jordan Professor Emeritus from Illinois and Distinguished Amrita Professor from Amrita, wherever I am in this global world.

I always believed in the power of education—transcending the boundaries of national origin, race, and religion—to assure the future of the world. Throughout my life, I have been involved in education, as a student, professor, researcher, teacher, author, and administrator. The sheer enjoyment of my work led me to coining the word "grattitude," in 2005, in answer to people wondering if I would ever retire from my job at Illinois. "Grattitude" is a word combining "gratitude" and "attitude," and meaning an "attitude of gratitude." In my journey, I feel grattitude for the opportunity I have been given to help facilitate the education of the wonderful youth from countries all over the world, through my books, teaching, and international activities. I have learned that engaging in an activity with "grattitude" yields immediate enjoyment. I conclude this story of "gratitude and grattitude" with the following poem:

To the students from all around the world
And to the students all over the world
EMpowered by the Jordan name
And inspired by the Amrita name
I offer to you this book on EM
Beginning with this poem which I call PoEM
If you are wondering why you should study EM
Let me tell you about it by means of this PoEM
First you should know that the beauty of EM
Lies in the nature of its compact formalism
Through a set of four wonderful EMantras
Familiarly known as Maxwell's equations
They might be like mere four lines of mathematics to you
But in them lie a wealth of phenomena that surround you
Based on them are numerous devices
That provide you everyday services
Without the principles of Maxwell's equations
Surely we would all have been in the dark ages
Because there would be no such thing as electrical power
Nor would there be electronic communication or computer
Which are typical of the important applications of ECE
And so you see, EM is fundamental to the study of ECE.

So, you are curious about learning EM
Let us proceed further with this PoEM
First you should know that **E** *means electric field*
And furthermore that **B** *stands for magnetic field*

*Now, the static **E** and **B** fields may be independent*
*But the dynamic **E** and **B** fields are interdependent*
Causing them to be simultaneous
And to coexist in any given space
Which makes EM very illuminating
And modern day life most interesting
*For it is the interdependence of **E** and **B** fields*
That is responsible for electromagnetic waves
In your beginning courses you might have learnt circuit theory
It is all an approximation of electromagnetic field theory
So you see they put the cart before the horse
But it is okay to do that and still make sense
Because at low frequencies circuit approximations are fine
But at high frequencies electromagnetic effects are prime
So, whether you are an electrical engineer
Or you happen to be a computer engineer
Whether you are interested in high frequency electronics
Or maybe high-speed computer communication networks
You see, electromagnetic effects are prime
Studying the fundamentals of EM is sublime.

If you still have a ProblEM with EM,
Because it is full of abstract mathematics,
I say, my dear ECE student who dislikes electromagnetics
Because you complain it is full of abstract mathematics
I want you to know that it is the power of mathematics
That enabled Maxwell's prediction through his equations
Of the physical phenomenon of electromagnetic radiation
Even before its finding by Hertz through experimentation
In fact it was this accomplishment
That partly resulted in the entitlement
For the equations to be known after Maxwell
Whereas in reality they are not his laws after all
For example the first one among the four of them
Is Faraday's Law expressed in mathematical form
You see, mathematics is a compact means
For representing the underlying physics
Therefore do not despair when you see mathematical derivations
Throughout your textbook on the Fundamentals of Electromagnetics
Instead look through the derivations to understand the concepts
Realizing that mathematics is only a means to extend the physics
Think of yourself as riding the horse of mathematics
To conquer the new frontier of electromagnetics
Let you and me together go on the ride
As I take you through the steps in stride, with grattitude!

N. Narayana Rao

Preface to the Indian Edition

FROM AN INDIAMERICAN WITH GRATITUDE AND GRATTITUDE

Here is a little poem for Mother
My mother, your mother, our mother
The mother of a billion people on her land
The mother of millions of people outside her land
Mother India, my citizenship is American
But the blood you sent me with is Indian
So, as they say, am I an Indian American?
Or, am I an American of Indian origin?
I may be known as an Indian American
Or they may call me an American of Indian origin
But mother, I feel more like an IndiAmerican!
With the "Indian" fused into "American"
And I shall always be an IndiAmerican!
As my "Indian" is inseparable from my "American"
Or, for that matter, from any other "an!"

The above poem, which I call the "IndiAmerican Poem," is a slightly modified version of a poem I composed for the Republic Day in January 2006 and sent to the then-President, Bharat Ratna, Dr A. P. J. Abdul Kalam. Dr Kalam was kind enough to send me a brief letter, stating: *Dear Prof. Rao, I was going through the poem composed by you, "Mother, I am an IndiAmerican!" It is indeed very inspiring. My best wishes to you. Yours Sincerely, A. P. J. Abdul Kalam.* The IndiAmerican Poem appeared in the Preface of the Indian Edition of "Elements of Engineering Electromagnetics, Sixth Edition," published by Pearson Education in July 2006, and "dedicated to 'the young minds that will take this country to the greatest heights,' in the words of the President of India, Bharat Ratna, Dr A. P. J. Abdul Kalam, a fellow alumnus of the Madras Institute of Technology."

As stated in the section on "Gratitude and Grattitude," in which the word, "grattitude," is explained, I always believed in the power of education—transcending the boundaries of national origin, race, and religion—to assure the future of the world. When I left India for the United States of America in 1958, there were only a few technological institutions in a given state in India. I came with the solid foundation laid at my alma mater, the Madras Institute of Technology (MIT), which to me is not just the MIT, the "Madras Institute of Technology" but the MIT, the "Mother Institute of Technology." I acquired more education at my alma mater in the United States of America, the University of Washington (UW), and prospered in my profession of engineering education at the University of Illinois at Urbana-Champaign (UIUC). For all of this, I am grateful to my two Lands, the land of my birth, India, for the foundation, and the land of my work, America, for the prosperity. I would now like to share with you a fascinating and heart-warming story because it has a bearing with my belief in the power of education, before proceeding on to items of direct relevance to the subject matter of this book.

AMERICA, CHINA, AND INDIA

On April 18, 2006, while I was staying in the Le Meridien Hotel in New Delhi on one of my trips to the motherland, along with my wife, I received an email from one of my former colleagues at UIUC and an emeritus faculty member, on some department business. Along with it, he forwarded an e-mail from Da Hsuan Feng, then Vice President for Research and Economic Development and Professor of Physics at the University of Texas at Dallas, and now Senior Executive Vice President at National Chung Kung University in Tainan, Taiwan. Da Hsuan Feng was born in New Delhi in 1945 to Chinese parents, when his father, Paul Kuo-Jin Feng, was stationed as a foreign correspondent.

Paul Kuo-Jin Feng died tragically when Da Hsuan Feng was a young boy so that he knew little about what his father did, according to him. On the afternoon of April 18, 2006, it suddenly occurred to him that it might be interesting to google "Paul Feng." And he did just that and the results absolutely astounded him! It returned two websites, both of which were the same. What came to him on his screen stunned him so much that for a moment he was almost paralyzed! From an article, "Forging an Asian identity," by Manoj Das in The Hindu of January 7, 2001, he read the following: Speaking to Mr. Paul Feng of the Central News Agency on January 20, 1946, Pundit Jawaharlal Nehru, the first Prime Minister of India, said, "If China and India hold together, the future of Asia is assured." This holding together need not be confined to diplomacy; it can, by all means, be a psychological force that can work wonders in the realms of creativity...

Da Hsuan Feng was stunned that 60 years ago then, his father was THE person who heard the profound statement—"If China and India hold together, the future of Asia is assured"—from Pundit Nehru. I was in high school in 1946, and from 1947, when I joined college in Madras, until I left India for the United States, I had several occasions to see the revered Prime Minister. Since I received the e-mail when I was in India in New Delhi in a hotel in a room overlooking the Parliament building and the Presidential mansion, where I met President Kalam three times in the preceding five months, the statement felt even more profound to me.

As I responded to my colleague with a copy to Da Hsuan Feng on April 20, 2006, just before my departure back to the United States after a very successful trip, from New Delhi: "Now, 60 years after that, that statement is even more significant. I would go further and say, with utmost reverence to Pundit Nehru, that today, if America, China, and India hold together, the future of the world is assured. In no other sphere of endeavor is the 'holding together' more important than in the sphere of education, because that is where future leaders are cultivated. It is in this regard that UIUC comes into the picture. And the reason I am here in India is on the matter of developing relations in India...."

It is my hope that this book and its original U.S. edition and its international versions serve as a contribution, however small, in this regard by propagating the knowledge on the fundamentals of electromagnetics to students and academics throughout the world, transcending the boundaries of national origin, race, and religion.

INDIAN EDITION VERSUS THE U.S. EDITION

Differences in Terminology

This Indian edition bears the title, *Fundamentals of Electromagnetics for Engineering*, whereas the original U.S. edition is entitled, *Fundamentals of Electromagnetics for Electrical and Computer Engineering*, reflecting the differences in terminology associated with the use of the term, "electrical," in the Indian and American systems of engineering education. As is commonly known, in engineering education in the United States of America, the term "electrical" refers to the totality of what are known in India by the terms, "electrical," "electronics," and "communications." In Indian engineering education, following the British system, the use of "electrical" is generally confined to the areas of power and energy systems and control. Because this book serves the needs of all three of the engineering disciplines, "electrical," "electronics," and "communications," as well as the "computer engineering" discipline, in the Indian system, it is titled, *Fundamentals of Electromagnetics for Engineering*.

The Terminology, "Electrical and Computer Engineering"

Because the term "electrical and computer engineering" is used in the rest of the front matter of this book, it is of interest to elaborate upon this terminology. In engineering departments in the United States educational institutions, electrical and computer engineering is generally one academic department, although not in all institutions. The name, ECEDHA, Electrical and Computer Engineering Department Heads Association, reflects this situation. In the College of Engineering at the University of Illinois at UIUC, the Department of Electrical and Computer Engineering (ECE) offers two undergraduate programs leading to the Bachelor of Science degrees: Electrical Engineering and Computer Engineering. The following two paragraphs, taken from the descriptions of these two undergraduate programs, provide an understanding of the scope of these disciplines.

Scope of the Fields of Electrical Engineering and Computer Engineering

"A list of the twenty greatest engineering achievements of the twentieth century compiled by the National Academy of Engineering includes ten achievements primarily related to the field of *electrical engineering*: electrification, electronics, radio and television, computers, telephone, Internet, imaging, household appliances, health technologies, and laser and fiber optics. The remaining achievements in the list—automobile, airplane, water supply and distribution, agricultural mechanization, air conditioning and refrigeration, highways, spacecraft, petroleum/petrochemical technologies, nuclear technologies, and high-performance materials—also require knowledge of electrical engineering to differing degrees. In the twenty-first century, the discipline of electrical engineering continues to be one of the primary drivers of change and progress in technology and standards of living around the globe."

"*Computer engineering* is a discipline that applies principles of physics and mathematics to the design, implementation, and analysis of computer and communication systems. The discipline is broad, spanning topics as diverse as radio communications, coding and encryption, computer architecture, testing and analysis of computer and communication systems, vision, and robotics. A defining characteristic of the discipline is its grounding in physical aspects of computer and communication systems. Computer engineering concerns itself with development of devices that exploit physical phenomena to store and process information, with the design of hardware that incorporates such devices, and with software that takes advantage of this hardware's characteristics. It addresses problems in design, testing, and evaluation of system properties, such as reliability, and security. It is an exciting area to work in, one that has immediate impact on the technology that shapes society today."

The Illinois Curricula

Because of the fundamental nature of the subject of electromagnetics (EM) for electrical engineering as well as for computer engineering, a course on it is required in both programs. For the *electrical engineering* program, the core curriculum focuses on fundamental electrical engineering knowledge: circuits, systems, electromagnetics, solid state electronics, computer engineering, and design. A rich set of elective courses permits students to select from collections of courses in seven areas of electrical and computer engineering: bioengineering, acoustics, and magnetic resonance engineering; circuits and signal processing; communication and control; computer engineering; electromagnetics, optics, and remote sensing; microelectronics and quantum electronics; power and energy systems. For the *computer engineering* program, the core curriculum focuses on fundamental computer engineering knowledge: circuits, systems, electromagnetics, computer engineering, solid state electronics, and computer science. A rich set of elective courses permits students to concentrate in any subdiscipline of computer engineering including computer systems; electronic circuits; networks; engineering applications; software, languages, and theory; and algorithms and mathematical tools.

We shall hereafter use the term, "electrical and computer engineering," to mean the totality of the three engineering disciplines, "electrical," "electronics," and "communications," and the "computer engineering" discipline, in the Indian system.

ON THE APPROACH USED IN THIS BOOK FOR TEACHING ELECTROMAGNETICS

Electromagnetics Courses and Textbooks

In the context of a course offering or the title of a textbook, the term, "electromagnetics," is freely interchanged with "electromagnetic theory," "electricity and magnetism," "electric and magnetic fields," "electromagnetic fields," "electromagnetic fields and waves," and so on. Introductory textbooks on engineering electromagnetics can be classified broadly into three categories:

1. One-semester textbooks based on a traditional approach of covering essentially electrostatics and magnetostatics, and culminating in Maxwell's equations and some discussion of their applications.

2. Two-semester textbooks, with the first half or more covering electrostatics and magnetostatics, as in category 1, and the remainder devoted to topics associated with electromagnetic waves.

3. One- or two-semester textbooks that deviate from the traditional approach, with the degree and nature of the deviation dependent on the author.

This book, a one-semester book, belongs to category 3. The teaching of the course, using books in categories 1 and 2 is heavy on static electric and magnetic fields prior to introducing Maxwell's equations for time-varying fields and wave propagation. As explained in the preface, this book deviates from that approach, with the deviation originating in 1977, by introducing Maxwell's equations for time-varying fields at the outset.

About Electromagnetics

By the very nature of the word, *electromagnetics* implies having to do with a phenomenon involving both electric and magnetic fields and furthermore coupled. This is indeed the case when the situation is dynamic, that is, time-varying, because time-varying electric and magnetic fields are interdependent, with one field producing the other. In other words, a time-varying electric field or a time-varying magnetic field cannot exist alone; the two fields coexist in time and space, with the space-variation of one field governed by the time-variation of the second field. This is the essence of Faraday's law and Ampere's circuital law, the first two of the four Maxwell's equations.

Faraday's law says that a time-varying magnetic field gives rise to an electric field, the space-variation of which is related to the time-variation of the magnetic field. Ampere's circuital law tells us that a time-varying electric field produces a magnetic field, the space variation of which is related to the time-variation of the electric field. Thus, if one time-varying field is generated, it produces the second one, which, in turn, gives rise to the first one, and so on, which is the phenomenon of electromagnetic wave propagation, characterized by time delay of propagation of signals. In addition, Ampere's circuital law tells us that an electric current produces a magnetic field, so that a time-varying current source results in a time-varying magnetic field, beginning the process of one field generating the second.

Only when the fields are not changing with time, that is, for the static case, they are independent; a static electric field or a static magnetic field can exist alone, with the exception of one case in which there is a one-way coupling, electric field resulting in magnetic field, but not the other way. Thus, in the entire frequency spectrum, except for dc, all electrical phenomena are, in the strictest sense, governed by interdependent electric and magnetic fields, or electromagnetic fields. However, at low frequencies, an approximation, known as the qusistatic approximation, can be made in which the time-varying fields in a physical structure are approximated to have the same spatial variations as the static fields in the structure obtained by setting the source frequency equal to zero. Thus, although the actual situation in the structure is one of electromagnetic wave nature, it is approximated by a dynamic but not wavelike nature. As the frequency becomes higher and higher, this approximation violates the actual situation more and more, and it becomes increasingly necessary to consider the wave solution.

Historical Development of Technologies and Teaching of Electromagnetics

Historically, the development of major technologies based on Maxwell's equations occurred in the sequence of electrically and magnetically based technologies (electromechanics and electrical power) in the nineteenth century; electronics hardware and software in the twentieth century; and photonics technologies, entering into the twenty-first century. The teaching of electromagnetics evolved following this sequence, that is, beginning with a course on electrostatics, magnetostatics, energy and forces, and in some cases quasistatic fields, followed by Maxwell's equations for time-varying fields and an introduction to electromagnetic waves. This course was then followed by one or more courses on transmission lines, electromagnetic waves, waveguides and antennas. In fact, when I joined the University of Illinois at Urbana-Champaign in 1965, a three-semester sequence of courses in electromagnetics (Electric and Magnetic Fields, Transmission Lines, and Electromagnetic Fields) for a total of nine credit-hours, were required in the Electrical Engineering (EE) curriculum of a total of 145 credit-hours. The treatment of material in this manner in the first course was largely due to development and application of electrical machinery. The course on transmission lines was to extend the low frequency (lumped) circuit theory, into the high frequency regime by beginning with the introduction of transmission line as a distributed circuit. The third course was then based on full electromagnetic wave solution of Maxwell's equations.

Principal Drawback of the Traditional, Inductive, Approach

In 1973, the Computer Engineering curriculum (CompE) was introduced, and the total number of credit-hours were decreased to 124 credit-hours (now 128 credit-hours) for both curricula. Only one course in electromagnetics, the first of the above three-course sequence, was required in both curricula, with the remaining courses moved to elective status. Soon, this was found to be unsuitable, because the course was based on the inductive approach, that is, an approach consisting of developing general principles from particular facts, which in this case was developing complete set of Maxwell's equations beginning with the particular laws of static fields. Since much time was taken up for the coverage of static fields before getting to the complete set of Maxwell's equations, the time was cut short for the more useful material, centered on electromagnetic waves. This principal drawback of the traditional approach of teaching electromagnetics was unnecessarily aggravating for both curricula, because students were already exposed to the traditional treatment in a prerequisite physics course on electricity and magnetism. It was further aggravating for the Computer Engineering curriculum, because generally that was the only EM course they would take, whereas the EE students would take one or more of the remaining courses as electives.

Elimination of the Drawback by Using the Deductive Approach and Implementation

To resolve this problem, without having separate courses for the two curricula (EE and CompE), a new completely revised course was created based on the deductive approach, that is, an approach in which one begins with the general principles that are accepted as true and then applies it to particular cases, which in this case was beginning with the complete set of Maxwell's equations for time-varying fields and then developing their applications, as well as considering special cases of static and quasistatic fields.

Electromagnetic theory, which is extremely elegant in structure and uniquely rigorous in formulation, is a truly fascinating subject. For those interested in getting enlightened from this cultural aspect, the traditional, inductive, approach is the best, which is what I have done with the publication of my first book, *Basic Electromagnetics with Applications*, as early as in 1972, and with which I continue to be fascinated. Practical considerations described above made me stretch this fascination for the benefit of the modern day curricula and write my second book, *Elements of Engineering Electromagnetics*, using the deductive approach. The first edition of this book was published in 1977, only five years after the publication of the first book, "breaking with the tradition." At that time, it was a one-semester textbook. From 1977 to date, this deductive approach served the needs of the Illinois curricula well by imparting to students the elements of engineering electromagnetics that (a) constitute the foundation for preparing the EE majors to take follow-on courses, and (b) represent the essentials for the CompE majors taking this course only. The subsequent editions of the book were expanded for a two-semester coverage, but retaining the deductive approach in the beginning chapters for the first course and adding material in the later chapters for a follow-on elective course for EE students. Thus, the principal drawback of the traditional approach has been eliminated and shown to be successful by the experience of over more than 30 years.

Using the Deductive Approach for Further Benefit in This Book

A significant secondary drawback of the traditional treatment is that, generally, transmission lines are introduced from the circuit equivalent point of view and then followed by the treatment of electromagnetic waves. This creates a lack of appreciation of the fact that transmission lines are actually physical structures in which electromagnetic waves propagate, except that they are represented by (distributed) circuit equivalents, let alone that (lumped) circuit elements themselves arise from approximations of dynamic field solutions for frequencies low enough that wave propagation effects are negligible. From the very outset, including even in the 1972 book, I have always introduced electromagnetic waves first and then only transmission lines from the field aspect, deriving from the field equations the circuit equations and the equivalent circuit. An understanding based on this approach has become increasingly important with the applications extending farther and farther in frequency along the spectrum.

In this book, I have further stretched my fascination with the subject of electromagnetics by taking the usefulness of the deductive approach one step further by following it with the thread of statics-quasistatics-waves. In this treatment, first electromagnetic wave propagation is introduced by solving Maxwell's equations for time-varying fields. This provides appreciation of the fact that regardless of how low the frequency is, as long as it is nonzero, the phenomenon is one of electromagnetic waves, resulting from the interdependence of time-varying electric and magnetic fields. Then, the thread of statics-quasistatics-waves is used to bring out the frequency-behavior of physical structures.

The thread of statics-quasistatics-waves consists of first studying static fields by setting the time derivatives in Maxwell's equations to zero, and introducing the (lumped) circuit elements, familiar in circuit theory, through the different classifications of static fields. Then, quasistatic fields are studied as low-frequency extensions of static fields, as approximations to exact solutions of Maxwell's equations for time-varying fields. For a given physical structure, this is done by beginning with a time-varying field having the same spatial characteristics as that of the static field solution for the structure, and obtaining field solutions to Maxwell's equations containing terms up to and including the first power (which is the lowest power) in frequency, leading to the result that circuit representation for the input behavior of the structure is the same as the circuit representation for the static case. The thread is then continued by discussing that beyond this "quasistatic approximation," the situation requires exact solution by simultaneous solution of Maxwell's equations, leading to waves and the "distributed circuit" concept of a transmission line.

Later on, during the discussion of transmission line analysis, the condition for the validity of the quasistatic approximation, which is that the dimensions of the physical structure are small compared to the wavelength corresponding to the frequency of operation, is derived, by considering the approximation of the exact solution for low frequencies. In this manner, it becomes quite clear that, along the frequency spectrum, the quasistatic behavior approached from the static (zero frequency) limit as an extension of the static behavior to dynamic behavior of first order in frequency is the same as the low-frequency behavior approached from the other (higher frequencies) side, by approximating the exact dynamic solution for low frequencies. This very important concept is not always clearly understood or appreciated when the inductive approach is employed.

WHY IS EM FUNDAMENTAL TO ELECTRICAL AND COMPUTER ENGINEERING?

IEEE Logo

Very often, the question is asked why one should study EM as a required course for all branches of electrical and computer engineering. In 1963, the American Institute of Electrical Engineers (AIEE) and the Institute of Radio Engineers (IRE) were merged into the Institute of Electrical and Electronics Engineers (IEEE). The IEEE is a global nonprofit organization with over 375,000 members. It is "the world's leading professional association for the advancement of technology." The IEEE logo or badge is a merger of the badges of the two parent organizations. It contains a vertical arrow surrounded by a circular arrow, within a kite-shaped border. No letters clutter the badge because a badge without letters can be read in any language. The AIEE badge had the kite shape which was meant to symbolize the kite from Benjamin Franklin's famous kite experiment to study electricity. The IRE badge had the two arrows that symbolize the right hand rule of electromagnetism. Alternatively, the vertical arrow can be thought of as representing one of the two fields, electric or magnetic, and the circular arrow surrounding it representing the second field, produced by it, so that together they represent an electromagnetic field.

Whether this logo of IEEE was intended to be a recognition of the fact that electromagnetics is fundamental to all of electrical and computer engineering, it is a fact that all electrical phenomena are governed by the laws of electromagnetics, and hence, the study of electromagnetics is essential to all branches of electrical and computer engineering, and indirectly impacts many other branches. In the following, I present the descriptions of the relevance of EM first in general and then by considering certain specific areas of ECE, listed in alphabetical order. (This material is taken, with some modification from the Indian Edition of *Elements of Engineering Electromagnetics, Sixth Edition*, Pearson Education, 2006.)

General

Many laws in circuit theory are derived from the laws of EM. The increased clock rates of computers dictate that the manipulation of electrical signals in computer circuits and chips be more and more in accordance with their electromagnetic nature, requiring a fundamental understanding of EM. EM includes the study of antennas, wireless communication systems, and radar technologies. In turn, these technologies are supported by microwave engineering, which is an important branch of EM. Traditionally, the understanding of EM phenomena has been aided by mathematical modeling, where solutions to simplified models are sought for the understanding of complex phenomena. The branch of mathematical modeling in EM has now been replaced by computational electromagnetics where solutions to complex models can be sought efficiently. The use of laws of EM also extends into the realms of remote sensing, subsurface sensing, optics, power systems, EM sources at all frequencies, terahertz systems, and many other areas. Understanding of electric fields is important for understanding the operating principles of many semiconductor and nanotechnology devices. Electrical signals are conveyed as electromagnetic waves, and hence, communications, control, and signal processing are indirectly influenced by our understanding of the laws of EM. EM is also important in biomedical engineering, nondestructive testing, electromagnetic compatibility and interference analysis, microelectromechanical systems, and many more areas.

Computer Systems

Computer systems and digital electronics are based on a hierarchy of abstractions and approximations that manage the amount of complexity an engineer must consider at any given time. At first glance, these abstractions might seem to make understanding EM less important for a student or an engineer whose interests lie in the digital domain. However, this is not the case. While the fields, vectors, and mathematical expressions that describe EM structures are somewhat removed from the Boolean logic, microprocessor instruction sets, and programming languages of computer systems, it is essential that computer engineers have both a qualitative and a quantitative understanding of EM in order to evaluate which approximations and abstractions are appropriate to any particular design. Choosing approximations that neglect important factors can lead to designs that fail when implemented in hardware, and including unimportant effects in calculations can significantly increase the amount of effort required to design a system and/or obscure the impact of important parameters. A computer engineer or digital system designer must be able to consider EM effects in order to build systems that meet their design requirements. As technology advances, such cases will become more and more common, if for no other reason than the fact that designers are continually driven to push the limits of a given integrated circuit fabrication technology in order to outperform their competition. To be successful, an engineer must be not only a master of his or her specialty, but an expert in all of the areas of electrical engineering that impact that specialty, including EM.

DSL Systems

Hundreds of millions of digital subscriber line (DSL) broadband access connections are now in use around the globe. Such DSLs use the copper telephone-line twisted pair at or near its fundamental data-carrying limits to effect the broadband service. A twisted pair transmission line can be divided into a series of incrementally small circuits that are characterized by fundamental passive circuit elements of resistance (R), inductance (L), capacitance (C), and conductance (G), sometimes known as the RLCG parameters. These parameters often vary as a function of frequency also, and so such models can be repeated at a set of frequencies over a band of use or interest, which is typically 500 kHz to up to as much as 30 MHz in DSL systems. EM theory and, in particular, the basic Maxwell's equations essentially allow the construction of these incremental circuits and their cascade, allowing calculation of the various transfer functions and impedances and then characterize the achievable data rates of the DSL. Of significant interest is the subsequent use of such theory to model an entire binder of copper-twisted pairs using vector/matrix generalizations of the simple isolated transmission lines. The modeling of "crosstalk" between the individual lines is important to understanding the limits of transmission of all the lines within the binder and their mutual effects upon one another. EM theory fundamentally allows such characterization and the calculation of the impact of the various transmission lines upon one another. EM theory is, thus, fundamental to understanding of and design thereupon of DSL systems.

Electrical Power and Energy Systems

The use of electricity for generation, transport, and conversion of energy is a dominant factor in the global economy. EM theory is an essential basis for understanding the devices, methods, and systems used for electrical energy. Both electric and magnetic fields are *defined* in terms of the forces they produce. A strong grasp of fields is essential to the study of electromechanics—the use of fields to create forces and motion to do useful work. In electromechanics, engineers design and use magnetic field arrangements to create electric machines, transformers, inductors, and related devices that are central to electric power systems. In microelectromechanical systems (MEMS), engineers use both magnetic and electric fields for motion control at size scales down to nanometers. At the opposite end of the size scale, electric fields must be managed carefully in the enormous power transmission grid that supplies energy to cities and towns around the world. Today's transmission towers carry up to a million volts and thousands of amps on each conductor. The lines they carry can be millions of meters long. EM theory is a vital tool for the design and operation of these lines and the many devices needed to connect to them. All engineering study related to electrical energy and power relies on key concepts from EM theory. In the future, society needs more efficient energy processing, expanded use of alternative energy resources, more sophisticated control capabilities in the power grid, and better industrial processes. EM represents an essential and fundamental background that underlies future advances in energy systems.

Electromagnetic Compatibility

We live in an increasingly "spectrally rich" EM environment. The potential for a product to *interfere with* other devices, or to be *interfered with* by neighboring electronic systems, is ever increasing. Coupled with this is that increasingly stringent RF emission regulatory standards must be satisfied before a product may be marketed. The electrical or computer engineer of today must know how to design "electromagnetically compatible" (EMC) systems that perform their intended function even in the presence of unintended EM radiation from nearby electronic equipment. Likewise, he or she must know how to design systems that do not themselves pollute the EM spectrum further. The only way this can be done is through a solid understanding of electric fields, magnetic fields, electromagnetic wave propagation, signal-coupling mechanisms, and filtering, shielding, and grounding techniques.

High-frequency Electronics

As information technology continues into the realm of ever higher frequencies, circuits and devices must be designed with an ever keener awareness of EM. At frequencies higher than a few gigahertz, electronic devices can no longer be treated as simple lumped components. Rather, they become enmeshed in a complex web of interconnected phenomena—all of them determined by the laws of EM. High frequency means short wavelength, and as wavelength diminishes to the point where it is comparable to integrated circuit dimensions, EM phenomena called *transmission line effects* become critical. These effects include conductor loss, dielectric loss, and radiation loss. They are a signal's worst enemies. The radio-frequency or microwave circuit designer must lay out transmission lines to achieve optimal matching conditions among parts of the circuit and to limit the signal attenuation caused by transmission line effects.

Integrated Circuit (IC) Design

Electrical signals move from one part of an IC to another according to the laws of EM. Unwanted coupling of electrical signals from different parts of an IC can be explained, and solved, only through recourse to fundamental knowledge about EM. Engineers who specialize in both communications and circuits must bring their EM knowledge to bear on their system design. If the components are not electromagnetically compatible, the system will not function.

Lasers, Fiber Optics, and Optoelectronics

During the past few decades, the invention of lasers and low-loss optical fibers has revolutionized the use of optical communication technologies for high-speed Internet. Although stimulated absorption and emission of photons may require a quantum mechanical description of the photon-electron interaction, classical EM plays a crucial role in understanding the system because light follows the theory of EM waves for most of the guided wave phenomena. In the study of optical communication systems, four important areas of devices are required: generation of light (semiconductor lasers, light-emitting diodes or LEDs, and erbium-doped fiber optical amplifiers), modulation of light (electro-optical modulators, electro-absorption modulators, and optical phase modulators), propagation of light (optical fibers, and optical dielectric waveguides), and detection of light (semiconductor photodetectors). The growing demand for ultra-high-bandwidth Internet technologies requires researchers and engineers to develop novel devices for the generation, propagation, modulation, and detection of light. Knowing EM is a necessity because the wave nature of light plays a vital role in all the above devices.

Medical and Biomedical Imaging

Light, and its interactions with biological tissues and cells, has the potential to provide helpful diagnostic information about structure and function. The study of EM is essential to understanding the properties of light, its propagation through tissue, scattering and absorption effects, and changes in the state of polarization. The spectroscopic (wavelength) content of light provides a new dimension of diagnostic information since many of the constituents of biological tissue, such as hemoglobin in blood, melanin in skin, and ubiquitous water, have wavelength-dependent optical properties over the visible and near-infrared EM spectrum. The study of EM has direct relevance to understanding how light interacts with tissue, and novel technology for medical and biological imaging can be developed based on these EM principles.

Mobile Communication

In 1864, Maxwell predicted the existence of the EM waves by logically examining the known experimental laws: Faraday's law, Ampere's law, Gauss' law and the charge conservation law. Maxwell's prediction was verified by Hertz in 1887 when he propagated an electric spark across his laboratory. Within a few years of Hertz's experiment, Marconi demonstrated the potential application of EM waves for communication by successfully propagating a telegraphic signal across the Atlantic. He coined the term "wireless," when he established his Wireless Telegraph and Signal Company in 1897, and wireless communication took off. For many years, radio signals bouncing off the ionosphere became the main carrier of the global communication networks, connecting people and institutions across the continents. In the 1960s, when the world moved into the space age, satellite communication was introduced which offered faster, better and more reliable services. With this new development, the future of wireless communication was considered very promising. However, without much warning, the optical fiber came along. Broader bandwidth, more secure communication and lower costs of the optical systems made satellite communication a less attractive choice. The world seemed to be moving back to cable communication. For the two decades in the 1970s and 80s, wireless almost became obsolete. Then mobile communication appeared and all of a sudden, thanks to the miniaturization of the devices, we are in the era of personal communication. Wireless is back. New applications are coming out almost every month. It now seems that people's communication needs can no longer be satisfied by mobility alone. They require ubiquity which most likely can only be provided by an innovative wireless environment.

Remote Sensing

The attribute of EM waves to carry energy and information from one point in space (the sender) to another point (the receiver) also makes them a good tool for probing something from a distance. Radar was invented in the 1930s using precisely this property of radio waves (a subset of EM waves). Later, this new application of EM waves developed into a thriving new discipline called "remote sensing." New active and passive devices and systems have been invented to improve remote sensing capabilities. Nowadays data from various remote sensing techniques and equipments provide people with the necessary information to monitor the status of the global environment, information vital to our pursuit of the sustainable development of human society.

Signal Processing

The vast majority of signals processed in high-tech systems and components are EM waves. Engineers must know how to model signal propagation in the physical medium of interest, be it optical fiber, coaxial cable, twisted pair wires, or in the air. Communication system designers employ this knowledge to design algorithms and architectures for transmitting data reliably over a noisy channel.

CONCLUSION

In this preface to the Indian Edition, I have used the word, "fascinating," several times. My dictionary defines "fascinating" as "captivating; enchanting; charming." Whether it is captivating, enchanting, or charming, fascination with the subject of electromagnetics has led me to recognizing a parallel between the guiding principles of my life, the set of four principles from the Upanishads, and the guiding equations of my work, the set of four Maxwell's equations:

Matrudevo bhava!	⇒	*Revere the mother as God!*
Pitrudevo bhava!	⇒	*Revere the father as God!*
Acharyadevo bhava!	⇒	*Revere the preceptor as God*
Atithidevo bhava!	⇒	*Revere the guest as God!*

$$\oint_C \mathbf{E} \cdot d\mathbf{l} = -\frac{d}{dt}\int_S \mathbf{B} \cdot d\mathbf{S}$$

$$\oint_C \mathbf{H} \cdot d\mathbf{l} = \int_S \mathbf{J} \cdot d\mathbf{S} + \frac{d}{dt}\int_S \mathbf{D} \cdot d\mathbf{S}$$

$$\oint_S \mathbf{D} \cdot d\mathbf{S} = \int_V \rho \, dv$$

$$\oint_S \mathbf{B} \cdot d\mathbf{S} = 0$$

Besides the fact that both sets are four in number, the first two of the principles from the Upanishads have to do with mother and father, whose interdependence is the essence of life, whereas the first two of the four Maxwell's equations have to do with the interdependence of dynamic electric and magnetic fields, which is the essence of electromagnetic waves. Whether you will find the subject of electromagnetics fascinating or not, I hope you will find this aspect of electromagnetics, which I call the "Mahatmyam (greatness) of Maxwell's equations," fascinating!

N. NARAYANA RAO

Vectors and Fields

Electromagnetics deals with the study of electric and magnetic *fields*. It is at once apparent that we need to familiarize ourselves with the concept of a *field*, and in particular with *electric* and *magnetic* fields. These fields are vector quantities and their behavior is governed by a set of laws known as *Maxwell's equations*. The mathematical formulation of Maxwell's equations and their subsequent application in our study of the fundamentals of electromagnetics require that we first learn the basic rules pertinent to mathematical manipulations involving vector quantities. With this goal in mind, we shall devote this chapter to vectors and fields.

We shall first study certain simple rules of vector algebra without the implication of a coordinate system and then introduce the Cartesian coordinate system, which is the coordinate system employed for the most part of our study in this book. After learning the vector algebraic rules, we shall turn our attention to a discussion of scalar and vector fields, static as well as time-varying, by means of some familiar examples. We shall devote particular attention to sinusoidally time-varying fields, scalar as well as vector, and to the phasor technique of dealing with sinusoidally time-varying quantities. With this general introduction to vectors and fields, we shall then devote the remainder of the chapter to an introduction of the electric and magnetic field concepts, from considerations of the experimental laws of Coulomb and Ampere.

1.1 VECTOR ALGEBRA

In the study of elementary physics we come across several quantities such as mass, temperature, velocity, acceleration, force, and charge. Some of these quantities have associated with them not only a magnitude but also a direction in space, whereas others are characterized by magnitude only. The former class of quantities are known as *vectors*, and the latter class of quantities are known as *scalars*. Mass, temperature, and charge are scalars, whereas velocity, acceleration, and force are vectors. Other examples are voltage and current for scalars and electric and magnetic fields for vectors.

Vector quantities are represented by boldface roman type symbols, for example, **A**, in order to distinguish them from scalar quantities, which are represented by lightface italic type symbols, for example, *A*. Graphically, a vector, say **A**, is represented by a straight line with an arrowhead pointing in the direction of **A** and having a length proportional to the magnitude of **A**, denoted |**A**| or simply *A*. Figures 1.1(a)–(d) show four vectors drawn to the same scale. If the top of the page represents north, then vectors **A** and **B** are directed eastward, with the magnitude of **B** being twice that of **A**. Vector **C** is directed toward the northeast and has a magnitude three times that of **A**. Vector **D** is directed toward the southwest and has a magnitude equal to that of **C**. Since **C** and **D** are equal in magnitude but opposite in direction, one is the negative of the other. It is important to note that the lengths of the lines are not associated with the physical quantity *distance*, unless the vector quantity represents distance; they are associated with the magnitudes of the physical quantity that the vector represents, such as velocity, acceleration, or force.

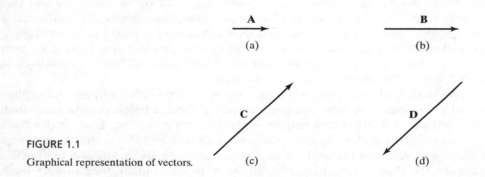

FIGURE 1.1

Graphical representation of vectors.

Since a vector may have in general an arbitrary orientation in three dimensions, we need to define a set of three reference directions at each and every point in space in terms of which we can describe vectors drawn at that point. It is convenient to choose these three reference directions to be mutually orthogonal as, for example, east, north, and upward or the three contiguous edges of a rectangular room. Thus, let us consider three mutually orthogonal reference directions and direct *unit vectors* along the three directions as shown, for example, in Figure 1.2(a). A unit vector has magnitude unity. We shall represent a unit vector by the symbol **a** and use a subscript to denote its direction. We shall denote the three directions by subscripts 1, 2, and 3. We note that for a fixed orientation of $\mathbf{a}_1$, two combinations are possible for the orientations of $\mathbf{a}_2$ and $\mathbf{a}_3$, as shown in Figures 1.2(a) and (b). If we take a right-hand screw and turn it from $\mathbf{a}_1$ to $\mathbf{a}_2$ through the 90°-angle, it progresses in the direction of $\mathbf{a}_3$ in Figure 1.2(a) but opposite to the direction of $\mathbf{a}_3$ in Figure 1.2(b). Alternatively, a left-hand screw when turned from $\mathbf{a}_1$ to $\mathbf{a}_2$ in Figure 1.2(b) will progress in the direction of $\mathbf{a}_3$. Hence, the set of unit vectors in Figure 1.2(a) corresponds to a right-handed system, whereas the set in Figure 1.2(b) corresponds to a left-handed system. We shall work consistently with the right-handed system.

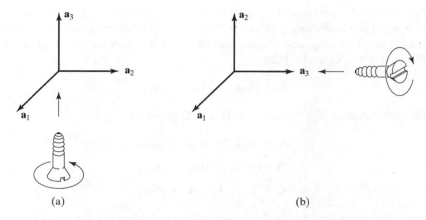

FIGURE 1.2

(a) Set of three orthogonal unit vectors in a right-handed system. (b) Set of three orthogonal unit vectors in a left-handed system.

A vector of magnitude different from unity along any of the reference directions can be represented in terms of the unit vector along that direction. Thus, $4\mathbf{a}_1$ represents a vector of magnitude 4 units in the direction of $\mathbf{a}_1$, $6\mathbf{a}_2$ represents a vector of magnitude 6 units in the direction of $\mathbf{a}_2$, and $-2\mathbf{a}_3$ represents a vector of magnitude 2 units in the direction opposite to that of $\mathbf{a}_3$, as shown in Figure 1.3. Two vectors are added by placing the beginning of the second vector at the tip of the first vector and then drawing the sum vector from the beginning of the first vector to the tip of the second vector. Thus to add $4\mathbf{a}_1$ and $6\mathbf{a}_2$, we simply slide $6\mathbf{a}_2$ without changing its direction until its beginning coincides with the tip of $4\mathbf{a}_1$ and then draw the vector ($4\mathbf{a}_1 + 6\mathbf{a}_2$) from the beginning of $4\mathbf{a}_1$ to the tip of $6\mathbf{a}_2$, as shown in Figure 1.3. By adding $-2\mathbf{a}_3$ to this vector ($4\mathbf{a}_1 + 6\mathbf{a}_2$) in a similar manner, we obtain the vector ($4\mathbf{a}_1 + 6\mathbf{a}_2 - 2\mathbf{a}_3$), as shown in Figure 1.3. We note that the magnitude of ($4\mathbf{a}_1 + 6\mathbf{a}_2$) is $\sqrt{4^2 + 6^2}$, or 7.211, and that the magnitude of ($4\mathbf{a}_1 + 6\mathbf{a}_2 - 2\mathbf{a}_3$) is $\sqrt{4^2 + 6^2 + 2^2}$, or 7.483. Conversely to the

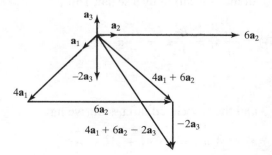

FIGURE 1.3

Graphical addition of vectors.

foregoing discussion, a vector **A** at a given point is simply the superposition of three vectors $A_1\mathbf{a}_1$, $A_2\mathbf{a}_2$, and $A_3\mathbf{a}_3$ that are the projections of **A** onto the reference directions at that point. A_1, A_2, and A_3 are known as the components of **A** along the 1, 2, and 3 directions, respectively. Thus,

$$\mathbf{A} = A_1\mathbf{a}_1 + A_2\mathbf{a}_2 + A_3\mathbf{a}_3 \tag{1.1}$$

We now consider three vectors **A**, **B**, and **C** given by

$$\mathbf{A} = A_1\mathbf{a}_1 + A_2\mathbf{a}_2 + A_3\mathbf{a}_3 \tag{1.2a}$$

$$\mathbf{B} = B_1\mathbf{a}_1 + B_2\mathbf{a}_2 + B_3\mathbf{a}_3 \tag{1.2b}$$

$$\mathbf{C} = C_1\mathbf{a}_1 + C_2\mathbf{a}_2 + C_3\mathbf{a}_3 \tag{1.2c}$$

at a point and discuss several algebraic operations involving vectors as follows.

Vector Addition and Subtraction

Since a given pair of like components of two vectors are parallel, addition of two vectors consists simply of adding the three pairs of like components of the vectors. Thus,

$$\mathbf{A} + \mathbf{B} = (A_1\mathbf{a}_1 + A_2\mathbf{a}_2 + A_3\mathbf{a}_3) + (B_1\mathbf{a}_1 + B_2\mathbf{a}_2 + B_3\mathbf{a}_3)$$

$$= (A_1 + B_1)\mathbf{a}_1 + (A_2 + B_2)\mathbf{a}_2 + (A_3 + B_3)\mathbf{a}_3 \tag{1.3}$$

Vector subtraction is a special case of addition. Thus,

$$\mathbf{B} - \mathbf{C} = \mathbf{B} + (-\mathbf{C}) = (B_1\mathbf{a}_1 + B_2\mathbf{a}_2 + B_3\mathbf{a}_3) + (-C_1\mathbf{a}_1 - C_2\mathbf{a}_2 - C_3\mathbf{a}_3)$$

$$= (B_1 - C_1)\mathbf{a}_1 + (B_2 - C_2)\mathbf{a}_2 + (B_3 - C_3)\mathbf{a}_3 \tag{1.4}$$

Multiplication and Division by a Scalar

Multiplication of a vector **A** by a scalar m is the same as repeated addition of the vector. Thus,

$$m\mathbf{A} = m(A_1\mathbf{a}_1 + A_2\mathbf{a}_2 + A_3\mathbf{a}_3) = mA_1\mathbf{a}_1 + mA_2\mathbf{a}_2 + mA_3\mathbf{a}_3 \tag{1.5}$$

Division by a scalar is a special case of multiplication by a scalar. Thus,

$$\frac{\mathbf{B}}{n} = \frac{1}{n}(\mathbf{B}) = \frac{B_1}{n}\mathbf{a}_1 + \frac{B_2}{n}\mathbf{a}_2 + \frac{B_3}{n}\mathbf{a}_3 \tag{1.6}$$

Magnitude of a Vector

From the construction of Figure 1.3 and the associated discussion, we have

$$|\mathbf{A}| = |A_1\mathbf{a}_1 + A_2\mathbf{a}_2 + A_3\mathbf{a}_3| = \sqrt{A_1^2 + A_2^2 + A_3^2} \tag{1.7}$$

Unit Vector Along A

The unit vector $\mathbf{a}_A$ has a magnitude equal to unity but its direction is the same as that of $\mathbf{A}$. Hence,

$$\mathbf{a}_A = \frac{\mathbf{A}}{|\mathbf{A}|} = \frac{A_1 \mathbf{a}_1 + A_2 \mathbf{a}_2 + A_3 \mathbf{a}_3}{\sqrt{A_1^2 + A_2^2 + A_3^2}}$$

$$= \frac{A_1}{\sqrt{A_1^2 + A_2^2 + A_3^2}} \mathbf{a}_1 + \frac{A_2}{\sqrt{A_1^2 + A_2^2 + A_3^2}} \mathbf{a}_2 + \frac{A_3}{\sqrt{A_1^2 + A_2^2 + A_3^2}} \mathbf{a}_3 \qquad (1.8)$$

Scalar or Dot Product of Two Vectors

The scalar or dot product of two vectors $\mathbf{A}$ and $\mathbf{B}$ is a scalar quantity equal to the product of the magnitudes of $\mathbf{A}$ and $\mathbf{B}$ and the cosine of the angle between $\mathbf{A}$ and $\mathbf{B}$. It is represented by a boldface dot between $\mathbf{A}$ and $\mathbf{B}$. Thus if α is the angle between $\mathbf{A}$ and $\mathbf{B}$, then

$$\mathbf{A} \cdot \mathbf{B} = |\mathbf{A}||\mathbf{B}| \cos \alpha = AB \cos \alpha \qquad (1.9)$$

For the unit vectors $\mathbf{a}_1, \mathbf{a}_2, \mathbf{a}_3$, we have

$$\mathbf{a}_1 \cdot \mathbf{a}_1 = 1 \qquad \mathbf{a}_1 \cdot \mathbf{a}_2 = 0 \qquad \mathbf{a}_1 \cdot \mathbf{a}_3 = 0 \qquad (1.10a)$$

$$\mathbf{a}_2 \cdot \mathbf{a}_1 = 0 \qquad \mathbf{a}_2 \cdot \mathbf{a}_2 = 1 \qquad \mathbf{a}_2 \cdot \mathbf{a}_3 = 0 \qquad (1.10b)$$

$$\mathbf{a}_3 \cdot \mathbf{a}_1 = 0 \qquad \mathbf{a}_3 \cdot \mathbf{a}_2 = 0 \qquad \mathbf{a}_3 \cdot \mathbf{a}_3 = 1 \qquad (1.10c)$$

By noting that $\mathbf{A} \cdot \mathbf{B} = A(B \cos \alpha) = B(A \cos \alpha)$, we observe that the dot product operation consists of multiplying the magnitude of one vector by the scalar obtained by projecting the second vector onto the first vector as shown in Figures 1.4(a) and (b). The dot product operation is commutative since

$$\mathbf{B} \cdot \mathbf{A} = BA \cos \alpha = AB \cos \alpha = \mathbf{A} \cdot \mathbf{B} \qquad (1.11)$$

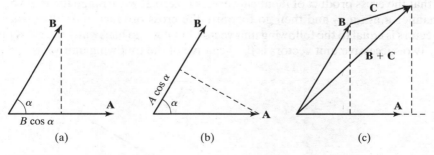

FIGURE 1.4

(a) and (b) For showing that the dot product of two vectors $\mathbf{A}$ and $\mathbf{B}$ is the product of the magnitude of one vector and the projection of the second vector onto the first vector. (c) For proving the distributive property of the dot product operation.

The distributive property also holds for the dot product, as can be seen from the construction of Figure 1.4(c), which illustrates that the projection of $(\mathbf{B} + \mathbf{C})$ onto $\mathbf{A}$ is equal to the sum of the projections of $\mathbf{B}$ and $\mathbf{C}$ onto $\mathbf{A}$. Thus,

$$\mathbf{A} \cdot (\mathbf{B} + \mathbf{C}) = \mathbf{A} \cdot \mathbf{B} + \mathbf{A} \cdot \mathbf{C} \tag{1.12}$$

Using this property, and the relationships (1.10a)–(1.10c), we have

$$
\begin{aligned}
\mathbf{A} \cdot \mathbf{B} &= (A_1\mathbf{a}_1 + A_2\mathbf{a}_2 + A_3\mathbf{a}_3) \cdot (B_1\mathbf{a}_1 + B_2\mathbf{a}_2 + B_3\mathbf{a}_3) \\
&= A_1\mathbf{a}_1 \cdot B_1\mathbf{a}_1 + A_1\mathbf{a}_1 \cdot B_2\mathbf{a}_2 + A_1\mathbf{a}_1 \cdot B_3\mathbf{a}_3 \\
&\quad + A_2\mathbf{a}_2 \cdot B_1\mathbf{a}_1 + A_2\mathbf{a}_2 \cdot B_2\mathbf{a}_2 + A_2\mathbf{a}_2 \cdot B_3\mathbf{a}_3 \\
&\quad + A_3\mathbf{a}_3 \cdot B_1\mathbf{a}_1 + A_3\mathbf{a}_3 \cdot B_2\mathbf{a}_2 + A_3\mathbf{a}_3 \cdot B_3\mathbf{a}_3 \\
&= A_1 B_1 + A_2 B_2 + A_3 B_3
\end{aligned}
\tag{1.13}
$$

Thus, the dot product of two vectors is the sum of the products of the like components of the two vectors.

Vector or Cross Product of Two Vectors

The vector or cross product of two vectors $\mathbf{A}$ and $\mathbf{B}$ is a vector quantity whose magnitude is equal to the product of the magnitudes of $\mathbf{A}$ and $\mathbf{B}$ and the sine of the smaller angle α between $\mathbf{A}$ and $\mathbf{B}$ and whose direction is the direction of advance of a right-hand screw as it is turned from $\mathbf{A}$ to $\mathbf{B}$ through the angle α, as shown in Figure 1.5. It is represented by a boldface cross between $\mathbf{A}$ and $\mathbf{B}$. Thus if $\mathbf{a}_N$ is the unit vector in the direction of advance of the right-hand screw, then

$$\mathbf{A} \times \mathbf{B} = |\mathbf{A}||\mathbf{B}| \sin \alpha \, \mathbf{a}_N = AB \sin \alpha \, \mathbf{a}_N \tag{1.14}$$

For the unit vectors $\mathbf{a}_1, \mathbf{a}_2, \mathbf{a}_3$, we have

$$\mathbf{a}_1 \times \mathbf{a}_1 = 0 \qquad \mathbf{a}_1 \times \mathbf{a}_2 = \mathbf{a}_3 \qquad \mathbf{a}_1 \times \mathbf{a}_3 = -\mathbf{a}_2 \tag{1.15a}$$

$$\mathbf{a}_2 \times \mathbf{a}_1 = -\mathbf{a}_3 \qquad \mathbf{a}_2 \times \mathbf{a}_2 = 0 \qquad \mathbf{a}_2 \times \mathbf{a}_3 = \mathbf{a}_1 \tag{1.15b}$$

$$\mathbf{a}_3 \times \mathbf{a}_1 = \mathbf{a}_2 \qquad \mathbf{a}_3 \times \mathbf{a}_2 = -\mathbf{a}_1 \qquad \mathbf{a}_3 \times \mathbf{a}_3 = 0 \tag{1.15c}$$

Note that the cross product of identical vectors is zero. If we arrange the unit vectors in the manner $\mathbf{a}_1\mathbf{a}_2\mathbf{a}_3\mathbf{a}_1\mathbf{a}_2$ and then go forward, the cross product of any two successive unit vectors is equal to the following unit vector, but if we go backward, the cross product of any two successive unit vectors is the negative of the following unit vector.

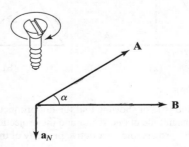

FIGURE 1.5

The cross product operation $\mathbf{A} \times \mathbf{B}$.

The cross product operation is not commutative, since

$$\mathbf{B} \times \mathbf{A} = |\mathbf{B}||\mathbf{A}| \sin \alpha \, (-\mathbf{a}_N) = -AB \sin \alpha \, \mathbf{a}_N = -\mathbf{A} \times \mathbf{B} \tag{1.16}$$

The distributive property holds for the cross product (we shall prove this later in this section) so that

$$\mathbf{A} \times (\mathbf{B} + \mathbf{C}) = \mathbf{A} \times \mathbf{B} + \mathbf{A} \times \mathbf{C} \tag{1.17}$$

Using this property and the relationships (1.15a)–(1.15c), we obtain

$$\begin{aligned}
\mathbf{A} \times \mathbf{B} &= (A_1\mathbf{a}_1 + A_2\mathbf{a}_2 + A_3\mathbf{a}_3) \times (B_1\mathbf{a}_1 + B_2\mathbf{a}_2 + B_3\mathbf{a}_3) \\
&= A_1\mathbf{a}_1 \times B_1\mathbf{a}_1 + A_1\mathbf{a}_1 \times B_2\mathbf{a}_2 + A_1\mathbf{a}_1 \times B_3\mathbf{a}_3 \\
&\quad + A_2\mathbf{a}_2 \times B_1\mathbf{a}_1 + A_2\mathbf{a}_2 \times B_2\mathbf{a}_2 + A_2\mathbf{a}_2 \times B_3\mathbf{a}_3 \\
&\quad + A_3\mathbf{a}_3 \times B_1\mathbf{a}_1 + A_3\mathbf{a}_3 \times B_2\mathbf{a}_2 + A_3\mathbf{a}_3 \times B_3\mathbf{a}_3 \\
&= A_1B_2\mathbf{a}_3 - A_1B_3\mathbf{a}_2 - A_2B_1\mathbf{a}_3 + A_2B_3\mathbf{a}_1 \\
&\quad + A_3B_1\mathbf{a}_2 - A_3B_2\mathbf{a}_1 \\
&= (A_2B_3 - A_3B_2)\mathbf{a}_1 + (A_3B_1 - A_1B_3)\mathbf{a}_2 \\
&\quad + (A_1B_2 - A_2B_1)\mathbf{a}_3
\end{aligned} \tag{1.18}$$

This can be expressed in determinant form in the manner

$$\mathbf{A} \times \mathbf{B} = \begin{vmatrix} \mathbf{a}_1 & \mathbf{a}_2 & \mathbf{a}_3 \\ A_1 & A_2 & A_3 \\ B_1 & B_2 & B_3 \end{vmatrix} \tag{1.19}$$

A triple cross product involves three vectors in two cross product operations. Caution must be exercised in evaluating a triple cross product since the order of evaluation is important, that is, $\mathbf{A} \times (\mathbf{B} \times \mathbf{C})$ is not equal to $(\mathbf{A} \times \mathbf{B}) \times \mathbf{C}$. This can be illustrated by means of a simple example involving unit vectors. Thus if $\mathbf{A} = \mathbf{a}_1$, $\mathbf{B} = \mathbf{a}_1$, and $\mathbf{C} = \mathbf{a}_2$, then

$$\mathbf{A} \times (\mathbf{B} \times \mathbf{C}) = \mathbf{a}_1 \times (\mathbf{a}_1 \times \mathbf{a}_2) = \mathbf{a}_1 \times \mathbf{a}_3 = -\mathbf{a}_2$$

whereas

$$(\mathbf{A} \times \mathbf{B}) \times \mathbf{C} = (\mathbf{a}_1 \times \mathbf{a}_1) \times \mathbf{a}_2 = 0 \times \mathbf{a}_2 = 0$$

Scalar Triple Product

The scalar triple product involves three vectors in a dot product operation and a cross product operation as, for example, $\mathbf{A} \cdot \mathbf{B} \times \mathbf{C}$. It is not necessary to include parentheses, since this quantity can be evaluated in only one manner, that is, by evaluating $\mathbf{B} \times \mathbf{C}$ first and then dotting the resulting vector with $\mathbf{A}$. It is meaningless to try to evaluate the dot product first since it results in a scalar quantity and hence we cannot proceed any further. From (1.13) and (1.19), we have

$$\mathbf{A} \cdot \mathbf{B} \times \mathbf{C} = (A_1\mathbf{a}_1 + A_2\mathbf{a}_2 + A_3\mathbf{a}_3) \cdot \begin{vmatrix} \mathbf{a}_1 & \mathbf{a}_2 & \mathbf{a}_3 \\ B_1 & B_2 & B_3 \\ C_1 & C_2 & C_3 \end{vmatrix} = \begin{vmatrix} A_1 & A_2 & A_3 \\ B_1 & B_2 & B_3 \\ C_1 & C_2 & C_3 \end{vmatrix} \tag{1.20}$$

Since the value of the determinant on the right side of (1.20) remains unchanged if the rows are interchanged in a cyclical manner,

$$\mathbf{A} \cdot \mathbf{B} \times \mathbf{C} = \mathbf{B} \cdot \mathbf{C} \times \mathbf{A} = \mathbf{C} \cdot \mathbf{A} \times \mathbf{B} \tag{1.21}$$

We shall now show that the distributive law holds for the cross product operation by using (1.21). Thus, let us consider $\mathbf{A} \times (\mathbf{B} + \mathbf{C})$. Then, if $\mathbf{D}$ is any arbitrary vector, we have

$$\mathbf{D} \cdot \mathbf{A} \times (\mathbf{B} + \mathbf{C}) = (\mathbf{B} + \mathbf{C}) \cdot (\mathbf{D} \times \mathbf{A}) = \mathbf{B} \cdot (\mathbf{D} \times \mathbf{A}) + \mathbf{C} \cdot (\mathbf{D} \times \mathbf{A})$$
$$= \mathbf{D} \cdot \mathbf{A} \times \mathbf{B} + \mathbf{D} \cdot \mathbf{A} \times \mathbf{C} = \mathbf{D} \cdot (\mathbf{A} \times \mathbf{B} + \mathbf{A} \times \mathbf{C}) \tag{1.22}$$

where we have used the distributive property of the dot product operation. Since (1.22) holds for any $\mathbf{D}$, it follows that

$$\mathbf{A} \times (\mathbf{B} + \mathbf{C}) = \mathbf{A} \times \mathbf{B} + \mathbf{A} \times \mathbf{C}$$

Example 1.1

Given three vectors

$$\mathbf{A} = \mathbf{a}_1 + \mathbf{a}_2$$
$$\mathbf{B} = \mathbf{a}_1 + 2\mathbf{a}_2 - 2\mathbf{a}_3$$
$$\mathbf{C} = \mathbf{a}_2 + 2\mathbf{a}_3$$

let us carry out several of the vector algebraic operations.

(a) $\mathbf{A} + \mathbf{B} = (\mathbf{a}_1 + \mathbf{a}_2) + (\mathbf{a}_1 + 2\mathbf{a}_2 - 2\mathbf{a}_3) = 2\mathbf{a}_1 + 3\mathbf{a}_2 - 2\mathbf{a}_3$

(b) $\mathbf{B} - \mathbf{C} = (\mathbf{a}_1 + 2\mathbf{a}_2 - 2\mathbf{a}_3) - (\mathbf{a}_2 + 2\mathbf{a}_3) = \mathbf{a}_1 + \mathbf{a}_2 - 4\mathbf{a}_3$

(c) $4\mathbf{C} = 4(\mathbf{a}_2 + 2\mathbf{a}_3) = 4\mathbf{a}_2 + 8\mathbf{a}_3$

(d) $|\mathbf{B}| = |\mathbf{a}_1 + 2\mathbf{a}_2 - 2\mathbf{a}_3| = \sqrt{(1)^2 + (2)^2 + (-2)^2} = 3$

(e) $\mathbf{a}_B = \dfrac{\mathbf{B}}{|\mathbf{B}|} = \dfrac{\mathbf{a}_1 + 2\mathbf{a}_2 - 2\mathbf{a}_3}{3} = \dfrac{1}{3}\mathbf{a}_1 + \dfrac{2}{3}\mathbf{a}_2 - \dfrac{2}{3}\mathbf{a}_3$

(f) $\mathbf{A} \cdot \mathbf{B} = (\mathbf{a}_1 + \mathbf{a}_2) \cdot (\mathbf{a}_1 + 2\mathbf{a}_2 - 2\mathbf{a}_3) = (1)(1) + (1)(2) + (0)(-2) = 3$

(g) $\mathbf{A} \times \mathbf{B} = \begin{vmatrix} \mathbf{a}_1 & \mathbf{a}_2 & \mathbf{a}_3 \\ 1 & 1 & 0 \\ 1 & 2 & -2 \end{vmatrix} = (-2 - 0)\mathbf{a}_1 + (0 + 2)\mathbf{a}_2 + (2 - 1)\mathbf{a}_3$

$$= -2\mathbf{a}_1 + 2\mathbf{a}_2 + \mathbf{a}_3$$

(h) $(\mathbf{A} \times \mathbf{B}) \times \mathbf{C} = \begin{vmatrix} \mathbf{a}_1 & \mathbf{a}_2 & \mathbf{a}_3 \\ -2 & 2 & 1 \\ 0 & 1 & 2 \end{vmatrix} = 3\mathbf{a}_1 + 4\mathbf{a}_2 - 2\mathbf{a}_3$

(i) $\mathbf{A} \cdot \mathbf{B} \times \mathbf{C} = \begin{vmatrix} 1 & 1 & 0 \\ 1 & 2 & -2 \\ 0 & 1 & 2 \end{vmatrix} = (1)(6) + (1)(-2) + (0)(1) = 4$

1.2   CARTESIAN COORDINATE SYSTEM

In the previous section we introduced the technique of expressing a vector at a point in space in terms of its component vectors along a set of three mutually orthogonal directions defined by three mutually orthogonal unit vectors at that point. Now, in order to relate vectors at one point in space to vectors at another point in space, we must define the set of three reference directions at each and every point in space. To do this in a systematic manner, we need to use a coordinate system. Although there are several different coordinate systems, we shall use for the most part of our study the simplest of these, namely, the Cartesian coordinate system, also known as the *rectangular coordinate system*, to keep the geometry simple and yet sufficient to learn the fundamentals of electromagnetics. We shall, however, find it necessary in a few cases to resort to the use of cylindrical and spherical coordinate systems. Hence, a discussion of these coordinate systems is included in Appendix A. In this section we introduce the Cartesian coordinate system.

The Cartesian coordinate system is defined by a set of three mutually orthogonal planes, as shown in Figure 1.6(a). The point at which the three planes intersect is known as the origin O. The origin is the reference point relative to which we locate any other point in space. Each pair of planes intersects in a straight line. Hence, the three planes define a set of three straight lines that form the coordinate axes. These coordinate axes are denoted as the x-, y-, and z-axes. Values of x, y, and z are measured from the origin and hence the coordinates of the origin are $(0, 0, 0)$, that is, $x = 0$, $y = 0$, and $z = 0$. Directions in which values of x, y, and z increase along the respective coordinate axes are indicated by arrowheads. The same set of three directions is used to erect a set of three unit vectors, denoted $\mathbf{a}_x$, $\mathbf{a}_y$, and $\mathbf{a}_z$, as shown in Figure 1.6(a), for the purpose of describing vectors drawn at the origin. Note that the positive x-, y-, and z-directions are chosen such that they form a right-handed system, that is, a system for which $\mathbf{a}_x \times \mathbf{a}_y = \mathbf{a}_z$.

On one of the three planes, namely, the yz-plane, the value of x is constant and equal to zero, its value at the origin, since movement on this plane does not require any movement in the x-direction. Similarly, on the zx-plane the value of y is constant and equal to zero, and on the xy-plane the value of z is constant and equal to zero. Any point other than the origin is now given by the intersection of three planes obtained by incrementing the values of the coordinates by appropriate amounts. For example, by displacing the $x = 0$ plane by 2 units in the positive x-direction, the $y = 0$ plane by 5 units in the positive y-direction, and the $z = 0$ plane by 4 units in the positive z-direction, we obtain the planes $x = 2$, $y = 5$, and $z = 4$, respectively, which intersect at the point $(2, 5, 4)$, as shown in Figure 1.6(b). The intersections of pairs of these planes define three straight lines along which we can erect the unit vectors $\mathbf{a}_x$, $\mathbf{a}_y$, and $\mathbf{a}_z$ toward the directions of increasing values of x, y, and z, respectively, for the purpose of describing vectors drawn at that point. These unit vectors are parallel to the corresponding unit vectors drawn at the origin, as can be seen from Figure 1.6(b). The same is true for any point in space in the Cartesian coordinate system. Thus, each one of the three unit vectors in the Cartesian coordinate system has the same direction at all points and hence it is uniform. This behavior does not, however, hold for all unit vectors in the cylindrical and spherical coordinate systems.

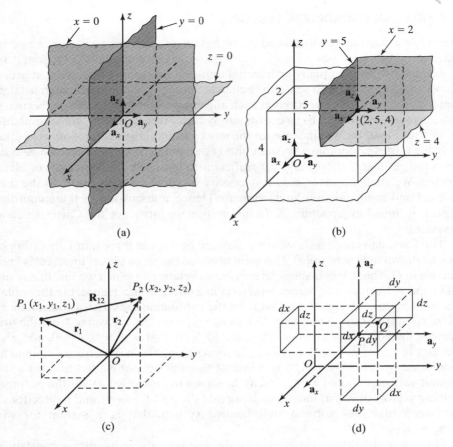

FIGURE 1.6

Cartesian coordinate system. (a) The three orthogonal planes defining the coordinate system. (b) Unit vectors at an arbitrary point. (c) Vector from one arbitrary point to another arbitrary point. (d) Differential lengths, surfaces, and volume formed by incrementing the coordinates.

It is now a simple matter to apply what we have learned in Section 1.1 to vectors in Cartesian coordinates. All we need to do is to replace the subscripts 1, 2, and 3 for the unit vectors and the components along the unit vectors by the subscripts x, y, and z, respectively, and also utilize the property that $\mathbf{a}_x$, $\mathbf{a}_y$, and $\mathbf{a}_z$ are uniform vectors. Thus let us, for example, obtain the expression for the vector $\mathbf{R}_{12}$ drawn from point $P_1(x_1, y_1, z_1)$ to point $P_2(x_2, y_2, z_2)$, as shown in Figure 1.6(c). To do this, we note that the position vector $\mathbf{r}_1$ drawn from the origin to the point P_1 is given by

$$\mathbf{r}_1 = x_1\mathbf{a}_x + y_1\mathbf{a}_y + z_1\mathbf{a}_z \tag{1.23}$$

and that the position vector $\mathbf{r}_2$ drawn from the origin to the point P_2 is given by

$$\mathbf{r}_2 = x_2\mathbf{a}_x + y_2\mathbf{a}_y + z_2\mathbf{a}_z \tag{1.24}$$

The position vector is so called because it defines the position of the point in space relative to the origin. Since, from the rule for vector addition, $\mathbf{r}_1 + \mathbf{R}_{12} = \mathbf{r}_2$, we obtain

$$\mathbf{R}_{12} = \mathbf{r}_2 - \mathbf{r}_1 = (x_2 - x_1)\mathbf{a}_x + (y_2 - y_1)\mathbf{a}_y + (z_2 - z_1)\mathbf{a}_z \qquad (1.25)$$

In our study of electromagnetic fields, we have to work with line integrals, surface integrals, and volume integrals. As in elementary calculus, these involve differential lengths, surfaces, and volumes, obtained by incrementing the coordinates by infinitesimal amounts. Since in the Cartesian coordinate system the three coordinates represent lengths, the differential length elements obtained by incrementing one coordinate at a time, keeping the other two coordinates constant, are $dx\,\mathbf{a}_x$, $dy\,\mathbf{a}_y$, and $dz\,\mathbf{a}_z$ for the x-, y-, and z-coordinates, respectively, as shown in Figure 1.6(d), at an arbitrary point $P(x, y, z)$. The three differential length elements form the contiguous edges of a rectangular box in which the corner Q diagonally opposite to P has the coordinates $(x + dx, y + dy, z + dz)$. The differential length vector $d\mathbf{l}$ from P to Q is simply the vector sum of the three differential length elements. Thus,

$$d\mathbf{l} = dx\,\mathbf{a}_x + dy\,\mathbf{a}_y + dz\,\mathbf{a}_z \qquad (1.26)$$

The box has six differential surfaces with each surface defined by two of the three length elements, as shown by the projections onto the coordinate planes in Figure 1.6(d). The orientation of a differential surface dS is specified by a unit vector normal to it, that is, a unit vector perpendicular to any two vectors tangential to the surface. Unless specified, the normal vector can be drawn toward any one of the two sides of a given surface. Thus, the differential surfaces formed by the pairs of differential length elements are

$$\pm dS\,\mathbf{a}_z = \pm dx\,dy\,\mathbf{a}_z = \pm dx\,\mathbf{a}_x \times dy\,\mathbf{a}_y \qquad (1.27a)$$

$$\pm dS\,\mathbf{a}_x = \pm dy\,dz\,\mathbf{a}_x = \pm dy\,\mathbf{a}_y \times dz\,\mathbf{a}_z \qquad (1.27b)$$

$$\pm dS\,\mathbf{a}_y = \pm dz\,dx\,\mathbf{a}_y = \pm dz\,\mathbf{a}_z \times dx\,\mathbf{a}_x \qquad (1.27c)$$

Finally, the differential volume dv formed by the three differential lengths is simply the volume of the box, that is,

$$dv = dx\,dy\,dz \qquad (1.28)$$

We shall now briefly review some elementary analytic geometrical details that will be useful in our study of electromagnetics. An arbitrary surface is defined by an equation of the form

$$f(x, y, z) = 0 \qquad (1.29)$$

In particular, the equation for a plane surface making intercepts a, b, and c on the x-, y-, and z-axes, respectively, is given by

$$\frac{x}{a} + \frac{y}{b} + \frac{z}{c} = 1 \qquad (1.30)$$

Since a curve is the intersection of two surfaces, an arbitrary curve is defined by a pair of equations

$$f(x, y, z) = 0 \quad \text{and} \quad g(x, y, z) = 0 \qquad (1.31)$$

Alternatively, a curve is specified by a set of three parametric equations

$$x = x(t), \qquad y = y(t), \qquad z = z(t) \qquad (1.32)$$

where t is an independent parameter. For example, a straight line passing through the origin and making equal angles with the positive x-, y-, and z-axes is given by the pair of equations $y = x$ and $z = x$, or by the set of three parametric equations $x = t$, $y = t$, and $z = t$.

Example 1.2

Let us find a unit vector normal to the plane

$$5x + 2y + 4z = 20$$

By writing the given equation for the plane in the form

$$\frac{x}{4} + \frac{y}{10} + \frac{z}{5} = 1$$

we identify the intercepts made by the plane on the x-, y-, and z-axes to be 4, 10, and 5, respectively. The portion of the plane lying in the first octant of the coordinate system is shown in Figure 1.7.

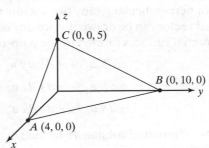

FIGURE 1.7
The plane surface $5x + 2y + 4z = 20$.

To find a unit vector normal to the plane, we consider two vectors lying in the plane and evaluate their cross product. Thus considering the vectors $\mathbf{R}_{AB}$ and $\mathbf{R}_{AC}$, we have from (1.25),

$$\mathbf{R}_{AB} = (0 - 4)\mathbf{a}_x + (10 - 0)\mathbf{a}_y + (0 - 0)\mathbf{a}_z = -4\mathbf{a}_x + 10\mathbf{a}_y$$

$$\mathbf{R}_{AC} = (0 - 4)\mathbf{a}_x + (0 - 0)\mathbf{a}_y + (5 - 0)\mathbf{a}_z = -4\mathbf{a}_x + 5\mathbf{a}_z$$

The cross product of $\mathbf{R}_{AB}$ and $\mathbf{R}_{AC}$ is then given by

$$\mathbf{R}_{AB} \times \mathbf{R}_{AC} = \begin{vmatrix} \mathbf{a}_x & \mathbf{a}_y & \mathbf{a}_z \\ -4 & 10 & 0 \\ -4 & 0 & 5 \end{vmatrix} = 50\mathbf{a}_x + 20\mathbf{a}_y + 40\mathbf{a}_z$$

This vector is perpendicular to both $\mathbf{R}_{AB}$ and $\mathbf{R}_{AC}$ and hence to the plane. Finally, the required unit vector is obtained by dividing $\mathbf{R}_{AB} \times \mathbf{R}_{AC}$ by its magnitude. Thus, it is equal to

$$\frac{50\mathbf{a}_x + 20\mathbf{a}_y + 40\mathbf{a}_z}{|50\mathbf{a}_x + 20\mathbf{a}_y + 40\mathbf{a}_z|} = \frac{5\mathbf{a}_x + 2\mathbf{a}_y + 4\mathbf{a}_z}{\sqrt{25 + 4 + 16}} = \frac{1}{3\sqrt{5}}(5\mathbf{a}_x + 2\mathbf{a}_y + 4\mathbf{a}_z)$$

1.3   SCALAR AND VECTOR FIELDS

Before we take up the task of studying electromagnetic fields, we must understand what is meant by a *field*. A field is associated with a region in space and we say that a field exists in the region if there is a physical phenomenon associated with points in that region. For example, in everyday life we are familiar with the earth's gravitational field. We do not "see" the field in the same manner as we see light rays, but we know of its existence in the sense that objects are acted upon by the gravitational force of the earth. In a broader context, we can talk of the field of any physical quantity as being a description, mathematical or graphical, of how the quantity varies from one point to another in the region of the field and with time. We can talk of scalar or vector fields depending on whether the quantity of interest is a scalar or a vector. We can talk of static or time-varying fields depending on whether the quantity of interest is independent of or changing with time.

We shall begin our discussion of fields with some simple examples of scalar fields. Thus, let us consider the case of the conical pyramid shown in Figure 1.8(a). A description of the height of the pyramidal surface versus position on its base is an example of a scalar field involving two variables. Choosing the origin to be the projection of the vertex of the cone onto the base and setting up an xy-coordinate system to locate points on the base, we obtain the height field as a function of x and y to be

$$h(x, y) = 6 - 2\sqrt{x^2 + y^2} \tag{1.33}$$

Although we are able to depict the height variation of points on the conical surface graphically by using the third coordinate for h, we will have to be content with the visualization of the height field by a set of constant-height contours on the xy-plane if only two coordinates were available, as in the case of a two-dimensional space. For the field under consideration, the constant-height contours are circles in the xy-plane centered at the origin and equally spaced for equal increments of the height value, as shown in Figure 1.8(a).

For an example of a scalar field in three dimensions, let us consider a rectangular room and the distance field of points in the room from one corner of the room, as shown in Figure 1.8(b). For convenience, we choose this corner to be the origin O and set up a Cartesian coordinate system with the three contiguous edges meeting at that point as the coordinate axes. Each point in the room is defined by a set of values for the three coordinates x, y, and z. The distance r from the origin to that point is $\sqrt{x^2 + y^2 + z^2}$. Thus, the distance field of points in the room from the origin is given by

$$r(x, y, z) = \sqrt{x^2 + y^2 + z^2} \tag{1.34}$$

Since the three coordinates are already used up for defining the points in the field region, we have to visualize the distance field by means of a set of constant-distance surfaces. A constant-distance surface is a surface for which points on it correspond to a particular constant value of r. For the case under consideration, the constant-distance surfaces are spherical surfaces centered at the origin and are equally spaced for equal increments in the value of the distance, as shown in Figure 1.8(b).

The fields we have discussed thus far are static fields. A simple example of a time-varying scalar field is provided by the temperature field associated with points in a room, especially when it is being heated or cooled. Just as in the case of the distance

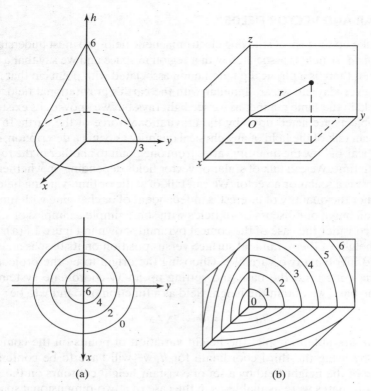

FIGURE 1.8

(a) A conical pyramid lying above the xy-plane, and a set of constant-height contours for the conical surface. (b) A rectangular room, and a set of constant-distance surfaces depicting the distance field of points in the room from one corner of the room.

field of Figure 1.8(b), we set up a three-dimensional coordinate system and to each set of three coordinates corresponding to the location of a point in the room, we assign a number to represent the temperature T at that point. Since the temperature at that point, however, varies with time t, this number is a function of time. Thus, we describe mathematically the time-varying temperature field in the room by a function $T(x, y, z, t)$. For any given instant of time, we can visualize a set of constant-temperature or isothermal surfaces corresponding to particular values of T as representing the temperature field for that value of time. For a different instant of time, we will have a different set of isothermal surfaces for the same values of T. Thus, we can visualize the time-varying temperature field in the room by a set of isothermal surfaces continuously changing their shapes as though in a motion picture.

The foregoing discussion of scalar fields may now be extended to vector fields by recalling that a vector quantity has associated with it a direction in space in addition to magnitude. Hence, in order to describe a vector field we attribute to each point in the field region a vector that represents the magnitude and direction of the physical quantity under consideration at that point. Since a vector at a given point can be expressed as the sum of its components along the set of unit vectors at that point, a mathematical description of the vector field involves simply the descriptions of the three component scalar fields. Thus for a vector field $\mathbf{F}$ in the Cartesian coordinate system, we have

$$\mathbf{F}(x, y, z, t) = F_x(x, y, z, t)\mathbf{a}_x + F_y(x, y, z, t)\mathbf{a}_y + F_z(x, y, z, t)\mathbf{a}_z \quad (1.35)$$

Similar expressions hold in the cylindrical and spherical coordinate systems. We should, however, note that two of the unit vectors in the cylindrical coordinate system and all the unit vectors in the spherical coordinate system are themselves functions of the coordinates.

To illustrate the graphical description of a vector field, let us consider the linear velocity vector field associated with points on a circular disk rotating about its center with a constant angular velocity ω rad/s. We know that the magnitude of the linear velocity of a point on the disk is then equal to the product of the angular velocity ω and the radial distance r of the point from the center of the disk. The direction of the linear velocity is tangential to the circle drawn through that point and concentric with the disk. Hence, we may depict the linear velocity field by drawing at several points on the disk vectors that are tangential to the circles concentric with the disk and passing through those points, and whose lengths are proportional to the radii of the circles, as shown in Figure 1.9(a), where the points are carefully selected in order to reveal the circular symmetry of the field with respect to the center of the disk. We, however, find that this method of representation of the vector field results in a congested sketch of vectors. Hence, we may simplify the sketch by omitting the vectors and simply placing arrowheads along the circles, giving us a set of *direction lines*, also known as *stream lines* and *flux lines*, which simply represent the direction of the field at points on them. We note that for the field under consideration the direction lines are also contours of constant magnitude of the velocity, and hence by increasing the density of the direction lines as r increases, we can indicate the magnitude variation, as shown in Figure 1.9(b).

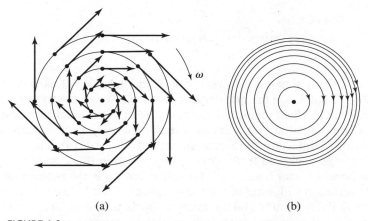

(a)   (b)

FIGURE 1.9

(a) Linear velocity vector field associated with points on a rotating disk.
(b) Same as (a) except that the vectors are omitted, and the density of direction lines is used to indicate the magnitude variation.

1.4 SINUSOIDALLY TIME-VARYING FIELDS

In our study of electromagnetic fields we will be particularly interested in fields that vary sinusoidally with time. Hence, we shall devote this section to a discussion of sinusoidally time-varying fields. Let us first consider a scalar sinusoidal function of time. Such a function is given by an expression of the form $A \cos (\omega t + \phi)$ where A is the peak amplitude of the sinusoidal variation, $\omega = 2\pi f$ is the radian frequency, f is the linear frequency, and $(\omega t + \phi)$ is the phase. In particular, the phase of the function for $t = 0$ is ϕ. A plot of this function versus t, shown in Figure 1.10, illustrates how the function changes periodically between positive and negative values. If we now have a sinusoidally time-varying scalar field, we can visualize the field quantity varying sinusoidally with time at each point in the field region with the amplitude and phase governed by the spatial dependence of the field quantity. Thus, for example, the field $Ae^{-\alpha z} \cos (\omega t - \beta z)$, where A, α, and β are positive constants, is characterized by sinusoidal time variations with amplitude decreasing exponentially with z and the phase at any given time decreasing linearly with z.

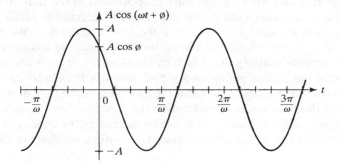

FIGURE 1.10

Sinusoidally time-varying scalar function $A \cos (\omega t + \phi)$.

For a sinusoidally time-varying vector field, the behavior of each component of the field may be visualized in the manner just discussed. If we now fix our attention on a particular point in the field region, we can visualize the sinusoidal variation with time of a particular component at that point by a vector changing its magnitude and direction as shown, for example, for the x-component in Figure 1.11(a). Since the tip of the vector simply moves back and forth along a line, which in this case is parallel to the x-axis, the component vector is said to be *linearly polarized* in the x-direction. Similarly, the sinusoidal variation with time of the y-component of the field can be visualized by a vector changing its magnitude and direction as shown in Figure 1.11(b), not necessarily with the same amplitude and phase as those of the x-component. Since the tip of the vector moves back and forth parallel to the y-axis, the y-component is said to be linearly polarized in the y-direction. In the same manner, the z-component is linearly polarized in the z-direction.

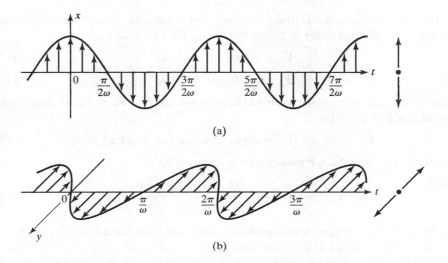

FIGURE 1.11

(a) Time variation of a linearly polarized vector in the x-direction. (b) Time variation of a linearly polarized vector in the y-direction.

If two components sinusoidally time-varying vectors have arbitrary amplitudes but are in phase or phase opposition as, for example,

$$\mathbf{F}_1 = F_1 \cos(\omega t + \phi)\,\mathbf{a}_x \tag{1.36a}$$

$$\mathbf{F}_2 = \pm F_2 \cos(\omega t + \phi)\,\mathbf{a}_y \tag{1.36b}$$

then the sum vector $\mathbf{F} = \mathbf{F}_1 + \mathbf{F}_2$ is linearly polarized in a direction making an angle

$$\alpha = \tan^{-1}\frac{F_y}{F_x} = \pm\tan^{-1}\frac{F_2}{F_1}$$

with the x-direction, as shown in the series of sketches in Figure 1.12 for the in-phase case illustrating the time history of the magnitude and direction of $\mathbf{F}$ over an interval of one period.

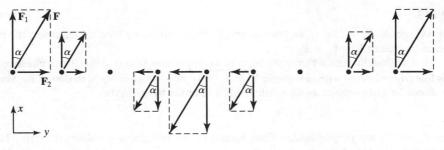

FIGURE 1.12

The sum vector of two linearly polarized vectors in phase is a linearly polarized vector.

If two component sinusoidally time-varying vectors have equal amplitudes, differ in direction by 90°, and differ in phase by $\pi/2$, as, for example,

$$\mathbf{F}_1 = F_0 \cos(\omega t + \phi)\, \mathbf{a}_x \tag{1.37a}$$

$$\mathbf{F}_2 = F_0 \sin(\omega t + \phi)\, \mathbf{a}_y \tag{1.37b}$$

then, to determine the *polarization* of the sum vector $\mathbf{F} = \mathbf{F}_1 + \mathbf{F}_2$, we note that the magnitude of $\mathbf{F}$ is given by

$$|\mathbf{F}| = |F_0 \cos(\omega t + \phi)\, \mathbf{a}_x + F_0 \sin(\omega t + \phi)\, \mathbf{a}_y| = F_0 \tag{1.38}$$

and that the angle α which $\mathbf{F}$ makes with $\mathbf{a}_x$ is given by

$$\alpha = \tan^{-1}\frac{F_y}{F_x} = \tan^{-1}\left[\frac{F_0 \sin(\omega t + \phi)}{F_0 \cos(\omega t + \phi)}\right] = \omega t + \phi \tag{1.39}$$

Thus, the sum vector rotates with constant magnitude F_0 and at a rate of ω rad/s so that its tip describes a circle. The sum vector is then said to be *circularly polarized*. The series of sketches in Figure 1.13 illustrates the time history of the magnitude and direction of $\mathbf{F}$ over an interval of one period.

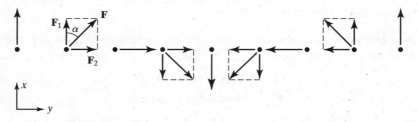

FIGURE 1.13

Circular polarization.

For the general case in which two component sinusoidally time-varying vectors differ in amplitude, direction, and phase by arbitrary amounts, the sum vector is *elliptically polarized*, that is, its tip describes an ellipse.

Example 1.3

Given two vectors $\mathbf{F}_1 = (3\mathbf{a}_x - 4\mathbf{a}_z)\cos\omega t$ and $\mathbf{F}_2 = 5\mathbf{a}_y \sin\omega t$, we wish to determine the polarization of the vector $\mathbf{F} = \mathbf{F}_1 + \mathbf{F}_2$.

We note that the vector $\mathbf{F}_1$, consisting of two components (x and z) that are in phase opposition, is linearly polarized with amplitude $\sqrt{3^2 + (-4)^2}$ or 5, which is equal to that of $\mathbf{F}_2$. Since $\mathbf{F}_1$ varies as $\cos\omega t$ and $\mathbf{F}_2$ varies as $\sin\omega t$, they differ in phase by $\pi/2$. Also,

$$\mathbf{F}_1 \cdot \mathbf{F}_2 = (3\mathbf{a}_x - 4\mathbf{a}_z) \cdot 5\mathbf{a}_y = 0$$

so that $\mathbf{F}_1$ and $\mathbf{F}_2$ are perpendicular. Thus $\mathbf{F}_1$ and $\mathbf{F}_2$ are two linearly polarized vectors having equal amplitudes but differing in direction by 90° and differing in phase by $\pi/2$. Hence, $\mathbf{F} = \mathbf{F}_1 + \mathbf{F}_2$ is circularly polarized.

In the remainder of this section we shall briefly review the phasor technique which, as the student may have already learned in sinusoidal steady-state circuit analysis, is very useful in carrying out mathematical manipulations involving sinusoidally time-varying quantities. Let us consider the simple problem of adding the two quantities $10 \cos \omega t$ and $10 \sin (\omega t - 30°)$. To illustrate the basis behind the phasor technique, we carry out the following steps:

$$
\begin{aligned}
10 \cos \omega t + 10 \sin (\omega t - 30°) &= 10 \cos \omega t + 10 \cos (\omega t - 120°) \\
&= \text{Re}[10e^{j\omega t}] + \text{Re}[10e^{j(\omega t - 2\pi/3)}] \\
&= \text{Re}[10e^{j0}e^{j\omega t}] + \text{Re}[10e^{-j2\pi/3}e^{j\omega t}] \\
&= \text{Re}[(10e^{j0} + 10e^{-j2\pi/3})e^{j\omega t}] \\
&= \text{Re}[10e^{-j\pi/3}e^{j\omega t}] \\
&= \text{Re}[10e^{j(\omega t - \pi/3)}] \\
&= 10 \cos (\omega t - 60°)
\end{aligned}
\tag{1.40}
$$

where Re stands for *real part of*, and the addition of the two complex numbers $10e^{j0}$ and $10e^{-j2\pi/3}$ is performed by locating them in the complex plane and then using the parallelogram law of addition of complex numbers, as shown in Figure 1.14. Alternatively, the complex numbers may be expressed in terms of their real and imaginary parts and then added up for conversion into exponential form in the manner

$$
\begin{aligned}
10e^{j0} + 10e^{-j2\pi/3} &= (10 + j0) + (-5 - j8.66) \\
&= 5 - j8.66 = \sqrt{5^2 + 8.66^2}\, e^{-j \tan^{-1} 8.66/5} \\
&= 10e^{-j\pi/3}
\end{aligned}
\tag{1.41}
$$

In practice, we do not write all of the steps shown in (1.40). First, we express all functions in their cosine forms and then recognize the phasor corresponding to each cosine function as the complex number having the magnitude equal to the amplitude

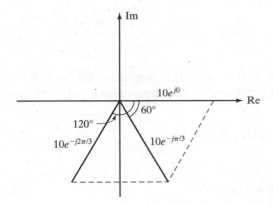

FIGURE 1.14

Addition of two complex numbers.

of the cosine function and phase angle equal to the phase angle of the cosine function for $t = 0$. For the above example, the complex numbers $10e^{j0}$ and $10e^{-j2\pi/3}$ are the phasors corresponding to $10 \cos \omega t$ and $10 \sin (\omega t - 30°)$, respectively. Then we add the phasors and from the sum phasor write down the required cosine function. Thus, the steps involved are as shown in Figure 1.15.

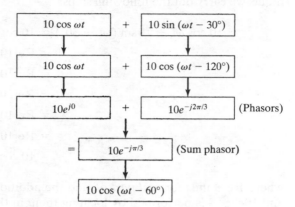

FIGURE 1.15

Block diagram of steps involved in the application of phasor technique to the addition of two sinusoidally time-varying functions.

The same technique is adopted for solving differential equations by recognizing, for example, that

$$\frac{d}{dt} [A \cos (\omega t + \theta)] = -A\omega \sin (\omega t + \theta) = A\omega \cos (\omega t + \theta + \pi/2)$$

and hence the phasor for $\dfrac{d}{dt} [A \cos (\omega t + \theta)]$ is

$$A\omega e^{j(\theta + \pi/2)} = A\omega e^{j\pi/2} e^{j\theta} = j\omega A e^{j\theta}$$

or $j\omega$ times the phasor for $A \cos (\omega t + \theta)$. Thus, the differentiation operation is replaced by $j\omega$ for converting the differential equation into an algebraic equation involving phasors. To illustrate this, let us consider the differential equation

$$10^{-3} \frac{di}{dt} + i = 10 \cos 1000t \qquad (1.42)$$

The solution for this is of the form $i = I_0 \cos (\omega t + \theta)$. Recognizing that $\omega = 1000$ and replacing d/dt by $j1000$ and all time functions by their phasors, we obtain the corresponding algebraic equation as

$$10^{-3}(j1000\bar{I}) + \bar{I} = 10e^{j0} \qquad (1.43)$$

or

$$\bar{I}(1 + j1) = 10e^{j0} \qquad (1.44)$$

where the overbar above I indicates the complex nature of the quantity. Solving (1.44) for $\bar{I}$, we obtain

$$\bar{I} = \frac{10e^{j0}}{1 + j1} = \frac{10e^{j0}}{\sqrt{2}e^{j\pi/4}} = 7.07e^{-j\pi/4} \tag{1.45}$$

and finally

$$i = 7.07 \cos\left(1000t - \frac{\pi}{4}\right) \tag{1.46}$$

1.5   THE ELECTRIC FIELD

Basic to our study of the fundamentals of electromagnetics is an understanding of the concepts of electric and magnetic fields. Hence, we shall devote this and the following section to an introduction of the electric and magnetic fields. From our study of Newton's law of gravitation in elementary physics, we are familiar with the gravitational force field associated with material bodies by virtue of their physical property known as *mass*. Newton's experiments showed that the gravitational force of attraction between two bodies of masses m_1 and m_2 separated by a distance R, which is very large compared to their sizes, is equal to $m_1 m_2 G/R^2$ where G is the constant of universal gravitation. In a similar manner, a force field known as the *electric field* is associated with bodies that are *charged*. A material body may be charged positively or negatively or may possess no net charge. In the International System of Units that we shall use throughout this book, the unit of charge is coulomb, abbreviated C. The charge of an electron is -1.60219×10^{-19} C. Alternatively, approximately 6.24×10^{18} electrons represent a charge of one negative coulomb.

Experiments conducted by Coulomb showed that the following hold for two charged bodies that are very small in size compared to their separation so that they can be considered as *point charges*:

1. The magnitude of the force is proportional to the product of the magnitudes of the charges.
2. The magnitude of the force is inversely proportional to the square of the distance between the charges.
3. The magnitude of the force depends on the medium.
4. The direction of the force is along the line joining the charges.
5. Like charges repel; unlike charges attract.

For free space, the constant of proportionality is $1/4\pi\epsilon_0$ where ϵ_0 is known as the permittivity of free space, having a value 8.854×10^{-12} or approximately equal to $10^{-9}/36\pi$. Thus, if we consider two point charges Q_1 C and Q_2 C separated R m in free space, as shown in Figure 1.16, then the forces F_1 and F_2 experienced by Q_1 and Q_2, respectively, are given by

$$\mathbf{F}_1 = \frac{Q_1 Q_2}{4\pi\epsilon_0 R^2}\mathbf{a}_{21} \tag{1.47a}$$

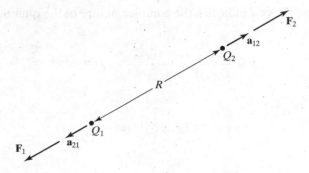

FIGURE 1.16

Forces experienced by two point charges Q_1 and Q_2.

and

$$\mathbf{F}_2 = \frac{Q_2 Q_1}{4\pi\epsilon_0 R^2}\mathbf{a}_{12} \tag{1.47b}$$

where $\mathbf{a}_{21}$ and $\mathbf{a}_{12}$ are unit vectors along the line joining Q_1 and Q_2, as shown in Figure 1.16. Equations (1.47a) and (1.47b) represent Coulomb's law. Since the units of force are newtons, we note that ϵ_0 has the units (coulomb)^2 per (newton-meter^2). These are commonly known as *farads per meter*, where a farad is (coulomb)^2 per newton-meter.

In the case of the gravitational field of a material body, we define the gravitational field intensity as the force per unit mass experienced by a small test mass placed in that field. In a similar manner, the force per unit charge experienced by a small test charge placed in an electric field is known as the *electric field intensity*, denoted by the symbol **E**. Alternatively, if in a region of space, a test charge q experiences a force **F**, then the region is said to be characterized by an electric field of intensity **E** given by

$$\mathbf{E} = \frac{\mathbf{F}}{q} \tag{1.48}$$

The unit of electric field intensity is newton per coulomb, or more commonly volt per meter, where a volt is newton-meter per coulomb. The test charge should be so small that it does not alter the electric field in which it is placed. Ideally, **E** is defined in the limit that q tends to zero, that is,

$$\mathbf{E} = \lim_{q \to 0} \frac{\mathbf{F}}{q} \tag{1.49}$$

Equation (1:49) is the defining equation for the electric field intensity irrespective of the source of the electric field. Just as one body by virtue of its mass is the source of a gravitational field acting upon other bodies by virtue of their masses, a charged body is the source of an electric field acting upon other charged bodies. We will, however, learn in Chapter 2 that there exists another source for the electric field, namely, a time-varying magnetic field.

Returning now to Coulomb's law and letting one of the two charges in Figure 1.16, say Q_2, be a small test charge q, we have

$$\mathbf{F}_2 = \frac{Q_1 q}{4\pi\epsilon_0 R^2}\mathbf{a}_{12} \tag{1.50}$$

The electric field intensity $\mathbf{E}_2$ at the test charge due to the point charge Q_1 is then given by

$$\mathbf{E}_2 = \frac{\mathbf{F}_2}{q} = \frac{Q_1}{4\pi\epsilon_0 R^2}\mathbf{a}_{12} \tag{1.51}$$

Generalizing this result by making R a variable, that is, by moving the test charge around in the medium, writing the expression for the force experienced by it, and dividing the force by the test charge, we obtain the electric field intensity $\mathbf{E}$ of a point charge Q to be

$$\mathbf{E} = \frac{Q}{4\pi\epsilon_0 R^2}\mathbf{a}_R \tag{1.52}$$

where R is the distance from the point charge to the point at which the field intensity is to be computed and $\mathbf{a}_R$ is the unit vector along the line joining the two points under consideration and directed away from the point charge. The electric field intensity due to a point charge is thus directed everywhere radially away from the point charge and its constant-magnitude surfaces are spherical surfaces centered at the point charge, as shown in Figure 1.17.

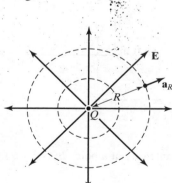

FIGURE 1.17

Direction lines and constant-magnitude surfaces of electric field due to a point charge.

If we now have several point charges $Q_1, Q_2, \ldots$, as shown in Figure 1.18, the force experienced by a test charge situated at a point P is the vector sum of the forces experienced by the test charge due to the individual charges. It then follows that the

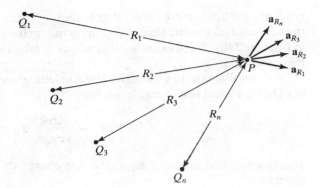

FIGURE 1.18

A collection of point charges and unit
vectors along the directions of their
electric fields at a point P.

electric field intensity at point P is the superposition of the electric field intensities due
to the individual charges, that is,

$$\mathbf{E} = \frac{Q_1}{4\pi\epsilon_0 R_1^2}\mathbf{a}_{R_1} + \frac{Q_2}{4\pi\epsilon_0 R_2^2}\mathbf{a}_{R_2} + \cdots + \frac{Q_n}{4\pi\epsilon_0 R_n^2}\mathbf{a}_{R_n} \qquad (1.53)$$

Let us now consider an example.

Example 1.4

Figure 1.19 shows eight point charges situated at the corners of a cube. We wish to find the electric field intensity at each point charge, due to the remaining seven point charges.

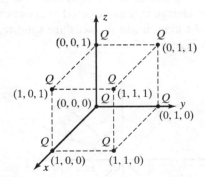

FIGURE 1.19

A cubical arrangement of point charges.

First, we note from (1.52) that the electric field intensity at a point $B(x_2, y_2, z_2)$ due to a
point charge Q at point $A(x_1, y_1, z_1)$ is given by

$$\mathbf{E}_B = \frac{Q}{4\pi\epsilon_0(AB)^2}\mathbf{a}_{AB} = \frac{Q}{4\pi\epsilon_0(AB)^2}\frac{\mathbf{R}_{AB}}{(AB)} = \frac{Q(\mathbf{R}_{AB})}{4\pi\epsilon_0(AB)^3}$$

$$= \frac{Q}{4\pi\epsilon_0}\frac{(x_2 - x_1)\mathbf{a}_x + (y_2 - y_1)\mathbf{a}_y + (z_2 - z_1)\mathbf{a}_z}{[(x_2 - x_1)^2 + (y_2 - y_1)^2 + (z_2 - z_1)^2]^{3/2}} \qquad (1.54)$$

where we have used $\mathbf{R}_{AB}$ to denote the vector from A to B. Let us now consider the point $(1, 1, 1)$. Applying (1.54) to each of the charges at the seven other points and using (1.53), we obtain the electric field intensity at the point $(1, 1, 1)$ to be

$$
\begin{aligned}
\mathbf{E}_{(1,1,1)} &= \frac{Q}{4\pi\epsilon_0}\left[\frac{\mathbf{a}_x}{(1)^{3/2}} + \frac{\mathbf{a}_y}{(1)^{3/2}} + \frac{\mathbf{a}_z}{(1)^{3/2}} + \frac{\mathbf{a}_y + \mathbf{a}_z}{(2)^{3/2}} + \frac{\mathbf{a}_z + \mathbf{a}_x}{(2)^{3/2}} \right. \\
&\quad \left. + \frac{\mathbf{a}_x + \mathbf{a}_y}{(2)^{3/2}} + \frac{\mathbf{a}_x + \mathbf{a}_y + \mathbf{a}_z}{(3)^{3/2}}\right] \\
&= \frac{Q}{4\pi\epsilon_0}\left(1 + \frac{1}{\sqrt{2}} + \frac{1}{3\sqrt{3}}\right)(\mathbf{a}_x + \mathbf{a}_y + \mathbf{a}_z) \\
&= \frac{3.29Q}{4\pi\epsilon_0}\left(\frac{\mathbf{a}_x + \mathbf{a}_y + \mathbf{a}_z}{\sqrt{3}}\right)
\end{aligned}
$$

Noting that $(\mathbf{a}_x + \mathbf{a}_y + \mathbf{a}_z)/\sqrt{3}$ is the unit vector directed from $(0, 0, 0)$ to $(1, 1, 1)$, we find the electric field intensity at $(1, 1, 1)$ to be directed diagonally away from $(0, 0, 0)$, with a magnitude equal to $\dfrac{3.29Q}{4\pi\epsilon_0}$ N/C. From symmetry considerations, it then follows that the electric field intensity at each point charge, due to the remaining seven point charges, has a magnitude $\dfrac{3.29Q}{4\pi\epsilon_0}$ N/C, and it is directed away from the corner opposite to that charge.

The foregoing illustration of the computation of the electric field intensity due to a multitude of point charges may be extended to the computation of the field intensity for a continuous charge distribution by dividing the region in which the charge exists into elemental lengths, surfaces, or volumes depending on whether the charge is distributed along a line, over a surface, or in a volume, and treating the charge in each elemental length, surface, or volume as a point charge and then applying superposition. We shall include some of the simpler cases in the problems for the interested reader.

Let us now consider the motion of a cloud of electrons, distributed uniformly with density N, under the influence of a time-varying electric field of intensity

$$
\mathbf{E} = E_0 \cos \omega t \, \mathbf{a}_x \tag{1.55}
$$

Each electron experiences a force given by

$$
\mathbf{F} = e\mathbf{E} = eE_0 \cos \omega t \, \mathbf{a}_x \tag{1.56}
$$

where e is the charge of the electron. The equation of motion of the electron is then given by

$$
m\frac{d\mathbf{v}}{dt} = eE_0 \cos \omega t \, \mathbf{a}_x \tag{1.57}
$$

where m is the mass of the electron and $\mathbf{v}$ is its velocity. Solving (1.57) for $\mathbf{v}$, we obtain

$$
\mathbf{v} = \frac{eE_0}{m\omega} \sin \omega t \, \mathbf{a}_x + \mathbf{C} \tag{1.58}
$$

where $\mathbf{C}$ is the constant of integration. Assuming an initial condition of $\mathbf{v} = 0$ for $t = 0$ gives us $\mathbf{C} = 0$, reducing (1.58) to

$$
\mathbf{v} = \frac{eE_0}{m\omega} \sin \omega t \, \mathbf{a}_x = -\frac{|e|E_0}{m\omega} \sin \omega t \, \mathbf{a}_x \tag{1.59}
$$

The motion of the electron cloud gives rise to current flow. To find the current crossing an infinitesimal surface of area ΔS oriented such that the normal vector to the surface makes an angle α with the x direction as shown in Figure 1.20, let us for instance consider an infinitesimal time interval Δt when v_x is negative. The number of electrons crossing the area ΔS from its right side to its left side in this time interval is the same as that which exists in a column of length $|v_x|\Delta t$ and cross-sectional area $\Delta S \cos \alpha$ to the right of the area under consideration. Thus, the negative charge ΔQ crossing the area ΔS in time Δt to its left side is given by

$$\Delta Q = (\Delta S \cos \alpha)(|v_x|\Delta t)Ne$$
$$= Ne|v_x|\Delta S \cos \alpha \, \Delta t \tag{1.60}$$

The current ΔI flowing through the area ΔS from its left side to its right side is then given by

$$\Delta I = \frac{|\Delta Q|}{\Delta t} = N|e||v_x|\Delta S \cos \alpha$$
$$= \frac{N|e|^2}{m\omega} E_0 \sin \omega t \, \Delta S \cos \alpha$$
$$= \frac{Ne^2}{m\omega} E_0 \sin \omega t \, \mathbf{a}_x \cdot \Delta S \, \mathbf{a}_n \tag{1.61}$$

where $\mathbf{a}_n$ is the unit vector normal to the area ΔS, as shown in Figure 1.20.

FIGURE 1.20

For finding the current crossing an infinitesimal area in a moving cloud of electrons.

We can now talk of a current density vector $\mathbf{J}$, associated with the current flow. The current density vector has a mgnitude equal to the current per unit area and a direction normal to the area when the area is oriented in order to maximize the current crossing it. The current crossing ΔS is maximized when $\alpha = 0$, that is, when the area is oriented such that $\mathbf{a}_n = \mathbf{a}_x$. The current per unit area is then equal to $\frac{Ne^2}{m\omega} E_0 \sin \omega t$. Thus, the current density vector is given by

$$\mathbf{J} = \frac{Ne^2}{m\omega} E_0 \sin \omega t \, \mathbf{a}_x$$
$$= Ne\mathbf{v} \tag{1.62}$$

Finally, by substituting (1.62) back into (1.61), we note that the current crossing any area $\Delta \mathbf{S} = \Delta S \, \mathbf{a}_n$ is simply equal to $\mathbf{J} \cdot \Delta \mathbf{S}$.

1.6   THE MAGNETIC FIELD

In the preceding section we presented an experimental law known as Coulomb's law having to do with the electric force associated with two charged bodies, and we introduced the electric field intensity vector as the force per unit charge experienced by a test charge placed in the electric field. In this section we present another experimental law known as *Ampere's law of force*, analogous to Coulomb's law, and use it to introduce the magnetic field concept.

Ampere's law of force is concerned with *magnetic* forces associated with two loops of wire carrying currents by virtue of motion of charges in the loops. Figure 1.21 shows two loops of wire carrying currents I_1 and I_2 and each of which is divided into a large number of elements having infinitesimal lengths. The total force experienced by a loop is the vector sum of forces experienced by the infinitesimal current elements comprising the loop. The force experienced by each of these current elements is the vector sum of the forces exerted on it by the infinitesimal current elements comprising the second loop. If the number of elements in loop 1 is m and the number of elements in loop 2 is n, then there are $m \times n$ pairs of elements. A pair of magnetic forces is associated with each pair of these elements just as a pair of electric forces is associated with a pair of point charges. Thus, if we consider an element $d\mathbf{l}_1$ in loop 1 and an element $d\mathbf{l}_2$ in loop 2, then the forces $d\mathbf{F}_1$ and $d\mathbf{F}_2$ experienced by the elements $d\mathbf{l}_1$ and $d\mathbf{l}_2$, respectively, are given by

$$d\mathbf{F}_1 = I_1\, d\mathbf{l}_1 \times \left(\frac{kI_2\, d\mathbf{l}_2 \times \mathbf{a}_{21}}{R^2} \right) \tag{1.63a}$$

$$d\mathbf{F}_2 = I_2\, d\mathbf{l}_2 \times \left(\frac{kI_1\, d\mathbf{l}_1 \times \mathbf{a}_{12}}{R^2} \right) \tag{1.63b}$$

where $\mathbf{a}_{21}$ and $\mathbf{a}_{12}$ are unit vectors along the line joining the two current elements, R is the distance between them, and k is a constant of proportionality that depends on the medium. For free space, k is equal to $\mu_0/4\pi$, where μ_0 is known as the permeability of free space, having a value $4\pi \times 10^{-7}$. From (1.63a) or (1.63b), we note that the units of μ_0 are newtons per ampere squared. These are commonly known as *henrys per meter* where a henry is a newton-meter per ampere squared.

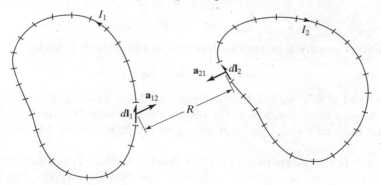

FIGURE 1.21

Two loops of wire carrying currents I_1 and I_2.

Equations (1.63a) and (1.63b) represent Ampere's force law as applied to a pair of current elements. Some of the features evident from these equations are as follows:

1. The magnitude of the force is proportional to the product of the two currents and to the product of the lengths of the two current elements.

2. The magnitude of the force is inversely proportional to the square of the distance between the current elements.

3. To determine the direction of the force acting on the current element dl_1, we first find the cross product $dl_2 \times \mathbf{a}_{21}$ and then cross dl_1 into the resulting vector. Similarly, to determine the direction of the force acting on the current element dl_2, we first find the cross product $dl_1 \times \mathbf{a}_{12}$ and then cross dl_2 into the resulting vector. For the general case of arbitrary orientations of dl_1 and dl_2, these operations yield $d\mathbf{F}_{12}$ and $d\mathbf{F}_{21}$ which are not equal and opposite. This is not a violation of Newton's third law since isolated current elements do not exist without sources and sinks of charges at their ends. Newton's third law, however, must and does hold for complete current loops.

The forms of (1.63a) and (1.63b) suggest that each current element is acted upon by a field which is due to the other current element. By definition, this field is the magnetic field and is characterized by a quantity known as the *magnetic flux density vector*, denoted by the symbol **B**. Thus, we note from (1.63b) that the magnetic flux density at the element dl_2 due to the element dl_1 is given by

$$\mathbf{B}_1 = \frac{\mu_0}{4\pi} \frac{I_1 \, d\mathbf{l}_1 \times \mathbf{a}_{12}}{R^2} \tag{1.64}$$

and that this flux density acting upon dl_2 results in a force on it given by

$$d\mathbf{F}_2 = I_2 \, d\mathbf{l}_2 \times \mathbf{B}_1 \tag{1.65}$$

Similarly, we note from (1.63a) that the magnetic flux density at the element dl_1 due to the element dl_2 is given by

$$\mathbf{B}_2 = \frac{\mu_0}{4\pi} \frac{I_2 \, d\mathbf{l}_2 \times \mathbf{a}_{21}}{R^2} \tag{1.66}$$

and that this flux density acting upon dl_1 results in a force on it given by

$$d\mathbf{F}_1 = I_1 \, d\mathbf{l}_1 \times \mathbf{B}_2 \tag{1.67}$$

From (1.65) and (1.67), we see that the units of **B** are newtons per ampere-meter, commonly known as *webers/meter*^{2} (or tesla), where a weber is a newton-meter per ampere. The units of webers per unit area give the character of flux density to the quantity **B**.

Although **B** has the character of a flux density, whereas **E** has the character of a field intensity, they are the fundamental field vectors, because together they define the force acting on a charge in a region of electric and magnetic fields, as we shall learn later in this section. We will introduce the electric flux density and the magnetic field intensity vectors in Chapter 2.

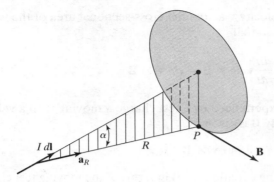

FIGURE 1.22

Magnetic flux density due to an
infinitesimal current element.

Generalizing (1.64) and (1.66), we obtain the magnetic flux density due to an infinitesimal current element of length $d\mathbf{l}$ and carrying current I to be

$$\mathbf{B} = \frac{\mu_0}{4\pi} \frac{I \, d\mathbf{l} \times \mathbf{a}_R}{R^2} \qquad (1.68)$$

where R is the distance from the current element to the point at which the flux density is to be computed and $\mathbf{a}_R$ is the unit vector along the line joining the current element and the point under consideration and directed away from the current element as shown in Figure 1.22. Equation (1.68) is known as the *Biot-Savart law* and is analogous to the expression for the electric field intensity due to a point charge. The Biot-Savart law tells us that the magnitude of $\mathbf{B}$ at a point P is proportional to the current I, the element length dl, and the sine of the angle α between the current element and the line joining it to the point P, and is inversely proportional to the square of the distance from the current element to the point P. Hence, the magnetic flux density is zero at points along the axis of the current element. The direction of $\mathbf{B}$ at point P is normal to the plane containing the current element and the line joining the current element to P, as given by the cross product operation $d\mathbf{l} \times \mathbf{a}_R$, that is, right circular to the axis of the wire. As a numerical example, for a current element $0.01\mathbf{a}_z$ m situated at the origin and carrying current 2 A, the magnetic flux density at the point $(0, 1, 1)$ has a magnitude $10^{-9}/\sqrt{2}$ Wb/m^2 and is directed in the $-\mathbf{a}_x$-direction. The magnetic field due to a given current distribution can be found by dividing the current distribution into a number of infinitesimal current elements, applying the Biot-Savart law to find the magnetic field due to each current element, and then using superposition. We shall include some simple cases in the problems for the interested reader.

Turning our attention now to (1.65) and (1.67) and generalizing, we say that an infinitesimal current element of length $d\mathbf{l}$ and current I placed in a magnetic field of flux density $\mathbf{B}$ experiences a force $d\mathbf{F}$ given by

$$d\mathbf{F} = I \, d\mathbf{l} \times \mathbf{B} \qquad (1.69)$$

Alternatively, if a current element experiences a force in a region of space, then the region is said to be characterized by a magnetic field. Since current is due to flow of charges, (1.69) can be formulated in terms of the moving charge causing the flow of current. Thus, if the time taken by the charge dq contained in the length $d\mathbf{l}$ of the current element to flow with a velocity **v** across the cross-sectional area of the wire is dt, then $I = dq/dt$, and $d\mathbf{l} = \mathbf{v}\,dt$ so that

$$d\mathbf{F} = \frac{dq}{dt}\mathbf{v}\,dt \times \mathbf{B} = dq\,\mathbf{v} \times \mathbf{B} \qquad (1.70)$$

It then follows that the force **F** experienced by a test charge q moving with a velocity **v** in a magnetic field of flux density **B** is given by

$$\mathbf{F} = q\mathbf{v} \times \mathbf{B} \qquad (1.71)$$

We may now obtain a defining equation for **B** in terms of the moving test charge. To do this, we note from (1.71) that the magnetic force is directed normally to both **v** and **B**, as shown in Figure 1.23, and that its magnitude is equal to $qvB \sin \delta$, where δ is the angle between **v** and **B**. A knowledge of the force **F** acting on a test charge moving with an arbitrary velocity **v** provides only the value of $B \sin \delta$. To find **B**, we must determine the maximum force qvB that occurs for δ equal to 90° by trying out several directions of **v**, keeping its magnitude constant. Thus, if this maximum force is $\mathbf{F}_m$ and it occurs for a velocity $v\mathbf{a}_m$, then

$$\mathbf{B} = \frac{\mathbf{F}_m \times \mathbf{a}_m}{qv} \qquad (1.72)$$

As in the case of defining the electric field intensity, we assume that the test charge does not alter the magnetic field in which it is placed. Ideally, **B** is defined in the limit that qv tends to zero, that is,

$$\mathbf{B} = \mathop{\text{Lim}}_{qv \to 0} \frac{\mathbf{F}_m \times \mathbf{a}_m}{qv} \qquad (1.73)$$

Equation (1.73) is the defining equation for the magnetic flux density irrespective of the source of the magnetic field. We have learned in this section that an electric current or a charge in motion is a source of the magnetic field. We will learn in Chapter 2 that there exists another source for the magnetic field, namely, a time-varying electric field.

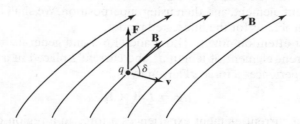

FIGURE 1.23

Force experienced by a test charge q moving with a velocity **v** in a magnetic field **B**.

We can now combine (1.48) and (1.71) to write the expression for the total force acting on a test charge q moving with a velocity $\mathbf{v}$ in a region characterized by an electric field of intensity $\mathbf{E}$ and a magnetic field of flux density $\mathbf{B}$ as

$$\mathbf{F} = q\mathbf{E} + q\mathbf{v} \times \mathbf{B} = q(\mathbf{E} + \mathbf{v} \times \mathbf{B}) \tag{1.74}$$

Equation (1.74) is known as the *Lorentz force equation*. We shall now consider an example.

Example 1.5

The forces experienced by a test charge q for three different velocities at a point in a region characterized by electric and magnetic fields are given by

$$\mathbf{F}_1 = q[E_0\mathbf{a}_x + (E_0 - v_0B_0)\mathbf{a}_y] \qquad \text{for } \mathbf{v}_1 = v_0\mathbf{a}_x$$

$$\mathbf{F}_2 = q[(E_0 + v_0B_0)\mathbf{a}_x + E_0\mathbf{a}_y] \qquad \text{for } \mathbf{v}_2 = v_0\mathbf{a}_y$$

$$\mathbf{F}_3 = q[E_0\mathbf{a}_x + E_0\mathbf{a}_y] \qquad \text{for } \mathbf{v}_3 = v_0\mathbf{a}_z$$

where v_0, E_0, and B_0 are constants. Find $\mathbf{E}$ and $\mathbf{B}$ at the point.

From Lorentz force equation, we have

$$q\mathbf{E} + qv_0\mathbf{a}_x \times \mathbf{B} = q[E_0\mathbf{a}_x + (E_0 - v_0B_0)\mathbf{a}_y] \tag{1.75a}$$

$$q\mathbf{E} + qv_0\mathbf{a}_y \times \mathbf{B} = q[(E_0 + v_0B_0)\mathbf{a}_x + E_0\mathbf{a}_y] \tag{1.75b}$$

$$q\mathbf{E} + qv_0\mathbf{a}_z \times \mathbf{B} = q[E_0\mathbf{a}_x + E_0\mathbf{a}_y] \tag{1.75c}$$

Eliminating $\mathbf{E}$ by subtracting (1.75a) from (1.75b) and (1.75c) from (1.75b), we obtain

$$(\mathbf{a}_y - \mathbf{a}_x) \times \mathbf{B} = B_0(\mathbf{a}_x + \mathbf{a}_y) \tag{1.76a}$$

$$(\mathbf{a}_y - \mathbf{a}_z) \times \mathbf{B} = B_0\mathbf{a}_x \tag{1.76b}$$

It follows from these two equations that $\mathbf{B}$ is perpendicular to both $(\mathbf{a}_x + \mathbf{a}_y)$ and $\mathbf{a}_x$. Hence it is equal to $C(\mathbf{a}_x + \mathbf{a}_y) \times \mathbf{a}_x$ or $-C\mathbf{a}_z$ where C is to be determined. To do this, we substitute $\mathbf{B} = -C\mathbf{a}_z$ in (1.76a) to obtain

$$(\mathbf{a}_y - \mathbf{a}_x) \times (-C\mathbf{a}_z) = B_0(\mathbf{a}_x + \mathbf{a}_y)$$

$$-C(\mathbf{a}_x + \mathbf{a}_y) = B_0(\mathbf{a}_x + \mathbf{a}_y)$$

or $C = -B_0$. Thus, we get

$$\mathbf{B} = B_0\mathbf{a}_z$$

Substituting this result in (1.75c), we obtain

$$\mathbf{E} = E_0(\mathbf{a}_x + \mathbf{a}_y)$$

SUMMARY

We first learned in this chapter several rules of vector algebra that are necessary for our study of the fundamentals of electromagnetics by considering vectors expressed in terms of their components along three mutually orthogonal directions. To carry out the manipulations involving vectors at different points in space in a systematic manner, we then introduced the Cartesian coordinate system and discussed the application of the vector algebraic rules to vectors in the Cartesian coordinate system. To summarize these rules, we consider three vectors

$$\mathbf{A} = A_x \mathbf{a}_x + A_y \mathbf{a}_y + A_z \mathbf{a}_z$$
$$\mathbf{B} = B_x \mathbf{a}_x + B_y \mathbf{a}_y + B_z \mathbf{a}_z$$
$$\mathbf{C} = C_x \mathbf{a}_x + C_y \mathbf{a}_y + C_z \mathbf{a}_z$$

in a right-handed Cartesian coordinate system, that is, with $\mathbf{a}_x \times \mathbf{a}_y = \mathbf{a}_z$. We then have

$$\mathbf{A} + \mathbf{B} = (A_x + B_x)\mathbf{a}_x + (A_y + B_y)\mathbf{a}_y + (A_z + B_z)\mathbf{a}_z$$
$$\mathbf{B} - \mathbf{C} = (B_x - C_x)\mathbf{a}_x + (B_y - C_y)\mathbf{a}_y + (B_z - C_z)\mathbf{a}_z$$
$$m\mathbf{A} = mA_x \mathbf{a}_x + mA_y \mathbf{a}_y + mA_z \mathbf{a}_z$$
$$\frac{\mathbf{B}}{n} = \frac{B_x}{n}\mathbf{a}_x + \frac{B_y}{n}\mathbf{a}_y + \frac{B_z}{n}\mathbf{a}_z$$
$$|\mathbf{A}| = \sqrt{A_x^2 + A_y^2 + A_z^2}$$
$$\mathbf{a}_A = \frac{A_x}{\sqrt{A_x^2 + A_y^2 + A_z^2}}\mathbf{a}_x + \frac{A_y}{\sqrt{A_x^2 + A_y^2 + A_z^2}}\mathbf{a}_y + \frac{A_z}{\sqrt{A_x^2 + A_y^2 + A_z^2}}\mathbf{a}_z$$
$$\mathbf{A} \cdot \mathbf{B} = A_x B_x + A_y B_y + A_z B_z$$
$$\mathbf{A} \times \mathbf{B} = \begin{vmatrix} \mathbf{a}_x & \mathbf{a}_y & \mathbf{a}_z \\ A_x & A_y & A_z \\ B_x & B_y & B_z \end{vmatrix}$$
$$\mathbf{A} \cdot \mathbf{B} \times \mathbf{C} = \begin{vmatrix} A_x & A_y & A_z \\ B_x & B_y & B_z \\ C_x & C_y & C_z \end{vmatrix}$$

Other useful expressions are

$$d\mathbf{l} = dx\,\mathbf{a}_x + dy\,\mathbf{a}_y + dz\,\mathbf{a}_z$$
$$d\mathbf{S} = \pm dx\,dy\,\mathbf{a}_z, \qquad \pm dy\,dz\,\mathbf{a}_x, \qquad \pm dz\,dx\,\mathbf{a}_y$$
$$dv = dx\,dy\,dz$$

As a prelude to the introduction of electric and magnetic fields, we discussed the concepts of scalar and vector fields, static and time-varying, by means of some simple examples, such as the height of points on a conical surface above its base, the temperature field of points in a room, and the velocity vector field associated with points on a disk rotating about its center. We learned about the visualization of fields by means of constant-magnitude contours or surfaces and in addition by means of direction lines in the case of vector fields. Particular attention was devoted to sinusoidally time-varying fields. Polarization of vector fields as a means of describing how the orientation of a vector at a point changes with time was discussed. The phasor technique as a means of facilitating mathematical operations involving sinusoidally time-varying quantities was reviewed.

Having obtained the necessary background vector algebraic tools and physical field concepts, we then introduced the electric and magnetic fields from considerations of experimental laws known as Coulomb's law and Ampere's force law, having to do with the electric forces between two point charges, and the magnetic forces between two current elements, respectively. From these laws, we deduced the expressions for the electric field intensity $\mathbf{E}$ due to a point charge Q and the magnetic flux density $\mathbf{B}$ due to a current element $I\,d\mathbf{l}$. These expressions are

$$\mathbf{E} = \frac{Q}{4\pi\epsilon_0 R^2}\,\mathbf{a}_R$$

$$\mathbf{B} = \frac{\mu_0 I\,d\mathbf{l} \times \mathbf{a}_R}{4\pi R^2}$$

where ϵ_0 and μ_0 are the permittivity and the permeability, respectively, of free space, R is the distance from the source to the point, say P, at which the field is to be computed, and $\mathbf{a}_R$ is the unit vector directed from the source toward the point P. We learned that the electric field is a force field acting on charges merely by virtue of the property of charge. The electric force is given simply by

$$\mathbf{F} = q\mathbf{E}$$

On the other hand, the magnetic field exerts forces only on moving charges, or current elements, as given by

$$\mathbf{F} = dq\,\mathbf{v} \times \mathbf{B} = I d\mathbf{l} \times \mathbf{B}$$

Combining the electric and magnetic field concepts, we finally introduced the Lorentz force equation for the force exerted on a charge q moving with a velocity $\mathbf{v}$ in a region of electric and magnetic fields $\mathbf{E}$ and $\mathbf{B}$, respectively, as

$$\mathbf{F} = q(\mathbf{E} + \mathbf{v} \times \mathbf{B})$$

REVIEW QUESTIONS

1.1. Give some examples of scalars.

1.2. Give some examples of vectors.

1.3. State all conditions for which $\mathbf{A} \cdot \mathbf{B} = 0$.

1.4. State all conditions for which $\mathbf{A} \times \mathbf{B} = 0$.

1.5. What is the significance of $\mathbf{A} \cdot \mathbf{B} \times \mathbf{C} = 0$?

1.6. Is it necessary for the reference vectors $\mathbf{a}_1$, $\mathbf{a}_2$, and $\mathbf{a}_3$ to be an orthogonal set?

1.7. State whether $\mathbf{a}_1$, $\mathbf{a}_2$, and $\mathbf{a}_3$ directed westward, northward, and downward, respectively, is a right-handed or a left-handed set.

1.8. What is the particular advantageous characteristic associated with the unit vectors in the Cartesian coordinate system?

1.9. How do you find a vector perpendicular to a plane?

1.10. How do you find the perpendicular distance from a point to a plane?

1.11. What is the total distance around the circumference of a circle of radius 1 m? What is the total vector distance around the circle?

1.12. What is the total surface area of a cube of sides 1 m? Assuming the normals to the surfaces to be directed outward of the cubical volume, what is the total vector surface area of the cube?

1.13. Describe briefly your concept of a scalar field and illustrate with an example.

1.14. Describe briefly your concept of a vector field and illustrate with an example.

1.15. How do you depict pictorially the gravitational field of the earth?

1.16. A sinusoidally time-varying vector is expressed in terms of its components along the x-, y-, and z-axes. What is the polarization of each of the components?

1.17. What are the conditions for the sum of two linearly polarized sinusoidally time-varying vectors to be circularly polarized?

1.18. What is the polarization for the general case of the sum of two sinusoidally time-varying linearly polarized vectors having arbitrary amplitudes, phase angles, and directions?

1.19. Considering the second hand on your watch to be a vector, state its polarization. What is the frequency?

1.20. What is a phasor?

1.21. Is there any relationship between a phasor and a vector? Explain.

1.22. Describe the phasor technique of adding two sinusoidal functions of time.

1.23. Describe the phasor technique of solving a differential equation for the sinusoidal steady-state solution.

1.24. State Coulomb's law. To what law in mechanics is Coulomb's law analogous?

1.25. What is the definition of the electric field intensity?

1.26. What are the units of the electric field intensity?

1.27. What is the permittivity of free space? What are its units?

1.28. Describe the electric field due to a point charge.

1.29. How do you find the electric field intensity due to a continuous charge distribution?

1.30. How is current density defined? What are its units?

1.31. For a current flowing on a sheet, how would you define the current density at a point on the sheet? What are the units?

1.32. State Ampere's force law as applied to current elements.

1.33. Why is it not necessary for Newton's third law to hold for current elements?

1.34. What is the permeability of free space? What are its units?

1.35. Describe the magnetic field due to a current element.

1.36. How is the magnetic flux density defined in terms of force on a current element?

1.37. How is the magnetic flux density defined in terms of force on a moving charge?

1.38. What are the units of the magnetic flux density?

1.39. State Lorentz force equation.

1.40. If it is assumed that there is no electric field, the magnetic field at a point can be found from the knowledge of forces exerted on a moving test charge for two noncollinear velocities. Explain.

PROBLEMS

1.1. A bug starts at a point and travels 1 m northward, $\frac{1}{2}$ m eastward, $\frac{1}{4}$ m southward, $\frac{1}{8}$ m westward, $\frac{1}{16}$ m northward, and so on, making a 90°-turn to the right and halving the distance each time. (a) What is the total distance traveled by the bug? (b) Find the final position of the bug relative to its starting location. (c) Find the straight-line distance from the starting location to the final position.

1.2. Solve the following equations for $\mathbf{A}, \mathbf{B}$, and $\mathbf{C}$:

$$\mathbf{A} + \mathbf{B} + \mathbf{C} = 2\mathbf{a_1} + 3\mathbf{a_2} + 2\mathbf{a_3}$$
$$2\mathbf{A} + \mathbf{B} - \mathbf{C} = \mathbf{a_1} + 3\mathbf{a_2}$$
$$\mathbf{A} - 2\mathbf{B} + 3\mathbf{C} = 4\mathbf{a_1} + 5\mathbf{a_2} + \mathbf{a_3}$$

1.3. Show that $(\mathbf{A} + \mathbf{B}) \cdot (\mathbf{A} - \mathbf{B}) = A^2 - B^2$ and that $(\mathbf{A} + \mathbf{B}) \times (\mathbf{A} - \mathbf{B}) = 2\mathbf{B} \times \mathbf{A}$. Verify the above for $\mathbf{A} = 3\mathbf{a_1} - 5\mathbf{a_2} + 4\mathbf{a_3}$ and $\mathbf{B} = \mathbf{a_1} + \mathbf{a_2} - 2\mathbf{a_3}$.

1.4. Given $\mathbf{A} = -2\mathbf{a_1} + \mathbf{a_2}$, $\mathbf{B} = \mathbf{a_1} - 2\mathbf{a_2} + \mathbf{a_3}$,   and $\mathbf{C} = 3\mathbf{a_1} + 2\mathbf{a_2} + \mathbf{a_3}$,   find $\mathbf{A} \times (\mathbf{B} \times \mathbf{C}) + \mathbf{B} \times (\mathbf{C} \times \mathbf{A}) + \mathbf{C} \times (\mathbf{A} \times \mathbf{B})$.

1.5. Show that $\frac{1}{2}|\mathbf{A} \times \mathbf{B}|$ is equal to the area of the triangle having $\mathbf{A}$ and $\mathbf{B}$ as two of its sides. Then find the area of the triangle formed by the points $(1, 2, 1), (-3, -4, 5)$, and $(2, -1, -3)$.

1.6. Show that $\mathbf{A} \cdot \mathbf{B} \times \mathbf{C}$ is the volume of the parallelepiped having $\mathbf{A}, \mathbf{B}$, and $\mathbf{C}$ as three of its contiguous edges. Then find the volume if $\mathbf{A} = 4\mathbf{a_x}, \mathbf{B} = 2\mathbf{a_x} + \mathbf{a_y} + 3\mathbf{a_z}$, and $\mathbf{C} = 2\mathbf{a_y} + 6\mathbf{a_z}$. Comment on your result.

1.7. Given $\mathbf{a_x} \times \mathbf{A} = -\mathbf{a_y} + 2\mathbf{a_z}$ and $\mathbf{a_y} \times \mathbf{A} = \mathbf{a_x} - 2\mathbf{a_z}$, find A.

1.8. Find the component of the vector drawn from $(5, 0, 3)$ to $(3, 3, 2)$ along the direction of the vector drawn from $(6, 2, 4)$ to $(3, 3, 6)$.

1.9. Find the unit vector normal to the plane $4x - 5y + 3z = 60$. Then find the distance from the origin to the plane.

1.10. Write the expression for the differential length vector $d\mathbf{l}$ at the point $(1, 2, 8)$ on the straight line $y = 2x, z = 4y$, and having the projection dx on the x-axis.

1.11. Write the expression for the differential length vector $d\mathbf{l}$ at the point $(4, 4, 2)$ on the curve $x = y = z^2$ and having the projection dz on the z-axis.

1.12. Write the expression for the differential surface vector $d\mathbf{S}$ at the point $(1, 1, \frac{1}{2})$ on the plane $x + 2z = 2$ and having the projection $dx\, dy$ on the xy-plane.

1.13. Find two differential length vectors tangential to the surface $y = x^2$ at the point $(2, 4, 1)$ and then find a unit vector normal to the surface at that point.

1.14. A hemispherical bowl of radius 2 m lies with its base on the xy-plane and with its center at the origin. Write the expression for the scalar field, describing the height of points on the bowl as a function of x and y.

1.15. A number equal to the sum of its coordinates is assigned to each point in a rectangular room having three of its contiguous edges as the coordinate axes. Draw a sketch of the constant-magnitude surfaces for the number field generated in this manner.

1.16. Write the expression for the vector distance of a point in a rectangular room from one corner of the room, choosing the three edges meeting at that point as the coordinate axes. Describe the vector distance field associated with the points in the room.

1.17. For the rotating disk of Figure 1.9, write the expression for the linear velocity vector field associated with the points on the disk; use an xy-coordinate system with the origin at the center of the disk.

1.18. Given $f(z, t) = 10 \cos(2\pi \times 10^7 t - 0.1\pi z)$, (a) draw sketches of f versus z for $t = 0, \frac{1}{8} \times 10^{-7}, \frac{1}{4} \times 10^{-7}, \frac{3}{8} \times 10^{-7}$, and $\frac{1}{2} \times 10^{-7}$ s, and (b) draw sketches of f versus t for $z = 0, 2.5, 5, 7.5$, and 10 m. From your sketches of part (a), what can you say about the function $f(z, t)$?

1.19. Repeat Problem 1.18 for $f(z, t) = 10 \cos(2\pi \times 10^7 t + 0.1\pi z)$.

1.20. Repeat Problem 1.18 for $f(z, t) = 10 \cos 2\pi \times 10^7 t \cos 0.1\pi z$.

1.21. For each of the following vector fields, find the polarization:

(a) $1 \cos(\omega t + 30°) \mathbf{a}_x + \sqrt{2} \cos(\omega t + 30°) \mathbf{a}_y$

(b) $1 \cos(\omega t + 30°) \mathbf{a}_x + 1 \cos(\omega t - 60°) \mathbf{a}_y$

(c) $1 \cos(\omega t + 30°) \mathbf{a}_x + \sqrt{2} \cos(\omega t - 60°) \mathbf{a}_y$

1.22. Determine the polarization of the sum vector obtained by adding the two vector fields

$$\mathbf{F}_1 = (-\sqrt{3}\mathbf{a}_x + \mathbf{a}_y) \cos \omega t$$

$$\mathbf{F}_2 = \left(\frac{1}{2}\mathbf{a}_x + \frac{\sqrt{3}}{2}\mathbf{a}_y - \sqrt{3}\mathbf{a}_z \right) \sin \omega t$$

1.23. For the vector field $1 \cos \omega t \, \mathbf{a}_x + \sqrt{2} \sin \omega t \, \mathbf{a}_y$, draw sketches similar to those of Figures 1.12 and 1.13 and describe the polarization.

1.24. Find $10 \cos(\omega t - 30°) + 10 \cos(\omega t + 210°)$ by using the phasor technique.

1.25. Find $3 \cos(\omega t + 60°) - 4 \cos(\omega t + 150°)$ by using the phasor technique.

1.26. Solve the differential equation $5 \times 10^{-6} \dfrac{di}{dt} + 12i = 13 \cos 10^6 t$ by using the phasor technique.

1.27. Two point charges each of mass m and charge q are suspended by strings of length l from a common point. Find the value of q for which the angle made by the strings at the common point is 90°.

1.28. Point charges Q and $-Q$ are situated at $(0, 0, 1)$ and $(0, 0, -1)$, respectively. Find the approximate electric field intensity at (a) $(0, 0, 100)$, and (b) $(100, 0, 0)$.

1.29. For the point charge configuration of Example 1.4, find $\mathbf{E}$ at the point $(2, 2, 2)$.

1.30. A line charge consists of charge distributed along a line just as graphite in a pencil lead. We then talk of line charge density, or charge per unit length, having the units C/m. Obtain a series expression for the electric field intensity at $(0, 1, 0)$ for a line charge situated along the z-axis between $(0, 0, -1)$ and $(0, 0, 1)$ with uniform density 10^{-3} C/m by dividing the line into 100 equal segments. Consider the charge in each segment to be a point charge located at the center of the segment, and use superposition.

1.31. Repeat Problem 1.30, but assume the line charge density to be $10^{-3} |z|$ C/m.

1.32. Charge is distributed uniformly with density 10^{-3} C/m on a circular ring of radius 2 m lying in the xy-plane and centered at the origin. Obtain the electric field intensity at the point $(0, 0, 1)$ by using the procedure described in Problem 1.30.

1.33. A surface charge consists of charge distributed on a surface just as paint on a table top. We then talk of surface charge density, or charge per unit area, having the units C/m². Obtain a series expression for the electric field intensity at $(0, 0, 1)$ for a surface charge of uniform density 10^{-3} C/m² situated within the square on the xy-plane having the corners $(1, 1, 0), (-1, 1, 0), (-1, -1, 0)$, and $(1, -1, 0)$ by dividing the square into 10,000 equal areas. Consider the charge in each area as a point charge located at the center of the area, and use superposition.

1.34. Repeat Problem 1.33, but assume the surface charge density to be $10^{-3} |xy^2|$ C/m².

1.35. For an electron cloud of uniform density $N = 10^{12}\,\text{m}^{-3}$ oscillating under the influence of an electric field $\mathbf{E} = 10^{-3}\cos 2\pi \times 10^7\,t\,\mathbf{a}_x$ V/m, find (a) the current density, and (b) the current crossing the surface $0.01(\mathbf{a}_x + \mathbf{a}_y)\,\text{m}^2$.

1.36. An object of mass m and charge q, suspended by a spring of spring constant k is acted upon by the earth's gravitational field and an electric field $E_0 \cos \omega t$ parallel to the gravitational field. Obtain the steady-state solution for the velocity of the object.

1.37. Find $d\mathbf{F}_1$ and $d\mathbf{F}_2$ for $I_1\,d\mathbf{l}_1 = I_1\,dx\,\mathbf{a}_x$ located at the origin and $I_2\,d\mathbf{l}_2 = I_2\,dy\,\mathbf{a}_y$ located at $(0,1,0)$.

1.38. For an infinitesimal current element $I\,dx\,(\mathbf{a}_x + 2\mathbf{a}_y + 2\mathbf{a}_z)$ located at the point $(1,0,0)$, find the magnetic flux density at (a) the point $(0,1,1)$ and (b) the point $(2,2,2)$.

1.39. A square loop of wire of sides 0.01 m lies in the xy-plane, with its sides parallel to the x- and y-axes and with its center at the origin. It carries a current of 1 A in the clockwise sense as seen along the positive z-axis. Find the approximate magnetic flux density at (a) $(0,0,1)$ and (b) $(0,1,0)$.

1.40. A straight wire along the z-axis carries current I in the positive z-direction. Consider the portion of the wire lying between $(0,0,-1)$ and $(0,0,1)$. By dividing this portion into 100 equal segments and using superposition, obtain a series expression for $\mathbf{B}$ at $(0,1,0)$.

1.41. A circular loop of wire of radius 2 m is situated in the xy-plane and with its center at the origin. It carries a current I in the clockwise sense as seen along the positive z-axis. Find $\mathbf{B}$ at $(0,0,1)$ by dividing the loop into a large number of equal infinitesimal segments and by using superposition.

1.42. Obtain the expression for the orbital frequency for an electron moving in a circular orbit normal to a uniform magnetic field of flux density B_0 Wb/m². Compute its value for B_0 equal to 5×10^{-5}.

1.43. A magnetic field $\mathbf{B} = B_0(\mathbf{a}_x + 2\mathbf{a}_y - 4\mathbf{a}_z)$ exists at a point. What should be the electric field at that point if the force experienced by a test charge moving with a velocity $\mathbf{v} = v_0(3\mathbf{a}_x - \mathbf{a}_y + 2\mathbf{a}_z)$ is to be zero?

1.44. The forces experienced by a test charge q at a point in a region of electric and magnetic fields are given as follows for three different velocities of the test charge:

$$\mathbf{F}_1 = 0 \qquad \text{for } \mathbf{v} = v_0\mathbf{a}_x$$
$$\mathbf{F}_2 = 0 \qquad \text{for } \mathbf{v} = v_0\mathbf{a}_y$$
$$\mathbf{F}_3 = -qE_0\mathbf{a}_z \quad \text{for } \mathbf{v} = v_0(\mathbf{a}_x + \mathbf{a}_y)$$

where v_0 and E_0 are constants. (a) Find $\mathbf{E}$ and $\mathbf{B}$ at that point. (b) Find the force experienced by the test charge for $\mathbf{v} = v_0(\mathbf{a}_x - \mathbf{a}_y)$.

2

Maxwell's Equations in Integral Form

In Chapter 1 we learned the simple rules of vector algebra and familiarized ourselves with the basic concepts of fields, particularly those associated with electric and magnetic fields. We now have the necessary background to introduce the additional tools required for the understanding of the various quantities associated with Maxwell's equations and then discuss Maxwell's equations. In particular, our goal in this chapter is to learn Maxwell's equations in integral form as a prerequisite to the derivation of their differential forms in the next chapter. Maxwell's equations in integral form govern the interdependence of certain field and source quantities associated with regions in space, that is, contours, surfaces, and volumes. The differential forms of Maxwell's equations, however, relate the characteristics of the field vectors at a given point to one another and to the source densities at that point.

Maxwell's equations in integral form are a set of four laws resulting from several experimental findings and a purely mathematical contribution. We shall, however, consider them as postulates and learn to understand their physical significance as well as their mathematical formulation. The source quantities involved in their formulation are charges and currents. The field quantities have to do with the line and surface integrals of the electric and magnetic field vectors. We shall therefore first introduce line and surface integrals and then consider successively the four Maxwell's equations in integral form.

2.1 THE LINE INTEGRAL

Let us consider in a region of electric field $\mathbf{E}$ the movement of a test charge q from the point A to the point B along the path C, as shown in Figure 2.1(a). At each and every point along the path the electric field exerts a force on the test charge and, hence, does a certain amount of work in moving the charge to another point an infinitesimal distance away. To find the total amount of work done from A to B, we divide the path into a number of infinitesimal segments $\Delta\mathbf{l}_1, \Delta\mathbf{l}_2, \Delta\mathbf{l}_3, \ldots, \Delta\mathbf{l}_n$, as shown in Figure 2.1(b), find the infinitesimal amount of work done for each segment and then add up the contributions from all the segments. Since the segments are infinitesimal in length, we can consider each of them to be straight and the electric field at all points within a segment to be the same and equal to its value at the start of the segment.

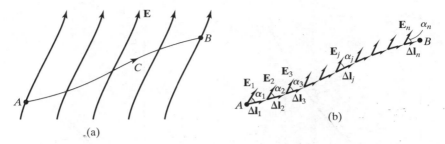

FIGURE 2.1

For evaluating the total amount of work done in moving a test charge along a path C from point A to point B in a region of electric field.

If we now consider one segment, say the jth segment, and take the component of the electric field for that segment along the length of that segment, we obtain the result $E_j \cos \alpha_j$, where α_j is the angle between the direction of the electric field vector $\mathbf{E}_j$ at the start of that segment and the direction of that segment. Since the electric field intensity has the meaning of force per unit charge, the electric force along the direction of the jth segment is then equal to $qE_j \cos \alpha_j$, where q is the value of the test charge. To obtain the work done in carrying the test charge along the length of the jth segment, we then multiply this electric force component by the length Δl_j of that segment. Thus for the jth segment, we obtain the result for the work done by the electric field as

$$\Delta W_j = qE_j \cos \alpha_j \, \Delta l_j \tag{2.1}$$

If we do this for all the infinitesimal segments and add up all the contributions, we get the total work done by the electric field in moving the test charge from A to B as

$$
\begin{aligned}
W_A^B &= \Delta W_1 + \Delta W_2 + \Delta W_3 + \cdots + \Delta W_n \\
&= qE_1 \cos \alpha_1 \, \Delta l_1 + qE_2 \cos \alpha_2 \, \Delta l_2 + qE_3 \cos \alpha_3 \, \Delta l_3 \\
&\quad + \cdots + qE_n \cos \alpha_n \, \Delta l_n \\
&= q \sum_{j=1}^{n} E_j \cos \alpha_j \, \Delta l_j
\end{aligned}
\tag{2.2}
$$

In vector notation we make use of the dot product operation between two vectors to write this quantity as

$$W_A^B = q \sum_{j=1}^{n} \mathbf{E}_j \cdot \Delta \mathbf{l}_j \tag{2.3}$$

Example 2.1

Let us consider the electric field given by

$$\mathbf{E} = y\mathbf{a}_y$$

and determine the work done by the field in carrying 3 μC of charge from the point $A(0,0,0)$ to the point $B(1,1,0)$ along the parabolic path $y = x^2, z = 0$ shown in Figure 2.2(a).

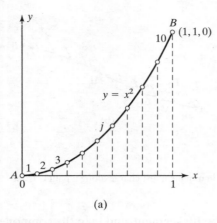 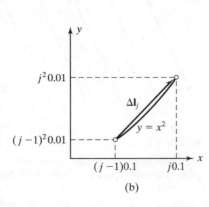

FIGURE 2.2

(a) Division of the path $y = x^2$ from $A\,(0, 0, 0)$ to $B\,(1, 1, 0)$ into ten segments. (b) The length vector corresponding to the jth segment of part (a) approximated as a straight line.

For convenience, we shall divide the path into ten segments having equal widths along the x direction, as shown in Figure 2.2(a). We shall number the segments $1, 2, 3, \ldots, 10$. The coordinates of the starting and ending points of the jth segment are as shown in Figure 2.2(b). The electric field at the start of the jth segment is given by

$$\mathbf{E}_j = (j - 1)^2\,0.01\mathbf{a}_y$$

The length vector corresponding to the jth segment, approximated as a straight line connecting its starting and ending points, is

$$\Delta\mathbf{l}_j = 0.1\mathbf{a}_x + [j^2 - (j - 1)^2]\,0.01\mathbf{a}_y$$

$$= 0.1\mathbf{a}_x + (2j - 1)0.01\mathbf{a}_y$$

The required work is then given by

$$W_A^B = 3 \times 10^{-6} \sum_{j=1}^{10} \mathbf{E}_j \cdot \Delta\mathbf{l}_j$$

$$= 3 \times 10^{-6} \sum_{j=1}^{10} [(j - 1)^2 0.01\mathbf{a}_y] \cdot [0.1\mathbf{a}_x + (2j - 1)0.01\mathbf{a}_y]$$

$$= 3 \times 10^{-10} \sum_{j=1}^{10} (j - 1)^2(2j - 1)$$

$$= 3 \times 10^{-10}[0 + 3 + 20 + 63 + 144 + 275 + 468 + 735$$

$$+ 1088 + 1539]$$

$$= 3 \times 10^{-10} \times 4335 = 1.3005\ \mu\mathbf{J}$$

The result that we have obtained in Example 2.1, for W_A^B, is approximate since we divided the path from A to B into a finite number of segments. By dividing it into larger and larger numbers of segments, we can obtain more and more accurate results. In fact, the problem can be conveniently formulated for a computer solution and by varying the number of segments from a small value to a large value, the convergence of the result can be verified. The value to which the result converges is that for which $n = \infty$. The summation in (2.3) then becomes an integral, which represents exactly the work done by the field and is given by

$$W_A^B = q \int_A^B \mathbf{E} \cdot d\mathbf{l} \tag{2.4}$$

The integral on the right side of (2.4) is known as the *line integral of* $\mathbf{E}$ *from A to B.*

Example 2.2

We shall illustrate the evaluation of the line integral by computing the exact value of the work done by the electric field in Example 2.1.

To do this, we note that at any arbitrary point $(x, y, 0)$ on the curve $y = x^2$, $z = 0$, the infinitesimal length vector tangential to the curve is given by

$$\begin{aligned}
d\mathbf{l} &= dx\, \mathbf{a}_x + dy\, \mathbf{a}_y \\
&= dx\, \mathbf{a}_x + d(x^2)\, \mathbf{a}_y \\
&= dx\, \mathbf{a}_x + 2x\, dx\, \mathbf{a}_y
\end{aligned}$$

The value of $\mathbf{E} \cdot d\mathbf{l}$ at the point $(x, y, 0)$ is

$$\begin{aligned}
\mathbf{E} \cdot d\mathbf{l} &= y\mathbf{a}_y \cdot (dx\, \mathbf{a}_x + dy\, \mathbf{a}_y) \\
&= x^2 \mathbf{a}_y \cdot (dx\, \mathbf{a}_x + 2x\, dx\, \mathbf{a}_y) \\
&= 2x^3\, dx
\end{aligned}$$

Thus, the required work is given by

$$\begin{aligned}
W_A^B &= q \int_A^B \mathbf{E} \cdot d\mathbf{l} = 3 \times 10^{-6} \int_{(0,0,0)}^{(1,1,0)} 2x^3\, dx \\
&= 3 \times 10^{-6} \left[\frac{2x^4}{4} \right]_{x=0}^{x=1} = 1.5\ \mu\mathbf{J}
\end{aligned}$$

Dividing both sides of (2.4) by q, we note that the line integral of $\mathbf{E}$ from A to B has the physical meaning of work per unit charge done by the field in moving the test charge from A to B. This quantity is known as the *voltage between A and B* and is denoted by the symbol $[V]_A^B$, having the units of volts. Thus,

$$[V]_A^B = \int_A^B \mathbf{E} \cdot d\mathbf{l} \tag{2.5}$$

When the path under consideration is a closed path, as shown in Figure 2.3, the line integral is written with a circle associated with the integral sign in the manner $\oint_C \mathbf{E} \cdot d\mathbf{l}$. The line integral of a vector around a closed path is known as the *circulation* of that vector. In particular, the line integral of $\mathbf{E}$ around a closed path is the work per unit charge done by the field in moving a test charge around the closed path. It is the voltage around the closed path and is also known as the *electromotive force*. We shall now consider an example of evaluating the line integral of a vector around a closed path.

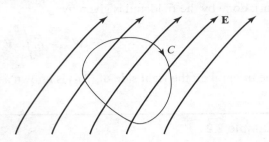

FIGURE 2.3

Closed path C in a region of electric field.

Example 2.3

Let us consider the force field

$$\mathbf{F} = x\mathbf{a}_y$$

and evaluate $\oint_C \mathbf{F} \cdot d\mathbf{l}$, where C is the closed path $ABCDA$ shown in Figure 2.4.

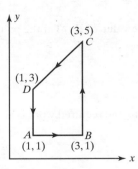

FIGURE 2.4

For evaluating the line integral of a vector field around a closed path.

Noting that

$$\oint_{ABCDA} \mathbf{F} \cdot d\mathbf{l} = \int_A^B \mathbf{F} \cdot d\mathbf{l} + \int_B^C \mathbf{F} \cdot d\mathbf{l} + \int_C^D \mathbf{F} \cdot d\mathbf{l} + \int_D^A \mathbf{F} \cdot d\mathbf{l} \tag{2.6}$$

we simply evaluate each of the line integrals on the right side of (2.6) and add them up to obtain the required quantity. Thus for the side AB,

$$y = 1, \qquad dy = 0, \qquad d\mathbf{l} = dx\,\mathbf{a}_x + (0)\mathbf{a}_y = dx\,\mathbf{a}_x$$

$$\mathbf{F} \cdot d\mathbf{l} = (x\mathbf{a}_y) \cdot (dx\,\mathbf{a}_x) = 0$$

$$\int_A^B \mathbf{F} \cdot d\mathbf{l} = 0$$

For the side *BC*,

$$x = 3, \qquad dx = 0, \qquad d\mathbf{l} = (0)\mathbf{a}_x + dy\, \mathbf{a}_y = dy\, \mathbf{a}_y$$

$$\mathbf{F} \cdot d\mathbf{l} = (3\mathbf{a}_y) \cdot (dy\, \mathbf{a}_y) = 3\, dy$$

$$\int_B^C \mathbf{F} \cdot d\mathbf{l} = \int_1^5 3\, dy = 12$$

For the side *CD*,

$$y = 2 + x, \qquad dy = dx, \qquad d\mathbf{l} = dx\, \mathbf{a}_x + dx\, \mathbf{a}_y$$

$$\mathbf{F} \cdot d\mathbf{l} = (x\mathbf{a}_y) \cdot (dx\, \mathbf{a}_x + dx\, \mathbf{a}_y) = x\, dx$$

$$\int_C^D \mathbf{F} \cdot d\mathbf{l} = \int_3^1 x\, dx = -4$$

For the side *DA*,

$$x = 1, \qquad dx = 0, \qquad d\mathbf{l} = (0)\mathbf{a}_x + dy\, \mathbf{a}_y$$

$$\mathbf{F} \cdot d\mathbf{l} = (\mathbf{a}_y) \cdot (dy\, \mathbf{a}_y) = dy$$

$$\int_D^A \mathbf{F} \cdot d\mathbf{l} = \int_3^1 dy = -2$$

Finally,

$$\oint_{ABCDA} \mathbf{F} \cdot d\mathbf{l} = 0 + 12 - 4 - 2 = 6$$

2.2 THE SURFACE INTEGRAL

Let us consider a region of magnetic field and an infinitesimal surface at a point in that region. Since the surface is infinitesimal, we can assume the magnetic flux density to be uniform on the surface, although it may be nonuniform over a wider region. If the surface is oriented normal to the magnetic field lines, as shown in Figure 2.5(a), then the magnetic flux crossing the surface is simply given by the product of the surface area and the magnetic flux density on the surface, that is, $B \, \Delta S$. If, however, the surface is oriented parallel to the magnetic field lines, as shown in Figure 2.5(b), there is no magnetic flux crossing the surface. If the surface is oriented in such a manner that the normal to the surface makes an angle α with the magnetic field lines, as shown in Figure 2.5(c), then the amount of magnetic flux crossing the surface can be determined by considering that the component of **B** normal to the surface is $B \cos \alpha$ and the component tangential to the surface is $B \sin \alpha$. The component of **B** normal to the surface results in a flux of $(B \cos \alpha) \, \Delta S$ crossing the surface, whereas the component tangential to the surface does not contribute at all to the flux crossing the surface. Thus, the magnetic flux crossing the surface in this case is $(B \cos \alpha) \, \Delta S$. We can obtain this result

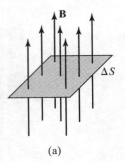

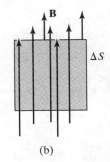

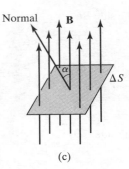

(a)   (b)   (c)

FIGURE 2.5

An infinitesimal surface ΔS in a magnetic field **B** oriented (a) normal to the field, (b) parallel to the field, and (c) with its normal making an angle α to the field.

alternatively by noting that the projection of the surface onto the plane normal to the magnetic field lines is $\Delta S \cos \alpha$.

Let us now consider a large surface S in the magnetic field region, as shown in Figure 2.6. The magnetic flux crossing this surface can be found by dividing the surface into a number of infinitesimal surfaces $\Delta S_1, \Delta S_2, \Delta S_3, \ldots, \Delta S_n$ and applying the result obtained above for each infinitesimal surface and adding up the contributions from all the surfaces. To obtain the contribution from the jth surface, we draw the normal vector to that surface and find the angle α_j between the normal vector and the magnetic flux density vector $\mathbf{B}_j$ associated with that surface. Since the surface is infinitesimal, we can assume $\mathbf{B}_j$ to be the value of **B** at the centroid of the surface and we can also erect the normal vector at that point. The contribution to the total magnetic flux from the jth infinitesimal surface is then given by

$$\Delta \psi_j = B_j \cos \alpha_j \, \Delta S_j \tag{2.7}$$

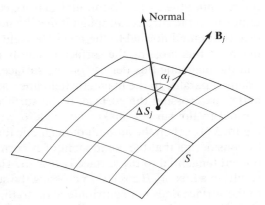

FIGURE 2.6

Division of a large surface S in a magnetic field region into a number of infinitesimal surfaces.

where the symbol ψ represents magnetic flux. The total magnetic flux crossing the surface S is then given by

$$
\begin{aligned}
[\psi]_S &= \Delta\psi_1 + \Delta\psi_2 + \Delta\psi_3 + \cdots + \Delta\psi_n \\
&= B_1 \cos \alpha_1 \, \Delta S_1 + B_2 \cos \alpha_2 \, \Delta S_2 + B_3 \cos \alpha_3 \, \Delta S_3 \\
&\quad + \cdots + B_n \cos \alpha_n \, \Delta S_n \\
&= \sum_{j=1}^{n} B_j \cos \alpha_j \, \Delta S_j
\end{aligned}
\tag{2.8}
$$

In vector notation we make use of the dot product operation between two vectors to write this quantity as

$$
[\psi]_S = \sum_{j=1}^{n} \mathbf{B}_j \cdot \Delta S_j \, \mathbf{a}_{nj}
\tag{2.9}
$$

where $\mathbf{a}_{nj}$ is the unit vector normal to the surface ΔS_j. In fact, by recalling that the infinitesimal surface can be considered as a vector quantity having magnitude equal to the area of the surface and direction normal to the surface, that is,

$$
\Delta \mathbf{S}_j = \Delta S_j \, \mathbf{a}_{nj}
\tag{2.10}
$$

we can write (2.9) as

$$
[\psi]_S = \sum_{j=1}^{n} \mathbf{B}_j \cdot \Delta \mathbf{S}_j
\tag{2.11}
$$

Example 2.4

Let us consider the magnetic field given by

$$
\mathbf{B} = 3xy^2 \mathbf{a}_z \text{ Wb/m}^2
$$

and determine the magnetic flux crossing the portion of the xy-plane lying between $x = 0$, $x = 1$, $y = 0$, and $y = 1$.

For convenience, we shall divide the surface into 25 equal areas, as shown in Figure 2.7(a). We shall designate the squares as $11, 12, \ldots, 15, 21, 22, \ldots, 55$, where the first digit represents the number of the square in the x-direction and the second digit represents the number of the square in the y-direction. The x- and y-coordinates of the midpoint of the ijth square are $(2i - 1)0.1$ and $(2j - 1)0.1$, respectively, as shown in Figure 2.7(b). The magnetic field at the center of the ijth square is then given by

$$
\mathbf{B}_{ij} = 3(2i - 1)(2j - 1)^2 0.001 \mathbf{a}_z
$$

Since we have divided the surface into equal areas and since all areas are in the xy-plane,

$$
\Delta \mathbf{S}_{ij} = 0.04 \, \mathbf{a}_z \qquad \text{for all } i \text{ and } j
$$

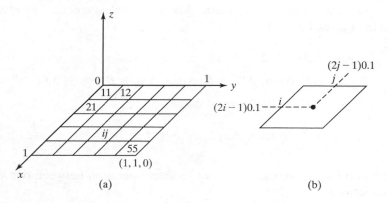

FIGURE 2.7

(a) Division of the portion of the xy-plane lying between $x = 0$, $x = 1$, $y = 0$, and $y = 1$ into 25 squares. (b) The area corresponding to the ijth square.

The required magnetic flux is then given by

$$[\psi]_S = \sum_{i=1}^{5} \sum_{j=1}^{5} \mathbf{B}_{ij} \cdot \Delta \mathbf{S}_{ij}$$

$$= \sum_{i=1}^{5} \sum_{j=1}^{5} 3(2i - 1)(2j - 1)^2 0.001\mathbf{a}_z \cdot 0.04\mathbf{a}_z$$

$$= 0.00012 \sum_{i=1}^{5} \sum_{j=1}^{5} (2i - 1)(2j - 1)^2$$

$$= 0.00012(1 + 3 + 5 + 7 + 9)(1 + 9 + 25 + 49 + 81)$$

$$= 0.495 \text{ Wb}$$

The result that we have obtained for $[\psi]_S$ in Example 2.4 is approximate since we have divided the surface S into a finite number of areas. By dividing it into larger and larger numbers of squares, we can obtain more and more accurate results. In fact, the problem can be conveniently formulated for a computer solution, and by varying the number of squares from a small value to a large value, the convergence of the result can be verified. The value to which the result converges is that for which the number of squares in each direction is infinity. The summation in (2.11) then becomes an integral that represents exactly the magnetic flux crossing the surface and is given by

$$[\psi]_S = \int_S \mathbf{B} \cdot d\mathbf{S} \tag{2.12}$$

where the symbol S associated with the integral sign denotes that the integration is performed over the surface S. The integral on the right side of (2.12) is known as the *surface integral of* **B** *over* S. The surface integral is a double integral since dS is equal to the product of two differential lengths. In fact, the work in Example 2.4 indicates that as i and j tend to infinity, the double summation becomes a double integral involving the variables of integration x and y.

Example 2.5

We shall illustrate the evaluation of the surface integral by computing the exact value of the magnetic flux in Example 2.4.

To do this, we note that at any arbitrary point (x, y) on the surface, the infinitesimal surface vector is given by

$$d\mathbf{S} = dx\, dy\, \mathbf{a}_z$$

The value of $\mathbf{B} \cdot d\mathbf{S}$ at the point (x, y) is

$$\mathbf{B} \cdot d\mathbf{S} = 3xy^2\, \mathbf{a}_z \cdot dx\, dy\, \mathbf{a}_z$$
$$= 3xy^2\, dx\, dy$$

Thus, the required magnetic flux is given by

$$[\psi]_S = \int_S \mathbf{B} \cdot d\mathbf{S}$$
$$= \int_{x=0}^{1} \int_{y=0}^{1} 3xy^2\, dx\, dy = 0.5\ \text{Wb}$$

When the surface under consideration is a closed surface, the surface integral is written with a circle associated with the integral sign in the manner $\oint_S \mathbf{B} \cdot d\mathbf{S}$. The surface integral of $\mathbf{B}$ over the closed surface S is simply the magnetic flux emanating from the volume bounded by the surface. We shall now consider an example of evaluating the closed surface integral.

Example 2.6

Let us consider the vector field

$$\mathbf{A} = (x + 2)\mathbf{a}_x + (1 - 3y)\mathbf{a}_y + 2z\mathbf{a}_z$$

and evaluate $\oint_S \mathbf{A} \cdot d\mathbf{S}$ where S is the surface of the cubical box bounded by the planes

$$x = 0, \quad x = 1$$
$$y = 0, \quad y = 1$$
$$z = 0, \quad z = 1$$

as shown in Figure 2.8.

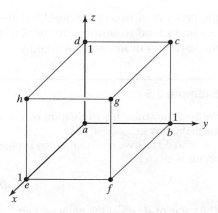

FIGURE 2.8

For evaluating the surface integral of a vector field over a closed surface.

Noting that

$$\oint_S \mathbf{A} \cdot d\mathbf{S} = \int_{abcd} \mathbf{A} \cdot d\mathbf{S} + \int_{efgh} \mathbf{A} \cdot d\mathbf{S} + \int_{aehd} \mathbf{A} \cdot d\mathbf{S} + \int_{bfgc} \mathbf{A} \cdot d\mathbf{S}$$

$$+ \int_{aefb} \mathbf{A} \cdot d\mathbf{S} + \int_{dhgc} \mathbf{A} \cdot d\mathbf{S} \tag{2.13}$$

we simply evaluate each of the surface integrals on the right side of (2.13) and add them up to obtain the required quantity. In doing so, we recognize that since the quantity we want is the flux of $\mathbf{A}$ out of the box, we should direct the normal vectors toward the outside of the box. Thus for the surface *abcd*,

$$x = 0, \qquad \mathbf{A} = 2\mathbf{a}_x + (1 - 3y)\mathbf{a}_y + 2z\mathbf{a}_z, \qquad d\mathbf{S} = -dy\, dz\, \mathbf{a}_x$$

$$\mathbf{A} \cdot d\mathbf{S} = -2\, dy\, dz$$

$$\int_{abcd} \mathbf{A} \cdot d\mathbf{S} = \int_{z=0}^{1} \int_{y=0}^{1} (-2)\, dy\, dz = -2$$

For the surface *efgh*,

$$x = 1, \qquad \mathbf{A} = 3\mathbf{a}_x + (1 - 3y)\mathbf{a}_y + 2z\mathbf{a}_z, \qquad d\mathbf{S} = dy\, dz\, \mathbf{a}_x$$

$$\mathbf{A} \cdot d\mathbf{S} = 3\, dy\, dz$$

$$\int_{efgh} \mathbf{A} \cdot d\mathbf{S} = \int_{z=0}^{1} \int_{y=0}^{1} 3\, dy\, dz = 3$$

For the surface *aehd*,

$$y = 0, \qquad \mathbf{A} = (x + 2)\mathbf{a}_x + 1\mathbf{a}_y + 2z\mathbf{a}_z, \qquad d\mathbf{S} = -dz\, dx\, \mathbf{a}_y$$

$$\mathbf{A} \cdot d\mathbf{S} = -dz\, dx$$

$$\int_{aehd} \mathbf{A} \cdot d\mathbf{S} = \int_{x=0}^{1} \int_{z=0}^{1} (-1)\, dz\, dx = -1$$

For the surface *bfgc*,

$$y = 1, \qquad \mathbf{A} = (x + 2)\mathbf{a}_x - 2\mathbf{a}_y + 2z\mathbf{a}_z, \qquad d\mathbf{S} = dz\, dx\, \mathbf{a}_y$$

$$\mathbf{A} \cdot d\mathbf{S} = -2\, dz\, dx$$

$$\int_{bfgc} \mathbf{A} \cdot d\mathbf{S} = \int_{x=0}^{1} \int_{z=0}^{1} (-2)\, dz\, dx = -2$$

For the surface *aefb*,

$$z = 0, \qquad \mathbf{A} = (x + 2)\mathbf{a}_x + (1 - 3y)\mathbf{a}_y + 0\mathbf{a}_z, \qquad d\mathbf{S} = -dx\, dy\, \mathbf{a}_z$$

$$\mathbf{A} \cdot d\mathbf{S} = 0$$

$$\int_{aefb} \mathbf{A} \cdot d\mathbf{S} = 0$$

For the surface *dhgc*,

$$z = 1, \qquad \mathbf{A} = (x + 2)\mathbf{a}_x + (1 - 3y)\mathbf{a}_y + 2\mathbf{a}_z, \qquad d\mathbf{S} = dx\, dy\, \mathbf{a}_z$$

$$\mathbf{A} \cdot d\mathbf{S} = 2\, dx\, dy$$

$$\int_{dhgc} \mathbf{A} \cdot d\mathbf{S} = \int_{y=0}^{1} \int_{x=0}^{1} 2\, dx\, dy = 2$$

Finally,

$$\oint_S \mathbf{A} \cdot d\mathbf{S} = -2 + 3 - 1 - 2 + 0 + 2 = 0$$

2.3 FARADAY'S LAW

In the previous sections we introduced the line and surface integrals. We are now ready to consider Maxwell's equations in integral form. The first equation, which we shall discuss in this section, is a consequence of an experimental finding by Michael Faraday in 1831 that time-varying magnetic fields give rise to electric fields and hence it is known as *Faraday's law*. Faraday discovered that when the magnetic flux enclosed by a loop of wire changes with time, a current is produced in the loop, indicating that a voltage or an *electromotive force*, abbreviated as emf, is induced around the loop. The variation of the magnetic flux can result from the time variation of the magnetic flux enclosed by a fixed loop or from a moving loop in a static magnetic field or from a combination of the two, that is, a moving loop in a time-varying magnetic field.

Thus far we have merely stated Faraday's finding without regard to the polarity of the induced emf around the loop or that of the magnetic flux enclosed by the loop. To clarify the point, let us consider a planar circular loop in the plane of the paper as shown in Figure 2.9. Then, we can talk of emf induced in the clockwise sense or in the counterclockwise sense. The emf induced in the clockwise sense is the line integral of $\mathbf{E}$ ($\oint \mathbf{E} \cdot d\mathbf{l}$) evaluated by traversing the loop in the clockwise direction, as shown in Figures 2.9(a) and 2.9(b). The emf induced in the counterclockwise sense is the line integral of $\mathbf{E}$ ($\oint \mathbf{E} \cdot d\mathbf{l}$) evaluated by traversing the loop in the counterclockwise direction, as shown in Figures 2.9(c) and 2.9(d). One is, of course, the negative of the other. Similarly, we can talk of enclosed magnetic flux directed into the paper or out of the paper. The enclosed magnetic flux into the paper is the surface integral of $\mathbf{B}$ ($\int \mathbf{B} \cdot d\mathbf{S}$) evaluated over the plane surface bounded by the loop and with the normal to the surface directed into the paper, as shown in Figures 2.9(a) and 2.9(c). The enclosed magnetic flux out of the paper is the surface integral of $\mathbf{B}$ ($\int \mathbf{B} \cdot d\mathbf{S}$) evaluated over the plane surface bounded by the loop and with the normal to the surface directed out of the paper, as shown in Figures 2.9(b) and 2.9(d). One is, of course, the negative of the other.

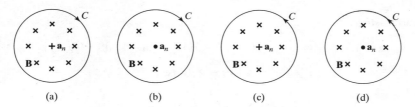

(a)   (b)   (c)   (d)

FIGURE 2.9

Four possible pairs of directions of traversal around a planar circular loop and normal to the surface bounded by the loop.

If we do not pay any attention to the polarities, we can write four equations relating the emf around the loop to the magnetic flux enclosed by the loop. These are

$$[\text{emf}]_{\text{clockwise}} = \frac{d}{dt} [\text{magnetic flux}]_{\text{into the paper}} \tag{2.14a}$$

$$[\text{emf}]_{\text{clockwise}} = \frac{d}{dt} [\text{magnetic flux}]_{\text{out of the paper}} \tag{2.14b}$$

$$[\text{emf}]_{\text{counterclockwise}} = \frac{d}{dt} [\text{magnetic flux}]_{\text{into the paper}} \tag{2.14c}$$

$$[\text{emf}]_{\text{counterclockwise}} = \frac{d}{dt} [\text{magnetic flux}]_{\text{out of the paper}} \tag{2.14d}$$

The fourth equation is, however, consistent with the first and the third equation is consistent with the second. Thus, we are left with a choice between the first and the second. Only one of them can be correct, since they provide contradictory results for the emf. Faraday's experiments showed that the second equation is the one that should be used. Alternatively, if we wish to work with clockwise-induced emf and magnetic flux into the paper (or with counterclockwise-induced emf and magnetic flux out of the paper), we must include a minus sign in front of the time derivative. This is, in fact, what is done conventionally. The convention is to use that normal to the surface which is directed toward the advancing direction of a right-hand screw when it is turned in the sense in which the loop is traversed, as shown in Figures 2.10(a) and 2.10(b). This is known as the *right-hand screw rule* and is applied consistently for all electromagnetic field laws. Hence, it is well worthwhile digesting it at this early stage.

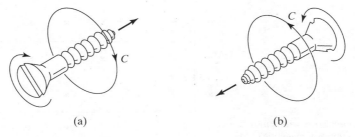

(a)   (b)

FIGURE 2.10

Right-hand screw rule convention employed in the formulation of electromagnetic field laws.

We can now express Faraday's law mathematically as

$$\oint_C \mathbf{E} \cdot d\mathbf{l} = -\frac{d}{dt}\int_S \mathbf{B} \cdot d\mathbf{S} \tag{2.15}$$

where S is a surface bounded by C. For the law to be unique, the surface S need not be a plane surface and can be any curved surface bounded by C, as depicted in Figure 2.11. This tells us that the magnetic flux through all possible surfaces bounded by C must be the same. We shall make use of this later. In fact, if C is not a planar loop, we cannot have a plane surface bounded by C. A further point of interest is that C need not represent a loop of wire but can be an imaginary closed path. It means that the time-varying magnetic flux induces an electric field in the region and this results in an emf around the closed path. If a wire is placed in the position occupied by the closed path, the emf will produce a current in the loop simply because the charges in the wire are constrained to move along the wire. Let us now consider some examples.

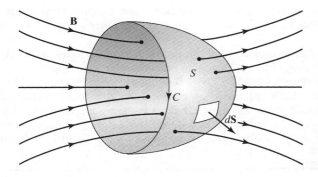

FIGURE 2.11

For illustrating Faraday's law.

Example 2.7

A rectangular loop of wire with three sides fixed and the fourth side movable is situated in a plane perpendicular to a uniform magnetic field $\mathbf{B} = B_0\mathbf{a}_z$, as illustrated in Figure 2.12. The movable side consists of a conducting bar moving with a velocity v_0 in the y-direction. It is desired to find the emf induced in the loop.

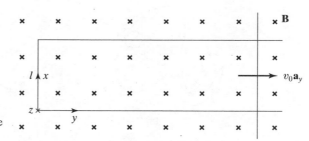

FIGURE 2.12

A rectangular loop of wire with a movable side situated in a uniform magnetic field.

Letting the position of the movable side at any time t be $y_0 + v_0t$, we obtain the magnetic flux enclosed by the loop and directed into the paper as

$$\psi = \text{(area of the loop)}B_0$$

$$= l(y_0 + v_0t)B_0$$

The emf induced in the loop in the clockwise sense is then given by

$$\oint \mathbf{E} \cdot d\mathbf{l} = -\frac{d}{dt}\psi$$

$$= -\frac{d}{dt}[l(y_0 + v_0t)B_0]$$

$$= -B_0lv_0$$

Thus, if the bar is moving to the right, the induced emf produces a current in the counterclockwise sense. Note that this polarity of the current is such that it gives rise to a magnetic field directed out of the paper inside the loop. The flux of this magnetic field is in opposition to the flux of the original magnetic field and hence tends to decrease it. This observation is in accordance with *Lenz's law*, which states that the induced emf is such that it acts to oppose the *change* in the magnetic flux producing it. The minus sign on the right side of Faraday's law ensures that Lenz's law is always satisfied.

It is also of interest to note that the induced emf can also be interpreted as due to the electric field induced in the moving bar by virtue of its motion perpendicular to the magnetic field. Thus, a charge Q in the bar experiences a force $\mathbf{F} = Q\mathbf{v} \times \mathbf{B}$ or $Qv_0\mathbf{a}_y \times B_0\mathbf{a}_z = Qv_0B_0\mathbf{a}_x$. To an observer moving with the bar, this force appears as an electric force due to an electric field $\mathbf{F}/Q = v_0B_0\mathbf{a}_x$. Viewed from inside the loop, this electric field is in the counterclockwise direction and hence the induced emf is v_0B_0l in that sense, as deduced above from Faraday's law. This concept of induced emf is known as the *motional emf concept*, which is employed widely in the study of electromechanics.

Example 2.8

A time-varying magnetic field is given by

$$\mathbf{B} = B_0 \cos \omega t \, \mathbf{a}_y$$

where B_0 is a constant. It is desired to find the induced emf around a rectangular loop in the xz-plane, as shown in Figure 2.13.

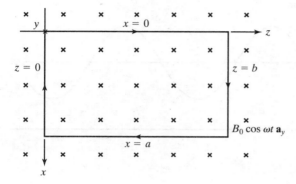

FIGURE 2.13

A rectangular loop in the xz-plane situated in a time-varying magnetic field.

The magnetic flux enclosed by the loop and directed into the paper is given by

$$\psi = \int_S \mathbf{B} \cdot d\mathbf{S} = \int_{z=0}^{b} \int_{x=0}^{a} B_0 \cos \omega t \, \mathbf{a}_y \cdot dx \, dz \, \mathbf{a}_y$$

$$= B_0 \cos \omega t \int_{z=0}^{b} \int_{x=0}^{a} dx \, dz = ab B_0 \cos \omega t$$

The induced emf in the clockwise sense is then given by

$$\oint_C \mathbf{E} \cdot d\mathbf{l} = -\frac{d}{dt} \int_S \mathbf{B} \cdot d\mathbf{S}$$

$$= -\frac{d}{dt} [ab B_0 \cos \omega t] = ab B_0 \omega \sin \omega t$$

The time variations of the magnetic flux enclosed by the loop and the induced emf around the loop are shown in Figure 2.14. It can be seen that when the magnetic flux enclosed by the loop is decreasing with time, the induced emf is positive, thereby producing a clockwise current if the loop were a wire. This polarity of the current gives rise to a magnetic field directed into the paper inside the loop and hence acts to increase the magnetic flux enclosed by the loop. When the magnetic flux enclosed by the loop is increasing with time, the induced emf is negative, thereby producing a counterclockwise current around the loop. This polarity of the current gives rise to a magnetic field directed out of the paper inside the loop and hence acts to decrease the magnetic flux enclosed by the loop. These observations are once again consistent with Lenz's law.

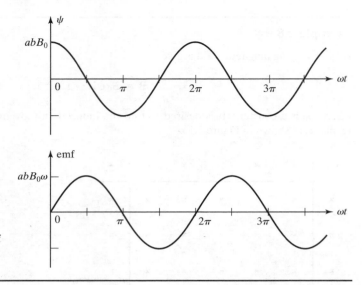

FIGURE 2.14

Time variations of magnetic flux ψ enclosed by the loop of Figure 2.13, and the resulting induced emf around the loop.

2.4 AMPERE'S CIRCUITAL LAW

In the previous section we introduced Faraday's law, one of Maxwell's equations, in integral form. In this section we introduce another of Maxwell's equations in integral form. This equation, known as *Ampere's circuital law*, is a combination of an experimental finding of Oersted that electric currents generate magnetic fields and a mathematical contribution of Maxwell that time-varying electric fields give rise to magnetic fields. It is this contribution of Maxwell that led to the prediction of electromagnetic wave propagation even before the phenomenon was discovered experimentally. In mathematical form, Ampere's circuital law is analogous to Faraday's law and is given by

$$\oint_C \frac{\mathbf{B}}{\mu_0} \cdot d\mathbf{l} = \int_S \mathbf{J} \cdot d\mathbf{S} + \frac{d}{dt} \int_S \epsilon_0 \mathbf{E} \cdot d\mathbf{S} \tag{2.16}$$

where S is any surface bounded by C, as shown in Figure 2.15. Here again, in order to evaluate the surface integrals on the right side of (2.16), we choose that normal to the surface which is directed toward the advancing direction of a right-hand screw when it is turned in the sense of C, just as in the case of Faraday's law. Also, both integrals on the right side of (2.16) must be evaluated on the same surface, whatever be the surface chosen.

The quantity $\mathbf{J}$ on the right side of (2.16) is the volume current density vector having the magnitude equal to the maximum value of current per unit area (A/m^2) at the point under consideration, as discussed in Section 1.5. Thus, the quantity $\int_S \mathbf{J} \cdot d\mathbf{S}$, being the surface integral of $\mathbf{J}$ over S, has the meaning of current due to flow of charges crossing the surface S bounded by C. It also includes line currents, that is, currents flowing along thin filamentary wires enclosed by C, and surface currents, that is, currents flowing along ribbon-like wires enclosed by C. Thus, $\int_S \mathbf{J} \cdot d\mathbf{S}$, although formulated in

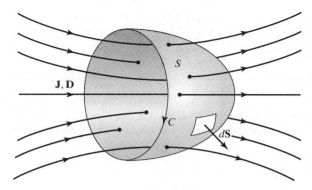

FIGURE 2.15

For illustrating Ampere's circuital law.

terms of the volume current density vector **J**, represents the algebraic sum of all the currents due to flow of charges across the surface S.

The quantity $\int_S \epsilon_0 \mathbf{E} \cdot d\mathbf{S}$ on the right side of (2.16) is the flux of the vector field $\epsilon_0 \mathbf{E}$ crossing the surface S. The vector $\epsilon_0 \mathbf{E}$ is known as the *displacement vector* or the *displacement flux density vector* and is denoted by the symbol **D**. By recalling from (1.52) that **E** has the units of (charge) per [(permittivity)(distance)^2], we note that the quantity **D** has the units of charge per unit area, or C/m^2. Hence, the quantity $\int_S \epsilon_0 \mathbf{E} \cdot d\mathbf{S}$, that is, the displacement flux has the units of charge, and the quantity $\frac{d}{dt} \int_S \epsilon_0 \mathbf{E} \cdot d\mathbf{S}$ has the units of $\frac{d}{dt}$ (charge) or current and is known as the *displacement current*. Physically, it is not a current in the sense that it does not represent the flow of charges, but mathematically it is equivalent to a current crossing the surface S.

The quantity $\oint_C \frac{\mathbf{B}}{\mu_0} \cdot d\mathbf{l}$ on the left side of (2.16) is the line integral of the vector field $\mathbf{B}/\mu_0$ around the closed path C. We learned in Section 2.1 that the quantity $\oint_C \mathbf{E} \cdot d\mathbf{l}$ has the physical meaning of work per unit charge associated with the movement of a test charge around the closed path C. The quantity $\oint_C \frac{\mathbf{B}}{\mu_0} \cdot d\mathbf{l}$ does not have a similar physical meaning. This is because magnetic force on a moving charge is directed perpendicular to the direction of motion of the charge as well as to the direction of the magnetic field and hence does not do work in the movement of the charge. The vector $\mathbf{B}/\mu_0$ is known as the *magnetic field intensity vector* and is denoted by the symbol **H**. By recalling from (1.68) that **B** has the units of [(permeability)(current)(length)] per [(distance)^2], we note that the quantity **H** has the units of current per unit distance, or A/m. This gives the units of current or A to $\oint_C \mathbf{H} \cdot d\mathbf{l}$. In analogy with the name *electromotive force* for $\oint_C \mathbf{E} \cdot d\mathbf{l}$, the quantity $\oint_C \mathbf{H} \cdot d\mathbf{l}$ is known as the *magnetomotive force*, abbreviated as mmf.

Replacing $\mathbf{B}/\mu_0$ and $\epsilon_0 \mathbf{E}$ in (2.16) by **H** and **D**, respectively, we rewrite Ampere's circuital law as

$$\oint_C \mathbf{H} \cdot d\mathbf{l} = \int_S \mathbf{J} \cdot d\mathbf{S} + \frac{d}{dt} \int_S \mathbf{D} \cdot d\mathbf{S} \tag{2.17}$$

In words, (2.17) states that "the magnetomotive force around a closed path C is equal to the total current, that is, the current due to actual flow of charges plus the displacement current bounded by C." When we say "the total current bounded by C," we mean "the total current crossing any given surface S bounded by C." This implies that the total current crossing all possible surfaces bounded by C must be the same since for a given C, $\oint_C \mathbf{H} \cdot d\mathbf{l}$ must have a unique value.

Example 2.9

An infinitely long, thin, straight wire situated along the z-axis carries a current I in the z-direction. It is desired to find $\oint_C \mathbf{H} \cdot d\mathbf{l}$ around a circle of radius a lying on the xy-plane and centered at the origin as shown in Figure 2.16.

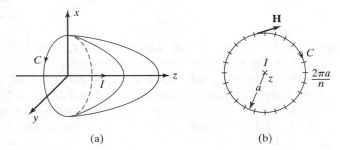

FIGURE 2.16

(a) For illustrating the uniqueness of a wire current enclosed by a closed path for an infinitely long, straight wire. (b) For finding the magnetic field due to the wire.

(a)　　　　　　　(b)

Let us consider the plane surface enclosed by C. The total current crossing the surface consists entirely of the current I carried by the wire. In fact, since the wire is infinitely long, the total current crossing any of the infinite number of surfaces bounded by C is equal to I. The situation is illustrated in Figure 2.16(a) for a few of the infinite number of surfaces. Thus, noting that the current I is bounded by C in the right-hand sense, and that it is uniquely given, we obtain

$$\oint_C \mathbf{H} \cdot d\mathbf{l} = I \tag{2.18}$$

We can proceed further and evaluate $\mathbf{H}$ at points on the circular path from symmetry considerations. In order for $\oint_C \mathbf{H} \cdot d\mathbf{l}$ to be nonzero, $\mathbf{H}$ must be directed (or have a component) tangential to the circular path and then, from symmetry considerations, it must have the same magnitude at all points on the circle, since the circle is centered at the wire. We, however, know from elementary considerations of the magnetic field due to a current element that $\mathbf{H}$ must be directed entirely tangential to the circular path. Thus, let us divide the circle into a large number of equal segments, say n, as shown in Figure 2.16(b). Since the length of each segment is $2\pi a/n$ and since $\mathbf{H}$ is parallel to the segment, $\mathbf{H} \cdot d\mathbf{l}$ for the segment is $(2\pi a/n)H$ and

$$\oint_C \mathbf{H} \cdot d\mathbf{l} = \frac{2\pi a}{n}H(\text{number of segments})$$

$$= \frac{2\pi a}{n}H \cdot n = 2\pi aH$$

From (2.18), we then have

$$2\pi aH = I$$

or

$$H = \frac{I}{2\pi a}$$

Thus, the magnetic field intensity due to the infinitely long wire is directed circular to the wire in the right-hand sense and has a magnitude $I/2\pi a$, where a is the distance of the point from the wire. The method we have discussed here is a standard procedure for the determination of the static magnetic field due to current distributions possessing certain symmetries. We shall include some cases in the problems for the interested reader.

If the wire of Example 2.9 is finitely long, say, extending from $-d$ to $+d$ on the z-axis, then, the construction of Figure 2.17 illustrates that for some surfaces the wire pierces through the surface, whereas for some other surfaces it does not. Thus, for this case, there is no unique value of the wire current alone that is enclosed by C. Hence, there must be a displacement current through the surfaces in addition to the wire current so that the total current enclosed by C is uniquely given. In fact, this displacement current is provided by the time-varying electric field due to charges accumulating at one end and depleting at the other end of the current-carrying wire. Thus, considering, for example, the surfaces S_1 and S_3 and setting the total currents through S_1 and S_3 to be equal, we have

$$\int_{S_1} \mathbf{J} \cdot d\mathbf{S} + \frac{d}{dt}\int_{S_1} \mathbf{D} \cdot d\mathbf{S} = \int_{S_3} \mathbf{J} \cdot d\mathbf{S} + \frac{d}{dt}\int_{S_3} \mathbf{D} \cdot d\mathbf{S} \qquad (2.19)$$

Now, since the wire pierces through S_1 in the right-hand sense,

$$\int_{S_1} \mathbf{J} \cdot d\mathbf{S} = I \qquad (2.20)$$

The wire does not pierce through S_3. Hence,

$$\int_{S_3} \mathbf{J} \cdot d\mathbf{S} = 0 \qquad (2.21)$$

Substituting (2.20) and (2.21) into (2.19), we get

$$I + \frac{d}{dt}\int_{S_1} \mathbf{D} \cdot d\mathbf{S} = 0 + \frac{d}{dt}\int_{S_3} \mathbf{D} \cdot d\mathbf{S} \qquad (2.22)$$

or

$$\frac{d}{dt}\int_{S_3} \mathbf{D} \cdot d\mathbf{S} - \frac{d}{dt}\int_{S_1} \mathbf{D} \cdot d\mathbf{S} = I \qquad (2.23)$$

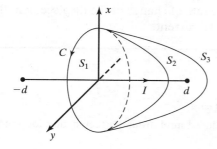

FIGURE 2.17

For illustrating that the wire current enclosed by a closed path is not unique for a finitely long wire.

Reversing the sense of evaluation of the surface integral of **D** over S_1 and changing the minus sign to a plus sign, we obtain

$$\frac{d}{dt} \oint_{S_3+S_1} \mathbf{D} \cdot d\mathbf{S} = I \tag{2.24}$$

Thus, the displacement current emanating from the closed surface $S_1 + S_3$ is equal to I.

Another example in which the wire current enclosed by C is not uniquely defined is shown in Figure 2.18, which is that of a simple circuit consisting of a capacitor driven by an alternating voltage source. Considering two surfaces S_1 and S_2, where S_1 cuts through the wire and S_2 passes between the plates of the capacitor, we have

$$\int_{S_1} \mathbf{J} \cdot d\mathbf{S} = I \tag{2.25}$$

and

$$\int_{S_2} \mathbf{J} \cdot d\mathbf{S} = 0 \tag{2.26}$$

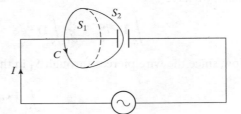

FIGURE 2.18

A capacitor circuit illustrating that the wire current enclosed by a closed path is not unique.

If we neglect fringing and assume that the electric field in the capacitor is contained entirely within the region between the plates, then

$$\int_{S_1} \mathbf{D} \cdot d\mathbf{S} = 0 \tag{2.27}$$

For $\oint_C \mathbf{H} \cdot d\mathbf{l}$ to be unique,

$$\int_{S_1} \mathbf{J} \cdot d\mathbf{S} + \frac{d}{dt} \int_{S_1} \mathbf{D} \cdot d\mathbf{S} = \int_{S_2} \mathbf{J} \cdot d\mathbf{S} + \frac{d}{dt} \int_{S_2} \mathbf{D} \cdot d\mathbf{S} \tag{2.28}$$

Substituting (2.25), (2.26), and (2.27) into (2.28), we obtain

$$\frac{d}{dt} \int_{S_2} \mathbf{D} \cdot d\mathbf{S} = I \tag{2.29}$$

Thus, the displacement current, that is, the time rate of change of the displacement flux between the capacitor plates, is equal to the wire current.

Example 2.10

A time-varying electric field is given by

$$\mathbf{E} = E_0 z \sin \omega t \, \mathbf{a}_x$$

where E_0 is a constant. It is desired to find the induced mmf around a rectangular loop in the yz-plane, as shown in Figure 2.19.

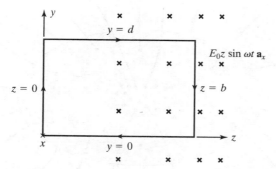

FIGURE 2.19

A rectangular loop in a time-varying electric field.

The total current here is composed entirely of displacement current. The displacement flux enclosed by the loop and directed into the paper is given by

$$\int_S \mathbf{D} \cdot d\mathbf{S} = \int_{z=0}^{b} \int_{y=0}^{d} \epsilon_0 E_0 z \sin \omega t \, \mathbf{a}_x \cdot dy \, dz \, \mathbf{a}_x$$

$$= \epsilon_0 E_0 \sin \omega t \int_{z=0}^{b} \int_{y=0}^{d} z \, dy \, dz$$

$$= \epsilon_0 \frac{b^2 d}{2} E_0 \sin \omega t$$

The induced mmf around C is then given by

$$\oint_C \mathbf{H} \cdot d\mathbf{l} = \frac{d}{dt} \int_S \mathbf{D} \cdot d\mathbf{S}$$

$$= \frac{d}{dt} \left(\epsilon_0 \frac{b^2 d}{2} E_0 \sin \omega t \right)$$

$$= \epsilon_0 \frac{b^2 d}{2} E_0 \omega \cos \omega t$$

2.5   GAUSS' LAW FOR THE ELECTRIC FIELD

In the previous two sections we learned two of the four Maxwell's equations. These two equations have to do with the line integrals of the electric and magnetic fields around closed paths. The remaining two Maxwell's equations are pertinent to the surface integrals of the electric and magnetic fields over closed surfaces. These are known as *Gauss' laws*.

Gauss' law for the electric field states that "the total displacement flux emanating from a closed surface S is equal to the total charge contained within the volume V bounded by that surface," as illustrated in Figure 2.20. This statement, although familiarly known as Gauss' law, has its origin in experiments conducted by Faraday. In mathematical form, Gauss' law for the electric field is given by

$$\oint_S \mathbf{D} \cdot d\mathbf{S} = \int_V \rho \, dv \tag{2.30}$$

where ρ is the volume charge density associated with points in the volume V.

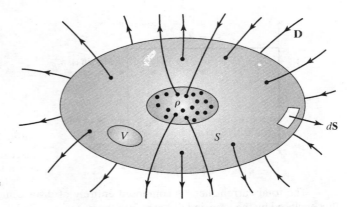

FIGURE 2.20

For illustrating Gauss' law for the electric field.

The volume charge density at a point is defined as the charge per unit volume (C/m^3) at that point in the limit that the volume shrinks to zero. Thus,

$$\rho = \lim_{\Delta v \to 0} \frac{\Delta Q}{\Delta v} \tag{2.31}$$

As an illustration of the computation of the charge contained in a given volume for a specified charge density, let us consider

$$\rho = (x + y + z) \, C/m^3$$

and the cubical volume V bounded by the planes $x = 0$, $x = 1$, $y = 0$, $y = 1$, $z = 0$, and $z = 1$. Then the charge Q contained within the cubical volume is given by

$$Q = \int_V \rho \, dv = \int_{x=0}^{1} \int_{y=0}^{1} \int_{z=0}^{1} (x + y + z) \, dx \, dy \, dz$$

$$= \int_{x=0}^{1} \int_{y=0}^{1} \left[xz + yz + \frac{z^2}{2} \right]_{z=0}^{1} dx \, dy$$

$$= \int_{x=0}^{1} \int_{y=0}^{1} \left(x + y + \frac{1}{2} \right) dx \, dy$$

$$= \int_{x=0}^{1} \left[xy + \frac{y^2}{2} + \frac{y}{2} \right]_{y=0}^{1} dx$$

$$= \int_{x=0}^{1} (x + 1) \, dx$$

$$= \left[\frac{x^2}{2} + x \right]_{x=0}^{1}$$

$$= \frac{3}{2} \, C$$

Although the quantity on the right side of (2.30), that is, the charge contained within the volume V bounded by the surface S associated with the quantity on the left side of (2.30), is formulated in terms of the volume charge density, it includes surface charges, line charges, and point charges enclosed by S. Thus it represents the algebraic sum of all the charges contained in the volume V. Let us now consider an example.

Example 2.11

A point charge Q is situated at the origin. It is desired to find $\oint_S \mathbf{D} \cdot d\mathbf{S}$ and $\mathbf{D}$ over the surface of a sphere of radius a centered at the origin.

According to Gauss' law for the electric field, the required displacement flux is given by

$$\oint_S \mathbf{D} \cdot d\mathbf{S} = Q \tag{2.32}$$

To evaluate $\mathbf{D}$ on the surface of the sphere, we note that in order for $\oint_S \mathbf{D} \cdot d\mathbf{S}$ to be nonzero, $\mathbf{D}$ must be directed normal to the spherical surface. From symmetry considerations, it must have the same magnitude at all points on the spherical surface, since the surface is centered at the origin. Thus, let us divide the spherical surface into a large number of infinitesimal areas, as shown in Figure 2.21. Since $\mathbf{D}$ is normal to each area, $\mathbf{D} \cdot d\mathbf{S}$ for each area is simply equal to $D\, dS$. Hence,

$$\oint_S \mathbf{D} \cdot d\mathbf{S} = D \int_S dS$$

$$= D \text{ (surface area of the sphere)}$$

$$= 4\pi a^2 D$$

From (2.32), we then have

$$4\pi a^2 D = Q$$

or

$$D = \frac{Q}{4\pi a^2}$$

Thus, the displacement flux density due to the point charge is directed away from the charge and has a magnitude $Q/4\pi a^2$ where a is the distance of the point from the charge. The method we have discussed here is a standard procedure for the determination of the static electric field due to charge distributions possessing certain symmetries. We shall include some cases in the problems for the interested reader.

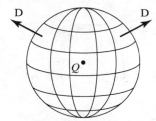

FIGURE 2.21

For evaluating the displacement flux density over the surface of a sphere centered at a point charge.

Gauss' law for the electric field is not independent of Ampere's circuital law if we recognize that, in view of conservation of electric charge, "the total current due to flow of charges emanating from a closed surface S is equal to the time rate of decrease of the charge within the volume V bounded by S," that is,

$$\oint_S \mathbf{J} \cdot d\mathbf{S} = -\frac{d}{dt} \int_V \rho \, dv$$

or

$$\oint_S \mathbf{J} \cdot d\mathbf{S} + \frac{d}{dt} \int_V \rho \, dv = 0 \tag{2.33}$$

This statement is known as the *law of conservation of charge*. In fact, it is this consideration that led to the mathematical contribution of Maxwell to Ampere's circuital law. Ampere's circuital law in its original form did not include the displacement current term which resulted in an inconsistency with (2.33) for time-varying fields.

Returning to the discussion of the dependency of Gauss' law on Ampere's circuital law through (2.33), let us consider the geometry of Figure 2.22, consisting of a closed path C and two surfaces S_1 and S_2, both of which are bounded by C. Applying Ampere's circuital law to C and S_1 and to C and S_2, we get

$$\oint_C \mathbf{H} \cdot d\mathbf{l} = \int_{S_1} \mathbf{J} \cdot d\mathbf{S}_1 + \frac{d}{dt} \int_{S_1} \mathbf{D} \cdot d\mathbf{S}_1 \tag{2.34a}$$

and

$$\oint_C \mathbf{H} \cdot d\mathbf{l} = -\int_{S_2} \mathbf{J} \cdot d\mathbf{S}_2 - \frac{d}{dt} \int_{S_2} \mathbf{D} \cdot d\mathbf{S}_2 \tag{2.34b}$$

respectively. Combining (2.34a) and (2.34b), we obtain

$$\oint_{S_1+S_2} \mathbf{J} \cdot d\mathbf{S} + \frac{d}{dt} \oint_{S_1+S_2} \mathbf{D} \cdot d\mathbf{S} = 0 \tag{2.35}$$

Now, using (2.33), we have

$$-\frac{d}{dt} \int_V \rho \, dv + \frac{d}{dt} \oint_S \mathbf{D} \cdot d\mathbf{S} = 0$$

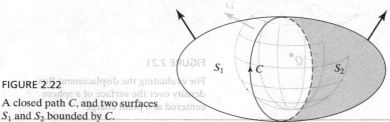

FIGURE 2.22

A closed path C, and two surfaces S_1 and S_2 bounded by C.

or

$$\frac{d}{dt}\left[\oint_S \mathbf{D}\cdot d\mathbf{S} - \int_V \rho\, dv\right] = 0 \tag{2.36}$$

where we have replaced $S_1 + S_2$ by S and where V is the volume enclosed by $S_1 + S_2$. Thus from (2.36), we get

$$\oint_S \mathbf{D}\cdot d\mathbf{S} - \int_V \rho\, dv = \text{constant with time} \tag{2.37}$$

Since there is no experimental evidence that the right side of (2.37) is nonzero, it follows that

$$\oint_S \mathbf{D}\cdot d\mathbf{S} = \int_V \rho\, dv$$

thereby giving Gauss' law for the electric field.

2.6   GAUSS' LAW FOR THE MAGNETIC FIELD

Gauss' law for the magnetic field states that "the total magnetic flux emanating from a closed surface S is equal to zero." In mathematical form, this is given by

$$\oint_S \mathbf{B}\cdot d\mathbf{S} = 0 \tag{2.38}$$

In physical terms, (2.38) signifies that magnetic charges do not exist and magnetic flux lines are closed. Whatever magnetic flux enters (or leaves) a certain part of a closed surface must leave (or enter) through the remainder of the closed surface, as illustrated in Figure 2.23.

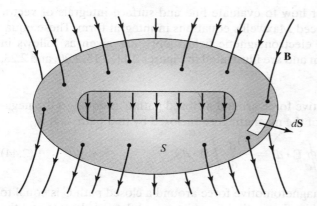

FIGURE 2.23

For illustrating Gauss' law for the magnetic field.

Equation (2.38) is not independent of Faraday's law. This can be shown by considering the geometry of Figure 2.22. Applying Faraday's law to C and S_1, we have

$$\oint_C \mathbf{E}\cdot d\mathbf{l} = -\frac{d}{dt}\int_{S_1} \mathbf{B}\cdot d\mathbf{S}_1 \tag{2.39}$$

where $d\mathbf{S}_1$ is directed out of the volume bounded by the closed surface $S_1 + S_2$. Applying Faraday's law to C and S_2, we have

$$\oint_C \mathbf{E} \cdot d\mathbf{l} = \frac{d}{dt} \int_{S_2} \mathbf{B} \cdot d\mathbf{S}_2 \qquad (2.40)$$

where $d\mathbf{S}_2$ is directed out of the volume bounded by $S_1 + S_2$. Combining (2.39) and (2.40), we obtain

$$-\frac{d}{dt} \int_{S_1} \mathbf{B} \cdot d\mathbf{S}_1 = \frac{d}{dt} \int_{S_2} \mathbf{B} \cdot d\mathbf{S}_2 \qquad (2.41)$$

or

$$\frac{d}{dt} \oint_{S_1+S_2} \mathbf{B} \cdot d\mathbf{S} = 0 \qquad (2.42)$$

or

$$\oint_{S_1+S_2} \mathbf{B} \cdot d\mathbf{S} = \text{constant with time} \qquad (2.43)$$

Since there is no experimental evidence that the right side of (2.43) is nonzero, it follows that

$$\oint_S \mathbf{B} \cdot d\mathbf{S} = 0$$

where we have replaced $S_1 + S_2$ by S.

SUMMARY

We first learned in this chapter how to evaluate line and surface integrals of vector quantities and then we introduced Maxwell's equations in integral form. These equations, which form the basis of electromagnetic field theory, are given as follows in words and in mathematical form and are illustrated in Figures 2.11, 2.15, 2.20, and 2.23, respectively.

Faraday's law.   The electromotive force around a closed path C is equal to the negative of the time rate of change of the magnetic flux enclosed by that path, that is,

$$\oint_C \mathbf{E} \cdot d\mathbf{l} = -\frac{d}{dt} \int_S \mathbf{B} \cdot d\mathbf{S} \qquad (2.44)$$

Ampere's circuital law.   The magnetomotive force around a closed path C is equal to the sum of the current enclosed by that path due to the actual flow of charges and the displacement current due to the time rate of change of the displacement flux enclosed by that path, that is,

$$\oint_C \mathbf{H} \cdot d\mathbf{l} = \int_S \mathbf{J} \cdot d\mathbf{S} + \frac{d}{dt} \int_S \mathbf{D} \cdot d\mathbf{S} \qquad (2.45)$$

Gauss' law for the electric field.   The displacement flux emanating from a closed surface S is equal to the charge enclosed by that surface, that is,

$$\oint_S \mathbf{D} \cdot d\mathbf{S} = \int_V \rho \, dv \tag{2.46}$$

Gauss' law for the magnetic field.   The magnetic flux emanating from a closed surface S is equal to zero, that is,

$$\oint_S \mathbf{B} \cdot d\mathbf{S} = 0 \tag{2.47}$$

The vectors $\mathbf{D}$ and $\mathbf{H}$, known as the displacement flux density and the magnetic field intensity vectors, respectively, are related to $\mathbf{E}$ and $\mathbf{B}$, known as the electric field intensity and the magnetic flux density vectors, respectively, in the manner

$$\mathbf{D} = \epsilon_0 \mathbf{E} \tag{2.48}$$

$$\mathbf{H} = \frac{\mathbf{B}}{\mu_0} \tag{2.49}$$

where ϵ_0 and μ_0 are the permittivity and the permeability of free space, respectively. In evaluating the right sides of (2.44) and (2.45), the normal vectors to th surfaces must be chosen such that they are directed in the right-hand sense, that is, toward the side of advance of a right-hand screw as it is turned around C, as shown in Figures 2.11 and 2.15. We have also learned that (2.47) is not independent of (2.44) and that (2.46) follows from (2.45) with the aid of the law of conservation of charge given by

$$\oint_S \mathbf{J} \cdot d\mathbf{S} + \frac{d}{dt} \int_V \rho \, dv = 0 \tag{2.50}$$

In words, (2.50) states that the sum of the current due to the flow of charges across a closed surface S and the time rate of increase of the charge within the volume V bounded by S is equal to zero. In (2.46), (2.47), and (2.50) the surface integrals must be evaluated in order to find the flux outward from the volume bounded by the surface.

Finally, we observe that time-varying electric and magnetic fields are interdependent, since according to Faraday's law (2.44), a time-varying magnetic field produces an electric field, whereas according to Ampere's circuital law (2.45), a time-varying electric field gives rise to a magnetic field. In addition, Ampere's circuital law tells us that an electric current generates a magnetic field. These properties from the basis for the phenomena of radiation and propagation of electromagnetic waves. To provide a simplified, qualitative explanation of radiation from an antenna, we begin with a piece of wire carrying a time-varying current, $I(t)$, as shown in Figure 2.24. Then, the time-varying current generates a time-varying magnetic field $\mathbf{H}(t)$, which surrounds the wire. Time-varying electric and magnetic fields, $\mathbf{E}(t)$ and $\mathbf{H}(t)$, are then produced in succession, as shown by two views in Figure 2.24, thereby giving rise to electromagnetic waves. Thus, just as water waves are produced when a rock is thrown in a pool of water, electromagnetic waves are radiated when a piece of wire in space is excited by a time-varying current.

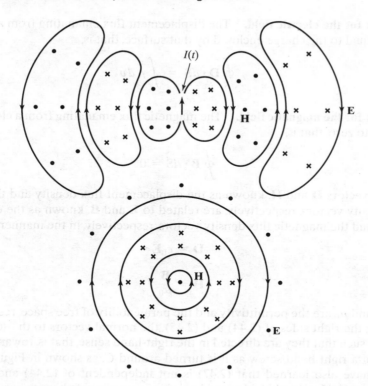

FIGURE 2.24

Two views of a simplified depiction of electromagnetic wave radiation
from a piece of wire carrying a time-varying current.

REVIEW QUESTIONS

2.1. How do you find the work done in moving a test charge by an infinitesimal distance in
an electric field?

2.2. What is the amount of work involved in moving a test charge normal to the electric field?

2.3. What is the physical interpretation of the line integral of **E** between two points A and B?

2.4. How do you find the approximate value of the line integral of a vector along a given path?

2.5. How do you find the exact value of the line integral?

2.6. What is the physical significance of the line integral of the earth's gravitational field
intensity?

2.7. What is the value of the line integral of the earth's gravitational field intensity around a
closed path?

2.8. How do you find the magnetic flux crossing an infinitesimal surface?

2.9. What is the magnetic flux crossing an infinitesimal surface oriented parallel to the magnetic flux density vector?

2.10. For what orientation of the infinitesimal surface relative to the magnetic flux density
vector is the magnetic flux crossing the surface a maximum?

2.11. How do you find the approximate value of the surface integral over a given surface?

2.12. How do you find the exact value of the surface integral?

2.13. Provide physical interpretations for the closed surface integrals of any two vectors of your choice.

2.14. State Faraday's law.

2.15. Why is it necessary to have the minus sign associated with the time rate of increase of magnetic flux on the right side of Faraday's law?

2.16. What is electromotive force?

2.17. What are the different ways in which an emf is induced around a loop?

2.18. To find the induced emf around a planar loop, is it necessary to consider the magnetic flux crossing the plane surface bounded by the loop?

2.19. Discuss briefly the motional emf concept.

2.20. What is Lenz's law?

2.21. How would you orient a loop antenna in order to obtain maximum signal from an incident electromagnetic wave which has its magnetic field linearly polarized in the north–south direction?

2.22. State three applications of Faraday's law.

2.23. State Ampere's circuital law.

2.24. What are the units of the magnetic field intensity vector?

2.25. What are the units of the displacement flux density vector?

2.26. What is displacement current? Give an example involving displacement current.

2.27. Why is it necessary to have the displacement current term on the right side of Ampere's circuital law?

2.28. When can you say that the current in a wire enclosed by a closed path is uniquely defined? Give two examples.

2.29. Give an example in which the current in a wire enclosed by a closed path is not uniquely defined.

2.30. Is it meaningful to consider two different surfaces bounded by a closed path to compute the two different currents on the right side of Ampere's circuital law to find $\oint \mathbf{H} \cdot d\mathbf{l}$ around the closed path?

2.31. Discuss briefly the application of Ampere's circuital law to determine the magnetic field due to current distributions.

2.32. State Gauss' law for the electric field.

2.33. How is volume charge density defined?

2.34. State the law of conservation of charge.

2.35. How is Gauss' law for the electric field derived from Ampere's circuital law?

2.36. Discuss briefly the application of Gauss' law for the electric field to determine the electric field due to charge distributions.

2.37. State Gauss' law for the magnetic field. How is it derived from Faraday's law?

2.38. What is the physical interpretation of Gauss' law for the magnetic field?

2.39. Summarize Maxwell's equations in integral form. Discuss the interdependence of time-varying electric and magnetic fields, with the aid of an example.

2.40. Which two of the Maxwell's equations are independent?

PROBLEMS

2.1. For the force field $\mathbf{F} = x^2\mathbf{a}_y$, find the approximate value of the line integral of $\mathbf{F}$ from the origin to the point $(1, 3, 0)$ along a straight line path by dividing the path into ten equal segments.

2.2. For the force field $\mathbf{F} = x^2\mathbf{a}_y$, obtain a series expression for the line integral of $\mathbf{F}$ from the origin to the point $(1, 3, 0)$ along a straight line path by dividing the path into n equal segments. Express the sum of the series in closed form and compute its value for values of n equal to $5, 10, 100$, and ∞.

2.3. For the force field $\mathbf{F} = x^2\mathbf{a}_y$, find the exact value of the line integral of $\mathbf{F}$ from the origin to the point $(1, 3, 0)$ along a straight line path.

2.4. Given $\mathbf{E} = y\mathbf{a}_x + x\mathbf{a}_y$, find $\int_{(0,0,0)}^{(1,1,0)} \mathbf{E} \cdot d\mathbf{l}$ along the following paths: (a) straight line path $y = x, z = 0$, (b) straight line path from $(0, 0, 0)$ to $(1, 0, 0)$, and then straight line path from $(1, 0, 0)$ to $(1, 1, 0)$, and (c) any path of your choice.

2.5. Show that for any closed path C, $\oint_C d\mathbf{l} = 0$ and hence show that for a uniform field $\mathbf{F}$, $\oint_C \mathbf{F} \cdot d\mathbf{l} = 0$.

2.6. Given $\mathbf{F} = y\mathbf{a}_x - x\mathbf{a}_y$, find $\oint_C \mathbf{F} \cdot d\mathbf{l}$ where C is the closed path in the xy-plane consisting of the following: the straight line path from $(0, 0, 0)$ to $(-1, 1, 0)$, the straight line path from $(-1, 1, 0)$ to $(0, \sqrt{2}, 0)$, the straight line path from $(0, \sqrt{2}, 0)$ to $(0, 1, 0)$, the circular path from $(0, 1, 0)$ to $(1, 0, 0)$ having its center at $(0, 0, 0)$, and the straight line path from $(1, 0, 0)$ to $(0, 0, 0)$.

2.7. Given $\mathbf{F} = xy\mathbf{a}_x + yz\mathbf{a}_y + zx\mathbf{a}_z$, find $\oint_C \mathbf{F} \cdot d\mathbf{l}$ where C is the closed path comprising the straight lines from $(0, 0, 0)$ to $(1, 1, 1)$, from $(1, 1, 1)$ to $(1, 1, 0)$, and from $(1, 1, 0)$ to $(0, 0, 0)$.

2.8. For the magnetic flux density vector $\mathbf{B} = x^2 e^{-y}\mathbf{a}_z$ Wb/m^2, find the approximate value of the magnetic flux crossing the portion of the xy-plane lying between $x = 0$, $x = 1$, $y = 0$, and $y = 1$, by dividing the area into 100 equal parts.

2.9. For the magnetic flux density vector $\mathbf{B} = x^2 e^{-y}\mathbf{a}_z$ Wb/m^2, obtain a series expression for the magnetic flux crossing the portion of the xy-plane lying between $x = 0$, $x = 1$, $y = 0$, and $y = 1$ by dividing the area into n^2 equal parts. Express the sum of the series in closed form and compute its value for values of n equal to $5, 10, 100$, and ∞.

2.10. For the magnetic flux density vector $\mathbf{B} = x^2 e^{-y}\mathbf{a}_z$ Wb/m^2, find the exact value of the magnetic flux crossing the portion of the xy-plane lying between $x = 0$, $x = 1$, $y = 0$, and $y = 1$ by evaluating the surface integral of $\mathbf{B}$.

2.11. Given $\mathbf{A} = x\mathbf{a}_x + y\mathbf{a}_y + z\mathbf{a}_z$, find $\int_S \mathbf{A} \cdot d\mathbf{S}$ where S is the hemispherical surface of radius 2 m lying above the xy-plane and having its center at the origin.

2.12. Show that for any closed surface S, $\oint_S d\mathbf{S} = 0$ and hence show that for a uniform field $\mathbf{A}$, $\oint_S \mathbf{A} \cdot d\mathbf{S} = 0$.

2.13. Given $\mathbf{J} = 3x\mathbf{a}_x + (y - 3)\mathbf{a}_y + (2 + z)\mathbf{a}_z$ A/m^2, find $\oint_S \mathbf{J} \cdot d\mathbf{S}$, that is, the current flowing out of the surface S of the rectangular box bounded by the planes $x = 0$, $x = 1$, $y = 0$, $y = 2$, $z = 0$, and $z = 3$.

2.14. Given $\mathbf{E} = (y\mathbf{a}_x - x\mathbf{a}_y) \cos \omega t$ V/m, find the time rate of decrease of the magnetic flux crossing toward the positive z-side and enclosed by the path in the xy-plane from $(0, 0, 0)$ to $(1, 0, 0)$ along $y = 0$, from $(1, 0, 0)$ to $(1, 1, 0)$ along $x = 1$, and from $(1, 1, 0)$ to $(0, 0, 0)$ along $y = x^3$.

2.15. A magnetic field is given in the xz-plane by $\mathbf{B} = \dfrac{B_0}{x}\mathbf{a}_y$ Wb/m^2, where B_0 is a constant. A rigid rectangular loop is situated in the xz-plane and with its corners at the points (x_0, z_0), $(x_0, z_0 + b)$, $(x_0 + a, z_0 + b)$, and $(x_0 + a, z_0)$. If the loop is moving in that plane with a velocity $\mathbf{v} = v_0\mathbf{a}_x$ m/s, where v_0 is a constant, find by using Faraday's law the induced emf around the loop in the sense defined by connecting the above specified points in succession. Discuss your result by using the motional emf concept.

2.16. Assuming the rectangular loop of Problem 2.15 to be stationary, find the induced emf around the loop if $\mathbf{B} = \dfrac{B_0}{x}\cos \omega t\, \mathbf{a}_y$ Wb/m^2.

2.17. Assuming the rectangular loop of Problem 2.15 to be moving with the velocity $\mathbf{v} = v_0\mathbf{a}_x$ m/s, find the induced emf around the loop if $\mathbf{B} = \dfrac{B_0}{x}\cos \omega t\, \mathbf{a}_y$ Wb/m^2.

2.18. For $\mathbf{B} = B_0 \cos \omega t\, \mathbf{a}_z$ Wb/m^2, find the induced emf around the closed path comprising the straight lines successively connecting the points $(0, 0, 0)$, $(1, 0, 0.01)$, $(1, 1, 0.02)$, $(0, 1, 0.03)$, $(0, 0, 0.04)$, and $(0, 0, 0)$.

2.19. Repeat Problem 2.18 for the closed path comprising the straight lines successively connecting the points $(0, 0, 0)$, $(1, 0, 0.01)$, $(1, 1, 0.02)$, $(0, 1, 0.03)$, $(0, 0, 0.04)$, $(1, 0, 0.05)$, $(1, 1, 0.06)$, $(0, 1, 0.07)$, $(0, 0, 0.08)$, and $(0, 0, 0)$, with a slight kink in the last straight line at the point $(0, 0, 0.04)$ to avoid touching the point.

2.20. A rigid rectangular loop of area A is situated normal to the xy-plane and symmetrically about the z-axis. It revolves around the z-axis at ω_1 rad/s in the sense defined by the curling of the fingers of the right hand when the z-axis is grabbed with the thumb pointed in the positive z-direction. Find the induced emf around the loop if $\mathbf{B} = B_0 \cos \omega_2 t\, \mathbf{a}_x$, where B_0 is a constant, and show that the induced emf has two frequency components $(\omega_1 + \omega_2)$ and $|\omega_1 - \omega_2|$.

2.21. For the revolving loop of Problem 2.20, find the induced emf around the loop if $\mathbf{B} = B_0(\cos \omega_1 t\, \mathbf{a}_x + \sin \omega_1 t\, \mathbf{a}_y)$.

2.22. For the revolving loop of Problem 2.20, find the induced emf around the loop if $\mathbf{B} = B_0(\cos \omega_1 t\, \mathbf{a}_x - \sin \omega_1 t\, \mathbf{a}_y)$.

2.23. A current I_1 flows from infinity to a point charge at the origin through a thin wire along the negative y-axis and a current I_2 flows from the point charge to infinity through another thin wire along the positive y-axis. From considerations of uniqueness of $\oint_C \mathbf{H} \cdot d\mathbf{l}$, find the displacement current emanating from (a) a spherical surface of radius 1 m and having its center at the point $(2, 2, 2)$ and (b) a spherical surface of radius 1 m and having its center at the origin.

2.24. A current density due to flow of charges is given by $\mathbf{J} = y \cos \omega t\, \mathbf{a}_y$ A/m^2. From consideration of uniqueness of $\oint_C \mathbf{H} \cdot d\mathbf{l}$, find the displacement current emanating from the cubical box bounded by the planes $x = 0$, $x = 1$, $y = 0$, $y = 1$, $z = 0$, and $z = 1$.

2.25. An infinitely long, cylindrical wire of radius a, having the z-axis as its axis, carries current in the positive z-direction with uniform density J_0 A/m^2. Find $\mathbf{H}$ both inside and outside the wire.

2.26. An infinitely long, hollow, cylindrical wire of inner radius a and outer radius b, having the z-axis as its axis, carries current in the positive z-direction with uniform density J_0 A/m^2. Find $\mathbf{H}$ everywhere.

2.27. An infinitely long, straight wire situated along the z-axis carries current I in the positive z-direction. What are the values of $\int_{(1,0,0)}^{(0,1,0)} \mathbf{H} \cdot d\mathbf{l}$ along (a) the circular path of radius 1 m and centered at the origin and (b) along a straight line path?

2.28. Given $\mathbf{D} = y\mathbf{a}_y$, find the charge contained in the volume of the wedge-shaped box defined by the planes $x = 0$, $x + z = 1$, $y = 0$, $y = 1$, and $z = 0$.

2.29. Given $\rho = xe^{-x^2}$ C/m^3, find the displacement flux emanating from the surface of the cubical box defined by the planes $x = 0$, $x = 1$, $y = 0$, $y = 1$, $z = 0$, and $z = 1$.

2.30. Charge is distributed uniformly along the z-axis with density ρ_{L0} C/m. Using Gauss' law for the electric field, find the electric field intensity due to the line charge.

2.31. Charge is distributed uniformly with density ρ_0 C/m^3 within a spherical volume of radius a m and having its center at the origin. Using Gauss' law for the electric field, find the electric field intensity both inside and outside the charge distribution.

2.32. A point charge Q C is situated at the origin. What are the values of the displacement flux crossing (a) the spherical surface $x^2 + y^2 + z^2 = 1$, $x > 0$, $y > 0$, and $z > 0$ and (b) the plane surface $x + y + z = 1$, $x > 0$, $y > 0$, and $z > 0$?

2.33. Given $\mathbf{J} = x\mathbf{a}_x$ A/m^2, find the time rate of increase of the charge contained in the cubical volume bounded by the planes $x = 0$, $x = 1$, $y = 0$, $y = 1$, $z = 0$, and $z = 1$.

2.34. Given $\mathbf{J} = x\mathbf{a}_x$ A/m^2, find the time rate of increase of the charge contained in the volume of the wedge-shaped box that is defined by the planes $x = 0$, $x + z = 1$, $y = 0$, $y = 1$, and $z = 0$.

2.35. Using the property that $\oint_S \mathbf{B} \cdot d\mathbf{S} = 0$, find the absolute value of $\int \mathbf{B} \cdot d\mathbf{S}$ over that portion of the surface $y = \sin x$ bounded by $x = 0$, $x = \pi$, $z = 0$, and $z = 1$, for $\mathbf{B} = y\mathbf{a}_x - x\mathbf{a}_y$.

2.36. Repeat Problem 2.35 for the plane rectangular surface having the vertices at $(0, 0, 0)$, $(0, 0, 1)$, $(1, 1, 1)$, and $(0, 1, 1)$.

Maxwell's Equations in Differential Form

In Chapter 2 we introduced Maxwell's equations in integral form. We learned that the quantities involved in the formulation of these equations are the scalar quantities, electromotive force, magnetomotive force, magnetic flux, displacement flux, charge, and current, which are related to the field vectors and source densities through line, surface, and volume integrals. Thus, the integral forms of Maxwell's equations, while containing all the information pertinent to the interdependence of the field and source quantities over a given region in space, do not permit us to study directly the interaction between the field vectors and their relationships with the source densities at individual points. It is our goal in this chapter to derive the differential forms of Maxwell's equations that apply directly to the field vectors and source densities at a given point.

We shall derive Maxwell's equations in differential form by applying Maxwell's equations in integral form to infinitesimal closed paths, surfaces, and volumes, in the limit that they shrink to points. We will find that the differential equations relate the spatial variations of the field vectors at a given point to their temporal variations and to the charge and current densities at that point. In this process we shall also learn two important operations in vector calculus, known as curl and divergence, and two related theorems, known as Stokes' and divergence theorems.

3.1 FARADAY'S LAW

We recall from the previous chapter that Faraday's law is given in integral form by

$$\oint_C \mathbf{E} \cdot d\mathbf{l} = -\frac{d}{dt}\int_S \mathbf{B} \cdot d\mathbf{S} \tag{3.1}$$

where S is any surface bounded by the closed path C. In the most general case, the electric and magnetic fields have all three components (x, y, and z) and are dependent on all three coordinates (x, y, and z) in addition to time (t). For simplicity, we shall, however, first consider the case in which the electric field has an x-component only, which is dependent only on the z-coordinate, in addition to time. Thus,

$$\mathbf{E} = E_x(z, t)\mathbf{a}_x \tag{3.2}$$

71

In other words, this simple form of time-varying electric field is everywhere directed in the x-direction and it is uniform in planes parallel to the xy-plane.

Let us now consider a rectangular path C of infinitesimal size lying in a plane parallel to the xz-plane and defined by the points (x, z), $(x, z + \Delta z)$, $(x + \Delta x, z + \Delta z)$, and $(x + \Delta x, z)$, as shown in Figure 3.1. According to Faraday's law, the emf around the closed path C is equal to the negative of the time rate of change of the magnetic flux enclosed by C. The emf is given by the line integral of $\mathbf{E}$ around C. Thus, evaluating the line integrals of $\mathbf{E}$ along the four sides of the rectangular path, we obtain

$$\int_{(x, z)}^{(x, z + \Delta z)} \mathbf{E} \cdot d\mathbf{l} = 0 \qquad \text{since } E_z = 0 \tag{3.3a}$$

$$\int_{(x, z + \Delta z)}^{(x + \Delta x, z + \Delta z)} \mathbf{E} \cdot d\mathbf{l} = [E_x]_{z + \Delta z} \, \Delta x \tag{3.3b}$$

$$\int_{(x + \Delta x, z + \Delta z)}^{(x + \Delta x, z)} \mathbf{E} \cdot d\mathbf{l} = 0 \quad \text{since } E_z = 0 \tag{3.3c}$$

$$\int_{(x + \Delta x, z)}^{(x, z)} \mathbf{E} \cdot d\mathbf{l} = -[E_x]_z \, \Delta x \tag{3.3d}$$

Adding up (3.3a)–(3.3d), we obtain

$$\oint_C \mathbf{E} \cdot d\mathbf{l} = [E_x]_{z + \Delta z} \, \Delta x - [E_x]_z \, \Delta x$$

$$= \{[E_x]_{z + \Delta z} - [E_x]_z\} \, \Delta x \tag{3.4}$$

In (3.3a)–(3.3d) and (3.4), $[E_x]_z$ and $[E_x]_{z + \Delta z}$ denote values of E_x evaluated along the sides of the path for which $z = z$ and $z = z + \Delta z$, respectively.

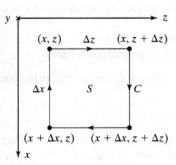

FIGURE 3.1

Infinitesimal rectangular path lying in a plane parallel to the xz-plane.

To find the magnetic flux enclosed by C, let us consider the plane surface S bounded by C. According to the right-hand screw rule, we must use the magnetic flux crossing S toward the positive y-direction, that is, into the page, since the path C is traversed in the clockwise sense. The only component of $\mathbf{B}$ normal to the area S is the y-component. Also, since the area is infinitesimal in size, we can assume B_y to be uniform over the area and equal to its value at (x, z). The required magnetic flux is then given by

$$\int_S \mathbf{B} \cdot d\mathbf{S} = [B_y]_{(x, z)}\, \Delta x\, \Delta z \tag{3.5}$$

Substituting (3.4) and (3.5) into (3.1) to apply Faraday's law to the rectangular path C under consideration, we get

$$\{[E_x]_{z+\Delta z} - [E_x]_z\}\, \Delta x = -\frac{d}{dt}\{[B_y]_{(x, z)}\, \Delta x\, \Delta z\}$$

or

$$\frac{[E_x]_{z+\Delta z} - [E_x]_z}{\Delta z} = -\frac{\partial [B_y]_{(x, z)}}{\partial t} \tag{3.6}$$

If we now let the rectangular path shrink to the point (x, z) by letting Δx and Δz tend to zero, we obtain

$$\underset{\substack{\Delta x \to 0 \\ \Delta z \to 0}}{\text{Lim}} \frac{[E_x]_{z+\Delta z} - [E_x]_z}{\Delta z} = -\underset{\substack{\Delta x \to 0 \\ \Delta z \to 0}}{\text{Lim}} \frac{\partial [B_y]_{(x, z)}}{\partial t}$$

or

$$\frac{\partial E_x}{\partial z} = -\frac{\partial B_y}{\partial t} \tag{3.7}$$

Equation (3.7) is Faraday's law in differential form for the simple case of $\mathbf{E}$ given by (3.2). It relates the variation of E_x with z (space) at a point to the variation of B_y with t (time) at that point. Since the above derivation can be carried out for any arbitrary point (x, y, z), it is valid for all points. It tells us in particular that a time-varying B_y at a point results in an E_x at that point having a differential in the z-direction. This is to be expected since if this is not the case, $\oint \mathbf{E} \cdot d\mathbf{l}$ around the infinitesimal rectangular path would be zero.

Example 3.1

Given $\mathbf{B} = B_0 \cos \omega t\, \mathbf{a}_y$ and it is known that $\mathbf{E}$ has an x-component only, let us find E_x.
From (3.6), we have

$$\frac{\partial E_x}{\partial z} = -\frac{\partial B_y}{\partial t} = -\frac{\partial}{\partial t}(B_0 \cos \omega t) = \omega B_0 \sin \omega t$$

$$E_x = \omega B_0 z \sin \omega t$$

We note that the uniform magnetic field gives rise to an electric field varying linearly with z.

Proceeding further, we can verify this result by evaluating $\oint \mathbf{E} \cdot d\mathbf{l}$ around the rectangular path of Example 2.8. This rectangular path is reproduced in Figure 3.2. The required line integral is given by

$$\oint_C \mathbf{E} \cdot d\mathbf{l} = \int_{z=0}^{b} [E_z]_{x=0}\, dz + \int_{x=0}^{a} [E_x]_{z=b}\, dx$$

$$+ \int_{z=b}^{0} [E_z]_{x=a}\, dz + \int_{x=a}^{0} [E_x]_{z=0}\, dx$$

$$= 0 + [\omega B_0 b \sin \omega t]a + 0 + 0$$

$$= ab B_0 \omega \sin \omega t$$

which agrees with the result of Example 2.8.

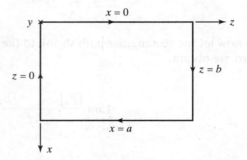

FIGURE 3.2

Rectangular path of Example 2.8.

We shall now proceed to generalize (3.7) for the arbitrary case of the electric field having all three components (x, y, and z), each of them depending on all three coordinates (x, y, and z), in addition to time (t), that is,

$$\mathbf{E} = E_x(x, y, z, t)\mathbf{a}_x + E_y(x, y, z, t)\mathbf{a}_y + E_z(x, y, z, t)\mathbf{a}_z \tag{3.8}$$

To do this, let us consider the three infinitesimal rectangular paths in planes parallel to the three mutually orthogonal planes of the Cartesian coordinate system, as shown in Figure 3.3. Evaluating $\oint \mathbf{E} \cdot d\mathbf{l}$ around the closed paths *abcda*, *adefa*, and *afgba*, we get

$$\oint_{abcda} \mathbf{E} \cdot d\mathbf{l} = [E_y]_{(x,z)}\, \Delta y + [E_z]_{(x,y+\Delta y)}\, \Delta z$$

$$- [E_y]_{(x,z+\Delta z)}\, \Delta y - [E_z]_{(x,y)}\, \Delta z \tag{3.9a}$$

$$\oint_{adefa} \mathbf{E} \cdot d\mathbf{l} = [E_z]_{(x,y)}\, \Delta z + [E_x]_{(y,z+\Delta z)}\, \Delta x$$

$$- [E_z]_{(x+\Delta x,y)}\, \Delta z - [E_x]_{(y,z)}\, \Delta x \tag{3.9b}$$

$$\oint_{afgba} \mathbf{E} \cdot d\mathbf{l} = [E_x]_{(y,\,z)}\,\Delta x + [E_y]_{(x+\Delta x,\,z)}\,\Delta y$$

$$- [E_x]_{(y+\Delta y,\,z)}\,\Delta x - [E_y]_{(x,\,z)}\,\Delta y \tag{3.9c}$$

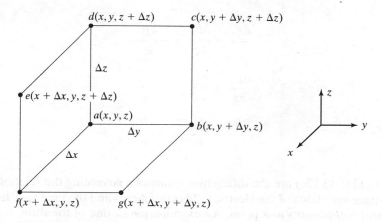

FIGURE 3.3

Infinitesimal rectangular paths in three mutually orthogonal planes.

In (3.9a)–(3.9c) the subscripts associated with the field components in the various terms on the right sides of the equations denote the value of the coordinates that remain constant along the sides of the closed paths corresponding to the terms. Now, evaluating $\int \mathbf{B} \cdot d\mathbf{S}$ over the surfaces *abcd*, *adef*, and *afgb*, keeping in mind the right-hand screw rule, we have

$$\int_{abcd} \mathbf{B} \cdot d\mathbf{S} = [B_x]_{(x,\,y,\,z)}\,\Delta y\,\Delta z \tag{3.10a}$$

$$\int_{adef} \mathbf{B} \cdot d\mathbf{S} = [B_y]_{(x,\,y,\,z)}\,\Delta z\,\Delta x \tag{3.10b}$$

$$\int_{afgb} \mathbf{B} \cdot d\mathbf{S} = [B_z]_{(x,\,y,\,z)}\,\Delta x\,\Delta y \tag{3.10c}$$

Applying Faraday's law to each of the three paths by making use of (3.9a)–(3.9c) and (3.10a)–(3.10c) and simplifying, we obtain

$$\frac{[E_z]_{(x,\,y+\Delta y)} - [E_z]_{(x,\,y)}}{\Delta y} - \frac{[E_y]_{(x,\,z+\Delta z)} - [E_y]_{(x,\,z)}}{\Delta z} = -\frac{\partial [B_x]_{(x,\,y,\,z)}}{\partial t} \tag{3.11a}$$

$$\frac{[E_x]_{(y,\,z+\Delta z)} - [E_x]_{(y,\,z)}}{\Delta z} - \frac{[E_z]_{(x+\Delta x,\,y)} - [E_z]_{(x,\,y)}}{\Delta x} = -\frac{\partial [B_y]_{(x,\,y,\,z)}}{\partial t} \tag{3.11b}$$

$$\frac{[E_y]_{(x+\Delta x,\, z)} - [E_y]_{(x,\, z)}}{\Delta x} - \frac{[E_x]_{(y+\Delta y,\, z)} - [E_x]_{(y,\, z)}}{\Delta y} = -\frac{\partial [B_z]_{(x,\, y,\, z)}}{\partial t} \qquad (3.11c)$$

If we now let all three paths shrink to the point a by letting Δx, Δy, and Δz tend to zero, (3.11a)–(3.11c) reduce to

$$\frac{\partial E_z}{\partial y} - \frac{\partial E_y}{\partial z} = -\frac{\partial B_x}{\partial t} \qquad (3.12a)$$

$$\frac{\partial E_x}{\partial z} - \frac{\partial E_z}{\partial x} = -\frac{\partial B_y}{\partial t} \qquad (3.12b)$$

$$\frac{\partial E_y}{\partial x} - \frac{\partial E_x}{\partial y} = -\frac{\partial B_z}{\partial t} \qquad (3.12c)$$

Equations (3.12a)–(3.12c) are the differential equations governing the relationships between the space variations of the electric field components and the time variations of the
magnetic field components at a point. An examination of one of the three equations is
sufficient to reveal the physical meaning of these relationships. For example, (3.12a) tells
us that a time-varying B_x at a point results in an electric field at that point having y- and
z-components such that their net right-lateral differential normal to the x-direction is
nonzero. The right-lateral differential of E_y normal to the x-direction is its derivative in
the $\mathbf{a}_y \times \mathbf{a}_x$, or $-\mathbf{a}_z$-direction, that is, $\partial E_y / \partial (-z)$ or $-\partial E_y / \partial z$. The right-lateral differential of E_z normal to the x-direction is its derivative in the $\mathbf{a}_z \times \mathbf{a}_x$, or $\mathbf{a}_y$-direction, that
is, $\partial E_z / \partial y$. Thus, the net right-lateral differential of the y- and z-components of the elec-
tric field normal to the x-direction is $(-\partial E_y / \partial z) + (\partial E_z / \partial y)$, or $(\partial E_z / \partial y - \partial E_y / \partial z)$. An
example in which the net right-lateral differential is zero, although the individual
derivatives are nonzero, is shown in Figure 3.4(a), whereas Figure 3.4(b) shows an example in which the net right-lateral differential is nonzero.

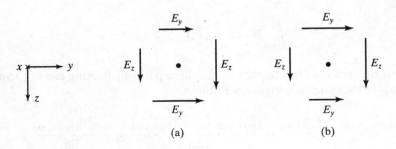

(a)   (b)

FIGURE 3.4

For illustrating (a) zero, and (b) nonzero net right-lateral differential of E_y
and E_z normal to the x-direction.

Equations (3.12a)–(3.12c) can be combined into a single vector equation as given by

$$\left(\frac{\partial E_z}{\partial y} - \frac{\partial E_y}{\partial z}\right)\mathbf{a}_x + \left(\frac{\partial E_x}{\partial z} - \frac{\partial E_z}{\partial x}\right)\mathbf{a}_y + \left(\frac{\partial E_y}{\partial x} - \frac{\partial E_x}{\partial y}\right)\mathbf{a}_z$$

$$= -\frac{\partial B_x}{\partial t}\mathbf{a}_x - \frac{\partial B_y}{\partial t}\mathbf{a}_y - \frac{\partial B_z}{\partial t}\mathbf{a}_z \qquad (3.13)$$

This can be expressed in determinant form as

$$\begin{vmatrix} \mathbf{a}_x & \mathbf{a}_y & \mathbf{a}_z \\ \dfrac{\partial}{\partial x} & \dfrac{\partial}{\partial y} & \dfrac{\partial}{\partial z} \\ E_x & E_y & E_z \end{vmatrix} = -\frac{\partial \mathbf{B}}{\partial t} \qquad (3.14)$$

or as

$$\left(\mathbf{a}_x \frac{\partial}{\partial x} + \mathbf{a}_y \frac{\partial}{\partial y} + \mathbf{a}_z \frac{\partial}{\partial z}\right) \times (E_x\mathbf{a}_x + E_y\mathbf{a}_y + E_z\mathbf{a}_z) = -\frac{\partial \mathbf{B}}{\partial t} \qquad (3.15)$$

The left side of (3.14) or (3.15) is known as the *curl of* **E**, denoted as $\nabla \times \mathbf{E}$ (del cross **E**), where ∇ (del) is the vector operator given by

$$\nabla = \mathbf{a}_x \frac{\partial}{\partial x} + \mathbf{a}_y \frac{\partial}{\partial y} + \mathbf{a}_z \frac{\partial}{\partial z} \qquad (3.16)$$

Thus, we have

$$\nabla \times \mathbf{E} = -\frac{\partial \mathbf{B}}{\partial t} \qquad (3.17)$$

Equation (3.17) is Maxwell's equation in differential form corresponding to Faraday's law. We shall discuss curl further in Section 3.3.

Example 3.2

Given $\mathbf{A} = y\mathbf{a}_x - x\mathbf{a}_y$, find $\nabla \times \mathbf{A}$.

From the determinant expansion for the curl of a vector, we have

$$\nabla \times \mathbf{A} = \begin{vmatrix} \mathbf{a}_x & \mathbf{a}_y & \mathbf{a}_z \\ \dfrac{\partial}{\partial x} & \dfrac{\partial}{\partial y} & \dfrac{\partial}{\partial z} \\ y & -x & 0 \end{vmatrix}$$

$$= \mathbf{a}_x\left[-\frac{\partial}{\partial z}(-x)\right] + \mathbf{a}_y\left[\frac{\partial}{\partial z}(y)\right] + \mathbf{a}_z\left[\frac{\partial}{\partial x}(-x) - \frac{\partial}{\partial y}(y)\right]$$

$$= -2\mathbf{a}_z$$

3.2 AMPERE'S CIRCUITAL LAW

In the previous section we derived the differential form of Faraday's law from its integral form. In this section we shall derive the differential form of Ampere's circuital law from its integral form in a completely analogous manner. We recall from Section 2.4 that Ampere's circuital law in integral form is given by

$$\oint_C \mathbf{H} \cdot d\mathbf{l} = \int_S \mathbf{J} \cdot d\mathbf{S} + \frac{d}{dt} \int_S \mathbf{D} \cdot d\mathbf{S} \tag{3.18}$$

where S is any surface bounded by the closed path C. For simplicity, we shall first consider the case in which the magnetic field has a y-component only, which is dependent only on the z-coordinate, in addition to time. Thus,

$$\mathbf{H} = H_y(z, t)\mathbf{a}_y \tag{3.19}$$

In other words, this simple form of the time-varying magnetic field is everywhere directed in the y-direction and is uniform in planes parallel to the xy-plane.

Let us now consider a rectangular path C of infinitesimal size lying in a plane parallel to the yz-plane and defined by the points (y, z), $(y, z + \Delta z)$, $(y + \Delta y, z + \Delta z)$, and $(y + \Delta y, z)$, as shown in Figure 3.5. According to Ampere's circuital law, the mmf around the closed path C is equal to the total current enclosed by C. The mmf is given by the line integral of $\mathbf{H}$ around C. Thus, evaluating the line integrals of $\mathbf{H}$ along the four sides of the rectangular path, we obtain

$$\oint_C \mathbf{H} \cdot d\mathbf{l} = \int_{(y, z)}^{(y+\Delta y, z)} \mathbf{H} \cdot d\mathbf{l} + \int_{(y+\Delta y, z)}^{(y+\Delta y, z+\Delta z)} \mathbf{H} \cdot d\mathbf{l}$$

$$+ \int_{(y+\Delta y, z+\Delta z)}^{(y, z+\Delta z)} \mathbf{H} \cdot d\mathbf{l} + \int_{(y, z+\Delta z)}^{(y, z)} \mathbf{H} \cdot d\mathbf{l}$$

$$= [H_y]_z \, \Delta y + 0 - [H_y]_{z+\Delta z} \, \Delta y + 0$$

$$= -\{[H_y]_{z+\Delta z} - [H_y]_z\} \, \Delta z \tag{3.20}$$

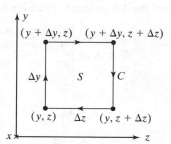

FIGURE 3.5

Infinitesimal rectangular path lying in a plane parallel to the yz-plane.

To find the total current enclosed by C, we consider the plane surface S bounded by C. According to the right-hand screw rule, we must find the current crossing S toward the positive x-direction, that is, into the page, since the path is traversed in the clockwise sense. This current consists of two parts:

$$\int_S \mathbf{J} \cdot d\mathbf{S} = [J_x]_{(y, z)} \Delta y \, \Delta z \tag{3.21a}$$

$$\frac{d}{dt} \int_S \mathbf{D} \cdot d\mathbf{S} = \frac{d}{dt}\{[D_x]_{(y, z)} \Delta y \, \Delta z\} = \frac{\partial [D_x]_{(y, z)}}{\partial t} \Delta y \, \Delta z \tag{3.21b}$$

where we have assumed that since the area is infinitesimal in size, J_x and D_x are uniform over the area and equal to their values at (y, z).

Substituting (3.20), (3.21a), and (3.21b) into (3.18) to apply Ampere's circuital law to the rectangular path C under consideration, we get

$$-\{[H_y]_{z+\Delta z} - [H_y]_z\} \, \Delta y = \left[J_x + \frac{\partial D_x}{\partial t} \right]_{(y, z)} \Delta y \, \Delta z$$

or

$$\frac{[H_y]_{z+\Delta z} - [H_y]_z}{\Delta z} = -\left[J_x + \frac{\partial D_x}{\partial t} \right]_{(y, z)} \tag{3.22}$$

If we now let the rectangular path shrink to the point (y, z) by letting Δy and Δz tend to zero, we obtain

$$\operatorname*{Lim}_{\substack{\Delta y \to 0 \\ \Delta z \to 0}} \frac{[H_y]_{z+\Delta z} - [H_y]_z}{\Delta z} = -\operatorname*{Lim}_{\substack{\Delta y \to 0 \\ \Delta z \to 0}} \left[J_x + \frac{\partial D_x}{\partial t} \right]_{(y, z)}$$

or

$$\frac{\partial H_y}{\partial z} = -J_x - \frac{\partial D_x}{\partial t} \tag{3.23}$$

Equation (3.23) is Ampere's circuital law in differential form for the simple case of $\mathbf{H}$ given by (3.19). It relates the variation of H_y with z (space) at a point to the current density J_x and to the variation of D_x with t (time) at that point. Since the above derivation can be carried out for any arbitrary point (x, y, z), it is valid at all points. It tells us in particular that a current density J_x or a time-varying D_x or a nonzero combination of the two quantities at a point results in an H_y at that point having a differential in the z-direction. This is to be expected since if this is not the case, $\oint \mathbf{H} \cdot d\mathbf{l}$ around the infinitesimal rectangular path would be zero.

Example 3.3

Given $\mathbf{E} = E_0 z \sin \omega t \, \mathbf{a}_x$ and it is known that $\mathbf{J}$ is zero and $\mathbf{B}$ has a y-component only, let us find B_y. From (3.23), we have

$$\frac{\partial H_y}{\partial z} = -J_x - \frac{\partial D_x}{\partial t} = 0 - \frac{\partial}{\partial t}(\epsilon_0 E_0 z \sin \omega t) = -\omega\epsilon_0 E_0 z \cos \omega t$$

$$H_y = -\omega\epsilon_0 E_0 \frac{z^2}{2} \cos \omega t$$

$$B_y = \mu_0 H_y = -\omega\mu_0\epsilon_0 E_0 \frac{z^2}{2} \cos \omega t$$

We note that the electric field varying linearly with z gives rise to a magnetic field proportional to z^2. In Example 3.1, however, an electric field varying linearly with z was found to result from a uniform magnetic field, according to Faraday's law in differential form. The inconsistency of these two results implies that neither the combination of E_x and B_y in Example 3.1 nor the combination of E_x and B_y in this example simultaneously satisfies the two Maxwell's equations in differential form given by (3.7) and (3.23). The pair of E_x and B_y in Example 3.1 satisfies only (3.7), whereas the pair of E_x and B_y in this example satisfies only (3.23). In the following chapter we shall find a pair of solutions for E_x and B_y that simultaneously satisfies the two Maxwell's equations.

Example 3.4

Let us consider the current distribution given by

$$\mathbf{J} = J_0 \mathbf{a}_x \quad \text{for } -a < z < a$$

as shown in Figure 3.6(a), where J_0 is a constant, and find the magnetic field everywhere.

Since the current density is independent of x and y, the field is also independent of x and y. Also, since the current density is not a function of time, the field is static. Hence, $(\partial D_x/\partial t) = 0$, and we have

$$\frac{\partial H_y}{\partial z} = -J_x$$

Integrating both sides with respect to z, we obtain

$$H_y = -\int_{-\infty}^{z} J_x \, dz + C$$

where C is the constant of integration.

The variation of J_x with z is shown in Figure 3.6(b). Integrating $-J_x$ with respect to z, that is, finding the area under the curve of Figure 3.6(b) as a function of z, and taking its negative, we obtain the result shown by the dashed curve in Figure 3.6(c) for $-\int_{-\infty}^{z} J_x \, dz$. From symmetry considerations, the field must be equal and opposite on either side of the current region $-a < z < a$. Hence, we choose the constant of integration C to be equal to $J_0 a$, thereby obtaining the final result for H_y as shown by the solid curve in Figure 3.6(c). Thus, the magnetic field intensity due to the current distribution is given by

$$\mathbf{H} = \begin{cases} J_0 a \mathbf{a}_y & \text{for } z < -a \\ -J_0 z \mathbf{a}_y & \text{for } -a < z < a \\ -J_0 a \mathbf{a}_y & \text{for } z > a \end{cases}$$

The magnetic flux density, $\mathbf{B}$, is equal to $\mu_0 \mathbf{H}$.

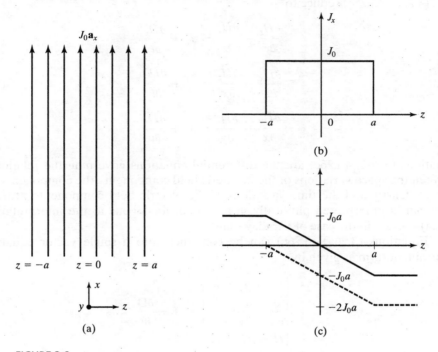

FIGURE 3.6

The determination of magnetic field due to a current distribution.

We now generalize (3.23) for the arbitrary case of a magnetic field having all three components, each of them depending on all three coordinates, in addition to t, that is,

$$\mathbf{H} = H_x(x, y, z, t)\mathbf{a}_x + H_y(x, y, z, t)\mathbf{a}_y + H_z(x, y, z, t)\mathbf{a}_z \qquad (3.24)$$

We do this in exactly the same manner as for the case of Faraday's law by considering the three infinitesimal rectangular paths shown in Figure 3.3. Applying Ampere's circuital law to each of the three paths and simplifying, we obtain

$$\frac{[H_z]_{(x, y+\Delta y)} - [H_z]_{(x, y)}}{\Delta y} - \frac{[H_y]_{(x, z+\Delta z)} - [H_y]_{(x, z)}}{\Delta z} = \left[J_x + \frac{\partial D_x}{\partial t} \right]_{(x, y, z)} \qquad (3.25a)$$

$$\frac{[H_x]_{(y, z+\Delta z)} - [H_x]_{(y, z)}}{\Delta z} - \frac{[H_z]_{(x+\Delta x, y)} - [H_z]_{(x, y)}}{\Delta x} = \left[J_y + \frac{\partial D_y}{\partial t} \right]_{(x, y, z)} \quad (3.25b)$$

$$\frac{[H_y]_{(x+\Delta x, z)} - [H_y]_{(x, z)}}{\Delta x} - \frac{[H_x]_{(y+\Delta y, z)} - [H_x]_{(y, z)}}{\Delta y} = \left[J_z + \frac{\partial D_z}{\partial t} \right]_{(x, y, z)} \quad (3.25c)$$

If we now let all three paths shrink to the point a by letting Δx, Δy, and Δz tend to zero, (3.25a)–(3.25c) reduce to

$$\frac{\partial H_z}{\partial y} - \frac{\partial H_y}{\partial z} = J_x + \frac{\partial D_x}{\partial t} \quad (3.26a)$$

$$\frac{\partial H_x}{\partial z} - \frac{\partial H_z}{\partial x} = J_y + \frac{\partial D_y}{\partial t} \quad (3.26b)$$

$$\frac{\partial H_y}{\partial x} - \frac{\partial H_x}{\partial y} = J_z + \frac{\partial D_z}{\partial t} \quad (3.26c)$$

Equations (3.26a)–(3.26c) are the differential equations governing the relationships between the space variations of the magnetic field components, the components of the current density and the time variations of the electric field components, at a point. They can be interpreted physically in a manner analogous to the interpretation of (3.12a)–(3.12c) in the case of Faraday's law.

Equations (3.26a)–(3.26c) can be combined into a single vector equation in determinant form as given by

$$\begin{vmatrix} \mathbf{a}_x & \mathbf{a}_y & \mathbf{a}_z \\ \dfrac{\partial}{\partial x} & \dfrac{\partial}{\partial y} & \dfrac{\partial}{\partial z} \\ H_x & H_y & H_z \end{vmatrix} = \mathbf{J} + \frac{\partial \mathbf{D}}{\partial t} \quad (3.27)$$

or

$$\nabla \times \mathbf{H} = \mathbf{J} + \frac{\partial \mathbf{D}}{\partial t} \quad (3.28)$$

Equation (3.28) is Maxwell's equation in differential form corresponding to Ampere's circuital law. The quantity $\partial \mathbf{D}/\partial t$ is known as the *displacement current density*. We shall discuss curl further in the following section.

3.3 CURL AND STOKES' THEOREM

In Sections 3.1 and 3.2 we derived the differential forms of Faraday's and Ampere's circuital laws from their integral forms. These differential forms involve a new vector quantity, namely, the *curl* of a vector. In this section we shall introduce the basic definition of curl and then present a physical interpretation of the curl. In order to do this, let us, for simplicity, consider Ampere's circuital law in differential form without the displacement current density term, that is,

$$\nabla \times \mathbf{H} = \mathbf{J} \tag{3.29}$$

We wish to express $\nabla \times \mathbf{H}$ at a point in the current region in terms of $\mathbf{H}$ at that point. If we consider an infinitesimal surface $\Delta \mathbf{S}$ at the point and take the dot product of both sides of (3.29) with $\Delta \mathbf{S}$, we get

$$(\nabla \times \mathbf{H}) \cdot \Delta \mathbf{S} = \mathbf{J} \cdot \Delta \mathbf{S} \tag{3.30}$$

But $\mathbf{J} \cdot \Delta \mathbf{S}$ is simply the current crossing the surface $\Delta \mathbf{S}$, and according to Ampere's circuital law in integral form without the displacement current term,

$$\oint_C \mathbf{H} \cdot d\mathbf{l} = \mathbf{J} \cdot \Delta \mathbf{S} \tag{3.31}$$

where C is the closed path bounding $\Delta \mathbf{S}$. Comparing (3.30) and (3.31), we have

$$(\nabla \times \mathbf{H}) \cdot \Delta \mathbf{S} = \oint_C \mathbf{H} \cdot d\mathbf{l}$$

or

$$(\nabla \times \mathbf{H}) \cdot \Delta S\, \mathbf{a}_n = \oint_C \mathbf{H} \cdot d\mathbf{l} \tag{3.32}$$

where $\mathbf{a}_n$ is the unit vector normal to ΔS and directed toward the side of advance of a right-hand screw as it is turned around C. Dividing both sides of (3.32) by ΔS, we obtain

$$(\nabla \times \mathbf{H}) \cdot \mathbf{a}_n = \frac{\oint_C \mathbf{H} \cdot d\mathbf{l}}{\Delta S} \tag{3.33}$$

The maximum value of $(\nabla \times \mathbf{H}) \cdot \mathbf{a}_n$, and hence that of the right side of (3.33), occurs when $\mathbf{a}_n$ is oriented parallel to $\nabla \times \mathbf{H}$, that is, when the surface ΔS is oriented normal to the current density vector $\mathbf{J}$. This maximum value is simply $|\nabla \times \mathbf{H}|$. Thus,

$$|\nabla \times \mathbf{H}| = \left[\frac{\oint_C \mathbf{H} \cdot d\mathbf{l}}{\Delta S} \right]_{\text{max}} \tag{3.34}$$

Since the direction of $\nabla \times \mathbf{H}$ is the direction of $\mathbf{J}$, or that of the unit vector normal to ΔS, we can then write

$$\nabla \times \mathbf{H} = \left[\frac{\oint_C \mathbf{H} \cdot d\mathbf{l}}{\Delta S} \right]_{\text{max}} \mathbf{a}_n \tag{3.35}$$

Equation (3.35) is only approximate since (3.32) is exact only in the limit that ΔS tends to zero. Thus,

$$\nabla \times \mathbf{H} = \lim_{\Delta S \to 0} \left[\frac{\oint_C \mathbf{H} \cdot d\mathbf{l}}{\Delta S} \right]_{\text{max}} \mathbf{a}_n \tag{3.36}$$

Equation (3.36) is the expression for $\nabla \times \mathbf{H}$ at a point in terms of $\mathbf{H}$ at that point. Although we have derived this for the $\mathbf{H}$ vector, it is a general result and, in fact, is often the starting point for the introduction of curl.

Equation (3.36) tells us that in order to find the curl of a vector at a point in that vector field, we first consider an infinitesimal surface at that point and compute the closed line integral or circulation of the vector around the periphery of this surface by orienting the surface such that the circulation is maximum. We then divide the circulation by the area of the surface to obtain the maximum value of the circulation per unit area. Since we need this maximum value of the circulation per unit area in the limit that the area tends to zero, we do this by gradually shrinking the area and making sure that each time we compute the circulation per unit area an orientation for the area that maximizes this quantity is maintained. The limiting value to which the maximum circulation per unit area approaches is the magnitude of the curl. The limiting direction to which the normal vector to the surface approaches is the direction of the curl. The task of computing the curl is simplified if we consider one component of the field at a time and compute the curl corresponding to that component since then it is sufficient if we always maintain the orientation of the surface normal to that component axis. In fact, this is what we did in Sections 3.1 and 3.2, which led us to the determinant form of curl.

We are now ready to discuss the physical interpretation of the curl. We do this with the aid of a simple device known as the *curl meter*. Although the curl meter may take several forms, we shall consider one consisting of a circular disc that floats in water with a paddle wheel attached to the bottom of the disc, as shown in Figure 3.7. A dot at the periphery on top of the disc serves to indicate any rotational motion of the curl meter about its axis, that is, the axis of the paddle wheel. Let us now consider a stream of rectangular cross section carrying water in the z-direction, as shown in Figure 3.7(a). Let us assume the velocity $\mathbf{v}$ of the water to be independent of height but increasing uniformly from a value of zero at the banks to a maximum value v_0 at the center, as shown in Figure 3.7(b), and investigate the behavior of the curl meter when it is placed vertically at different points in the stream. We assume that the size of the curl meter is vanishingly small so that it does not disturb the flow of water as we probe its behavior at different points.

Since exactly in midstream the blades of the paddle wheel lying on either side of the center line are hit by the same velocities, the paddle wheel does not rotate. The curl meter simply slides down the stream without any rotational motion, that is, with the dot on top of the disc maintaining the same position relative to the center of the disc, as shown in Figure 3.7(c). At a point to the left of the midstream the blades of the paddle wheel are hit by a greater velocity on the right side than on the left side so that the paddle wheel rotates in the counterclockwise sense. The curl meter rotates in the counterclockwise direction about its axis as it slides down the stream, as indicated by the changing position of the dot on top of the disc relative to the center of the disc, as shown in Figure 3.7(d). At a point to the right of midstream, the blades of the paddle wheel are hit by a greater velocity on the left side than on the right side so that the paddle wheel rotates in the clockwise sense. The curl meter rotates in the clockwise direction about its axis as it slides down the stream, as indicated by the changing position of the dot on top of the disc relative to the center of the disc, as shown in Figure 3.7(e).

To relate the foregoing discussion of the behavior of the curl meter with the curl of the velocity vector field of the water flow, we note that at a point in midstream, the

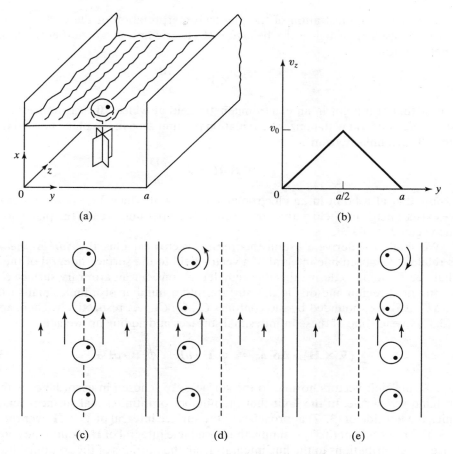

FIGURE 3.7

For explaining the physical interpretation of curl using the curl meter.

circulation of the velocity vector per unit area in the plane normal to the axis of the paddle wheel, that is, parallel to the surface of the stream, is zero and hence the component of the curl along that axis, that is, in the x-direction, is zero. At points on either side of midstream, however, the circulation per unit area is not zero in view of the velocity differential along the y-direction. Hence, the x-component of the curl is nonzero at these points. Furthermore, the x-component of the curl at points on the right side of midstream is opposite in sign to that on the left side of midstream, since the velocity differentials are opposite in sign. These properties are exactly similar to those of the rotational motion of the curl meter.

If we now pick up the curl meter and insert it in the water with its axis parallel to the surface of the stream, the curl meter does not rotate, because its blades are hit with the same force on either side of its axis. This behavior of the curl meter is akin to the property that the horizontal component of the curl of the velocity vector is zero, since the velocity differential along the x-direction is zero.

The foregoing illustration of the physical interpretation of the curl of a vector field can be used to visualize the behavior of electric and magnetic fields. Thus, for example, from

$$\nabla \times \mathbf{E} = -\frac{\partial \mathbf{B}}{\partial t}$$

we know that at a point in an electromagnetic field at which $\partial \mathbf{B}/\partial t$ is nonzero, there exists an electric field with nonzero circulation per unit area in the plane normal to the vector $\partial \mathbf{B}/\partial t$. Similarly, from

$$\nabla \times \mathbf{H} = \mathbf{J} + \frac{\partial \mathbf{D}}{\partial t}$$

we know that at a point in an electromagnetic field at which $\mathbf{J} + \partial \mathbf{D}/\partial t$ is nonzero, there exists a magnetic field with nonzero circulation per unit area in the plane normal to the vector $\mathbf{J} + \partial \mathbf{D}/\partial t$.

We shall now derive a useful theorem in vector calculus, the *Stokes' theorem*. This relates the closed line integral of a vector field to the surface integral of the curl of that vector field. To derive this theorem, let us consider an arbitrary surface S in a magnetic field region and divide this surface into a number of infinitesimal surfaces $\Delta S_1, \Delta S_2, \Delta S_3, \ldots$, bounded by the contours $C_1, C_2, C_3, \ldots$, respectively. Then, applying (3.32) to each one of these infinitesimal surfaces and adding up, we get

$$\sum_j (\nabla \times \mathbf{H})_j \cdot \Delta S_j \, \mathbf{a}_{nj} = \oint_{C_1} \mathbf{H} \cdot d\mathbf{l} + \oint_{C_2} \mathbf{H} \cdot d\mathbf{l} + \cdots \qquad (3.37)$$

where $\mathbf{a}_{nj}$ are unit vectors normal to the surfaces ΔS_j chosen in accordance with the right-hand screw rule. In the limit that the number of infinitesimal surfaces tends to infinity, the left side of (3.37) approaches to the surface integral of $\nabla \times \mathbf{H}$ over the surface S. The right side of (3.37) is simply the closed line integral of $\mathbf{H}$ around the contour C since the contributions to the line integrals from the portions of the contours interior to C cancel, as shown in Figure 3.8. Thus, we get

$$\int_S (\nabla \times \mathbf{H}) \cdot d\mathbf{S} = \oint_C \mathbf{H} \cdot d\mathbf{l} \qquad (3.38)$$

Equation (3.38) is Stokes' theorem. Although we have derived it by considering the $\mathbf{H}$ field, it is general and is applicable for any vector field.

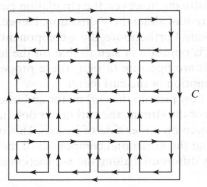

FIGURE 3.8

For deriving Stokes' theorem.

Example 3.5

Let us verify Stokes' theorem by considering

$$\mathbf{A} = y\mathbf{a}_x - x\mathbf{a}_y$$

and the closed path C shown in Figure 3.9.

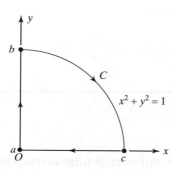

FIGURE 3.9

A closed path for verifying Stokes' theorem.

We first determine $\oint_C \mathbf{A} \cdot d\mathbf{l}$ by evaluating the line integrals along the three segments of the closed path. To do this, we first note that $\mathbf{A} \cdot d\mathbf{l} = y\,dx - x\,dy$. Then, from a to b, $x = 0$, $dx = 0$, $\mathbf{A} \cdot d\mathbf{l} = 0$

$$\int_a^b \mathbf{A} \cdot d\mathbf{l} = 0$$

From b to c, $x^2 + y^2 = 1$, $y = \sqrt{1 - x^2}$

$$2x\,dx + 2y\,dy = 0, \qquad dy = -\frac{x\,dx}{y} = -\frac{x}{\sqrt{1 - x^2}}\,dx$$

$$\mathbf{A} \cdot d\mathbf{l} = \sqrt{1 - x^2}\,dx + \frac{x^2\,dx}{\sqrt{1 - x^2}} = \frac{dx}{\sqrt{1 - x^2}}$$

$$\int_b^c \mathbf{A} \cdot d\mathbf{l} = \int_0^1 \frac{dx}{\sqrt{1 - x^2}} = \left[\sin^{-1} x \right]_0^1 = \frac{\pi}{2}$$

From c to a, $y = 0$, $dy = 0$, $\mathbf{A} \cdot d\mathbf{l} = 0$

$$\int_c^a \mathbf{A} \cdot d\mathbf{l} = 0$$

Thus,

$$\oint_C \mathbf{A} \cdot d\mathbf{l} = \int_a^b \mathbf{A} \cdot d\mathbf{l} + \int_b^c \mathbf{A} \cdot d\mathbf{l} + \int_c^a \mathbf{A} \cdot d\mathbf{l}$$

$$= 0 + \frac{\pi}{2} + 0 = \frac{\pi}{2}$$

Now, to evaluate $\oint_C \mathbf{A} \cdot d\mathbf{l}$ by using Stokes' theorem, we recall from Example 3.2 that

$$\nabla \times \mathbf{A} = \nabla \times (y\mathbf{a}_x - x\mathbf{a}_y) = -2\mathbf{a}_z$$

For the plane surface S enclosed by C,

$$d\mathbf{S} = -dx\,dy\,\mathbf{a}_z$$

Thus,

$$(\nabla \times \mathbf{A}) \cdot d\mathbf{S} = -2\mathbf{a}_z \cdot (-dx\,dy\,\mathbf{a}_z) = 2\,dx\,dy$$

$$\int_S (\nabla \times \mathbf{A}) \cdot d\mathbf{S} = \int_{x=0}^{1} \int_{y=0}^{\sqrt{1-x^2}} 2\,dx\,dy$$

$$= 2(\text{area enclosed by } C) = 2 \times \frac{\pi}{4} = \frac{\pi}{2}$$

thereby verifying Stokes' theorem.

3.4 GAUSS' LAW FOR THE ELECTRIC FIELD

Thus far we have derived Maxwell's equations in differential form corresponding to the two Maxwell's equations in integral form involving the line integrals of $\mathbf{E}$ and $\mathbf{H}$, that is, Faraday's law and Ampere's circuital law, respectively. The remaining two Maxwell's equations in integral form, namely, Gauss' law for the electric field and Gauss' law for the magnetic field, are concerned with the closed surface integrals of $\mathbf{D}$ and $\mathbf{B}$, respectively. We shall in this and the following sections derive the differential forms of these two equations.

We recall from Section 2.5 that Gauss' law for the electric field is given by

$$\oint_S \mathbf{D} \cdot d\mathbf{S} = \int_V \rho \, dv \tag{3.39}$$

where V is the volume enclosed by the closed surface S. To derive the differential form of this equation, let us consider a rectangular box of infinitesimal sides Δx, Δy, and Δz and defined by the six surfaces $x = x$, $x = x + \Delta x$, $y = y$, $y = y + \Delta y$, $z = z$, and $z = z + \Delta z$, as shown in Figure 3.10, in a region of electric field

$$\mathbf{D} = D_x(x, y, z, t)\mathbf{a}_x + D_y(x, y, z, t)\mathbf{a}_y + D_z(x, y, z, t)\mathbf{a}_z \tag{3.40}$$

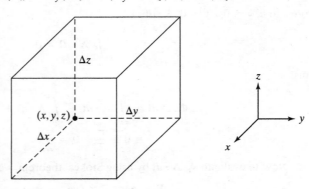

FIGURE 3.10

An infinitesimal rectangular box.

and charge of density $\rho(x, y, z, t)$. According to Gauss' law for the electric field, the displacement flux emanating from the box is equal to the charge enclosed by the box. The displacement flux is given by the surface integral of $\mathbf{D}$ over the surface of the box, which is comprised of six plane surfaces. Thus, evaluating the displacement flux emanating out of the box over each of the six plane surfaces of the box, we have

$$\int \mathbf{D} \cdot d\mathbf{S} = -[D_x]_x \, \Delta y \, \Delta z \qquad \text{for the surface } x = x \qquad (3.41\text{a})$$

$$\int \mathbf{D} \cdot d\mathbf{S} = [D_x]_{x+\Delta x} \, \Delta y \, \Delta z \qquad \text{for the surface } x = x + \Delta x \qquad (3.41\text{b})$$

$$\int \mathbf{D} \cdot d\mathbf{S} = -[D_y]_y \, \Delta z \, \Delta x \qquad \text{for the surface } y = y \qquad (3.41\text{c})$$

$$\int \mathbf{D} \cdot d\mathbf{S} = [D_y]_{y+\Delta y} \, \Delta z \, \Delta x \qquad \text{for the surface } y = y + \Delta y \qquad (3.41\text{d})$$

$$\int \mathbf{D} \cdot d\mathbf{S} = -[D_z]_z \, \Delta x \, \Delta y \qquad \text{for the surface } z = z \qquad (3.41\text{e})$$

$$\int \mathbf{D} \cdot d\mathbf{S} = [D_z]_{z+\Delta z} \, \Delta x \, \Delta y \qquad \text{for the surface } z = z + \Delta z \qquad (3.41\text{f})$$

Adding up (3.41a)–(3.41f), we obtain the total displacement flux emanating from the box to be

$$\oint_S \mathbf{D} \cdot d\mathbf{S} = \{[D_x]_{x+\Delta x} - [D_x]_x\} \, \Delta y \, \Delta z$$
$$+ \{[D_y]_{y+\Delta y} - [D_y]_y\} \, \Delta z \, \Delta x$$
$$+ \{[D_z]_{z+\Delta z} - [D_z]_z\} \, \Delta x \, \Delta y \qquad (3.42)$$

Now the charge enclosed by the rectangular box is given by

$$\int_V \rho \, dv = \rho(x, y, z, t) \cdot \Delta x \, \Delta y \, \Delta z = \rho \, \Delta x \, \Delta y \, \Delta z \qquad (3.43)$$

where we have assumed ρ to be uniform throughout the volume of the box and equal to its value at (x, y, z), since the box is infinitesimal in volume.

Substituting (3.42) and (3.43) into (3.39) to apply Gauss' law for the electric field to the surface of the box under consideration, we get

$$\{[D_x]_{x+\Delta x} - [D_x]_x\} \, \Delta y \, \Delta z + \{[D_y]_{y+\Delta y} - [D_y]_y\} \, \Delta z \, \Delta x$$
$$+ \{[D_z]_{z+\Delta z} - [D_z]_z\} \, \Delta x \, \Delta y = \rho \, \Delta x \, \Delta y \, \Delta z$$

or

$$\frac{[D_x]_{x+\Delta x} - [D_x]_x}{\Delta x} + \frac{[D_y]_{y+\Delta y} - [D_y]_y}{\Delta y} + \frac{[D_z]_{z+\Delta z} - [D_z]_z}{\Delta z} = \rho \qquad (3.44)$$

If we now let the box shrink to the point (x, y, z) by letting Δx, Δy, and Δz tend to zero, we obtain

$$\lim_{\Delta x \to 0} \frac{[D_x]_{x+\Delta x} - [D_x]_x}{\Delta x} + \lim_{\Delta y \to 0} \frac{[D_y]_{y+\Delta y} - [D_y]_y}{\Delta y}$$

$$+ \lim_{\Delta z \to 0} \frac{[D_z]_{z+\Delta z} - [D_z]_z}{\Delta z} = \lim_{\substack{\Delta x \to 0 \\ \Delta y \to 0 \\ \Delta z \to 0}} \rho$$

or

$$\frac{\partial D_x}{\partial x} + \frac{\partial D_y}{\partial y} + \frac{\partial D_z}{\partial z} = \rho \tag{3.45}$$

Equation (3.45) tells us that the net longitudinal differential of the components of **D**, that is, the algebraic sum of the derivatives of the components of **D** along their respective directions is equal to the charge density at that point. Conversely, a charge density at a point results in an electric field, having components of **D** such that their net longitudinal differential is nonzero. An example in which the net longitudinal differential is zero although some of the individual derivatives are nonzero is shown in Figure 3.11(a). Figure 3.11(b) shows an example in which the net longitudinal differential is nonzero. Equation (3.45) can be written in vector notation as

$$\left(\mathbf{a}_x \frac{\partial}{\partial x} + \mathbf{a}_y \frac{\partial}{\partial y} + \mathbf{a}_z \frac{\partial}{\partial z} \right) \cdot (D_x \mathbf{a}_x + D_y \mathbf{a}_y + D_z \mathbf{a}_z) = \rho \tag{3.46}$$

The left side of (3.46) is known as the *divergence of* **D**, denoted as $\nabla \cdot \mathbf{D}$ (del dot **D**). Thus, we have

$$\nabla \cdot \mathbf{D} = \rho \tag{3.47}$$

Equation (3.47) is Maxwell's equation in differential form corresponding to Gauss' law for the electric field. We shall discuss divergence further in Section 3.6.

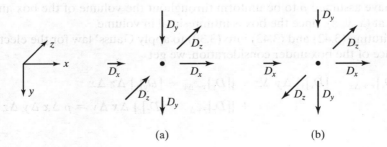

(a)   (b)

FIGURE 3.11

For illustrating (a) zero, and (b) nonzero net longitudinal differential of the components of **D**.

Example 3.6

Given $\mathbf{A} = 3x\mathbf{a}_x + (y - 3)\mathbf{a}_y + (2 - z)\mathbf{a}_z$, find $\nabla \cdot \mathbf{A}$.

From the expansion for the divergence of a vector, we have

$$\nabla \cdot \mathbf{A} = \left(\mathbf{a}_x \frac{\partial}{\partial x} + \mathbf{a}_y \frac{\partial}{\partial y} + \mathbf{a}_z \frac{\partial}{\partial z}\right) \cdot [3x\mathbf{a}_x + (y - 3)\mathbf{a}_y + (2 - z)\mathbf{a}_z]$$

$$= \frac{\partial}{\partial x}(3x) + \frac{\partial}{\partial y}(y - 3) + \frac{\partial}{\partial z}(2 - z)$$

$$= 3 + 1 - 1 = 3$$

Example 3.7

Let us consider the charge distribution given by

$$\rho = \begin{cases} -\rho_0 & \text{for } -a < x < 0 \\ \rho_0 & \text{for } 0 < x < a \end{cases}$$

as shown in Figure 3.12(a), where ρ_0 is a constant, and find the electric field everywhere.

Since the charge density is independent of y and z, the field is also independent of y and z, thereby giving us $\partial D_y/\partial y = \partial D_z/\partial z = 0$ and reducing Gauss' law for the electric field to

$$\frac{\partial D_x}{\partial x} = \rho$$

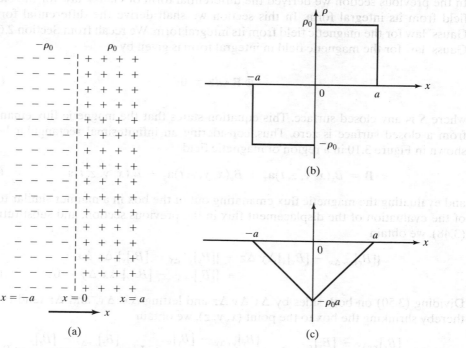

(a)

(b)

(c)

FIGURE 3.12

The determination of electric field due to a charge distribution.

Integrating both sides with respect to x, we obtain

$$D_x = \int_{-\infty}^{x} \rho \, dx + C$$

where C is the constant of integration.

The variation of ρ with x is shown in Figure 3.12(b). Integrating ρ with respect to x, that is, finding the area under the curve of Figure 3.12(b) as a function of x, we obtain the result shown in Figure 3.12(c) for $\int_{-\infty}^{x} \rho \, dx$. The constant of integration C is zero since the symmetry of equal and opposite fields on the two sides of the charge distribution, considered as a superposition of a series of thin slabs of charge, is already satisfied by the plot of Figure 3.12(c). Thus, the displacement flux density due to the charge distribution is given by

$$\mathbf{D} = \begin{cases} 0 & \text{for } x < -a \\ -\rho_0(x + a)\mathbf{a}_x & \text{for } -a < x < 0 \\ \rho_0(x - a)\mathbf{a}_x & \text{for } 0 < x < a \\ 0 & \text{for } x > a \end{cases}$$

The electric field intensity, $\mathbf{E}$, is equal to $\mathbf{D}/\epsilon_0$.

3.5 GAUSS' LAW FOR THE MAGNETIC FIELD

In the previous section we derived the differential form of Gauss' law for the electric field from its integral form. In this section we shall derive the differential form of Gauss' law for the magnetic field from its integral form. We recall from Section 2.6 that Gauss' law for the magnetic field in integral form is given by

$$\oint_S \mathbf{B} \cdot d\mathbf{S} = 0 \tag{3.48}$$

where S is any closed surface. This equation states that the magnetic flux emanating from a closed surface is zero. Thus, considering an infinitesimal rectangular box as shown in Figure 3.10 in a region of magnetic field

$$\mathbf{B} = B_x(x, y, z, t)\mathbf{a}_x + B_y(x, y, z, t)\mathbf{a}_y + B_z(x, y, z, t)\mathbf{a}_z \tag{3.49}$$

and evaluating the magnetic flux emanating out of the box in a manner similar to that of the evaluation of the displacement flux in the previous section, and substituting in (3.48), we obtain

$$\{[B_x]_{x+\Delta x} - [B_x]_x\}\Delta y \, \Delta z + \{[B_y]_{y+\Delta y} - [B_y]_y\} \Delta z \, \Delta x$$
$$+ \{[B_z]_{z+\Delta z} - [B_z]_z\} \Delta x \, \Delta y = 0 \tag{3.50}$$

Dividing (3.50) on both sides by $\Delta x \, \Delta y \, \Delta z$ and letting Δx, Δy, and Δz tend to zero, thereby shrinking the box to the point (x, y, z), we obtain

$$\lim_{\Delta x \to 0} \frac{[B_x]_{x+\Delta x} - [B_x]_x}{\Delta x} + \lim_{\Delta y \to 0} \frac{[B_y]_{y+\Delta y} - [B_y]_y}{\Delta y} + \lim_{\Delta z \to 0} \frac{[B_z]_{z+\Delta z} - [B_z]_z}{\Delta z} = 0$$

or

$$\frac{\partial B_x}{\partial x} + \frac{\partial B_y}{\partial y} + \frac{\partial B_z}{\partial z} = 0 \tag{3.51}$$

Equation (3.51) tells us that the net longitudinal differential of the components of **B** is zero. In vector form it is given by

$$\nabla \cdot \mathbf{B} = 0 \tag{3.52}$$

Equation (3.52) is Maxwell's equation in differential form corresponding to Gauss' law for the magnetic field. We shall discuss divergence further in the following section.

Example 3.8

Determine if the vector $\mathbf{A} = y\mathbf{a}_x - x\mathbf{a}_y$ can represent a magnetic field **B**.

  From (3.52), we note that a given vector can be realized as a magnetic field **B** if its divergence is zero. For $\mathbf{A} = y\mathbf{a}_x - x\mathbf{a}_y$,

$$\nabla \cdot \mathbf{A} = \frac{\partial}{\partial x}(y) + \frac{\partial}{\partial y}(-x) + \frac{\partial}{\partial z}(0) = 0$$

Hence, the given vector can represent a magnetic field **B**.

3.6 DIVERGENCE AND THE DIVERGENCE THEOREM

In Sections 3.4 and 3.5 we derived the differential forms of Gauss' laws for the electric and magnetic fields from their integral forms. These differential forms involve a new quantity, namely, the *divergence* of a vector. The divergence of a vector is a scalar as compared to the vector nature of the curl of a vector. In this section we shall introduce the basic definition of divergence and then present a physical interpretation for the divergence. In order to do this, let us consider Gauss' law for the electric field in differential form, that is,

$$\nabla \cdot \mathbf{D} = \rho \tag{3.53}$$

We wish to express $\nabla \cdot \mathbf{D}$ at a point in the charge region in terms of **D** at that point. If we consider an infinitesimal volume Δv at the point and multiply both sides of (3.53) by Δv, we get

$$(\nabla \cdot \mathbf{D}) \, \Delta v = \rho \, \Delta v \tag{3.54}$$

But $\rho \, \Delta v$ is simply the charge contained in the volume Δv, and according to Gauss' law for the electric field in integral form,

$$\oint_S \mathbf{D} \cdot d\mathbf{S} = \rho \, \Delta v \tag{3.55}$$

where S is the closed surface bounding Δv. Comparing (3.54) and (3.55), we have

$$(\boldsymbol{\nabla} \cdot \mathbf{D}) \, \Delta v = \oint_S \mathbf{D} \cdot d\mathbf{S} \tag{3.56}$$

Dividing both sides of (3.56) by Δv, we obtain

$$\boldsymbol{\nabla} \cdot \mathbf{D} = \frac{\oint_S \mathbf{D} \cdot d\mathbf{S}}{\Delta v} \tag{3.57}$$

Equation (3.57) is only approximate since (3.56) is exact only in the limit that Δv tends to zero. Thus,

$$\boldsymbol{\nabla} \cdot \mathbf{D} = \underset{\Delta v \to 0}{\text{Lim}} \frac{\oint_S \mathbf{D} \cdot d\mathbf{S}}{\Delta v} \tag{3.58}$$

Equation (3.58) is the expression for $\boldsymbol{\nabla} \cdot \mathbf{D}$ at a point in terms of $\mathbf{D}$ at that point. Although we have derived this for the $\mathbf{D}$ vector, it is a general result and, in fact, is often the starting point for the introduction of divergence.

Equation (3.58) tells us that in order to find the divergence of a vector at a point in that vector field, we first consider an infinitesimal volume at that point and compute the surface integral of the vector over the surface bounding that volume, that is, the outward flux of the vector field emanating from that volume. We then divide the flux by the volume to obtain the flux per unit volume. Since we need this flux per unit volume in the limit that the volume tends to zero, we do this by gradually shrinking the volume. The limiting value to which the flux per unit volume approaches is the value of the divergence of the vector field at the point to which the volume is shrunk.

We are now ready to discuss the physical interpretation of the divergence. To simplify this task, we shall consider the differential form of the law of conservation of charge given in integral form by (2.39), or

$$\oint_S \mathbf{J} \cdot d\mathbf{S} = -\frac{d}{dt} \int_V \rho \, dv \tag{3.59}$$

where S is the surface bounding the volume V. Applying (3.59) to an infinitesimal volume Δv, we have

$$\oint_S \mathbf{J} \cdot d\mathbf{S} = -\frac{d}{dt}(\rho \, \Delta v) = -\frac{\partial \rho}{\partial t} \Delta v$$

or

$$\frac{\oint_S \mathbf{J} \cdot d\mathbf{S}}{\Delta v} = -\frac{\partial \rho}{\partial t} \tag{3.60}$$

Now taking the limit on both sides of (3.60) as Δv tends to zero, we obtain

$$\underset{\Delta v \to 0}{\text{Lim}} \frac{\oint \mathbf{J} \cdot d\mathbf{S}}{\Delta v} = \underset{\Delta v \to 0}{\text{Lim}} -\frac{\partial \rho}{\partial t} \tag{3.61}$$

or

$$\nabla \cdot \mathbf{J} = -\frac{\partial \rho}{\partial t} \qquad (3.62)$$

or

$$\nabla \cdot \mathbf{J} + \frac{\partial \rho}{\partial t} = 0 \qquad (3.63)$$

Equation (3.63), which is the differential form of the law of conservation of charge, is familiarly known as the *continuity equation*. It tells us that the divergence of the current density vector at a point is equal to the time rate of decrease of the charge density at that point.

Let us now investigate three different cases: (a) positive value, (b) negative value, and (c) zero value of the time rate of decrease of the charge density at a point, that is, the divergence of the current density vector at that point. We shall do this with the aid of a simple device that we shall call the *divergence meter*. The divergence meter can be imagined to be a tiny, elastic balloon enclosing the point and that expands when hit by charges streaming outward from the point and contracts when acted upon by charges streaming inward toward the point. For case (a), that is, when the time rate of decrease of the charge density at the point is positive, there is a net amount of charge streaming out of the point in a given time, resulting in a net current flow outward from the point that will make the imaginary balloon expand. For case (b), that is, when the time rate of decrease of the charge density at the point is negative or the time rate of increase of the charge density is positive, there is a net amount of charge streaming toward the point in a given time, resulting in a net current flow toward the point and the imaginary balloon will contract. For case (c), that is, when the time rate of decrease of the charge density at the point is zero, the balloon will remain unaffected since the charge is streaming out of the point at exactly the same rate as it is streaming into the point. These three cases are illustrated in Figures 3.13(a), (b), and (c), respectively.

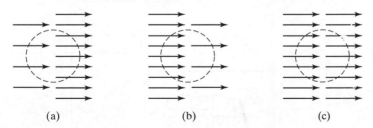

(a)   (b)   (c)

FIGURE 3.13

For explaining the physical interpretation of divergence using the divergence meter.

Generalizing the foregoing discussion to the physical interpretation of the divergence of any vector field at a point, we can imagine the vector field to be a velocity field of streaming charges acting upon the divergence meter and obtain in most cases a qualitative picture of the divergence of the vector field. If the divergence meter expands, the divergence is positive and a source of the flux of the vector field exists at that point. If the divergence meter contracts, the divergence is negative and a sink of the flux of the vector field exists at that point. If the divergence meter remains unaffected, the divergence is zero, and neither a source nor a sink of the flux of the vector field exists at that point. Alternatively, there can exist at the point pairs of sources and sinks of equal strengths.

We shall now derive a useful theorem in vector calculus, *the divergence theorem*. This relates the closed surface integral of the vector field to the volume integral of the divergence of that vector field. To derive this theorem, let us consider an arbitrary volume V in an electric field region and divide this volume into a number of infinitesimal volumes $\Delta v_1, \Delta v_2, \Delta v_3, \ldots$, bounded by the surfaces $S_1, S_2, S_3, \ldots$, respectively. Then, applying (3.56) to each one of these infinitesimal volumes and adding up, we get

$$\sum_j (\nabla \cdot \mathbf{D})_j \, \Delta v_j = \oint_{S_1} \mathbf{D} \cdot d\mathbf{S} + \oint_{S_2} \mathbf{D} \cdot d\mathbf{S} + \cdots \tag{3.64}$$

In the limit that the number of the infinitesimal volumes tends to infinity, the left side of (3.64) approaches to the volume integral of $\nabla \cdot \mathbf{D}$ over the volume V. The right side of (3.64) is simply the closed surface integral of $\mathbf{D}$ over S, since the contribution to the surface integrals from the portions of the surfaces interior to S cancel, as shown in Figure 3.14. Thus, we get

$$\int_V (\nabla \cdot \mathbf{D}) \, dv = \oint_S \mathbf{D} \cdot d\mathbf{S} \tag{3.65}$$

Equation (3.65) is the divergence theorem. Although we have derived it by considering the $\mathbf{D}$ field, it is general and is applicable for any vector field.

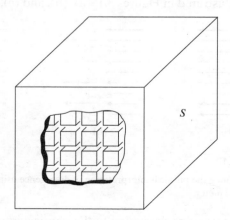

FIGURE 3.14

For deriving the divergence theorem.

Example 3.9

Let us verify the divergence theorem by considering

$$\mathbf{A} = 3x\mathbf{a}_x + (y - 3)\mathbf{a}_y + (2 - z)\mathbf{a}_z$$

and the closed surface of the box bounded by the planes $x = 0$, $x = 1$, $y = 0$, $y = 2$, $z = 0$, and $z = 3$.

We first determine $\oint_S \mathbf{A} \cdot d\mathbf{S}$ by evaluating the surface integrals over the six surfaces of the rectangular box. Thus for the surface $x = 0$,

$$\mathbf{A} = (y - 3)\mathbf{a}_y + (2 - z)\mathbf{a}_z, \qquad d\mathbf{S} = -dy\,dz\,\mathbf{a}_x$$

$$\mathbf{A} \cdot d\mathbf{S} = 0$$

$$\int \mathbf{A} \cdot d\mathbf{S} = 0$$

For the surface $x = 1$,

$$\mathbf{A} = 3\mathbf{a}_x + (y - 3)\mathbf{a}_y + (2 - z)\mathbf{a}_z, \qquad d\mathbf{S} = dy\,dz\,\mathbf{a}_x$$

$$\mathbf{A} \cdot d\mathbf{S} = 3\,dy\,dz$$

$$\int \mathbf{A} \cdot d\mathbf{S} = \int_{z=0}^{3}\int_{y=0}^{2} 3\,dy\,dz = 18$$

For the surface $y = 0$,

$$\mathbf{A} = 3x\mathbf{a}_x - 3\mathbf{a}_y + (2 - z)\mathbf{a}_z, \qquad d\mathbf{S} = -dz\,dx\,\mathbf{a}_y$$

$$\mathbf{A} \cdot d\mathbf{S} = 3\,dz\,dx$$

$$\int \mathbf{A} \cdot d\mathbf{S} = \int_{x=0}^{1}\int_{z=0}^{3} 3\,dz\,dx = 9$$

For the surface $y = 2$,

$$\mathbf{A} = 3x\mathbf{a}_x - \mathbf{a}_y + (2 - z)\mathbf{a}_z, \qquad d\mathbf{S} = dz\,dx\,\mathbf{a}_y$$

$$\mathbf{A} \cdot d\mathbf{S} = -dz\,dx$$

$$\int \mathbf{A} \cdot d\mathbf{S} = \int_{x=0}^{1}\int_{z=0}^{3} -dz\,dx = -3$$

For the surface $z = 0$,

$$\mathbf{A} = 3x\mathbf{a}_x + (y - 3)\mathbf{a}_y + 2\mathbf{a}_z, \qquad d\mathbf{S} = -dx\,dy\,\mathbf{a}_z$$

$$\mathbf{A} \cdot d\mathbf{S} = -2\,dx\,dy$$

$$\int \mathbf{A} \cdot d\mathbf{S} = \int_{y=0}^{2}\int_{x=0}^{1} -2\,dx\,dy = -4$$

For the surface $z = 3$,

$$\mathbf{A} = 3x\mathbf{a}_x + (y - 3)\mathbf{a}_y - \mathbf{a}_z, \qquad d\mathbf{S} = dx\,dy\,\mathbf{a}_z$$

$$\mathbf{A} \cdot d\mathbf{S} = -dx\,dy$$

$$\int \mathbf{A} \cdot d\mathbf{S} = \int_{y=0}^{2}\int_{x=0}^{1} -dx\,dy = -2$$

Thus,

$$\oint_S \mathbf{A} \cdot d\mathbf{S} = 0 + 18 + 9 - 3 - 4 - 2 = 18$$

Now, to evaluate $\oint_S \mathbf{A} \cdot d\mathbf{S}$ by using the divergence theorem, we recall from Example 3.6 that

$$\nabla \cdot \mathbf{A} = \nabla \cdot [3x\mathbf{a}_x + (y - 3)\mathbf{a}_y + (2 - z)\mathbf{a}_z] = 3$$

For the volume enclosed by the rectangular box,

$$\int (\nabla \cdot \mathbf{A})\, dv = \int_{z=0}^{3} \int_{y=0}^{2} \int_{x=0}^{1} 3\, dx\, dy\, dz = 18$$

thereby verifying the divergence theorem.

SUMMARY

We have in this chapter derived the differential forms of Maxwell's equations from their integral forms, which we introduced in the previous chapter. For the general case of electric and magnetic fields having all three components (x, y, z), each of them dependent on all coordinates (x, y, z), and time (t), Maxwell's equations in differential form are given as follows in words and in mathematical form.

Faraday's law. The curl of the electric field intensity is equal to the negative of the time derivative of the magnetic flux density, that is,

$$\nabla \times \mathbf{E} = -\frac{\partial \mathbf{B}}{\partial t} \tag{3.66}$$

Ampere's circuital law. The curl of the magnetic field intensity is equal to the sum of the current density due to flow of charges and the displacement current density, which is the time derivative of the displacement flux density, that is,

$$\nabla \times \mathbf{H} = \mathbf{J} + \frac{\partial \mathbf{D}}{\partial t} \tag{3.67}$$

Gauss' law for the electric field. The divergence of the displacement flux density is equal to the charge density, that is,

$$\nabla \cdot \mathbf{D} = \rho \tag{3.68}$$

Gauss' law for the magnetic field. The divergence of the magnetic flux density is equal to zero, that is,

$$\nabla \cdot \mathbf{B} = 0 \tag{3.69}$$

Auxiliary to (3.66)–(3.69), the continuity equation is given by

$$\mathbf{\nabla} \cdot \mathbf{J} + \frac{\partial \rho}{\partial t} = 0 \tag{3.70}$$

This equation, which is the differential form of the law of conservation of charge, states that the sum of the divergence of the current density due to flow of charges and the time derivative of the charge density is equal to zero. Also, we recall that

$$\mathbf{D} = \epsilon_0 \mathbf{E} \tag{3.71}$$

$$\mathbf{H} = \frac{\mathbf{B}}{\mu_0} \tag{3.72}$$

which relate $\mathbf{D}$ and $\mathbf{H}$ to $\mathbf{E}$ and $\mathbf{B}$, respectively, for free space.

We have learned that the basic definitions of curl and divergence, which have enabled us to discuss their physical interpretations with the aid of the curl and divergence meters, are

$$\mathbf{\nabla} \times \mathbf{A} = \lim_{\Delta S \to 0} \left[\frac{\oint_C \mathbf{A} \cdot d\mathbf{l}}{\Delta S} \right]_{\max} \mathbf{a}_n$$

$$\mathbf{\nabla} \cdot \mathbf{A} = \lim_{\Delta v \to 0} \frac{\oint_S \mathbf{A} \cdot d\mathbf{S}}{\Delta v}$$

Thus, the curl of a vector field at a point is a vector whose magnitude is the circulation of that vector field per unit area with the area oriented so as to maximize this quantity and in the limit that the area shrinks to the point. The direction of the vector is normal to the area in the aforementioned limit and in the right-hand sense. The divergence of a vector field at a point is a scalar quantity equal to the net outward flux of that vector field per unit volume in the limit that the volume shrinks to the point. In Cartesian coordinates the expansions for curl and divergence are

$$\mathbf{\nabla} \times \mathbf{A} = \begin{vmatrix} \mathbf{a}_x & \mathbf{a}_y & \mathbf{a}_z \\ \dfrac{\partial}{\partial x} & \dfrac{\partial}{\partial y} & \dfrac{\partial}{\partial z} \\ A_x & A_y & A_z \end{vmatrix}$$

$$= \left(\frac{\partial A_z}{\partial y} - \frac{\partial A_y}{\partial z} \right) \mathbf{a}_x + \left(\frac{\partial A_x}{\partial z} - \frac{\partial A_z}{\partial x} \right) \mathbf{a}_y + \left(\frac{\partial A_y}{\partial x} - \frac{\partial A_x}{\partial y} \right) \mathbf{a}_z$$

$$\mathbf{\nabla} \cdot \mathbf{A} = \frac{\partial A_x}{\partial x} + \frac{\partial A_y}{\partial y} + \frac{\partial A_z}{\partial z}$$

Thus, Maxwell's equations in differential form relate the spatial variations of the field vectors at a point to their temporal variations and to the charge and current densities at that point.

We have also learned two theorems associated with curl and divergence. These are the Stokes' theorem and the divergence theorem given, respectively, by

$$\oint_C \mathbf{A} \cdot d\mathbf{l} = \int_S (\nabla \times \mathbf{A}) \cdot d\mathbf{S}$$

and

$$\oint_S \mathbf{A} \cdot d\mathbf{S} = \int_V (\nabla \cdot \mathbf{A}) \, dv$$

Stokes' theorem enables us to replace the line integral of a vector around a closed path by the surface integral of the curl of that vector over any surface bounded by that closed path, and vice versa. The divergence theorem enables us to replace the surface integral of a vector over a closed surface by the volume integral of the divergence of that vector over the volume bounded by the closed surface, and vice versa.

In Chapter 2 we learned that all Maxwell's equations in integral form are not independent. Since Maxwell's equations in differential form are derived from their integral forms, it follows that the same is true for these equations. In fact, by noting that (see Problem 3.32),

$$\nabla \cdot \nabla \times \mathbf{A} \equiv 0 \tag{3.73}$$

and applying it to (3.66), we obtain

$$\nabla \cdot \left(-\frac{\partial \mathbf{B}}{\partial t} \right) = \nabla \cdot \nabla \times \mathbf{E} = 0$$

$$\frac{\partial}{\partial t} (\nabla \cdot \mathbf{B}) = 0$$

$$\nabla \cdot \mathbf{B} = \text{constant with time} \tag{3.74}$$

Similarly, applying (3.73) to (3.67), we obtain

$$\nabla \cdot \left(\mathbf{J} + \frac{\partial \mathbf{D}}{\partial t} \right) = \nabla \cdot \nabla \times \mathbf{H} = 0$$

$$\nabla \cdot \mathbf{J} + \frac{\partial}{\partial t} (\nabla \cdot \mathbf{D}) = 0$$

Using (3.70), we then have

$$-\frac{\partial \rho}{\partial t} + \frac{\partial}{\partial t} (\nabla \cdot \mathbf{D}) = 0$$

$$\frac{\partial}{\partial t} (\nabla \cdot \mathbf{D} - \rho) = 0$$

$$\nabla \cdot \mathbf{D} - \rho = \text{constant with time} \tag{3.75}$$

Since for any given point in space, the constants on the right sides of (3.74) and (3.75) can be made equal to zero at some instant of time, it follows that they are zero forever, giving us (3.69) and (3.68), respectively. Thus (3.69) follows from (3.66), whereas (3.68) follows from (3.67) with the aid of (3.70).

Finally, for the simple, special case in which

$$\mathbf{E} = E_x(z, t)\mathbf{a}_x$$
$$\mathbf{H} = H_y(z, t)\mathbf{a}_y$$

the two Maxwell's curl equations reduce to

$$\frac{\partial E_x}{\partial z} = -\frac{\partial B_y}{\partial t} \tag{3.76}$$

$$\frac{\partial H_y}{\partial z} = -J_x - \frac{\partial D_x}{\partial t} \tag{3.77}$$

In fact, we derived these equations first and then the general equations (3.66) and (3.67). We will be using (3.76) and (3.77) in the following chapters to study the phenomenon of electromagnetic wave propagation resulting from the interdependence between the space-variations and time-variations of the electric and magnetic fields.

In fact, Maxwell's equations in differential form lend themselves well for a qualitative discussion of the interdependence of time-varying electric and magnetic fields giving rise to the phenomenon of electromagnetic wave propagation. Recognizing that the operations of curl and divergence involve partial derivatives with respect to space coordinates, we observe that time-varying electric and magnetic fields coexist in space, with the spatial variation of the electric field governed by the temporal variation of the magnetic field in accordance with (3.66), and the spatial variation of the magnetic field governed by the temporal variation of the electric field in addition to the current density in accordance with (3.67). Thus, if in (3.67) we begin with a time-varying current source represented by $\mathbf{J}$, or a time-varying electric field represented by $\partial \mathbf{D}/\partial t$, or a combination of the two, then one can visualize that a magnetic field is generated in accordance with (3.67), which in turn generates an electric field in accordance with (3.66), which in turn contributes to the generation of the magnetic field in accordance with (3.67), and so on, as depicted in Figure 3.15. Note that $\mathbf{J}$ and ρ are coupled, since they must satisfy (3.70). Also, the magnetic field automatically satisfies (3.69), since (3.69) is not independent of (3.66).

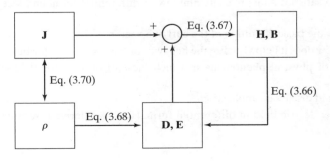

FIGURE 3.15

Generation of interdependent electric and magnetic fields, beginning with sources $\mathbf{J}$ and ρ.

The process depicted is exactly the phenomenon of electromagnetic waves propagating with a velocity (and other characteristics) determined by the parameters of the medium. In free space, the waves propagate unattenuated with the velocity $1/\sqrt{\mu_0 \varepsilon_0}$, familiarly represented by the symbol c, as we shall learn in Chapter 4. If either the term $\partial \mathbf{B}/\partial t$ in (3.66) or the term $\partial \mathbf{D}/\partial t$ in (1.28) is not present, then wave propagation would not occur. As already stated, it was through the addition of the term $\partial \mathbf{D}/\partial t$ in (3.67) that Maxwell predicted electromagnetic wave propagation before it was confirmed experimentally.

REVIEW QUESTIONS

3.1. State Faraday's law in differential form for the simple case of $\mathbf{E} = E_x(z, t)\mathbf{a}_x$. How is it derived from Faraday's law in integral form?

3.2. Discuss the physical interpretation of Faraday's law in differential form for the simple case of $\mathbf{E} = E_x(z, t)\mathbf{a}_x$.

3.3. State Faraday's law in differential form for the general case of an arbitrary electric field. How is it derived from its integral form?

3.4. What is meant by the net right-lateral differential of the x- and y-components of a vector normal to the z-direction?

3.5. Give an example in which the net right-lateral differential of E_y and E_z normal to the x-direction is zero, although the individual derivatives are nonzero.

3.6. If at a point in space B_y varies with time but B_x and B_z do not, what can we say about the components of $\mathbf{E}$ at that point?

3.7. What is the determinant expansion for the curl of a vector?

3.8. What is the significance of the curl of a vector being equal to zero?

3.9. State Ampere's circuital law in differential form for the simple case of $\mathbf{H} = H_y(z, t)\mathbf{a}_y$. How is it derived from Ampere's circuital law in integral form?

3.10. Discuss the physical interpretation of Ampere's circuital law in differential form for the simple case of $\mathbf{H} = H_y(z, t)\mathbf{a}_y$.

3.11. State Ampere's circuital law in differential form for the general case of an arbitrary magnetic field. How is it derived from its integral form?

3.12. What is the significance of a nonzero net right-lateral differential of H_x and H_y normal to the z-direction at a point in space?

3.13. If a pair of $\mathbf{E}$ and $\mathbf{B}$ at a point satisfies Faraday's law in differential form, does it necessarily follow that it also satisfies Ampere's circuital law in differential form, and vice versa?

3.14. State and briefly discuss the basic definition of the curl of a vector.

3.15. What is a curl meter? How does it help visualize the behavior of the curl of a vector field?

3.16. Provide two examples of physical phenomena in which the curl of a vector field is nonzero.

3.17. State Stokes' theorem and discuss its application.

3.18. State Gauss' law for the electric field in differential form. How is it derived from its integral form?

3.19. What is meant by the net longitudinal differential of the components of a vector field?

3.20. Give an example in which the net longitudinal differential of the components of a vector is zero, although the individual derivatives are nonzero.

3.21. What is the expansion for the divergence of a vector?

3.22. State Gauss' law for the magnetic field in differential form. How is it derived from its integral form?

3.23. How can you determine if a given vector can represent a magnetic field?

3.24. State and briefly discuss the basic definition of the divergence of a vector.

3.25. What is a divergence meter? How does it help visualize the behavior of the divergence of a vector field?

3.26. Provide two examples of physical phenomena in which the divergence of a vector field is nonzero.

3.27. State the continuity equation and discuss its physical interpretation.

3.28. Distinguish between the physical interpretations of the divergence and the curl of a vector field by means of examples.

3.29. State the divergence theorem and discuss its application.

3.30. What is the divergence of the curl of a vector?

3.31. Summarize Maxwell's equations in differential form.

3.32. Are all Maxwell's equations in differential form independent? If not, which of them are independent?

3.33. Provide a qualitative explanation of the phenomenon of electromagnetic wave propagation based on Maxwell's equations in differential form.

PROBLEMS

3.1. Given $\mathbf{B} = B_0 z \cos \omega t \, \mathbf{a}_y$ and it is known that $\mathbf{E}$ has only an x-component, find $\mathbf{E}$ by using Faraday's law in differential form. Then verify your result by applying Faraday's law in integral form to the rectangular closed path, in the xz-plane, defined by $x = 0$, $x = a$, $z = 0$, and $z = b$.

3.2. Assuming $\mathbf{E} = E_y(z, t)\mathbf{a}_y$ and considering a rectangular closed path in the yz-plane, carry out the derivation of Faraday's law in differential form similar to that in the text.

3.3. Find the curls of the following vector fields:
(a) $zx\mathbf{a}_x + xy\mathbf{a}_y + yz\mathbf{a}_z$; (b) $ye^{-x}\mathbf{a}_x - e^{-x}\mathbf{a}_y$.

3.4. For $\mathbf{A} = xy^2\mathbf{a}_x + x^2\mathbf{a}_y$, (a) find the net right-lateral differential of A_x and A_y normal to the z-direction at the point $(2, 1, 0)$, and (b) find the locus of the points at which the net right-lateral differential of A_x and A_y normal to the z-direction is zero.

3.5. Given $\mathbf{E} = 10 \cos (6\pi \times 10^8 t - 2\pi z) \, \mathbf{a}_x$ V/m, find $\mathbf{B}$ by using Faraday's law in differential form.

3.6. Show that the curl of $\left(\mathbf{a}_x \dfrac{\partial}{\partial x} + \mathbf{a}_y \dfrac{\partial}{\partial y} + \mathbf{a}_z \dfrac{\partial}{\partial z} \right) f$, that is, ∇f, where f is any scalar function of x, y, and z, is zero. Then find the scalar function for which $\nabla f = y\mathbf{a}_x + x\mathbf{a}_y$.

3.7. Given $\mathbf{E} = E_0 z^2 \sin \omega t \, \mathbf{a}_x$ and it is known that $\mathbf{J}$ is zero and $\mathbf{B}$ has only a y-component, find $\mathbf{B}$ by using Ampere's circuital law in differential form. Then find $\mathbf{E}$ from $\mathbf{B}$ by using Faraday's law in differential form. Comment on your result.

3.8. Assuming $\mathbf{H} = H_x(z, t)\, \mathbf{a}_x$ and considering a rectangular closed path in the xz-plane, carry out the derivation of Ampere's circuital law in differential form similar to that in the text.

3.9. Given $\mathbf{B} = \dfrac{10^{-7}}{3}\cos(6\pi \times 10^8 t - 2\pi z)\,\mathbf{a}_y$ Wb/m^2 and it is known that $\mathbf{J} = 0$, find $\mathbf{E}$ by using Ampere's circuital law in differential form. Then find $\mathbf{B}$ from $\mathbf{E}$ by using Faraday's law in differential form. Comment on your result.

3.10. Assuming $\mathbf{J} = 0$, determine which of the following pairs of E_x and H_y simultaneously satisfy the two Maxwell's equations in differential form given by (3.7) and (3.23):

(a) $E_x = 10\cos 2\pi z \cos 6\pi \times 10^8 t \qquad H_y = \dfrac{1}{12\pi}\sin 2\pi z \sin 6\pi \times 10^8 t$

(b) $E_x = (t - z\sqrt{\mu_0 \epsilon_0}) \qquad H_y = \sqrt{\dfrac{\epsilon_0}{\mu_0}}\,(t - z\sqrt{\mu_0 \epsilon_0})$

(c) $E_x = z^2 \sin \omega t \qquad H_y = -\dfrac{\omega \epsilon_0}{3} z^3 \cos \omega t$

3.11. A current distribution is given by

$$\mathbf{J} = \begin{cases} -J_0\mathbf{a}_x & \text{for } -a < z < 0 \\ J_0\mathbf{a}_x & \text{for } 0 < z < a \end{cases}$$

where J_0 is a constant. Using Ampere's circuital law in differential form and symmetry considerations, find the magnetic field everywhere.

3.12. A current distribution is given by

$$\mathbf{J} = J_0\left(1 - \frac{|z|}{a}\right)\mathbf{a}_x \qquad \text{for } -a < z < a$$

where J_0 is a constant. Using Ampere's circuital law in differential form and symmetry considerations, find the magnetic field everywhere.

3.13. Assume that the velocity of water in the stream of Figure 3.7(a) decreases linearly from a maximum at the top surface to zero at the bottom surface, with the velocity at the top surface given by Figure 3.7(b). Discuss the curl of the velocity vector field with the aid of the curl meter.

3.14. For the vector field $\mathbf{r} = x\mathbf{a}_x + y\mathbf{a}_y + z\mathbf{a}_z$, discuss the behavior of the curl meter and verify your reasoning by evaluating the curl of $\mathbf{r}$.

3.15. Discuss the curl of the vector field $y\mathbf{a}_x - x\mathbf{a}_y$ with the aid of the curl meter.

3.16. Verify Stokes' theorem for the vector field $\mathbf{A} = y\mathbf{a}_x + z\mathbf{a}_y + x\mathbf{a}_z$ and the closed path comprising the straight lines from $(1, 0, 0)$ to $(0, 1, 0)$, from $(0, 1, 0)$ to $(0, 0, 1)$, and from $(0, 0, 1)$ to $(1, 0, 0)$.

3.17. Verify Stokes' theorem for the vector field $\mathbf{A} = e^{-y}\mathbf{a}_x - xe^{-y}\mathbf{a}_y$ and any closed path of your choice.

3.18. For the vector $\mathbf{A} = yz\mathbf{a}_x + zx\mathbf{a}_y + xy\mathbf{a}_z$, use Stokes' theorem to show that $\oint_C \mathbf{A} \cdot d\mathbf{l}$ is zero for any closed path C. Then evaluate $\int \mathbf{A} \cdot d\mathbf{l}$ from the origin to the point $(1, 1, 2)$ along the curve $x = \sqrt{2}\sin t, y = \sqrt{2}\sin t, z = (8/\pi)t$.

3.19. Find the divergences of the following vector fields:
(a) $3xy^2\mathbf{a}_x + 3x^2y\mathbf{a}_y + z^3\mathbf{a}_z$; (b) $2xy\mathbf{a}_x - y^2\mathbf{a}_y$.

3.20. For $\mathbf{A} = xy\mathbf{a}_x + yz\mathbf{a}_y + zx\mathbf{a}_z$, (a) find the net longitudinal differential of the components of $\mathbf{A}$ at the point $(1, 1, 1)$, and (b) find the locus of the points at which the net longitudinal differential of the components of $\mathbf{A}$ is zero.

3.21. For each of the following vectors, find the curl and the divergence and discuss your results: (a) $xy\mathbf{a}_x$; (b) $y\mathbf{a}_x$; (c) $x\mathbf{a}_x$; (d) $y\mathbf{a}_x + x\mathbf{a}_y$.

3.22. A charge distribution is given by

$$\rho = \rho_0\left(1 - \frac{|x|}{a}\right) \quad \text{for } -a < x < a$$

where ρ_0 is a constant. Using Gauss' law for the electric field in differential form and symmetry considerations, find the electric field everywhere.

3.23. A charge distribution is given by

$$\rho = \rho_0\frac{x}{a} \quad \text{for } -a < x < a$$

where ρ_0 is a constant. Using Gauss' law for the electric field in differential form and symmetry considerations, find the electric field everywhere.

3.24. Given $\mathbf{D} = x^2y\mathbf{a}_x - y^3\mathbf{a}_y$, find the charge density at (a) the point $(2, 1, 0)$ and (b) the point $(3, 2, 0)$.

3.25. Determine which of the following vectors can represent a magnetic flux density vector $\mathbf{B}$: (a) $y\mathbf{a}_x - x\mathbf{a}_y$; (b) $x\mathbf{a}_x + y\mathbf{a}_y$; (c) $z^3 \cos \omega t \ \mathbf{a}_y$.

3.26. Given $\mathbf{J} = e^{-x}\mathbf{a}_x$, find the time rate of decrease of the charge density at (a) the point $(0, 0, 0)$ and (b) the point $(1, 0, 0)$.

3.27. For the vector field $\mathbf{r} = x\mathbf{a}_x + y\mathbf{a}_y + z\mathbf{a}_z$, discuss the behavior of the divergence meter, and verify your reasoning by evaluating the divergence of $\mathbf{r}$.

3.28. Discuss the divergence of the vector field $y\mathbf{a}_x - x\mathbf{a}_y$ with the aid of the divergence meter.

3.29. Verify the divergence theorem for the vector field $\mathbf{A} = x\mathbf{a}_x + y\mathbf{a}_y + z\mathbf{a}_z$ and the closed surface bounding the volume within the hemisphere of radius unity above the xy-plane and centered at the origin.

3.30. Verify the divergence theorem for the vector field $\mathbf{A} = xy\mathbf{a}_x + yz\mathbf{a}_y + zx\mathbf{a}_z$ and the closed surface of the volume bounded by the planes $x = 0$, $x = 1$, $y = 0$, $y = 1$, $z = 0$, and $z = 1$.

3.31. For the vector $\mathbf{A} = y^2\mathbf{a}_y - 2yz\mathbf{a}_z$, use the divergence theorem to show that $\oint_S \mathbf{A} \cdot d\mathbf{S}$ is zero for any closed surface S. Then evaluate $\int \mathbf{A} \cdot d\mathbf{S}$ over the surface $x + y + z = 1$, $x > 0$, $y > 0$, $z > 0$.

3.32. Show that $\nabla \cdot \nabla \times \mathbf{A} = 0$ for any $\mathbf{A}$ in two ways: (a) by evaluating $\nabla \cdot \nabla \times \mathbf{A}$ in Cartesian coordinates, and (b) by using Stokes' and divergence theorems.

Wave Propagation in Free Space

4

In Chapters 2 and 3, we learned Maxwell's equations in integral form and in differential form. We now have the knowledge of the fundamental laws of electromagnetics that enable us to embark upon the study of their applications. Many of these applications are based on electromagnetic wave phenomena, and hence it is necessary to gain an understanding of the basic principles of wave propagation, which is our goal in this chapter. In particular, we shall consider wave propagation in free space. We shall then in the next chapter consider the interaction of the wave fields with materials to extend the application of Maxwell's equations to material media and discuss wave propagation in material media.

We shall employ an approach in this chapter that will enable us not only to learn how the coupling between space-variations and time-variations of the electric and magnetic fields, as indicated by Maxwell's equations, results in wave motion, but also to illustrate the basic principle of radiation of waves from an antenna, which will be treated in detail in Chapter 9. In this process, we will also learn several techniques of analysis pertinent to field problems. We shall augment our discussion of radiation and propagation of waves by considering such examples as the principle of an antenna array and polarization. Finally, we shall discuss power flow and energy storage associated with the wave motion and introduce the Poynting vector.

4.1 THE INFINITE PLANE CURRENT SHEET

In Chapter 3, we learned that the space-variations of the electric and magnetic field components are related to the time-variations of the magnetic and electric field components, respectively, through Maxwell's equations. This interdependence gives rise to the phenomenon of electromagnetic wave propagation. In the general case, electromagnetic wave propagation involves electric and magnetic fields having more than one component, each dependent on all three coordinates, in addition to time. However, a simple and very useful type of wave that serves as a building block in the study of electromagnetic waves consists of electric and magnetic fields that are perpendicular to each other and to the direction of propagation and are uniform in planes perpendicular to the direction of propagation. These waves are known as *uniform plane waves*. By orienting the coordinate axes such that the electric field is in the x-direction, the magnetic field is in the y-direction, and the direction of propagation is in the z-direction, as shown in Figure 4.1, we have

$$\mathbf{E} = E_x(z, t)\mathbf{a}_x \qquad (4.1)$$

$$\mathbf{H} = H_y(z, t)\mathbf{a}_y \qquad (4.2)$$

Uniform plane waves do not exist in practice because they cannot be produced by finite-sized antennas. At large distances from physical antennas and ground, however, the waves can be approximated as uniform plane waves. Furthermore, the principles of guiding of electromagnetic waves along transmission lines and waveguides and the principles of many other wave phenomena can be studied basically in terms of uniform plane waves. Hence, it is very important that we understand the principles of uniform plane wave propagation.

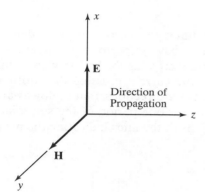

FIGURE 4.1

Directions of electric and magnetic fields and direction of propagation for a simple case of uniform plane wave.

In order to illustrate the phenomenon of interaction of electric and magnetic fields giving rise to uniform plane electromagnetic wave propagation, and the principle of radiation of electromagnetic waves from an antenna, we shall consider a simple, idealized, hypothetical source. This source consists of an infinite sheet lying in the xy-plane, as shown in Figure 4.2. On this infinite plane sheet a uniformly distributed current varying sinusoidally with time flows in the negative x-direction. Since the current is distributed on a surface, we talk of surface current density in order to express the current distribution mathematically. The surface current density, denoted by the symbol $\mathbf{J}_S$, is a vector quantity having the magnitude equal to the current per unit width (A/m) crossing an infinitesimally long line, on the surface, oriented so as to maximize the current. The direction of $\mathbf{J}_S$ is then normal to the line and toward the side of the current flow. In the present case, the surface current density is given by

$$\mathbf{J}_S = -J_{S0} \cos \omega t \, \mathbf{a}_x \qquad \text{for } z = 0 \qquad (4.3)$$

where J_{S0} is a constant and ω is the radian frequency of the sinusoidal time-variation of the current density.

Because of the uniformity of the surface current density on the infinite sheet, if we consider any line of width w parallel to the y-axis, as shown in Figure 4.2, the

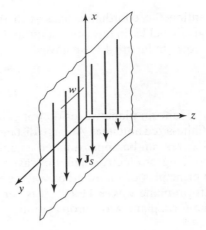

FIGURE 4.2

Infinite plane sheet in the xy-plane carrying surface
current of uniform density.

current crossing that line is simply given by w times the current density, that is,
$wJ_{S0} \cos \omega t$. If the current density is nonuniform, we have to perform an integration
along the width of the line in order to find the current crossing the line. In view of the
sinusoidal time-variation of the current density, the current crossing the width w
actually alternates between negative x- and positive x-directions, that is, downward
and upward. The time history of the current flow for one period of the sinusoidal
variation is illustrated in Figure 4.3, with the lengths of the lines indicating the magnitudes of the current.

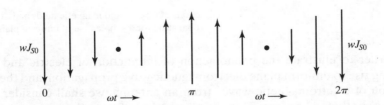

FIGURE 4.3

Time history of current flow across a line of width w parallel to the y-axis for
the current sheet of Figure 4.2.

4.2   MAGNETIC FIELD ADJACENT TO THE CURRENT SHEET

In the previous section, we introduced the infinite current sheet lying in the xy-plane
and upon which a surface current flows with density given by

$$\mathbf{J}_S = -J_{S0} \cos \omega t \, \mathbf{a}_x \qquad (4.4)$$

Our goal is to find the electromagnetic field due to this time-varying current distribution. In order to do this, we have to solve Faraday's and Ampere's circuital laws simultaneously. Since we have here only an x-component of the current density independent

of x and y, the equations of interest are

$$\frac{\partial E_x}{\partial z} = -\frac{\partial B_y}{\partial t} \tag{4.5}$$

$$\frac{\partial H_y}{\partial z} = -\left(J_x + \frac{\partial D_x}{\partial t}\right) \tag{4.6}$$

The quantity J_x on the right side of (4.6) represents volume current density, whereas we now have a surface current density. Furthermore, in the free space on either side of the current sheet the current density is zero and the differential equations reduce to

$$\frac{\partial E_x}{\partial z} = -\frac{\partial B_y}{\partial t} \tag{4.7}$$

$$\frac{\partial H_y}{\partial z} = -\frac{\partial D_x}{\partial t} \tag{4.8}$$

To obtain the solutions for E_x and H_y on either side of the current sheet, we therefore have to solve these two differential equations simultaneously.

To obtain a start on the solution, however, we need to consider the surface current distribution and find the magnetic field immediately adjacent to the current sheet. This is done by making use of Ampere's circuital law in integral form given by

$$\oint_C \mathbf{H} \cdot d\mathbf{l} = \int_S \mathbf{J} \cdot d\mathbf{S} + \frac{d}{dt} \int_S \mathbf{D} \cdot d\mathbf{S} \tag{4.9}$$

and applying it to a rectangular closed path $abcda$, as shown in Figure 4.4, with the sides ab and cd lying immediately adjacent to the current sheet, that is, touching the current sheet, and on either side of it. This choice of the rectangular path is not arbitrary but is intentionally chosen to achieve the task of finding the required magnetic field. First, we note from (4.6) that an x-directed current density gives rise to a

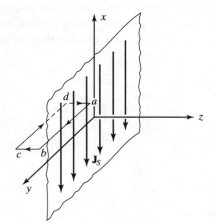

FIGURE 4.4

Rectangular path enclosing a portion of the current on the infinite plane current sheet.

magnetic field in the y-direction. At the source of the current, this magnetic field must also have a differential in the third direction, namely, the z-direction. In fact, from symmetry considerations, we can say that H_y on ab and cd must be equal in magnitude and opposite in direction.

If we now consider the line integral of $\mathbf{H}$ around the rectangular path $abcda$, we have

$$\int_{abcda} \mathbf{H} \cdot d\mathbf{l} = \int_a^b \mathbf{H} \cdot d\mathbf{l} + \int_b^c \mathbf{H} \cdot d\mathbf{l} + \int_c^d \mathbf{H} \cdot d\mathbf{l} + \int_d^a \mathbf{H} \cdot d\mathbf{l} \qquad (4.10)$$

The second and the fourth integrals on the right side of (4.10) are, however, equal to zero, since $\mathbf{H}$ is normal to the sides bc and da and furthermore bc and da are infinitesimally small. The first and third integrals on the right side of (4.10) are given by

$$\int_a^b \mathbf{H} \cdot d\mathbf{l} = [H_y]_{ab}(ab)$$

$$\int_c^d \mathbf{H} \cdot d\mathbf{l} = -[H_y]_{cd}(cd)$$

Thus,

$$\oint_{abcda} \mathbf{H} \cdot d\mathbf{l} = [H_y]_{ab}(ab) - [H_y]_{cd}(cd) = 2[H_y]_{ab}(ab) \qquad (4.11)$$

since $[H_y]_{cd} = -[H_y]_{ab}$.

We have just evaluated the left side of (4.9) for the particular problem under consideration here. To complete the task of finding the magnetic field adjacent to the current sheet, we now evaluate the right side of (4.9), which consists of two terms. The second term is, however, zero, since the area enclosed by the rectangular path is zero in view of the infinitesimally small thickness of the current sheet. The first term is not zero, since there is a current flowing on the sheet. Thus, the first term is simply equal to the current enclosed by the path $abcda$ in the right-hand sense, that is, the current crossing the width ab toward the negative x-direction. This is equal to the surface current density multiplied by the width ab, that is, $J_{S0} \cos \omega t \, (ab)$. Thus, substituting for the quantities on either side of (4.9), we have

$$2[H_y]_{ab}(ab) = J_{S0} \cos \omega t \, (ab)$$

or

$$[H_y]_{ab} = \frac{J_{S0}}{2} \cos \omega t \qquad (4.12)$$

It then follows that

$$[H_y]_{cd} = -\frac{J_{S0}}{2} \cos \omega t \qquad (4.13)$$

Thus, immediately adjacent to the current sheet the magnetic field intensity has a magnitude $\dfrac{J_{S0}}{2} \cos \omega t$ and is directed in the positive y-direction on the side $z > 0$ and in the negative y-direction on the side $z < 0$. This is illustrated in Figure 4.5. It is cautioned that this result is true only for points right next to the current sheet, since if we consider points at some distance from the current sheet, the second term on the right side of (4.9) will no longer be zero.

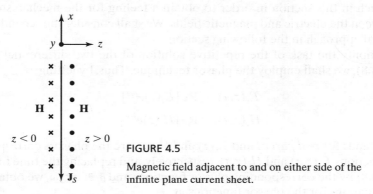

FIGURE 4.5

Magnetic field adjacent to and on either side of the infinite plane current sheet.

The technique we have used here for finding the magnetic field adjacent to the time-varying current sheet by using Ampere's circuital law in integral form is a standard procedure for finding the static electric and magnetic fields due to static charge and current distributions, possessing certain symmetries, by using Gauss' law for the electric field and Ampere's circuital law in integral forms, respectively, as we have already demonstrated in Chapter 2. Since for the static field case the terms involving time derivatives are zero, Ampere's circuital law simplifies to

$$\oint_C \mathbf{H} \cdot d\mathbf{l} = \int_S \mathbf{J} \cdot d\mathbf{S}$$

Hence, if the current distribution were not varying with time, then in order to compute the magnetic field we can choose a rectangular path of any width bc and it would still enclose the same current, namely, the current on the sheet. Thus, the magnetic field would be independent of the distance away from the sheet on either side of it. There are several problems in static fields that can be solved in this manner. We shall not discuss these here; instead, we shall include a few cases in the problems for the interested reader and shall continue with the derivation of the electromagnetic field due to our time-varying current sheet in the following section.

4.3 SUCCESSIVE SOLUTION OF MAXWELL'S EQUATIONS*

In the preceding section, we found the magnetic field adjacent to the infinite plane sheet of current introduced in Section 4.1. Now, to find the solutions for the fields everywhere on either side of the current sheet, let us first consider the region $z > 0$.

*This section may be omitted without loss of continuity.

In this region, the fields simultaneously satisfy the two differential equations (4.7) and (4.8) and with the constraint that the magnetic field at $z = 0$ is given by (4.12). To find the solutions for these differential equations, we have a choice of starting with the solution for H_y given by (4.12) and solving them successively and repeatedly in a step-by-step manner until the solutions satisfy both differential equations or of combining the two differential equations into one and then solving the single equation subject to the constraint at $z = 0$. Although it is somewhat longer and tedious, we shall use the first approach in this section in order to obtain a feeling for the mechanism of inter-action between the electric and magnetic fields. We shall consider the second and more conventional approach in the following section.

To simplify the task of the repetitive solution of the two differential equations (4.7) and (4.8), we shall employ the phasor technique. Thus, by letting

$$E_x(z, t) = \text{Re}[\bar{E}_x(z)e^{j\omega t}] \tag{4.14}$$

$$H_y(z, t) = \text{Re}[\bar{H}_y(z)e^{j\omega t}] \tag{4.15}$$

where Re stands for *real part of* and $\bar{E}_x(z)$ and $\bar{H}_y(z)$ are the phasors corresponding to the time functions $E_x(z, t)$ and $H_y(z, t)$, respectively, and replacing the time functions in (4.7) and (4.8) by the corresponding phasor functions and $\partial/\partial t$ by $j\omega$, we obtain the differential equations for the phasor functions as

$$\frac{\partial \bar{E}_x}{\partial z} = -j\omega \bar{B}_y = -j\omega\mu_0 \bar{H}_y \tag{4.16}$$

$$\frac{\partial \bar{H}_y}{\partial z} = -j\omega \bar{D}_x = -j\omega\epsilon_0 \bar{E}_x \tag{4.17}$$

We also note that since (4.12) can be written as

$$[H_y]_{ab} = \text{Re}\left(\frac{J_{S0}}{2}e^{j\omega t}\right)$$

the solution for the phasor $\bar{H}_y$ at $z = 0$ is given by

$$[\bar{H}_y]_{z=0} = \frac{J_{S0}}{2} \tag{4.18}$$

We start with (4.18) and solve (4.16) and (4.17) successively and repeatedly, and after obtaining the final solutions for $\bar{E}_x$ and $\bar{H}_y$, we put them in (4.14) and (4.15), respectively, to obtain the solutions for the real fields.

Thus, starting with (4.18) and substituting it in (4.16), we get

$$\frac{\partial \bar{E}_x}{\partial z} = -j\omega\mu_0 \frac{J_{S0}}{2}$$

Integrating both sides of this equation with respect to z, we have

$$\bar{E}_x = -j\omega\mu_0 \frac{J_{S0}z}{2} + \bar{C}$$

where $\bar{C}$ is the constant of integration. This constant of integration must, however, be equal to $[\bar{E}_x]_{z=0}$, since the first term on the right side tends to zero as $z \to 0$. Thus,

$$\bar{E}_x = -j\omega\mu_0 \frac{J_{S0}z}{2} + [\bar{E}_x]_{z=0} \tag{4.19}$$

Now, substituting (4.19) into (4.17), we obtain

$$\frac{\partial \bar{H}_y}{\partial z} = -j\omega\epsilon_0 \left\{ -j\omega\mu_0 \frac{J_{S0}z}{2} + [\bar{E}_x]_{z=0} \right\}$$

$$= -j\omega\epsilon_0 [\bar{E}_x]_{z=0} - \omega^2\mu_0\epsilon_0 \frac{J_{S0}z}{2}$$

$$\bar{H}_y = -j\omega\epsilon_0 z[\bar{E}_x]_{z=0} - \omega^2\mu_0\epsilon_0 \frac{J_{S0}z^2}{4} + [\bar{H}_y]_{z=0}$$

$$= -j\omega\epsilon_0 z[\bar{E}_x]_{z=0} - \omega^2\mu_0\epsilon_0 \frac{J_{S0}z^2}{4} + \frac{J_{S0}}{2}$$

$$= -j\omega\epsilon_0 z[\bar{E}_x]_{z=0} + \frac{J_{S0}}{2}\left(1 - \frac{\omega^2\mu_0\epsilon_0 z^2}{2} \right) \tag{4.20}$$

We have thus obtained a second-order solution for $\bar{H}_y$, which, however, does not satisfy (4.16) together with the solution for $\bar{E}_x$ given by (4.19). Hence, we must continue the step-by-step solution by substituting (4.20) into (4.16) and finding a higher-order solution for $\bar{E}_x$, and so on. Thus, by substituting (4.20) into (4.16), we get

$$\frac{\partial \bar{E}_x}{\partial z} = -j\omega\mu_0 \left\{ -j\omega\epsilon_0 z[\bar{E}_x]_{z=0} + \frac{J_{S0}}{2}\left(1 - \frac{\omega^2\mu_0\epsilon_0 z^2}{2} \right) \right\}$$

$$= -\omega^2\mu_0\epsilon_0 z[\bar{E}_x]_{z=0} - j\omega\mu_0 \frac{J_{S0}}{2}\left(1 - \frac{\omega^2\mu_0\epsilon_0 z^2}{2} \right)$$

$$\bar{E}_x = -\omega^2\mu_0\epsilon_0 \frac{z^2}{2}[\bar{E}_x]_{z=0} - j\omega\mu_0 \frac{J_{S0}}{2}\left(z - \frac{\omega^2\mu_0\epsilon_0 z^3}{6} \right) + [\bar{E}_x]_{z=0}$$

$$= [\bar{E}_x]_{z=0}\left(1 - \frac{\omega^2\mu_0\epsilon_0 z^2}{2} \right) - \frac{j\omega\mu_0 J_{S0}}{2}\left(z - \frac{\omega^2\mu_0\epsilon_0 z^3}{6} \right) \tag{4.21}$$

From (4.17), we then have

$$\frac{\partial \bar{H}_y}{\partial z} = -j\omega\epsilon_0 [\bar{E}_x]_{z=0}\left(1 - \frac{\omega^2\mu_0\epsilon_0 z^2}{2} \right) - \frac{\omega^2\mu_0\epsilon_0 J_{S0}}{2}\left(z - \frac{\omega^2\mu_0\epsilon_0 z^3}{6} \right)$$

$$\bar{H}_y = -j\omega\epsilon_0 [\bar{E}_x]_{z=0}\left(z - \frac{\omega^2\mu_0\epsilon_0 z^3}{6} \right)$$

$$\quad - \frac{\omega^2\mu_0\epsilon_0 J_{S0}}{2}\left(\frac{z^2}{2} - \frac{\omega^2\mu_0\epsilon_0 z^4}{24} \right) + [\bar{H}_y]_{z=0}$$

$$= -j\omega\epsilon_0 [\bar{E}_x]_{z=0}\left(z - \frac{\omega^2\mu_0\epsilon_0 z^3}{6} \right)$$

$$\quad + \frac{J_{S0}}{2}\left(1 - \frac{\omega^2\mu_0\epsilon_0 z^2}{2} + \frac{\omega^4\mu_0^2\epsilon_0^2 z^4}{24} \right) \tag{4.22}$$

Continuing in this manner, we will get infinite series expressions for $\bar{E}_x$ and $\bar{H}_y$ as follows:

$$\bar{E}_x = [\bar{E}_x]_{z=0}\left[1 - \frac{(\beta z)^2}{2!} + \frac{(\beta z)^4}{4!} - \cdots\right]$$
$$- j\frac{\eta_0 J_{S0}}{2}\left[\beta z - \frac{(\beta z)^3}{3!} + \frac{(\beta z)^5}{5!} - \cdots\right] \tag{4.23}$$

$$\bar{H}_y = -j\frac{1}{\eta_0}[\bar{E}_x]_{z=0}\left[\beta z - \frac{(\beta z)^3}{3!} + \frac{(\beta z)^5}{5!} - \cdots\right]$$
$$+ \frac{J_{S0}}{2}\left[1 - \frac{(\beta z)^2}{2!} + \frac{(\beta z)^4}{4!} - \cdots\right] \tag{4.24}$$

where we have introduced the notations

$$\beta = \omega\sqrt{\mu_0\epsilon_0} \tag{4.25}$$

$$\eta_0 = \sqrt{\frac{\mu_0}{\epsilon_0}} \tag{4.26}$$

It is left to the student to verify that the two expressions (4.23) and (4.24) simultaneously satisfy the two differential equations (4.16) and (4.17). Now, noting that

$$\cos \beta z = 1 - \frac{(\beta z)^2}{2!} + \frac{(\beta z)^4}{4!} - \cdots$$

$$\sin \beta z = \beta z - \frac{(\beta z)^3}{3!} + \frac{(\beta z)^5}{5!} + \cdots$$

and substituting into (4.23) and (4.24), we have

$$\bar{E}_x = [\bar{E}_x]_{z=0}\cos \beta z - j\frac{\eta_0 J_{S0}}{2}\sin \beta z \tag{4.27}$$

$$\bar{H}_y = -j\frac{1}{\eta_0}[\bar{E}_x]_{z=0}\sin \beta z + \frac{J_{S0}}{2}\cos \beta z \tag{4.28}$$

We now obtain the expressions for the real fields by putting (4.27) and (4.28) into (4.14) and (4.15), respectively. Thus,

$$E_x(z, t) = \text{Re}\left\{[\bar{E}_x]_{z=0}\cos \beta z\, e^{j\omega t} - j\frac{\eta_0 J_{S0}}{2}\sin \beta z\, e^{j\omega t}\right\}$$

$$= \cos \beta z\, \text{Re}\{[\bar{E}_x]_{z=0}e^{j\omega t}\} + \frac{\eta_0 J_{S0}}{2}\sin \beta z\, \text{Re}[e^{j(\omega t - \pi/2)}]$$

$$= \cos \beta z\, (C\cos \omega t + D\sin \omega t) + \frac{\eta_0 J_{S0}}{2}\sin \beta z\sin \omega t \tag{4.29}$$

$$H_y(z,t) = \text{Re}\left\{ -j\frac{1}{\eta_0} [\bar{E}_x]_{z=0} \sin \beta z \, e^{j\omega t} + \frac{J_{S0}}{2} \cos \beta z \, e^{j\omega t} \right\}$$

$$= \frac{1}{\eta_0} \sin \beta z \, \text{Re}\{[\bar{E}_x]_{z=0} \, e^{j(\omega t - \pi/2)}\} + \frac{J_{S0}}{2} \cos \beta z \, \text{Re}\,[e^{j\omega t}]$$

$$= \frac{1}{\eta_0} \sin \beta z \, (C \sin \omega t - D \cos \omega t) + \frac{J_{S0}}{2} \cos \beta z \cos \omega t \qquad (4.30)$$

where we have replaced the quantity $\text{Re}\{[\bar{E}_x]_{z=0} e^{j\omega t}\}$ by $(C \cos \omega t + D \sin \omega t)$, in which C and D are arbitrary constants to be determined. Making use of trigonometric identities and proceeding further, we write (4.29) and (4.30) as

$$E_x(z,t) = \frac{2C + \eta_0 J_{S0}}{4} \cos(\omega t - \beta z) + \frac{2C - \eta_0 J_{S0}}{4} \cos(\omega t + \beta z)$$

$$+ \frac{D}{2} \sin(\omega t - \beta z) + \frac{D}{2} \sin(\omega t + \beta z) \qquad (4.31)$$

$$H_y(z,t) = \frac{2C + \eta J_{S0}}{4\eta_0} \cos(\omega t - \beta z) - \frac{2C - \eta_0 J_{S0}}{4\eta_0} \cos(\omega t + \beta z)$$

$$+ \frac{D}{2\eta_0} \sin(\omega t - \beta z) - \frac{D}{2\eta_0} \sin(\omega t + \beta z) \qquad (4.32)$$

Equation (4.32) is the solution for H_y that together with the solution for E_x given by (4.31) satisfies the two differential equations (4.7) and (4.8) and that reduces to (4.12) for $z = 0$. Likewise, we can obtain the solutions for H_y and E_x for the region $z < 0$ by starting with $[H_y]_{z=0-}$ given by (4.13) and proceeding in a similar manner. We shall, however, proceed with the evaluation of the constants C and D in (4.31) and (4.32). In order to do this, we first have to understand the meanings of the functions $\cos(\omega t \mp \beta z)$ and $\sin(\omega t \mp \beta z)$. We shall do this in Section 4.5.

4.4 SOLUTION BY WAVE EQUATION

In Section 4.3, we found the solutions to the two simultaneous differential equations (4.7) and (4.8) by solving them successively and repeatedly in a step-by-step manner. In this section, we shall consider an alternative and more conventional method by combining the two equations into a single equation and then solving it. We recall that the two simultaneous differential equations to be satisfied in the free space on either side of the current sheet are

$$\frac{\partial E_x}{\partial z} = -\frac{\partial B_y}{\partial t} = -\mu_0 \frac{\partial H_y}{\partial t} \qquad (4.33)$$

$$\frac{\partial H_y}{\partial z} = -\frac{\partial D_x}{\partial t} = -\epsilon_0 \frac{\partial E_x}{\partial t} \qquad (4.34)$$

Differentiating (4.33) with respect to z and then substituting for $\partial H_y / \partial z$ from (4.34), we obtain

$$\frac{\partial^2 E_x}{\partial z^2} = -\mu_0 \frac{\partial}{\partial z}\left(\frac{\partial H_y}{\partial t}\right) = -\mu_0 \frac{\partial}{\partial t}\left(\frac{\partial H_y}{\partial z}\right) = -\mu_0 \frac{\partial}{\partial t}\left(-\epsilon_0 \frac{\partial E_x}{\partial t}\right)$$

or

$$\frac{\partial^2 E_x}{\partial z^2} = \mu_0 \epsilon_0 \frac{\partial^2 E_x}{\partial t^2} \tag{4.35}$$

We have thus eliminated H_y from (4.33) and (4.34) and obtained a single second-order partial differential equation involving E_x only.

Equation (4.35) is known as the *wave equation*. A technique of solving this equation is the *separation of variables* technique. Since it is a differential equation involving two variables, z and t, the technique consists of assuming that the required solution is the product of two functions, one of which is a function of z only and the second is a function of t only. Denoting these functions to be Z and T, respectively, we have

$$E_x(z, t) = Z(z)T(t) \tag{4.36}$$

Substituting (4.36) into (4.35) and dividing throughout by $\mu_0 \epsilon_0 Z(z)T(t)$, we obtain

$$\frac{1}{\mu_0 \epsilon_0 Z} \frac{d^2 Z}{dz^2} = \frac{1}{T} \frac{d^2 T}{dt^2} \tag{4.37}$$

In (4.37), the left side is a function of z only and the right side is a function of t only. In order for this to be satisfied, they both must be equal to a constant. Hence, setting them equal to a constant, say α^2, we have

$$\frac{d^2 Z}{dz^2} = \alpha^2 \mu_0 \epsilon_0 Z \tag{4.38a}$$

$$\frac{d^2 T}{dt^2} = \alpha^2 T \tag{4.38b}$$

We have thus obtained two ordinary differential equations involving separately the two variables z and t; hence, the technique is known as the *separation of variables* technique.

The constant α^2 in (4.38a) and (4.38b) is not arbitrary, since for the case of the sinusoidally time-varying current source the fields must also be sinusoidally time-varying with the same frequency, although not necessarily in phase with the source. Thus, the solution for $T(t)$ must be of the form

$$T(t) = A \cos \omega t + B \sin \omega t \tag{4.39}$$

where A and B are arbitrary constants to be determined. Substitution of (4.39) into (4.38b) gives us $\alpha^2 = -\omega^2$. The solution for (4.38a) is then given by

$$\begin{aligned} Z(z) &= A' \cos \omega \sqrt{\mu_0 \epsilon_0} z + B' \sin \omega \sqrt{\mu_0 \epsilon_0} z \\ &= A' \cos \beta z + B' \sin \beta z \end{aligned} \tag{4.40}$$

where A' and B' are arbitrary constants to be determined and we have defined

$$\beta = \omega\sqrt{\mu_0\epsilon_0} \tag{4.41}$$

The solution for E_x is then given by

$$
\begin{aligned}
E_x &= (A' \cos \beta z + B' \sin \beta z)(A \cos \omega t + B \sin \omega t) \\
&= C \cos \beta z \cos \omega t + D \cos \beta z \sin \omega t \\
&\quad + C' \sin \beta z \cos \omega t + D' \sin \beta z \sin \omega t
\end{aligned} \tag{4.42}
$$

The corresponding solution for H_y can be obtained by substituting (4.42) into one of the two equations (4.33) and (4.34). Thus, using (4.34), we get

$$
\begin{aligned}
\frac{\partial H_y}{\partial z} &= -\epsilon_0[-\omega C \cos \beta z \sin \omega t + \omega D \cos \beta z \cos \omega t \\
&\quad - \omega C' \sin \beta z \sin \omega t + \omega D' \sin \beta z \cos \omega t\,] \\
H_y &= \frac{\omega\epsilon_0}{\beta}[C \sin \beta z \sin \omega t - D \sin \beta z \cos \omega t \\
&\quad - C' \cos \beta z \sin \omega t + D' \cos \beta z \cos \omega t\,]
\end{aligned}
$$

Defining

$$\eta_0 = \frac{\beta}{\omega\epsilon_0} = \frac{\omega\sqrt{\mu_0\epsilon_0}}{\omega\epsilon_0} = \sqrt{\frac{\mu_0}{\epsilon_0}} \tag{4.43}$$

we have

$$
\begin{aligned}
H_y &= \frac{1}{\eta_0}[C \sin \beta z \sin \omega t - D \sin \beta z \cos \omega t \\
&\quad - C' \cos \beta z \sin \omega t + D' \cos \beta z \cos \omega t]
\end{aligned} \tag{4.44}
$$

Equation (4.44) is the general solution for H_y valid on both sides of the current sheet. In order to deduce the arbitrary constants, we first recall that the magnetic field adjacent to the current sheet is given by

$$
H_y = \begin{cases} \dfrac{J_{S0}}{2} \cos \omega t & \text{for } z = 0+ \\[2mm] -\dfrac{J_{S0}}{2} \cos \omega t & \text{for } z = 0- \end{cases} \tag{4.45}
$$

Thus, for $z > 0$,

$$\frac{1}{\eta_0}[-C' \sin \omega t + D' \cos \omega t] = \frac{J_{S0}}{2} \cos \omega t$$

or

$$C' = 0 \quad \text{and} \quad D' = \frac{\eta_0 J_{S0}}{2}$$

giving us

$$H_y = \frac{J_{S0}}{2} \cos \beta z \cos \omega t + \frac{1}{\eta_0} \sin \beta z \, (C \sin \omega t - D \cos \omega t) \qquad (4.46)$$

$$E_x = \frac{\eta_0 J_{S0}}{2} \sin \beta z \sin \omega t + \cos \beta z \, (C \cos \omega t + D \sin \omega t) \qquad (4.47)$$

Making use of trigonometric identities and proceeding further, we write (4.47) and (4.46) as

$$E_x(z, t) = \frac{2C + \eta_0 J_{S0}}{4} \cos(\omega t - \beta z) + \frac{2C - \eta_0 J_{S0}}{4} \cos(\omega t + \beta z)$$

$$+ \frac{D}{2} \sin(\omega t - \beta z) + \frac{D}{2} \sin(\omega t + \beta z) \qquad (4.48)$$

$$H_y(z, t) = \frac{2C + \eta_0 J_{S0}}{4\eta_0} \cos(\omega t - \beta z) - \frac{2C - \eta_0 J_{S0}}{4\eta_0} \cos(\omega t + \beta z)$$

$$+ \frac{D}{2\eta_0} \sin(\omega t - \beta z) - \frac{D}{2\eta_0} \sin(\omega t + \beta z) \qquad (4.49)$$

Equation (4.49) is the solution for H_y that together with the solution for E_x given by (4.48) satisfies the two differential equations (4.7) and (4.8) and that reduces to (4.12) for $z = 0$. Similarly, we can obtain the solutions for H_y and E_x for the region $z < 0$ by using the value of $[H_y]_{z=0-}$ to evaluate C' and D' in (4.44). We shall, however, proceed with the evaluation of the constants C and D in (4.48) and (4.49). In order to do this, we first have to understand the meanings of the functions $\cos(\omega t \mp \beta z)$ and $\sin(\omega t \mp \beta z)$. We shall do this in the following section.

4.5 UNIFORM PLANE WAVES

In the previous two sections, we derived the solutions for E_x and H_y, due to the infinite plane sheet of sinusoidally time-varying uniform current density, for the region $z > 0$. These solutions consist of the functions $\cos(\omega t \mp \beta z)$ and $\sin(\omega t \mp \beta z)$, which are dependent on both time and distance. Let us first consider the function $\cos(\omega t - \beta z)$. To understand the behavior of this function, we note that for a fixed value of time it varies in a cosinusoidal manner with the distance z. Let us therefore consider three values of time, $t = 0$, $t = \pi/4\omega$, and $t = \pi/2\omega$, and examine the sketches of this function versus z for these three times. By noting that

$$\text{for } t = 0, \quad \cos(\omega t - \beta z) = \cos(-\beta z) = \cos \beta z$$

$$\text{for } t = \frac{\pi}{4\omega}, \quad \cos(\omega t - \beta z) = \cos\left(\frac{\pi}{4} - \beta z\right)$$

$$\text{for } t = \frac{\pi}{2\omega}, \quad \cos(\omega t - \beta z) = \cos\left(\frac{\pi}{2} - \beta z\right) = \sin \beta z$$

we draw the sketches of the three functions as shown in Figure 4.6.

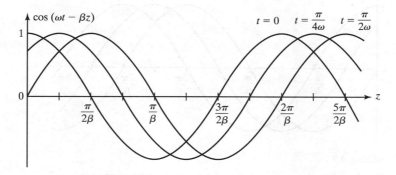

FIGURE 4.6

Sketches of the function $\cos(\omega t - \beta z)$ versus z for three values of t.

It is evident from Figure 4.6 that the sketch of the function for $t = \pi/4\omega$ is a replica of the function for $t = 0$ except that it is shifted by a distance of $\pi/4\beta$ toward the positive z-direction. Similarly, the sketch of the function for $t = \pi/2\omega$ is a replica of the function for $t = 0$ except that it is shifted by a distance of $\pi/2\beta$ toward the positive z-direction. Thus as time progresses, the function shifts bodily to the right, that is, toward increasing values of z. In fact, we can even find the velocity with which the function is traveling by dividing the distance moved by the time elapsed. This gives

$$\text{velocity} = \frac{\pi/\beta - \pi/2\beta}{\pi/2\omega - 0} = \frac{\omega}{\beta} = \frac{\omega}{\omega\sqrt{\mu_0\epsilon_0}}$$

$$= \frac{1}{\sqrt{\mu_0\epsilon_0}} = \frac{1}{\sqrt{4\pi \times 10^{-7} \times 10^{-9}/36\pi}}$$

$$= 3 \times 10^8 \, \text{m/s}$$

which is the velocity of light in free space, denoted c. Thus, the function $\cos(\omega t - \beta z)$ represents a *traveling wave* moving with a velocity ω/β toward the direction of increasing z. The wave is also known as the *positive going wave*, or *(+) wave*.

Similarly, by considering three values of time, $t = 0$, $t = \pi/4\omega$, and $t = \pi/2\omega$, for the function $\cos(\omega t + \beta z)$, we obtain the sketches shown in Figure 4.7. An examination of these sketches reveals that $\cos(\omega t + \beta z)$ represents a *traveling wave* moving with a velocity ω/β toward the direction of decreasing values of z. The wave is also known as the *negative going wave*, or *(−) wave*. Since the sine functions are cosine functions shifted in phase by $\pi/2$, it follows that $\sin(\omega t - \beta z)$ and $\sin(\omega t + \beta z)$ represent traveling waves moving in the positive and negative z-directions, respectively.

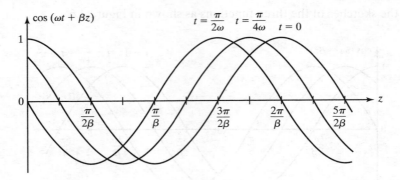

FIGURE 4.7

Sketches of the function $\cos(\omega t + \beta z)$ versus z for three values of t.

Returning to the solutions for E_x and H_y given by (4.31) and (4.32) or (4.48) and (4.49), we now know that these solutions consist of superpositions of traveling waves propagating away from and toward the current sheet. In the region $z > 0$, however, we have to rule out traveling waves propagating toward the current sheet, because such a situation requires a source of waves to the right of the sheet or an object that reflects the wave back toward the sheet. Thus, we have

$$D = 0$$

$$2C - \eta_0 J_{S0} = 0 \quad \text{or} \quad C = \frac{\eta_0 J_{S0}}{2}$$

which give us finally

$$\left.\begin{array}{l} E_x = \dfrac{\eta_0 J_{S0}}{2} \cos(\omega t - \beta z) \\[4mm] H_y = \dfrac{J_{S0}}{2} \cos(\omega t - \beta z) \end{array}\right\} \quad \text{for } z > 0 \qquad (4.50)$$

Having found the solutions for the fields in the region $z > 0$, we can now consider the solutions for the fields in the region $z < 0$. From our discussion of the functions $\cos(\omega t \mp \beta z)$, we know that these solutions must be of the form $\cos(\omega t + \beta z)$, since this function represents a traveling wave progressing in the negative z-direction, that is, away from the sheet in the region $z < 0$. Recalling that the magnetic field adjacent to the current sheet and to the left of it is given by

$$[H_y]_{z=0-} = -\frac{J_{S0}}{2} \cos \omega t$$

we get

$$H_y = -\frac{J_{S0}}{2} \cos(\omega t + \beta z) \qquad \text{for } z < 0 \qquad (4.51a)$$

The corresponding E_x can be obtained by simply substituting the result just obtained for H_y into one of the two differential equations (4.7) and (4.8). Thus using (4.7), we obtain

$$\frac{\partial E_x}{\partial z} = -\frac{\partial B_y}{\partial t} = -\frac{\mu_0 J_{S0}}{2}\omega \sin(\omega t + \beta z)$$

$$E_x = \frac{\mu_0 J_{S0}}{2}\frac{\omega}{\beta}\cos(\omega t + \beta z)$$

$$= \frac{\eta_0 J_{S0}}{2}\cos(\omega t + \beta z) \qquad \text{for } z < 0 \qquad (4.51b)$$

Combining (4.50) and (4.51), we find that the solution for the electromagnetic field due to the infinite plane current sheet in the xy-plane characterized by

$$\mathbf{J}_S = -J_{S0}\cos\omega t\,\mathbf{a}_x$$

is given by

$$\mathbf{E} = \frac{\eta_0 J_{S0}}{2}\cos(\omega t \mp \beta z)\,\mathbf{a}_x \qquad \text{for } z \gtrless 0 \qquad (4.52a)$$

$$\mathbf{H} = \pm\frac{J_{S0}}{2}\cos(\omega t \mp \beta z)\,\mathbf{a}_y \qquad \text{for } z \gtrless 0 \qquad (4.52b)$$

These results are illustrated in Figure 4.8, which shows sketches of the current density on the sheet and the distance-variation of the electric and magnetic fields on either side of the current sheet for a few values of t. It can be seen from these sketches that the phenomenon is one of electromagnetic waves *radiating* away from the current sheet to either side of it, in step with the time-variation of the current density on the sheet.

The solutions that we have just obtained for the fields due to the time-varying infinite plane current sheet are said to correspond to *uniform plane electromagnetic waves* propagating away from the current sheet to either side of it. The terminology arises from the fact that the fields are *uniform* (i.e., they do not vary with position) over the *planes z = constant*. Thus, the phase of the fields, that is, the quantity $(\omega t \pm \beta z)$, as well as the amplitudes of the fields, is uniform over the planes z = constant. The magnitude of the rate of change of phase with distance z for any fixed time is β. The quantity β is therefore known as the *phase constant*. Since the velocity of propagation of the wave, that is, ω/β, is the velocity with which a given constant phase progresses along the z-direction, that is, along the direction of propagation, it is known as the *phase velocity* and is denoted by the symbol v_p. Thus,

$$v_p = \frac{\omega}{\beta} \qquad (4.53)$$

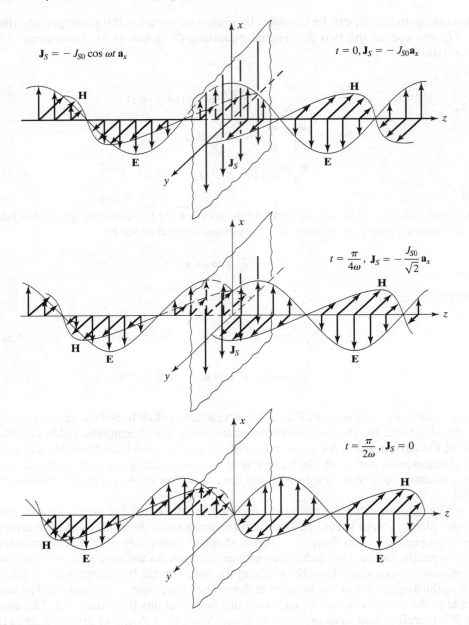

FIGURE 4.8

Time history of uniform plane electromagnetic wave radiating away from an infinite plane current sheet in free space.

The distance in which the phase changes by 2π radians for a fixed time is $2\pi/\beta$. This quantity is known as the *wavelength* and is denoted by the symbol λ. Thus,

$$\lambda = \frac{2\pi}{\beta} \tag{4.54}$$

Substituting (4.53) into (4.54), we obtain

$$\lambda = \frac{2\pi}{\omega/v_p} = \frac{v_p}{f}$$

or

$$\lambda f = v_p \tag{4.55}$$

Equation (4.55) is a simple relationship between the wavelength λ, which is a parameter governing the variation of the field with distance for a fixed time, and the frequency f, which is a parameter governing the variation of the field with time for a fixed value of z. Since for free space $v_p = 3 \times 10^8$ m/s, we have

$$\lambda \text{ in meters} \times f \text{ in Hz} = 3 \times 10^8$$
$$\lambda \text{ in meters} \times f \text{ in MHz} = 300 \tag{4.56}$$

Other properties of uniform plane waves evident from (4.52) are that the electric and magnetic fields have components lying in the planes of constant phase and perpendicular to each other and to the direction of propagation. In fact, the cross product of **E** and **H** results in a vector that is directed along the direction of propagation, as can be seen by noting that

$$\mathbf{E} \times \mathbf{H} = E_x \mathbf{a}_x \times H_y \mathbf{a}_y$$

$$= \pm \frac{\eta_0 J_{S0}^2}{4} \cos^2(\omega t \mp \beta z)\, \mathbf{a}_z \qquad \text{for } z \gtrless 0 \tag{4.57}$$

Finally, we note that the ratio of E_x to H_y is given by

$$\frac{E_x}{H_y} = \begin{cases} \eta_0 \text{ for } z > 0, \text{ that is, for the } (+) \text{ wave} \\ -\eta_0 \text{ for } z < 0, \text{ that is, for the } (-) \text{ wave} \end{cases} \tag{4.58}$$

The quantity η_0, which is equal to $\sqrt{\mu_0/\epsilon_0}$, is known as the *intrinsic impedance* of free space. Its value is given by

$$\eta_0 = \sqrt{\frac{(4\pi \times 10^{-7}) \text{ H/m}}{(10^{-9}/36\pi) \text{ F/m}}} = \sqrt{(144\pi^2 \times 10^2) \text{ H/F}}$$

$$= 120\pi\ \Omega = 377\ \Omega \tag{4.59}$$

Example 4.1

The electric field of a uniform plane wave is given by $\mathbf{E} = 10 \cos (3\pi \times 10^8 t - \pi z) \, \mathbf{a}_x$ V/m. Let us identify the various parameters associated with the uniform plane wave.

We recognize that

$$\omega = 3\pi \times 10^8 \text{ rad/s}$$

$$f = \frac{\omega}{2\pi} = 1.5 \times 10^8 \text{ Hz} = 150 \text{ MHz}$$

$$\beta = \pi \text{ rad/m}$$

$$\lambda = \frac{2\pi}{\beta} = 2 \text{ m}$$

$$v_p = \frac{\omega}{\beta} = \frac{3\pi \times 10^8}{\pi} = 3 \times 10^8 \text{ m/s}$$

Also, $\lambda f = v_p = 2 \times 1.5 \times 10^8 = 3 \times 10^8$ m/s. From (4.58), and since the given field represents a (+) wave,

$$\mathbf{H} = \frac{E_x}{\eta_0} \mathbf{a}_y = \frac{10}{377} \cos (3\pi \times 10^8 t - \pi z) \, \mathbf{a}_y \text{ A/m}$$

Example 4.2

An antenna array consists of two or more antenna elements spaced appropriately and excited with currents having the appropriate amplitudes and phases in order to obtain a desired radiation characteristic. To illustrate the principle of an antenna array, let us consider two infinite plane parallel current sheets, spaced $\lambda/4$ apart and carrying currents of equal amplitudes but out of phase by $\pi/2$, as given by the densities

$$\mathbf{J}_{S1} = -J_{S0} \cos \omega t \, \mathbf{a}_x \qquad z = 0$$

$$\mathbf{J}_{S2} = -J_{S0} \sin \omega t \, \mathbf{a}_x \qquad z = \frac{\lambda}{4}$$

and find the electric field due to the array of the two current sheets.

We apply the result given by (4.52) to each current sheet separately and then use superposition to find the required total electric field due to the array of the two current sheets. Thus, for the current sheet in the $z = 0$ plane, we have

$$\mathbf{E}_1 = \begin{cases} \dfrac{\eta_0 J_{S0}}{2} \cos (\omega t - \beta z) \, \mathbf{a}_x & \text{for } z > 0 \\[2ex] \dfrac{\eta_0 J_{S0}}{2} \cos (\omega t + \beta z) \, \mathbf{a}_x & \text{for } z < 0 \end{cases}$$

For the current sheet in the $z = \lambda/4$ plane, we have

$$\mathbf{E}_2 = \begin{cases} \dfrac{\eta_0 J_{S0}}{2} \sin\left[\omega t - \beta\left(z - \dfrac{\lambda}{4}\right)\right] \mathbf{a}_x & \text{for } z > \dfrac{\lambda}{4} \\[3mm] \dfrac{\eta_0 J_{S0}}{2} \sin\left[\omega t + \beta\left(z - \dfrac{\lambda}{4}\right)\right] \mathbf{a}_x & \text{for } z < \dfrac{\lambda}{4} \end{cases}$$

$$= \begin{cases} \dfrac{\eta_0 J_{S0}}{2} \sin\left(\omega t - \beta z + \dfrac{\pi}{2}\right) \mathbf{a}_x & \text{for } z > \dfrac{\lambda}{4} \\[3mm] \dfrac{\eta_0 J_{S0}}{2} \sin\left(\omega t + \beta z - \dfrac{\pi}{2}\right) \mathbf{a}_x & \text{for } z < \dfrac{\lambda}{4} \end{cases}$$

$$= \begin{cases} \dfrac{\eta_0 J_{S0}}{2} \cos\left(\omega t - \beta z\right) \mathbf{a}_x & \text{for } z > \dfrac{\lambda}{4} \\[3mm] -\dfrac{\eta_0 J_{S0}}{2} \cos\left(\omega t + \beta z\right) \mathbf{a}_x & \text{for } z < \dfrac{\lambda}{4} \end{cases}$$

Now, using superposition, we find the total electric field due to the two current sheets to be

$$\mathbf{E} = \mathbf{E}_1 + \mathbf{E}_2$$

$$= \begin{cases} \eta_0 J_{S0} \cos\left(\omega t - \beta z\right) \mathbf{a}_x & \text{for } z > \dfrac{\lambda}{4} \\[3mm] \eta_0 J_{S0} \sin \omega t \sin \beta z \, \mathbf{a}_x & \text{for } 0 < z < \dfrac{\lambda}{4} \\[3mm] 0 & \text{for } z < 0 \end{cases}$$

Thus, the total field is zero in the region $z < 0$, and hence there is no radiation toward that side of the array. In the region $z > \lambda/4$ the total field is twice that of the field due to a single sheet. The phenomenon is illustrated in Figure 4.9, which shows sketches of the individual fields E_{x1} and E_{x2} and the total field $E_x = E_{x1} + E_{x2}$ for a few values of t. The result that we have obtained here for the total field due to the array of two current sheets, spaced $\lambda/4$ apart and fed with currents of equal amplitudes but out of phase by $\pi/2$, is said to correspond to an *endfire* radiation pattern.

In Section 1.4, we introduced polarization of sinusoidally time-varying fields, which is of relevance here in wave propagation. To extend the discussion, in the case of circular and elliptical polarizations, since the circle or the ellipse can be traversed in one of two opposite senses relative to the direction of the wave propagation, we talk of right-handed or clockwise polarization and left-handed or counterclockwise polarization. The convention is that if in a given constant phase plane, the tip of the field vector of a circularly polarized wave rotates with time in the clockwise sense as seen looking along the direction of propagation of the wave, the wave is said to be right circularly polarized. If the tip of the field vector rotates in the counterclockwise sense, the wave is said to be left circularly polarized. Similar considerations hold for elliptically polarized waves, which arise due to the superposition of two linearly polarized waves in the general case.

For example, for a uniform plane wave propagating in the $+z$-direction and having the electric field,

$$\mathbf{E} = 10 \sin\left(3\pi \times 10^8 t - \pi z\right) \mathbf{a}_x + 10 \cos\left(3\pi \times 10^8 t - \pi z\right) \mathbf{a}_y \text{ V/m} \qquad (4.60)$$

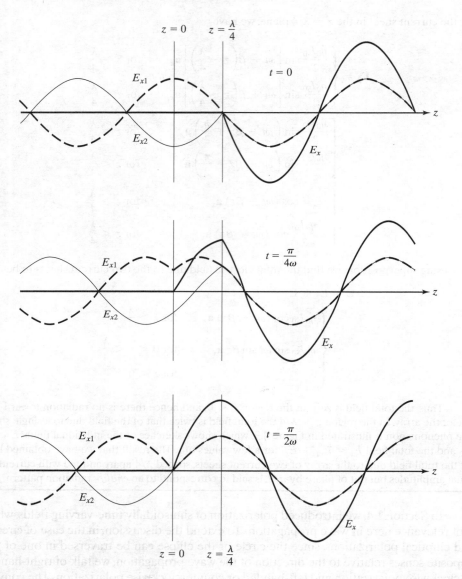

FIGURE 4.9

Time history of individual fields and the total field due to an array of two infinite plane parallel current sheets.

the two components of **E** are equal in amplitude, perpendicular, and out of phase by 90°. Therefore, the wave is circularly polarized. To determine if the polarization is right-handed or left-handed, we look at the electric field vectors in the $z = 0$ plane for two values of time, $t = 0$ and $t = \frac{1}{6} \times 10^{-8}$ s $(3\pi \times 10^8 t = \pi/2)$. These are shown in Figure 4.10. As time progresses, the tip of the vector rotates in the counterclockwise sense, as seen looking in the $+z$-direction. Hence, the wave is left circularly polarized.

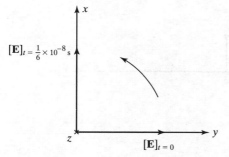

FIGURE 4.10

For the determination of the sense of circular polarization for the field of equation (4.60).

Thus far, we have considered a source of single frequency. We found that wave propagation in free space is characterized by a phase velocity v_p equal to c ($= 3 \times 10^8$ m/s) and intrinsic impedance η_0 ($= 377\ \Omega$), independent of frequency. Let us now consider a nonsinusoidal excitation for the current sheet. Then, since the propagation characteristics are the same for each frequency component of the nonsinusoidal excitation, the resulting fields at any given value of z will have the same shape as that of the source with time, that is, they propagate without change in shape with time. Thus, for an infinite plane current sheet of surface current density given by

$$\mathbf{J}_S(t) = -J_S(t)\mathbf{a}_x \qquad \text{for } z = 0 \tag{4.61}$$

the solution for the electromagnetic field is given by

$$\mathbf{E}(z, t) = \frac{\eta_0}{2}J_S\left(t \mp \frac{z}{v_p}\right)\mathbf{a}_x \qquad \text{for } z \gtrless 0 \tag{4.62a}$$

$$\mathbf{H}(z, t) = \pm\frac{1}{2}J_S\left(t \mp \frac{z}{v_p}\right)\mathbf{a}_y \qquad \text{for } z \gtrless 0 \tag{4.62b}$$

The time variation of the electric field component E_x in a given $z = $ constant plane is the same as the current density variation delayed by the time $|z|/v_p$ and multiplied by $\eta_0/2$. The time variation of the magnetic field component in a given $z = $ constant plane is the same as the current density variation delayed by $|z|/v_p$ and multiplied by $\pm\frac{1}{2}$, depending on $z \gtrless 0$. Using these properties, one can construct plots of the field components versus time for fixed values of z and versus z for fixed values of t.

Example 4.3

Let us consider the function $J_S(t)$ in (4.61) to be that given in Figure 4.11. We wish to find and sketch (a) E_x versus t for $z = 300$ m, (b) H_y versus t for $z = -450$ m, (c) E_x versus z for $t = 1\ \mu$s, and (d) H_y versus z for $t = 2.5\ \mu$s.

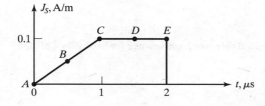

FIGURE 4.11

Plot of J_S versus t for Example 4.3.

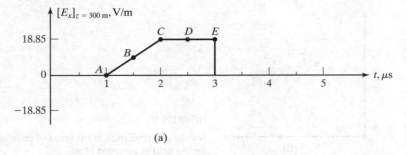

(a)

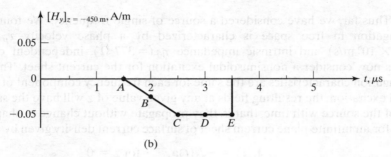

(b)

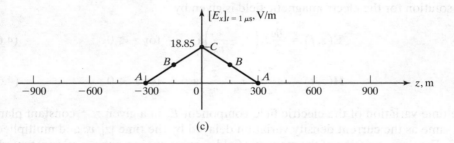

(c)

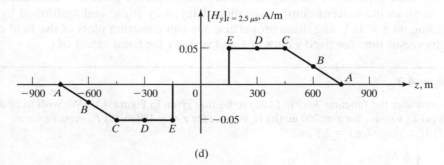

(d)

FIGURE 4.12

Plots of field components versus t for fixed values of z and versus z for fixed values of t for Example 4.3.

(a) Since $v_p = c = 3 \times 10^8$ m/s , the time delay corresponding to 300 m is 1 μs. Thus, the plot of E_x versus t for $z = 300$ m is the same as that of $J_S(t)$ multiplied by $\eta_0/2$, or 188.5, and delayed by 1 μs, as shown in Figure 4.12(a).

(b) The time delay corresponding to 450 m is 1.5 μs. Thus, the plot of H_y versus t for $z = -450$ m is the same as that of $J_S(t)$ multiplied by $-1/2$ and delayed by 1.5 μs, as shown in Figure 4.12(b).

(c) To sketch E_x versus z for a fixed value of t, say, t_1, we use the argument that a given value of E_x existing at the source at an earlier value of time, say, t_2, travels away from the source by the distance equal to $(t_1 - t_2)$ times v_p. Thus, at $t = 1$ μs, the values of E_x corresponding to points A and B in Figure 4.11 move to the locations $z = \pm300$ m and $z = \pm150$ m, respectively, and the value of E_x corresponding to point C exists right at the source. Hence, the plot of E_x versus z for $t = 1$ μs is as shown in Figure 4.12(c). Note that points beyond C in Figure 4.11 correspond to $t > 1$ μs, and therefore they do not appear in the plot of Figure 4.12(c).

(d) Using arguments as in part (c), we see that at $t = 2.5$ μs, the values of H_y corresponding to points A, B, C, D, and E in Figure 4.11 move to the locations $z = \pm750$ m, ±600 m, ±450 m, ±300 m, and ±150 m, respectively, as shown in Figure 4.12(d). Note that the plot is an odd function of z, since the factor by which J_{S0} is multiplied to obtain H_y is $\pm\frac{1}{2}$, depending on $z \lessgtr 0$.

4.6 POYNTING VECTOR AND ENERGY STORAGE

In the preceding section, we found the solution for the electromagnetic field due to an infinite plane current sheet situated in the $z = 0$ plane. For a surface current flowing in the negative x-direction, we found the electric field on the sheet to be directed in the positive x-direction. Since the current is flowing against the force due to the electric field, a certain amount of work must be done by the source of the current in order to maintain the current flow on the sheet. Let us consider a rectangular area of length Δx and width Δy on the current sheet, as shown in Figure 4.13. Since the current density is $J_{S0} \cos \omega t$, the charge crossing the width Δy in time dt is $dq = J_{S0} \Delta y \cos \omega t \, dt$. The force exerted on this charge by the electric field is given by

$$\mathbf{F} = dq \, \mathbf{E} = J_{S0} \Delta y \cos \omega t \, dt \, E_x \mathbf{a}_x \tag{4.63}$$

The amount of work required to be done against the electric field in displacing this charge by the distance Δx is

$$dw = F_x \Delta x = J_{S0} E_x \cos \omega t \, dt \, \Delta x \, \Delta y \tag{4.64}$$

Thus, the power supplied by the source of the current in maintaining the surface current over the area $\Delta x \, \Delta y$ is

$$\frac{dw}{dt} = J_{S0} E_x \cos \omega t \, \Delta x \, \Delta y \tag{4.65}$$

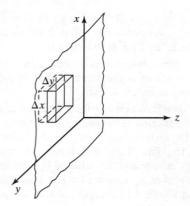

FIGURE 4.13

For the determination of power flow density associated with the electromagnetic field.

Recalling that E_x on the sheet is $\eta_0 \dfrac{J_{S0}}{2} \cos \omega t$, we obtain

$$\frac{dw}{dt} = \eta_0 \frac{J_{S0}^2}{2} \cos^2 \omega t \ \Delta x \ \Delta y \qquad (4.66)$$

We would expect the power given by (4.66) to be carried by the electromagnetic wave, half of it to either side of the current sheet. To investigate this, we note that the quantity $\mathbf{E} \times \mathbf{H}$ has the units of

$$\frac{\text{newtons}}{\text{coulomb}} \times \frac{\text{amperes}}{\text{meter}} = \frac{\text{newtons}}{\text{coulomb}} \times \frac{\text{coulomb}}{\text{second-meter}} \times \frac{\text{meter}}{\text{meter}}$$

$$= \frac{\text{newton-meters}}{\text{second}} \times \frac{1}{(\text{meter})^2} = \frac{\text{watts}}{(\text{meter})^2}$$

which represents power density. Let us then consider the rectangular box enclosing the area $\Delta x \ \Delta y$ on the current sheet and with its sides almost touching the current sheet on either side of it, as shown in Figure 4.13. Recalling that $\mathbf{E} \times \mathbf{H}$ is given by (4.57) and evaluating the surface integral of $\mathbf{E} \times \mathbf{H}$ over the surface of the rectangular box, we obtain the power flow out of the box as

$$\oint \mathbf{E} \times \mathbf{H} \cdot d\mathbf{S} = \eta_0 \frac{J_{S0}^2}{4} \cos^2 \omega t \ \mathbf{a}_z \cdot \Delta x \ \Delta y \ \mathbf{a}_z$$

$$+ \left(-\eta_0 \frac{J_{S0}^2}{4} \cos^2 \omega t \ \mathbf{a}_z \right) \cdot (-\Delta x \ \Delta y \ \mathbf{a}_z)$$

$$= \eta_0 \frac{J_{S0}^2}{2} \cos^2 \omega t \ \Delta x \ \Delta y \qquad (4.67)$$

This result is exactly equal to the power supplied by the current source as given by (4.66).

We now interpret the quantity **E** × **H** as the power flow density vector associated with the electromagnetic field. It is known as the *Poynting vector*, after J. H. Poynting, and is denoted by the symbol **P**. Although we have here introduced the Poynting vector by considering the specific case of the electromagnetic field due to the infinite plane current sheet, the interpretation that $\oint_S \mathbf{E} \times \mathbf{H} \cdot d\mathbf{S}$ is equal to the power flow out of the closed surface S is applicable in the general case.

Example 4.4

Far from a physical antenna, that is, at a distance of several wavelengths from the antenna, the radiated electromagnetic waves are approximately uniform plane waves with their constant phase surfaces lying normal to the radial directions away from the antenna, as shown for two directions in Figure 4.14. We wish to show from the Poynting vector and physical considerations that the electric and magnetic fields due to the antenna vary inversely proportional to the radial distance away from the antenna.

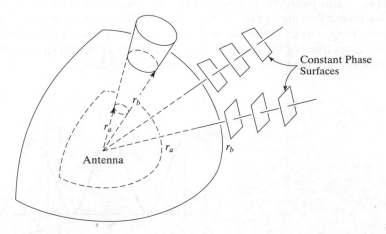

FIGURE 4.14

Radiation of electromagnetic waves far from a physical antenna.

From considerations of electric and magnetic fields of a uniform plane wave, the Poynting vector is directed everywhere in the radial direction indicating power flow radially away from the antenna and is proportional to the square of the magnitude of the electric field intensity. Let us now consider two spherical surfaces of radii r_a and r_b and centered at the antenna and insert a cone through these two surfaces such that the vertex is at the antenna, as shown in Figure 4.14. Then the power crossing the portion of the spherical surface of radius r_b inside the cone must be the same as the power crossing the portion of the spherical surface of radius r_a inside the cone. Since these surface areas are proportional to the square of the radius and since the surface integral of the Poynting vector gives the power, the Poynting vector must be inversely proportional to the square of the radius. This in turn means that the electric field intensity and hence the magnetic field intensity must be inversely proportional to the radius.

Thus, from these simple considerations we have established that far from a radiating antenna the electromagnetic field is inversely proportional to the radial distance away from the antenna. This reduction of the field intensity inversely proportional to the distance is known as the *free space reduction*. For example, let us consider communication from earth to the moon. The distance from the earth to the moon is approximately 38×10^4 km, or 38×10^7 m. Hence, the free space reduction factor for the field intensity is $10^{-7}/38$ or, in terms of decibels, the reduction is $20 \log_{10} 38 \times 10^7$, or 171.6 db.

Returning to the electromagnetic field due to the infinite plane current sheet, let us consider the region $z > 0$. The magnitude of the Poynting vector in this region is given by

$$P_z = E_x H_y = \eta_0 \frac{J_{S0}^2}{4} \cos^2 (\omega t - \beta z) \qquad (4.68)$$

The variation of P_z with z for $t = 0$ is shown in Figure 4.15. If we now consider a rectangluar box lying between $z = z$ and $z = z + \Delta z$ planes and having dimensions Δx and Δy in the x- and y-directions, respectively, we would in general obtain a nonzero result for the power flowing out of the box, since $\partial P_z/\partial z$ is not everywhere zero. Thus, there is some energy stored in the volume of the box. We then ask ourselves the question, "Where does this energy reside?" A convenient way of interpretation is to attribute the energy storage to the electric and magnetic fields.

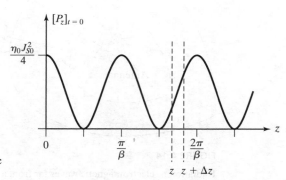

FIGURE 4.15

For the discussion of energy storage in electric and magnetic fields.

To discuss the energy storage in the electric and magnetic fields further, we evaluate the power flow out of the rectangular box. Thus,

$$\oint_S \mathbf{P} \cdot d\mathbf{S} = [P_z]_{z+\Delta z} \, \Delta x \, \Delta y - [P_z]_z \, \Delta x \, \Delta y$$

$$= \frac{[P_z]_{z+\Delta z} - [P_z]_z}{\Delta z} \, \Delta x \, \Delta y \, \Delta z$$

$$= \frac{\partial P_z}{\partial z} \, \Delta v \qquad (4.69)$$

where Δv is the volume of the box. Letting P_z equal $E_x H_y$ and using (4.7) and (4.8), we obtain

$$\oint_S \mathbf{P} \cdot d\mathbf{S} = \frac{\partial}{\partial z} [E_x H_y] \Delta v$$

$$= \left(H_y \frac{\partial E_x}{\partial z} + E_x \frac{\partial H_y}{\partial z} \right) \Delta v$$

$$= \left(-H_y \frac{\partial B_y}{\partial t} - E_x \frac{\partial D_x}{\partial t} \right) \Delta v$$

$$= -\mu_0 H_y \frac{\partial H_y}{\partial t} \Delta v - \epsilon_0 E_x \frac{\partial E_x}{\partial t} \Delta v$$

$$= -\frac{\partial}{\partial t} \left(\frac{1}{2} \mu_0 H_y^2 \Delta v \right) - \frac{\partial}{\partial t} \left(\frac{1}{2} \epsilon_0 E_x^2 \Delta v \right) \tag{4.70}$$

Equation (4.70), which is known as Poynting's theorem, tells us that the power flow out of the box is equal to the sum of the time rates of decrease of the quantities $\frac{1}{2}\epsilon_0 E_x^2 \Delta v$ and $\frac{1}{2}\mu_0 H_y^2 \Delta v$. These quantities are obviously the energies stored in the electric and magnetic fields, respectively, in the volume of the box. It then follows that the energy densities associated with the electric and magnetic fields are $\frac{1}{2}\epsilon_0 E_x^2$ and $\frac{1}{2}\mu_0 H_y^2$, respectively. It is left to the student to verify that the quantities $\frac{1}{2}\epsilon_0 E^2$ and $\frac{1}{2}\mu_0 H^2$ do indeed have the units J/m^3. Once again, although we have obtained these results by considering the particular case of the uniform plane wave, they hold in general.

Summarizing our discussion in this section, we have introduced the Poynting vector $\mathbf{P} = \mathbf{E} \times \mathbf{H}$ as the power flow density associated with the electromagnetic field characterized by the electric and magnetic fields, $\mathbf{E}$ and $\mathbf{H}$, respectively. The surface integral of $\mathbf{P}$ over a closed surface always gives the correct result for the power flow out of that surface. There is energy storage associated with the electric and magnetic fields with the energy densities given by

$$w_e = \frac{1}{2} \epsilon_0 E^2 \tag{4.71}$$

and

$$w_m = \frac{1}{2} \mu_0 H^2 \tag{4.72}$$

respectively.

SUMMARY

In this chapter, we studied the principles of uniform plane wave propagation in free space. Uniform plane waves are a building block in the study of electromagnetic wave propagation. They are the simplest type of solutions resulting from the coupling of the electric and magnetic fields in Maxwell's curl equations. We learned that uniform plane waves have their electric and magnetic fields perpendicular to each other and to the direction of propagation. The fields are *uniform* in the *planes* perpendicular to the direction of propagation.

We obtained the uniform plane wave solution to Maxwell's equations by considering an infinite plane current sheet in the xy-plane with uniform surface current density given by

$$\mathbf{J}_S = -J_{S0} \cos \omega t \, \mathbf{a}_x \, \text{A/m} \tag{4.73}$$

and deriving the electromagnetic field due to the current sheet to be given by

$$\mathbf{E} = \frac{\eta_0 J_{S0}}{2} \cos(\omega t \mp \beta z) \, \mathbf{a}_x \qquad \text{for } z \gtrless 0 \tag{4.74a}$$

$$\mathbf{H} = \pm \frac{J_{S0}}{2} \cos(\omega t \mp \beta z) \, \mathbf{a}_y \qquad \text{for } z \gtrless 0 \tag{4.74b}$$

In (4.74a) and (4.74b), $\cos(\omega t - \beta z)$ represents wave motion in the positive z-direction, whereas $\cos(\omega t + \beta z)$ represents wave motion in the negative z-direction. Thus, (4.74a) and (4.74b) correspond to waves propagating away from the current sheet to either side of it. Since the fields are independent of x and y, they represent uniform plane waves.

The quantity $\beta \, (= \omega \sqrt{\mu_0 \epsilon_0})$ is the phase constant, that is, the magnitude of the rate of change of phase with distance along the direction of propagation, for a fixed time. The phase velocity v_p, that is, the velocity with which a particular constant phase progresses along the direction of propagation, is given by

$$v_p = \frac{\omega}{\beta} \tag{4.75}$$

The wavelength λ, that is, the distance along the direction of propagation in which the phase changes by 2π radians, for a fixed time, is given by

$$\lambda = \frac{2\pi}{\beta} \tag{4.76}$$

The wavelength is related to the frequency f in a simple manner as given by

$$v_p = \lambda f \tag{4.77}$$

which follows from (4.75) and (4.76). The quantity $\eta_0 \, (= \sqrt{\mu_0/\epsilon_0})$ is the intrinsic impedance of free space. It is the ratio of the magnitude of $\mathbf{E}$ to the magnitude of $\mathbf{H}$ and has a value of $120\pi \, \Omega$.

In the process of deriving the electromagnetic field due to the infinite plane current sheet, we used two approaches and learned several useful techniques. These are discussed in the following:

1. The determination of the magnetic field adjacent to the current sheet by employing Ampere's circuital law in integral form: This is a common procedure used in the computation of static fields due to charge and current distributions possessing certain symmetries. In Chapter 5 we shall derive the *boundary conditions*, that is, the relationships between the fields on either side of an interface between two different media, by applying Maxwell's equations in integral form to closed paths and surfaces straddling the boundary as we have done here in the case of the current sheet.

2. The successive, step-by-step solution of the two Maxwell's curl equations, to obtain the final solution consistent with the two equations, starting with the solution obtained for the field adjacent to the current sheet: This technique provided us a feel for the phenomenon of *radiation* of electromagnetic waves resulting from the time-varying current distribution and the interaction between the electric and magnetic fields. We shall use this kind of approach and the knowledge gained on wave propagation to obtain in Chapter 9 the complete electromagnetic field due to an elemental antenna, which forms the basis for the study of physical antennas.

3. The solution of wave equation by the separation of variables technique: This is the standard technique employed in the solution of partial differential equations involving multiple variables.

4. The application of phasor technique for the solution of the differential equations: The phasor technique is a convenient tool for analyzing sinusoidal steady-state problems as we learned in Chapter 1.

We discussed (a) polarization of sinusoidally time-varying fields, as it pertains to uniform plane wave propagation, and (b) nonsinusoidal excitation giving rise to nonsinusoidal waves propagating in free space without change in shape, in view of phase velocity independent of frequency.

We also learned that there is power flow and energy storage associated with the wave propagation that accounts for the work done in maintaining the current flow on the sheet. The power flow density is given by the Poynting vector

$$\mathbf{P} = \mathbf{E} \times \mathbf{H}$$

and the energy densities associated with the electric and magnetic fields are given, respectively, by

$$w_e = \frac{1}{2}\,\epsilon_0 E^2$$

$$w_m = \frac{1}{2}\,\mu_0 H^2$$

The surface integral of the Poynting vector over a given closed surface gives the total power flow out of the volume bounded by that surface.

Finally, we have augmented our study of uniform plane wave propagation in free space by illustrating (a) the principle of an antenna array, and (b) the inverse distance dependence of the fields far from a physical antenna.

REVIEW QUESTIONS

4.1. What is a uniform plane wave?

4.2. Why is the study of uniform plane waves important?

4.3. How is the surface current density vector defined? Distinguish it from the volume current density vector.

4.4. How do you find the current crossing a given line on a sheet of surface current?

4.5. Why is it that Ampere's circuital law in integral form is used to find the magnetic field adjacent to the current sheet of Figure 4.2?

4.6. Why is the path chosen to evaluate the magnetic field in Figure 4.4 rectangular?

4.7. Outline the application of Ampere's circuital law in integral form to find the magnetic field adjacent to the current sheet of Figure 4.2.

4.8. Why is the displacement current enclosed by the rectangular path *abcda* in Figure 4.4 equal to zero?

4.9. How would you use Ampere's circuital law in differential form to find the magnetic field adjacent to the current sheet?

4.10. If the current density on the infinite plane current sheet of Figure 4.2 were directed in the positive y-direction, what would be the directions of the magnetic field adjacent to the current sheet and on either side of it?

4.11. Why are the results given by (4.12) and (4.13) for the magnetic field not valid for points at some distance from the current sheet?

4.12. Under what conditions would a result obtained for the magnetic field adjacent to the infinite plane current sheet of Figure 4.2 be valid at points distant from the current sheet?

4.13. Briefly outline the procedure involved in the successive solution of Maxwell's equations.

4.14. How does the technique of successive solution of Maxwell's equations reveal the interaction between the electric and magnetic fields giving rise to wave propagation?

4.15. State the wave equation for the case of $\mathbf{E} = E_x(z, t)\mathbf{a}_x$. How is it derived?

4.16. Briefly outline the separation of variables technique of solving the wave equation.

4.17. Discuss how the function $\cos(\omega t - \beta z)$ represents a traveling wave propagating in the positive z-direction.

4.18. Discuss how the function $\cos(\omega t + \beta z)$ represents a traveling wave propagating in the negative z-direction.

4.19. Discuss how the solution for the electromagnetic field given by (4.52) corresponds to that of a uniform plane wave.

4.20. Why is the quantity β in $\cos(\omega t - \beta z)$ known as the phase constant?

4.21. What is phase velocity? How is it related to the radian frequency and the phase constant of the wave?

4.22. Define wavelength. How is it related to the phase constant?

4.23. What is the relationship between frequency, wavelength, and phase velocity? What is the wavelength in free space for a frequency of 15 MHz?

4.24. What is the direction of propagation for a uniform plane wave having its electric field in the negative y-direction and its magnetic field in the positive z-direction?

4.25. What is the direction of the magnetic field for a uniform plane wave having its electric field in the positive z-direction and propagating in the positive x-direction?

4.26. What is intrinsic impedance? What is its value for free space?

4.27. Discuss the principle of an antenna array.

4.28. What should be the spacing and the relative phase angle of the current densities for an array of two infinite, plane, parallel current sheets of uniform densities, equal in magnitude, to confine their radiation to the region between the two sheets?

4.29. Discuss polarization of sinusoidally time-varying fields, as it is relevant to propagation of uniform plane waves.

4.30. Discuss the propagation of uniform plane waves arising from an infinite plane current sheet of nonsinusoidally time-varying surface current density.

4.31. Why is a certain amount of work involved in maintaining current flow on the sheet of Figure 4.2? How is this work accounted for?

4.32. What is a Poynting vector? What is its physical significance?

4.33. What is the physical interpretation of the surface integral of the Poynting vector over a closed surface?

4.34. Discuss how the fields far from a physical antenna vary inversely proportional to the distance from the antenna.

4.35. Discuss the interpretation of energy storage in the electric and magnetic fields of a uniform plane wave.

4.36. What are the energy densities associated with the electric and magnetic fields?

PROBLEMS

4.1. An infinite plane sheet lying in the $z = 0$ plane carries a current of uniform density $\mathbf{J}_S = -0.1\,\mathbf{a}_x$ A/m. Find the currents crossing the following straight lines: (a) from $(0,0,0)$ to $(0,2,0)$; (b) from $(0,0,0)$ to $(2,0,0)$; (c) from $(0,0,0)$ to $(2,2,0)$.

4.2. An infinite plane sheet lying in the $z = 0$ plane carries a current of nonuniform density $\mathbf{J}_S = -0.1e^{-|y|}\,\mathbf{a}_x$ A/m. Find the currents crossing the following straight lines: (a) from $(0,0,0)$ to $(0,1,0)$; (b) from $(0,0,0)$ to $(0,\infty,0)$; (c) from $(0,0,0)$ to $(1,1,0)$.

4.3. An infinite plane sheet lying in the $z = 0$ plane carries a current of uniform density

$$\mathbf{J}_S = (-0.1\cos\omega t\,\mathbf{a}_x + 0.1\sin\omega t\,\mathbf{a}_y)\ \text{A/m}$$

Find the currents crossing the following straight lines: (a) from $(0,0,0)$ to $(0,2,0)$; (b) from $(0,0,0)$ to $(2,0,0)$; (c) from $(0,0,0)$ to $(2,2,0)$.

4.4. An infinite plane sheet lying in the $z = 0$ plane carries a current of uniform density

$$\mathbf{J}_S = (-0.2\cos\omega t\,\mathbf{a}_x + 0.2\sin\omega t\,\mathbf{a}_y)\ \text{A/m}$$

Find the magnetic field intensities adjacent to the sheet and on either side of it. What is the polarization of the field?

4.5. An infinite plane sheet lying in the $z = 0$ plane carries a current of nonuniform density $\mathbf{J}_S = -0.2e^{-|y|}\cos\omega t\,\mathbf{a}_x$ A/m. Find the magnetic field intensities adjacent to the current sheet and on either side of it at (a) the point $(0,1,0)$ and (b) the point $(2,2,0)$.

4.6. Current flows with uniform density $\mathbf{J} = J_0\mathbf{a}_x$ A/m² in the region $|z| < a$. Using Ampere's circuital law in integral form and symmetry considerations, find $\mathbf{H}$ everywhere.

4.7. Current flows with nonuniform density $\mathbf{J} = J_0(1 - |z|/a)\mathbf{a}_x$ A/m² in the region $|z| < a$, where J_0 is a constant. Using Ampere's circuital law in integral form and symmetry considerations, find $\mathbf{H}$ everywhere.

4.8. For an infinite plane sheet of charge lying in the xy-plane with uniform surface charge density ρ_{S0} C/m², find the electric field intensity on both sides of the sheet by using Gauss' law for the electric field in integral form and symmetry considerations.

4.9. Charge is distributed with uniform density $\rho = \rho_0$ C/m³ in the region $|x| < a$. Using Gauss' law for the electric field in integral form and symmetry considerations, find $\mathbf{E}$ everywhere.

4.10. Charge is distributed with nonuniform density $\rho = \rho_0(1 - |x|/a)$ C/m³ in the region $|x| < a$, where ρ_0 is a constant. Using Gauss' law for the electric field in integral form and symmetry considerations, find $\mathbf{E}$ everywhere.

4.11. Verify that expressions (4.23) and (4.24) simultaneously satisfy the differential equations (4.16) and (4.17).

4.12. For the infinite plane current sheet in the $z = 0$ plane carrying surface current of density $\mathbf{J}_S = -J_{S0}t\, \mathbf{a}_x$ A/m, where J_{S0} is a constant, find the magnetic field adjacent to the current sheet. Then use the method of successive solution of Maxwell's equations to show that for $z > 0$,

$$E_x = \left(\frac{2C + \eta_0 J_{S0}}{4}\right)(t - z\sqrt{\mu_0\epsilon_0}) + \left(\frac{2C - \eta_0 J_{S0}}{4}\right)(t + z\sqrt{\mu_0\epsilon_0})$$

$$H_y = \left(\frac{2C + \eta_0 J_{S0}}{4\eta_0}\right)(t - z\sqrt{\mu_0\epsilon_0}) - \left(\frac{2C - \eta_0 J_{S0}}{4\eta_0}\right)(t + z\sqrt{\mu_0\epsilon_0})$$

where C is a constant.

4.13. For the infinite plane current sheet in the $z = 0$ plane carrying surface current of density $\mathbf{J}_S = -J_{S0}t^2\mathbf{a}_x$ A/m, where J_{S0} is a constant, find the magnetic field adjacent to the current sheet. Then use the method of successive solution of Maxwell's equations to show that for $z > 0$,

$$E_x = \left(\frac{2C + \eta_0 J_{S0}}{4}\right)(t - z\sqrt{\mu_0\epsilon_0})^2 + \left(\frac{2C - \eta_0 J_{S0}}{4}\right)(t + z\sqrt{\mu_0\epsilon_0})^2$$

$$H_y = \left(\frac{2C + \eta_0 J_{S0}}{4\eta_0}\right)(t - z\sqrt{\mu_0\epsilon_0})^2 - \left(\frac{2C - \eta_0 J_{S0}}{4\eta_0}\right)(t + z\sqrt{\mu_0\epsilon_0})^2$$

where C is a constant.

4.14. Verify that expressions (4.48) and (4.49) simultaneously satisfy the differential equations (4.7) and (4.8), and that (4.49) reduces to (4.12) for $z = 0+$.

4.15. Show that $(t - z\sqrt{\mu_0\epsilon_0})^2$ and $(t + z\sqrt{\mu_0\epsilon_0})^2$ are solutions of the wave equation. With the aid of sketches, discuss the nature of these functions.

4.16. For arbitrary time-variation of the fields, show that the solutions for the differential equations (4.33) and (4.34) are

$$E_x = Af(t - z\sqrt{\mu_0\epsilon_0}) + Bg(t + z\sqrt{\mu_0\epsilon_0})$$

$$H_y = \frac{1}{\eta_0}[Af(t - z\sqrt{\mu_0\epsilon_0}) - Bg(t + z\sqrt{\mu_0\epsilon_0})]$$

where A and B are arbitrary constants. Discuss the nature of the functions $f(t - z\sqrt{\mu_0\epsilon_0})$ and $g(t + z\sqrt{\mu_0\epsilon_0})$.

4.17. In Problems 4.12 and 4.13, evaluate the constant C and obtain the solutions for E_x and H_y in the region $z > 0$. Then write the solutions for E_x and H_y in the region $z < 0$.

4.18. The electric field intensity of a uniform plane wave is given by

$$\mathbf{E} = 37.7 \cos{(6\pi \times 10^8 t + 2\pi z)}\, \mathbf{a}_y \text{ V/m}.$$

Find (a) the frequency, (b) the wavelength, (c) the phase velocity, (d) the direction of propagation of the wave, and (e) the associated magnetic field intensity vector $\mathbf{H}$.

4.19. An infinite plane sheet lying in the $z = 0$ plane carries a surface current of density

$$\mathbf{J}_S = (-0.2 \cos 6\pi \times 10^8 t\, \mathbf{a}_x - 0.1 \cos 12\pi \times 10^8 t\, \mathbf{a}_x) \text{ A/m}$$

Find the expressions for the electric and magnetic fields on either side of the sheet.

4.20. An array is formed by two infinite plane parallel current sheets with the current densities given by

$$\mathbf{J}_{S1} = -J_{S0} \cos \omega t \, \mathbf{a}_x \qquad z = 0$$

$$\mathbf{J}_{S2} = -J_{S0} \cos \omega t \, \mathbf{a}_x \qquad z = \frac{\lambda}{2}$$

where J_{S0} is a constant. Find the electric field intensity in all three regions: (a) $z < 0$; (b) $0 < z < \lambda/2$; (c) $z > \lambda/2$.

4.21. Determine the spacing, relative amplitudes, and phase angles of current densities for an array of two infinite plane parallel current sheets required to obtain a radiation characteristic such that the field radiated to one side of the array is twice that of the field radiated to the other side of the array.

4.22. For two infinite plane parallel current sheets with the current densities given by

$$\mathbf{J}_{S1} = -J_{S0} \cos \omega t \, \mathbf{a}_x \qquad z = 0$$

$$\mathbf{J}_{S2} = -J_{S0} \cos \omega t \, \mathbf{a}_y \qquad z = \frac{\lambda}{2}$$

where J_{S0} is a constant, find the electric field in all three regions: (a) $z < 0$; (b) $0 < z < \lambda/2$; (c) $z > \lambda/2$. Discuss the polarization of the field in all three regions.

4.23. For each of the following fields, determine if the polarization is right- or left-circular.
(a) $E_0 \cos (\omega t - \beta y) \, \mathbf{a}_z + E_0 \sin (\omega t - \beta y) \, \mathbf{a}_x$
(b) $E_0 \cos (\omega t + \beta x) \, \mathbf{a}_y + E_0 \sin (\omega t + \beta x) \, \mathbf{a}_z$

4.24. For each of the following fields, determine if the polarization is right- or left-elliptical.
(a) $E_0 \cos (\omega t + \beta y) \, \mathbf{a}_x - 2E_0 \sin (\omega t + \beta y) \, \mathbf{a}_z$
(b) $E_0 \cos (\omega t - \beta x) \, \mathbf{a}_z - E_0 \sin (\omega t - \beta x + \pi/4) \, \mathbf{a}_y$

4.25. Express the following uniform plane wave electric field as a superposition of right- and left-circularly polarized fields: $E_0 \mathbf{a}_x \cos (\omega t + \beta z)$

4.26. Repeat Problem 4.25 for the following electric field: $E_0 \mathbf{a}_x \cos (\omega t - \beta z + \pi/3) - E_0 \mathbf{a}_y \cos (\omega t - \beta z + \pi/6)$

4.27. Write the expression for the electric field intensity of a sinusoidally time-varying uniform plane wave propagating in free space and having the following characteristics: (a) $f = 100$ MHz; (b) direction of propagation is the $+z$-direction; and (c) polarization is right circular with the electric field in the $z = 0$ plane at $t = 0$ having an x-component equal to E_0 and a y-component equal to $0.75E_0$.

4.28. An infinite plane sheet lying in the $z = 0$ plane carries a surface current of density $\mathbf{J}_S = -J_S(t) \mathbf{a}_x$, where $J_S(t)$ is the periodic function shown in Figure 4.16. Find and sketch (a) H_y versus t for $z = 0+$, (b) E_x versus t for $z = 150$ m, and (c) E_x versus z for $t = 1 \, \mu s$.

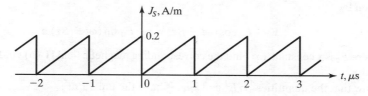

FIGURE 4.16

For Problem 4.28.

4.29. The time-variation of the electric field intensity E_x in the $z = 600$ m plane of a uniform plane wave propagating away from an infinite plane current sheet lying in the $z = 0$ plane is given by the periodic function shown in Figure 4.17. Find and sketch (a) E_x versus t for $z = 200$ m, (b) E_x versus z for $t = 0$, and (c) H_y versus z for $t = \frac{1}{3}$ μs.

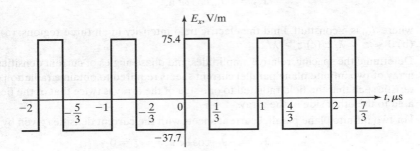

FIGURE 4.17

For Problem 4.29.

4.30. The time-variation of the electric field intensity E_x in the $z = 300$ m plane of a uniform plane wave propagating away from an infinite plane current sheet lying in the $z = 0$ plane is given by the aperiodic function shown in Figure 4.18. Find and sketch (a) E_x versus t for $z = 600$ m, (b) E_x versus z for $t = 1$ μs, and (c) H_y versus z for $t = 2$ μs.

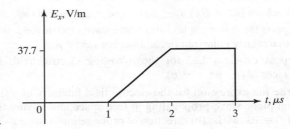

FIGURE 4.18

For Problem 4.30.

4.31. Show that the time-average value of the magnitude of the Poynting vector given by (4.68) is one-half its peak value. For an antenna radiating a time-average power of 150 kW, find the peak value of the electric field intensity at a distance of 100 km from the antenna. Assume the antenna to be radiating equally in all directions.

4.32. The electric field of a uniform plane wave propagating in the positive z-direction is given by

$$\mathbf{E} = E_0 \cos{(\omega t - \beta z)} \, \mathbf{a}_x + E_0 \sin{(\omega t - \beta z)} \, \mathbf{a}_y$$

where E_0 is a constant. (a) Find the corresponding magnetic field $\mathbf{H}$. (b) Find the Poynting vector.

4.33. Show that the quantities $\frac{1}{2}\epsilon_0 E^2$ and $\frac{1}{2}\mu_0 H^2$ have the units J/m³.

4.34. Show that the energy is stored equally in the electric and magnetic fields of a traveling wave.

Wave Propagation in Material Media

In Chapter 4, we introduced wave propagation in free space by considering the infinite plane current sheet of uniform, sinusoidally time-varying current density. We learned that the solution for the electromagnetic field due to the infinite plane current sheet represents uniform plane electromagnetic waves propagating away from the sheet to either side of it. With the knowledge of the principles of uniform plane wave propagation in free space, we are now ready to consider wave propagation in material media, which is our goal in this chapter. Materials contain charged particles that respond to applied electric and magnetic fields and give rise to currents, which modify the properties of wave propagation from those associated with free space.

We shall learn that there are three basic phenomena resulting from the interaction of the charged particles with the electric and magnetic fields. These are conduction, polarization, and magnetization. Although a given material may exhibit all three properties, it is classified as a conductor, a dielectric, or a magnetic material, depending on whether conduction, polarization, or magnetization is the predominant phenomenon. Thus, we shall introduce these three kinds of materials one at a time and develop a set of relations known as the constitutive relations that enable us to avoid the necessity of explicitly taking into account the interaction of the charged particles with the fields. We shall then use these constitutive relations together with Maxwell's equations to first discuss uniform plane wave propagation in a general material medium and then consider several special cases. Finally, we shall derive the *boundary conditions* and use them to study reflection and transmission of uniform plane waves at plane boundaries.

5.1 CONDUCTORS AND DIELECTRICS

We recall that the classical model of an atom postulates a tightly bound, positively charged nucleus surrounded by a diffuse cloud of electrons spinning and orbiting around the nucleus. In the absence of an applied electromagnetic field, the force of attraction between the positively charged nucleus and the negatively charged electrons is balanced by the outward centrifugal force to maintain stable electronic orbits. The electrons can be divided into *bound* electrons and *free* or *conduction* electrons.

The bound electrons can be displaced but not removed from the influence of the nucleus. The conduction electrons are constantly under thermal agitation, being released from the parent atom at one point and recaptured by another atom at a different point.

In the absence of an applied field, the motion of the conduction electrons is completely random; the average thermal velocity on a *macroscopic* scale, that is, over volumes large compared with atomic dimensions, is zero so that there is no net current and the electron cloud maintains a fixed position. With the application of an electromagnetic field, an additional velocity is superimposed on the random velocities, predominantly due to the electric force. This causes drift of the average position of the electrons in a direction opposite to that of the applied electric field. Due to the frictional mechanism provided by collisions of the electrons with the atomic lattice, the electrons, instead of accelerating under the influence of the electric field, drift with an average drift velocity proportional in magnitude to the applied electric field. This phenomenon is known as *conduction*, and the resulting current due to the electron drift is known as the *conduction current*.

In certain materials a large number of electrons may take part in the conduction process, but in certain other materials only a very few or negligible number of electrons may participate in conduction. The former class of materials is known as *conductors*, and the latter class is known as *dielectrics* or *insulators*. If the number of free electrons participating in conduction is N_e per cubic meter of the material, then the conduction current density is given by

$$\mathbf{J}_c = N_e e \mathbf{v}_d \tag{5.1}$$

where e is the charge of an electron, and $\mathbf{v}_d$ is the drift velocity of the electrons. The drift velocity varies from one conductor to another, depending on the average time between successive collisions of the electrons with the atomic lattice. It is related to the applied electric field in the manner

$$\mathbf{v}_d = -\mu_e \mathbf{E} \tag{5.2}$$

where μ_e is known as the *mobility* of the electron. Substituting (5.2) into (5.1), we obtain

$$\mathbf{J}_c = -\mu_e N_e e \mathbf{E} = \mu_e N_e |e| \mathbf{E} \tag{5.3}$$

Semiconductors are characterized by drift of *holes*, that is, vacancies created by detachment of electrons from covalent bonds, in addition to the drift of electrons. If N_e and N_h are the number of electrons and holes, respectively, per cubic meter of the material, and if μ_e and μ_h are the electron and hole mobilities, respectively, then the conduction current density in the semiconductor is given by

$$\mathbf{J}_c = (\mu_e N_e |e| + \mu_h N_h |e|)\mathbf{E} \tag{5.4}$$

Defining a quantity σ, known as the *conductivity* of the material, as given by

$$\sigma = \begin{cases} \mu_e N_e |e| & \text{for conductors} \\ \mu_e N_e |e| + \mu_h N_h |e| & \text{for semiconductors} \end{cases} \tag{5.5}$$

we obtain the simple and important relationship

$$\mathbf{J}_c = \sigma\mathbf{E} \tag{5.6}$$

for the conduction current density in a material. Equation (5.6) is known as Ohm's law applicable at a point from which follows the familiar form of Ohm's law used in circuit theory. The units of σ are siemens/meter where a siemen (S) is an ampere per volt. Values of σ for a few materials are listed in Table 5.1. In considering electromagnetic wave propagation in conducting media, the conduction current density given by (5.6) must be employed for the current density term on the right side of Ampere's circuital law. Thus, Maxwell's curl equation for $\mathbf{H}$ for a conducting medium is given by

$$\nabla \times \mathbf{H} = \mathbf{J}_c + \frac{\partial \mathbf{D}}{\partial t} = \sigma\mathbf{E} + \frac{\partial \mathbf{D}}{\partial t} \tag{5.7}$$

TABLE 5.1 Conductivities of Some Materials

Material	Conductivity S/m	Material	Conductivity S/m
Silver	6.1×10^7	Sea water	4
Copper	5.8×10^7	Intrinsic germanium	2.2
Gold	4.1×10^7	Intrinsic silicon	1.6×10^{-3}
Aluminum	3.5×10^7	Fresh water	10^{-3}
Tungsten	1.8×10^7	Distilled water	2×10^{-4}
Brass	1.5×10^7	Dry earth	10^{-5}
Solder	7.0×10^6	Bakelite	10^{-9}
Lead	4.8×10^6	Glass	$10^{-10} - 10^{-14}$
Constantin	2.0×10^6	Mica	$10^{-11} - 10^{-15}$
Mercury	1.0×10^6	Fused quartz	0.4×10^{-17}

While conductors are characterized by abundance of *conduction* or *free* electrons that give rise to conduction current under the influence of an applied electric field, in dielectric materials the *bound* electrons are predominant. Under the application of an external electric field, the bound electrons of an atom are displaced such that the centroid of the electron cloud is separated from the centroid of the nucleus. The atom is then said to be *polarized*, thereby creating an *electric dipole*, as shown in Figure 5.1(a). This kind of polarization is called *electronic polarization*. The schematic representation of an electric dipole is shown in Figure 5.1(b). The strength of the dipole is defined by the electric dipole moment $\mathbf{p}$ given by

$$\mathbf{p} = Q\mathbf{d} \tag{5.8}$$

where $\mathbf{d}$ is the vector displacement between the centroids of the positive and negative charges, each of magnitude Q.

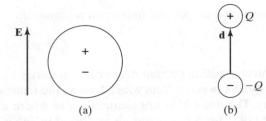

FIGURE 5.1

(a) An electric dipole. (b) Schematic representation of an electric dipole.

In certain dielectric materials, polarization may exist in the molecular structure of the material even under the application of no external electric field. The polarization of individual atoms and molecules, however, is randomly oriented, and hence the net polarization on a *macroscopic* scale is zero. The application of an external field results in torques acting on the *microscopic* dipoles, as shown in Figure 5.2, to convert the initially random polarization into a partially coherent one along the field, on a macroscopic scale. This kind of polarization is known as *orientational polarization.* A third kind of polarization, known as *ionic polarization*, results from the separation of positive and negative ions in molecules formed by the transfer of electrons from one atom to another in the molecule. Certain materials exhibit permanent polarization, that is, polarization even in the absence of an applied electric field. Electrets, when allowed to solidify in the applied electric field, become permanently polarized, and ferroelectric materials exhibit spontaneous, permanent polarization.

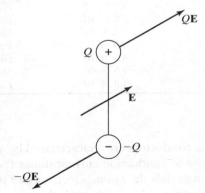

FIGURE 5.2

Torque acting on an electric dipole in an external electric field.

On a macroscopic scale, we define a vector **P**, called the *polarization vector*, as the *electric dipole moment per unit volume.* Thus, if N denotes the number of molecules per unit volume of the material, then there are $N \Delta v$ molecules in a volume Δv and

$$\mathbf{P} = \frac{1}{\Delta v} \sum_{j=1}^{N \Delta v} \mathbf{p}_j = N\mathbf{p} \tag{5.9}$$

where **p** is the average dipole moment per molecule. The units of **P** are coulomb-meter/meter^3 or coulombs per square meter. It is found that for many dielectric materials the polarization vector is related to the electric field **E** in the dielectric in the simple manner given by

$$\mathbf{P} = \epsilon_0 \chi_e \mathbf{E} \tag{5.10}$$

where χ_e, a dimensionless parameter, is known as the *electric susceptibility*. The quantity χ_e is a measure of the ability of the material to become polarized and differs from one dielectric to another.

To discuss the influence of polarization in the dielectric upon electromagnetic wave propagation in the dielectric medium, let us consider the case of the infinite plane current sheet of Figure 4.8, radiating uniform plane waves, except that now the space on either side of the current sheet is a dielectric medium instead of being free space. The electric field in the medium induces polarization. The polarization in turn acts together with other factors to govern the behavior of the electromagnetic field. For the case under consideration, the electric field is entirely in the x-direction and uniform in x and y. Thus, the induced electric dipoles are all oriented in the x-direction, on a macroscopic scale, with the dipole moment per unit volume given by

$$\mathbf{P} = P_x \mathbf{a}_x = \epsilon_0 \chi_e E_x \mathbf{a}_x \tag{5.11}$$

where E_x is understood to be a function of z and t.

If we now consider an infinitesimal surface of area $\Delta y \, \Delta z$ parallel to the yz-plane, we can write E_x associated with that infinitesimal area to be equal to $E_0 \cos \omega t$ where E_0 is a constant. The time history of the induced dipoles associated with that area can be sketched for one complete period of the current source, as shown in Figure 5.3. In view of the cosinusoidal variation of the electric field with time, the dipole moment of the individual dipoles varies in a cosinusoidal manner with maximum strength in the positive x-direction at $t = 0$, decreasing sinusoidally to zero strength at $t = \pi/2\omega$ and then reversing to the negative x-direction, increasing to maximum strength in that direction at $t = \pi/\omega$, and so on.

The arrangement can be considered as two plane sheets of equal and opposite time-varying charges displaced by the amount δ in the x-direction, as shown in Figure 5.4. To find the magnitude of either charge, we note that the dipole moment per unit volume is

$$P_x = \epsilon_0 \chi_e E_0 \cos \omega t \tag{5.12}$$

Since the total volume occupied by the dipoles is $\delta \, \Delta y \, \Delta z$, the total dipole moment associated with the dipoles is $\epsilon_0 \chi_e E_0 \cos \omega t \, (\delta \, \Delta y \, \Delta z)$. The dipole moment associated with two equal and opposite sheet charges is equal to the magnitude of either sheet charge multiplied by the displacement between the two sheets. Hence, we obtain the magnitude of either sheet charge to be $|\epsilon_0 \chi_e E_0 \cos \omega t \, \Delta y \, \Delta z|$. Thus, we have a situation in which a sheet charge $Q_1 = \epsilon_0 \chi_e E_0 \cos \omega t \, \Delta y \, \Delta z$ is above the surface and a sheet charge $Q_2 = -Q_1 = -\epsilon_0 \chi_e E_0 \cos \omega t \, \Delta y \, \Delta z$ is below the surface. This is equivalent to a current flowing across the surface, since the charges are varying with time.

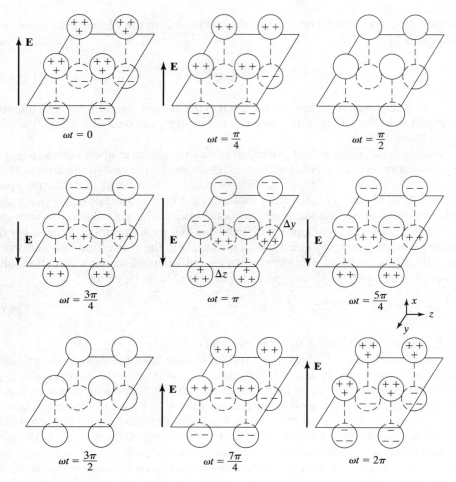

FIGURE 5.3

Time history of induced electric dipoles in a dielectric material under the influence of a sinusoidally time-varying electric field.

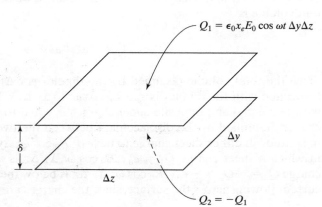

FIGURE 5.4

Two plane sheets of equal and opposite time-varying charges equivalent to the phenomenon depicted in Figure 5.3.

We call this current the *polarization current*, since it results from the time variation of the electric dipole moments induced in the dielectric due to polarization. The polarization current crossing the surface in the positive x-direction, that is, from below to above, is

$$I_{px} = \frac{dQ_1}{dt} = -\epsilon_0\chi_e E_0\omega \sin \omega t \, \Delta y \, \Delta z \qquad (5.13)$$

where the subscript p denotes polarization. By dividing I_{px} by $\Delta y \, \Delta z$ and letting the area tend to zero, we obtain the polarization current density associated with the points on the surface as

$$J_{px} = \lim_{\substack{\Delta y \to 0 \\ \Delta z \to 0}} \frac{I_{px}}{\Delta y \, \Delta z} = -\epsilon_0\chi_e E_0\omega \sin \omega t$$

$$= \frac{\partial}{\partial t}(\epsilon_0\chi_e E_0 \cos \omega t) = \frac{\partial P_x}{\partial t} \qquad (5.14)$$

or

$$\mathbf{J}_p = \frac{\partial \mathbf{P}}{\partial t} \qquad (5.15)$$

Although we have deduced this result by considering the special case of the infinite plane current sheet, it is valid in general.

In considering electromagnetic wave propagation in a dielectric medium, the polarization current density given by (5.15) must be included with the current density term on the right side of Ampere's circuital law. Thus, considering Ampere's circuital law in differential form for the general case given by (3.28), we have

$$\nabla \times \mathbf{H} = \mathbf{J} + \mathbf{J}_p + \frac{\partial}{\partial t}(\epsilon_0\mathbf{E}) \qquad (5.16)$$

Substituting (5.15) into (5.16), we get

$$\nabla \times \mathbf{H} = \mathbf{J} + \frac{\partial \mathbf{P}}{\partial t} + \frac{\partial}{\partial t}(\epsilon_0\mathbf{E})$$

$$= \mathbf{J} + \frac{\partial}{\partial t}(\epsilon_0\mathbf{E} + \mathbf{P}) \qquad (5.17)$$

In order to make (5.17) consistent with the corresponding equation for free space given by (3.28), we now revise the definition of the displacement vector $\mathbf{D}$ to read as

$$\mathbf{D} = \epsilon_0\mathbf{E} + \mathbf{P} \qquad (5.18)$$

Substituting for $\mathbf{P}$ by using (5.10), we obtain

$$\mathbf{D} = \epsilon_0\mathbf{E} + \epsilon_0\chi_e\mathbf{E}$$
$$= \epsilon_0(1 + \chi_e)\mathbf{E}$$
$$= \epsilon_0\epsilon_r\mathbf{E}$$
$$= \epsilon\mathbf{E} \qquad (5.19)$$

where we define

$$\epsilon_r = 1 + \chi_e \qquad (5.20)$$

and

$$\epsilon = \epsilon_0 \epsilon_r \tag{5.21}$$

The quantity ϵ_r is known as the *relative permittivity* or *dielectric constant* of the dielectric, and ϵ is the *permittivity* of the dielectric. The new definition for **D** permits the use of the same Maxwell's equations as for free space with ϵ_0 replaced by ϵ and without the need for explicitly considering the polarization current density. The permittivity ϵ takes into account the effects of polarization, and there is no need to consider them when we use ϵ for ϵ_0! The relative permittivity is an experimentally measurable parameter and its values for several dielectric materials are listed in Table 5.2.

TABLE 5.2   Relative Permittivities of Some Materials

Material	Relative Permittivity	Material	Relative Permittivity
Air	1.0006	Dry earth	5
Paper	2.0–3.0	Mica	6
Teflon	2.1	Neoprene	6.7
Polystyrene	2.56	Wet earth	10
Plexiglass	2.6–3.5	Ethyl alcohol	24.3
Nylon	3.5	Glycerol	42.5
Fused quartz	3.8	Distilled water	81
Bakelite	4.9	Titanium dioxide	100

Equation (5.19) governs the relationship between **D** and **E** for dielectric materials. Dielectrics for which ϵ is independent of the magnitude as well as the direction of **E** as indicated by (5.19) are known as *linear isotropic dielectrics*. For certain dielectric materials, each component of the polarization vector can be dependent on all components of the electric field intensity. For such materials, known as *anisotropic dielectric materials*, **D** is not in general parallel to **E**, and the relationship between these two quantities is expressed in the form of a matrix equation, as given by

$$\begin{bmatrix} D_x \\ D_y \\ D_z \end{bmatrix} = \begin{bmatrix} \epsilon_{xx} & \epsilon_{xy} & \epsilon_{xz} \\ \epsilon_{yx} & \epsilon_{yy} & \epsilon_{yz} \\ \epsilon_{zx} & \epsilon_{zy} & \epsilon_{zz} \end{bmatrix} \begin{bmatrix} E_x \\ E_y \\ E_z \end{bmatrix} \tag{5.22}$$

The square matrix in (5.22) is known as the *permittivity tensor* of the anisotropic dielectric.

Example 5.1

An anisotropic dielectric material is characterized by the permittivity tensor

$$[\epsilon] = \begin{bmatrix} 7\epsilon_0 & 2\epsilon_0 & 0 \\ 2\epsilon_0 & 4\epsilon_0 & 0 \\ 0 & 0 & 3\epsilon_0 \end{bmatrix}$$

Let us find **D** for several cases of **E**.

Substituting the given permittivity matrix in (5.22), we obtain

$$D_x = 7\epsilon_0 E_x + 2\epsilon_0 E_y$$

$$D_y = 2\epsilon_0 E_x + 4\epsilon_0 E_y$$

$$D_z = 3\epsilon_0 E_z$$

For $\mathbf{E} = E_0 \cos \omega t \, \mathbf{a}_z$, $\mathbf{D} = 3\epsilon_0 E_0 \cos \omega t \, \mathbf{a}_z$; $\mathbf{D}$ is parallel to $\mathbf{E}$.

For $\mathbf{E} = E_0 \cos \omega t \, \mathbf{a}_x$, $\mathbf{D} = 7\epsilon_0 E_0 \cos \omega t \, \mathbf{a}_x + 2\epsilon_0 E_0 \cos \omega t \, \mathbf{a}_y$; $\mathbf{D}$ is not parallel to $\mathbf{E}$.

For $\mathbf{E} = E_0 \cos \omega t \, \mathbf{a}_y$, $\mathbf{D} = 2\epsilon_0 E_0 \cos \omega t \, \mathbf{a}_x + 4\epsilon_0 E_0 \cos \omega t \, \mathbf{a}_y$; $\mathbf{D}$ is not parallel to $\mathbf{E}$.

For $\mathbf{E} = E_0 \cos \omega t \, (\mathbf{a}_x + 2\mathbf{a}_y)$, $\mathbf{D} = 11\epsilon_0 E_0 \cos \omega t \, \mathbf{a}_x + 10\epsilon_0 E_0 \cos \omega t \, \mathbf{a}_y$; $\mathbf{D}$ is not parallel to $\mathbf{E}$.

For $\mathbf{E} = E_0 \cos \omega t \, (2\mathbf{a}_x + \mathbf{a}_y)$, $\mathbf{D} = 16\epsilon_0 E_0 \cos \omega t \, \mathbf{a}_x + 8\epsilon_0 E_0 \cos \omega t \, \mathbf{a}_y = 8\epsilon_0 \mathbf{E}$; $\mathbf{D}$ is parallel to $\mathbf{E}$ and the dielectric behaves *effectively* in the same manner as an isotropic dielectric having the permittivity $8\epsilon_0$; that is, the *effective permittivity* of the anisotropic dielectric for this case is $8\epsilon_0$.

Thus, we find that in general $\mathbf{D}$ is not parallel to $\mathbf{E}$ but for certain polarizations of $\mathbf{E}$, $\mathbf{D}$ is parallel to $\mathbf{E}$. These polarizations are known as the characteristic polarizations.

5.2 MAGNETIC MATERIALS

The important characteristic of magnetic materials is *magnetization*. Magnetization is the phenomenon by means of which the orbital and spin motions of electrons are influenced by an external magnetic field. An electronic orbit is equivalent to a current loop, which is the magnetic analog of an electric dipole. The schematic representation of a magnetic dipole as seen from along its axis and from a point in its plane are shown in Figures 5.5(a) and 5.5(b), respectively. The strength of the dipole is defined by the magnetic dipole moment $\mathbf{m}$ given by

$$\mathbf{m} = I A \mathbf{a}_n \qquad (5.23)$$

where A is the area enclosed by the current loop and $\mathbf{a}_n$ is the unit vector normal to the plane of the loop and directed in the right-hand sense.

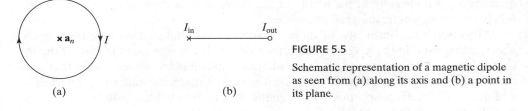

(a)

(b)

FIGURE 5.5

Schematic representation of a magnetic dipole as seen from (a) along its axis and (b) a point in its plane.

In many materials, the net magnetic moment of each atom is zero, that is, on the average, the magnetic dipole moments corresponding to the various electronic orbital and spin motions add up to zero. An external magnetic field has the effect of inducing a net dipole moment by changing the angular velocities of the electronic orbits, thereby magnetizing the material. This kind of magnetization, known as *diamagnetism*, is in fact prevalent in all materials. In certain materials known as *paramagnetic materials*, the individual atoms possess net nonzero magnetic moments even in the absence of an external magnetic field. These *permanent* magnetic moments of the individual atoms are, however, randomly oriented so that the net magnetization on a macroscoٟ.ic scale is zero. An applied magnetic field has the effect of exerting torques on the individual permanent dipoles, as shown in Figure 5.6, to convert, on a macroscopic scale, the initially random alignment into a partially coherent one along the magnetic field, that is, with the normal to the current loop directed along the magnetic field. This kind of magnetization is known as *paramagnetism*. Certain materials known as *ferromagnetic*, *antiferromagnetic*, and *ferrimagnetic* materials exhibit permanent magnetization, that is, magnetization even in the absence of an applied magnetic field.

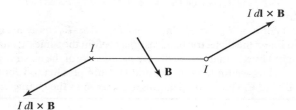

FIGURE 5.6

Torque acting on a magnetic dipole
in an external magnetic field.

On a macroscopic scale, we define a vector **M**, called the *magnetization vector*, as the *magnetic dipole moment per unit volume*. Thus, if N denotes the number of molecules per unit volume of the material, then there are $N \Delta v$ molecules in a volume Δv and

$$\mathbf{M} = \frac{1}{\Delta v} \sum_{j=1}^{N \Delta v} \mathbf{m}_j = N\mathbf{m} \qquad (5.24)$$

where **m** is the average dipole moment per molecule. The units of **M** are ampere-meter^2/meter^3 or amperes per meter. It is found that for many magnetic materials, the magnetization vector is related to the magnetic field **B** in the material in the simple manner given by

$$\mathbf{M} = \frac{\chi_m}{1 + \chi_m} \frac{\mathbf{B}}{\mu_0} \qquad (5.25)$$

where χ_m, a dimensionless parameter, is known as the *magnetic susceptibility*. The quantity χ_m is a measure of the ability of the material to become magnetized and differs from one magnetic material to another.

To discuss the influence of magnetization in the material on electromagnetic wave propagation in the magnetic material medium, let us consider the case of the infinite plane current sheet of Figure 4.8, radiating uniform plane waves, except that now the space on either side of the current sheet possesses magnetic material properties in addition to dielectric properties. The magnetic field in the medium induces magnetization. The magnetization in turn acts together with other factors to govern the behavior of the electromagnetic field. For the case under consideration, the magnetic field is entirely in the y-direction and uniform in x and y. Thus, the induced dipoles are all oriented with their axes in the y-direction, on a macroscopic scale, with the dipole moment per unit volume given by

$$\mathbf{M} = M_y \mathbf{a}_y = \frac{\chi_m}{1 + \chi_m} \frac{B_y}{\mu_0} \mathbf{a}_y \qquad (5.26)$$

where B_y is understood to be a function of z and t.

Let us now consider an infinitesimal surface of area $\Delta y \, \Delta z$ parallel to the yz-plane and the magnetic dipoles associated with the two areas $\Delta y \, \Delta z$ to the left and to the right of the center of this area, as shown in Figure 5.7(a). Since B_y is a function of z, we can assume the dipoles in the left area to have a different moment than the dipoles in the right area for any given time. If the dimension of an individual dipole is δ in the x-direction, then the total dipole moment associated with the dipoles in the left area is $[M_y]_{z-\Delta z/2} \, \delta \, \Delta y \, \Delta z$ and the total dipole moment associated with the dipoles in the right area is $[M_y]_{z+\Delta z/2} \, \delta \, \Delta y \, \Delta z$.

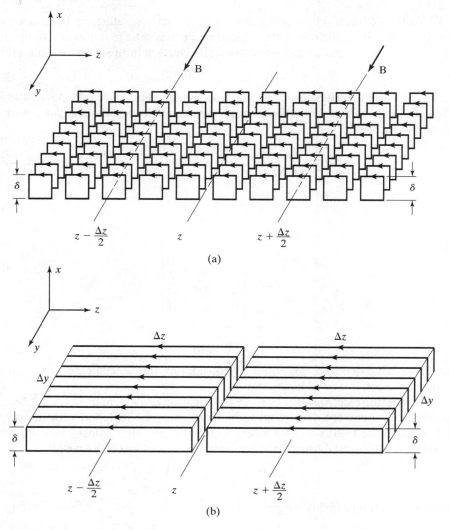

FIGURE 5.7

(a) Induced magnetic dipoles in a magnetic material. (b) Equivalent surface current loops.

The arrangement of dipoles can be considered to be equivalent to two rectangular surface current loops, as shown in Figure 5.7(b), with the left side current loop having a dipole moment $[M_y]_{z-\Delta z/2} \, \delta \, \Delta y \, \Delta z$ and the right side current loop having a dipole moment $[M_y]_{z+\Delta z/2} \, \delta \, \Delta y \, \Delta z$. Since the magnetic dipole moment of a rectangular surface current loop is simply equal to the product of the surface current and the cross-sectional area of the loop, the surface current associated with the left loop is $[M_y]_{z-\Delta z/2} \, \Delta y$ and the surface current associated with the right loop is $[M_y]_{z+\Delta z/2} \, \Delta y$. Thus, we have a situation in which a current equal to $[M_y]_{z-\Delta z/2} \, \Delta y$ is crossing the area $\Delta y \, \Delta z$ in the positive x-direction, and a current equal to $[M_y]_{z+\Delta z/2} \, \Delta y$ is crossing the same area in the negative x-direction. This is equivalent to a net current flowing across the surface.

We call this current the *magnetization current* since it results from the space variation of the magnetic dipole moments induced in the magnetic material due to magnetization. The net magnetization current crossing the surface in the positive x-direction is

$$I_{mx} = [M_y]_{z-\Delta z/2} \, \Delta y - [M_y]_{z+\Delta z/2} \, \Delta y \tag{5.27}$$

where the subscript m denotes magnetization. By dividing I_{mx} by $\Delta y \, \Delta z$ and letting the area tend to zero, we obtain the magnetization current density associated with the points on the surface as

$$J_{mx} = \lim_{\substack{\Delta y \to 0 \\ \Delta z \to 0}} \frac{I_{mx}}{\Delta y \, \Delta z} = \lim_{\Delta z \to 0} \frac{[M_y]_{z-\Delta z/2} - [M_y]_{z+\Delta z/2}}{\Delta z}$$

$$= -\frac{\partial M_y}{\partial z} \tag{5.28}$$

or

$$J_{mx}\mathbf{a}_x = \begin{vmatrix} \mathbf{a}_x & \mathbf{a}_y & \mathbf{a}_z \\ \dfrac{\partial}{\partial x} & \dfrac{\partial}{\partial y} & \dfrac{\partial}{\partial z} \\ 0 & M_y & 0 \end{vmatrix}$$

or

$$\mathbf{J}_m = \nabla \times \mathbf{M} \tag{5.29}$$

Although we have deduced this result by considering the special case of the infinite plane current sheet, it is valid in general.

In considering electromagnetic wave propagation in a magnetic material medium, the magnetization current density given by (5.29) must be included with the current density term on the right side of Ampere's circuital law. Thus, considering Ampere's circuital law in differential form for the general case given by (3.28), we have

$$\nabla \times \frac{\mathbf{B}}{\mu_0} = \mathbf{J} + \mathbf{J}_m + \frac{\partial \mathbf{D}}{\partial t} \tag{5.30}$$

Substituting (5.29) into (5.30), we get

$$\nabla \times \frac{\mathbf{B}}{\mu_0} = \mathbf{J} + \nabla \times \mathbf{M} + \frac{\partial \mathbf{D}}{\partial t}$$

or

$$\nabla \times \left(\frac{\mathbf{B}}{\mu_0} - \mathbf{M} \right) = \mathbf{J} + \frac{\partial \mathbf{D}}{\partial t} \tag{5.31}$$

In order to make (5.31) consistent with the corresponding equation for free space given by (3.28), we now revise the definition of the magnetic field intensity vector $\mathbf{H}$ to read as

$$\mathbf{H} = \frac{\mathbf{B}}{\mu_0} - \mathbf{M} \tag{5.32}$$

Substituting for $\mathbf{M}$ by using (5.25), we obtain

$$
\begin{aligned}
\mathbf{H} &= \frac{\mathbf{B}}{\mu_0} - \frac{\chi_m}{1 + \chi_m} \frac{\mathbf{B}}{\mu_0} \\
&= \frac{\mathbf{B}}{\mu_0(1 + \chi_m)} \\
&= \frac{\mathbf{B}}{\mu_0 \mu_r} \\
&= \frac{\mathbf{B}}{\mu}
\end{aligned}
\tag{5.33}
$$

where we define

$$\mu_r = 1 + \chi_m \tag{5.34}$$

and

$$\mu = \mu_0 \mu_r \tag{5.35}$$

The quantity μ_r is known as the *relative permeability* of the magnetic material, and μ is the *permeability* of the magnetic material. The new definition for $\mathbf{H}$ permits the use of the same Maxwell's equations as for free space with μ_0 replaced by μ and without the need for explicitly considering the magnetization current density. The permeability μ takes into account the effects of magnetization, and there is no need to consider them when we use μ for μ_0! For anisotropic magnetic materials, $\mathbf{H}$ is not in general parallel to $\mathbf{B}$ and the relationship between the two quantities is expressed in the form of a matrix equation, as given by

$$
\begin{bmatrix} B_x \\ B_y \\ B_z \end{bmatrix} = \begin{bmatrix} \mu_{xx} & \mu_{xy} & \mu_{xz} \\ \mu_{yx} & \mu_{yy} & \mu_{yz} \\ \mu_{zx} & \mu_{zy} & \mu_{zz} \end{bmatrix} \begin{bmatrix} H_x \\ H_y \\ H_z \end{bmatrix}
\tag{5.36}
$$

just as in the case of the relationship between $\mathbf{D}$ and $\mathbf{E}$ for anisotropic dielectric materials.

For many materials for which the relationship between $\mathbf{H}$ and $\mathbf{B}$ is linear, the relative permeability does not differ appreciably from unity, unlike the case of linear dielectric materials, for which the relative permittivity can be very large, as shown in Table 5.2. In fact, for diamagnetic materials, the magnetic susceptibility χ_m is a small negative number of the order -10^{-4} to -10^{-8}, whereas for paramagnetic materials, χ_m is a small positive number of the order 10^{-3} to 10^{-7}. Ferromagnetic materials, however, possess large values of relative permeability on the order of several hundreds, thousands, or more. The relationship between **B** and **H** for these materials is nonlinear, resulting in a nonunique value of μ_r for a given material. In fact, these materials are characterized by hysteresis, that is, the relationship between **B** and **H** dependent on the past history of the material.

A typical curve of B versus H, known as the *B–H curve* or the *hysteresis curve* for a ferromagnetic material, is shown in Figure 5.8. If we start with an unmagnetized sample of the material in which both B and H are initially zero, corresponding to point a in Figure 5.8, and then magnetize the material, the manner in which magnetization is built up initially to saturation is given by the portion ab of the curve. If the magnetization is now decreased gradually and then reversed in polarity, the curve does not retrace ab backward but instead follows along bcd until saturation is reached in the opposite direction at point e. A decrease in the magnetization back to zero followed by a reversal back to the original polarity brings the point back to b along the curve through the points f and g, thereby completing the loop. A continuous repetition of the process thereafter would simply make the point trace the hysteresis loop $bcdefgb$ repeatedly.

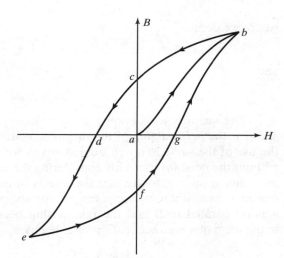

FIGURE 5.8

Hysteresis curve for a ferromagnetic material.

5.3 WAVE EQUATION AND SOLUTION

In the previous two sections, we introduced conductors, dielectrics, and magnetic materials. We found that conductors are characterized by conduction current, dielectrics are characterized by polarization current, and magnetic materials are characterized by magnetization current. The conduction current density is related to the electric field intensity through the conductivity σ of the conductor. To take into account the effects of polarization, we modified the relationship between **D** and **E** by introducing the permittivity ϵ of the dielectric. Similarly, to take into account the effects of magnetization, we modified the relationship between **H** and **B** by introducing the permeability μ of the magnetic material. The three pertinent relations, known as the *constitutive relations*, are

$$\mathbf{J}_c = \sigma \mathbf{E} \tag{5.37a}$$

$$\mathbf{D} = \epsilon \mathbf{E} \tag{5.37b}$$

$$\mathbf{H} = \frac{\mathbf{B}}{\mu} \tag{5.37c}$$

A given material may possess all three properties, although usually one of them is predominant. Hence, in this section we shall consider a material medium characterized by σ, ϵ, and μ. The Maxwell's curl equations for such a medium are

$$\nabla \times \mathbf{E} = -\frac{\partial \mathbf{B}}{\partial t} = -\mu \frac{\partial \mathbf{H}}{\partial t} \tag{5.38}$$

$$\nabla \times \mathbf{H} = \mathbf{J} + \frac{\partial \mathbf{D}}{\partial t} = \mathbf{J}_c + \frac{\partial \mathbf{D}}{\partial t} = \sigma \mathbf{E} + \epsilon \frac{\partial \mathbf{E}}{\partial t} \tag{5.39}$$

To discuss electromagnetic wave propagation in the material medium, let us consider the infinite plane current sheet of Figure 4.8, except that now the medium on either side of the sheet is a material instead of free space, as shown in Figure 5.9.

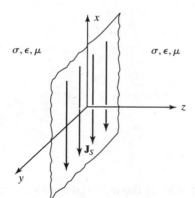

FIGURE 5.9

Infinite plane current sheet imbedded in a material medium.

The electric and magnetic fields for the simple case of the infinite plane current sheet in the $z = 0$ plane and carrying uniformly distributed current in the negative x-direction, as given by

$$\mathbf{J}_S = -J_{S0} \cos \omega t \, \mathbf{a}_x \tag{5.40}$$

are of the form

$$\mathbf{E} = E_x(z, t)\mathbf{a}_x \tag{5.41a}$$
$$\mathbf{H} = H_y(z, t)\mathbf{a}_y \tag{5.41b}$$

The corresponding simplified forms of the Maxwell's curl equations are

$$\frac{\partial E_x}{\partial z} = -\mu \frac{\partial H_y}{\partial t} \tag{5.42}$$

$$\frac{\partial H_y}{\partial z} = -\sigma E_x - \epsilon \frac{\partial E_x}{\partial t} \tag{5.43}$$

We shall make use of the phasor technique to solve these equations. Thus, letting

$$E_x(z, t) = \text{Re} \, [\bar{E}_x(z)e^{j\omega t}] \tag{5.44a}$$
$$H_y(z, t) = \text{Re} \, [\bar{H}_y(z)e^{j\omega t}] \tag{5.44b}$$

and replacing E_x and H_y in (5.42) and (5.43) by their phasors $\bar{E}_x$ and $\bar{H}_y$, respectively, and $\partial/\partial t$ by $j\omega$, we obtain the corresponding differential equations for the phasors $\bar{E}_x$ and $\bar{H}_y$ as

$$\frac{\partial \bar{E}_x}{\partial z} = -j\omega\mu\bar{H}_y \tag{5.45}$$

$$\frac{\partial \bar{H}_y}{\partial z} = -\sigma \bar{E}_x - j\omega\epsilon\bar{E}_x = -(\sigma + j\omega\epsilon)\bar{E}_x \tag{5.46}$$

Differentiating (5.45) with respect to z and using (5.46), we obtain

$$\frac{\partial^2 \bar{E}_x}{\partial z^2} = -j\omega\mu\frac{\partial \bar{H}_y}{\partial z} = j\omega\mu(\sigma + j\omega\epsilon)\bar{E}_x \tag{5.47}$$

Defining

$$\bar{\gamma} = \sqrt{j\omega\mu(\sigma + j\omega\epsilon)} \tag{5.48}$$

and substituting in (5.47), we have

$$\frac{\partial^2 \bar{E}_x}{\partial z^2} = \bar{\gamma}^2\bar{E}_x \tag{5.49}$$

Equation (5.49) is the wave equation for $\bar{E}_x$ in the material medium and its solution is given by

$$\bar{E}_x(z) = \bar{A}e^{-\bar{\gamma}z} + \bar{B}e^{\bar{\gamma}z} \tag{5.50}$$

where $\bar{A}$ and $\bar{B}$ are arbitrary constants. Noting that $\bar{\gamma}$ is a complex number and hence can be written as

$$\bar{\gamma} = \alpha + j\beta \tag{5.51}$$

and also writing $\bar{A}$ and $\bar{B}$ in exponential form as $Ae^{j\theta}$ and $Be^{j\phi}$, respectively, we have

$$\bar{E}_x(z) = Ae^{j\theta}e^{-\alpha z}e^{-j\beta z} + Be^{j\phi}e^{\alpha z}e^{j\beta z}$$

or

$$
\begin{aligned}
E_x(z, t) &= \text{Re}\,[\bar{E}_x(z)e^{j\omega t}] \\
&= \text{Re}\,[Ae^{j\theta}e^{-\alpha z}e^{-j\beta z}e^{j\omega t} + Be^{j\phi}e^{\alpha z}e^{j\beta z}e^{j\omega t}] \\
&= Ae^{-\alpha z}\cos{(\omega t - \beta z + \theta)} + Be^{\alpha z}\cos{(\omega t + \beta z + \phi)} \quad (5.52)
\end{aligned}
$$

We now recognize the two terms on the right side of (5.52) as representing uniform plane waves propagating in the positive z- and negative z-directions, respectively, with phase constant β, in view of the factors $\cos{(\omega t - \beta z + \theta)}$ and $\cos{(\omega t + \beta z + \phi)}$, respectively. They are, however, multiplied by the factors $e^{-\alpha z}$ and $e^{\alpha z}$, respectively. Hence, the peak amplitude of the field differs from one constant phase surface to another. Since there cannot be a positive going wave in the region $z < 0$, that is, to the left of the current sheet, and since there cannot be a negative going wave in the region $z > 0$, that is, to the right of the current sheet, the solution for the electric field is given by

$$E_x(z, t) = \begin{cases} Ae^{-\alpha z}\cos{(\omega t - \beta z + \theta)} & \text{for } z > 0 \\ Be^{\alpha z}\cos{(\omega t + \beta z + \phi)} & \text{for } z < 0 \end{cases} \quad (5.53)$$

To discuss how the peak amplitude of E_x varies with z on either side of the current sheet, we note that since σ, ϵ, and μ are all positive, the phase angle of $j\omega\mu(\sigma + j\omega\epsilon)$ lies between 90° and 180°, and hence the phase angle of $\bar{\gamma}$ lies between 45° and 90°, making α and β positive quantities. This means that $e^{-\alpha z}$ decreases with increasing value of z, that is, in the positive z-direction, and $e^{\alpha z}$ decreases with decreasing value of z, that is, in the negative z-direction. Thus, the exponential factors $e^{-\alpha z}$ and $e^{\alpha z}$ associated with the solutions for E_x in (5.53) have the effect of reducing the amplitude of the field, that is, attenuating it, as it propagates away from the sheet to either side of it. For this reason, the quantity α is known as the *attenuation constant*. The attenuation per unit length is equal to e^{α}. In terms of decibels, this is equal to $20\log_{10}e^{\alpha}$, or 8.686α db. The units of α are nepers per meter, abbreviated Np/m. The quantity $\bar{\gamma}$ is known as the *propagation constant*, since its real and imaginary parts, α and β, together determine the propagation characteristics, that is, attenuation and phase shift of the wave.

Returning now to the expression for $\bar{\gamma}$ given by (5.48), we can obtain the expressions for α and β by squaring it on both sides and equating the real and imaginary parts on both sides. Thus,

$$\bar{\gamma}^2 = (\alpha + j\beta)^2 = j\omega\mu(\sigma + j\omega\epsilon)$$

or

$$\alpha^2 - \beta^2 = -\omega^2\mu\epsilon \quad (5.54a)$$
$$2\alpha\beta = \omega\mu\sigma \quad (5.54b)$$

Now, squaring (5.54a) and (5.54b) and adding and then taking the square root, we obtain

$$\alpha^2 + \beta^2 = \omega^2\mu\epsilon\sqrt{1 + \left(\frac{\sigma}{\omega\epsilon}\right)^2} \quad (5.55)$$

From (5.54a) and (5.55), we then have

$$\alpha^2 = \frac{1}{2}\left[-\omega^2\mu\epsilon + \omega^2\mu\epsilon\sqrt{1 + \left(\frac{\sigma}{\omega\epsilon}\right)^2}\,\right]$$

$$\beta^2 = \frac{1}{2}\left[\omega^2\mu\epsilon + \omega^2\mu\epsilon\sqrt{1 + \left(\frac{\sigma}{\omega\epsilon}\right)^2}\,\right]$$

Since α and β are both positive, we finally get

$$\alpha = \frac{\omega\sqrt{\mu\epsilon}}{\sqrt{2}}\left[\sqrt{1 + \left(\frac{\sigma}{\omega\epsilon}\right)^2} - 1\right]^{1/2} \tag{5.56}$$

$$\beta = \frac{\omega\sqrt{\mu\epsilon}}{\sqrt{2}}\left[\sqrt{1 + \left(\frac{\sigma}{\omega\epsilon}\right)^2} + 1\right]^{1/2} \tag{5.57}$$

We note from (5.56) and (5.57) that α and β are both dependent on σ through the factor $\sigma/\omega\epsilon$. This factor, known as the *loss tangent*, is the ratio of the magnitude of the conduction current density $\sigma\bar{E}_x$ to the magnitude of the displacement current density $j\omega\epsilon\bar{E}_x$ in the material medium. In practice, the loss tangent is, however, not simply inversely proportional to ω, since both σ and ϵ are generally functions of frequency.

The phase velocity of the wave along the direction of propagation is given by

$$v_p = \frac{\omega}{\beta} = \frac{\sqrt{2}}{\sqrt{\mu\epsilon}}\left[\sqrt{1 + \left(\frac{\sigma}{\omega\epsilon}\right)^2} + 1\right]^{-1/2} \tag{5.58}$$

We note that the phase velocity is dependent on the frequency of the wave. Thus, waves of different frequencies travel with different phase velocities, that is, they undergo different rates of change of phase with z at any fixed time. This characteristic of the material medium gives rise to a phenomenon known as *dispersion*. The topic of dispersion is discussed in Section 8.3. The wavelength in the medium is given by

$$\lambda = \frac{2\pi}{\beta} = \frac{\sqrt{2}}{f\sqrt{\mu\epsilon}}\left[\sqrt{1 + \left(\frac{\sigma}{\omega\epsilon}\right)^2} + 1\right]^{-1/2} \tag{5.59}$$

Having found the solution for the electric field of the wave and discussed its general properties, we now turn to the solution for the corresponding magnetic field by substituting for $\bar{E}_x$ in (5.45). Thus,

$$\begin{aligned}
\bar{H}_y &= -\frac{1}{j\omega\mu}\frac{\partial\bar{E}_x}{\partial z} = \frac{\bar{\gamma}}{j\omega\mu}(\bar{A}e^{-\bar{\gamma}z} - \bar{B}e^{\bar{\gamma}z}) \\
&= \sqrt{\frac{\sigma + j\omega\epsilon}{j\omega\mu}}(\bar{A}e^{-\bar{\gamma}z} - \bar{B}e^{\bar{\gamma}z}) \\
&= \frac{1}{\eta}(\bar{A}e^{-\bar{\gamma}z} - \bar{B}e^{\bar{\gamma}z})
\end{aligned} \tag{5.60}$$

where

$$\overline{\eta} = \sqrt{\frac{j\omega\mu}{\sigma + j\omega\epsilon}} \tag{5.61}$$

is the intrinsic impedance of the medium. Writing

$$\overline{\eta} = |\overline{\eta}|e^{j\tau} \tag{5.62}$$

we obtain the solution for $H_y(z, t)$ as

$$
\begin{aligned}
H_y(z, t) &= \text{Re}\,[\overline{H}_y(z)e^{j\omega t}] \\
&= \text{Re}\left[\frac{1}{|\overline{\eta}|e^{j\tau}}Ae^{j\theta}e^{-\alpha z}e^{-j\beta z}e^{j\omega t} - \frac{1}{|\overline{\eta}|e^{j\tau}}Be^{j\phi}e^{\alpha z}e^{j\beta z}e^{j\omega t}\right] \\
&= \frac{A}{|\overline{\eta}|}e^{-\alpha z}\cos(\omega t - \beta z + \theta - \tau) - \frac{B}{|\overline{\eta}|}e^{\alpha z}\cos(\omega t + \beta z + \phi - \tau) \tag{5.63}
\end{aligned}
$$

Remembering that the first and second terms on the right side of (5.63) correspond to
$(+)$ and $(-)$ waves, respectively, and hence represent the solutions for the magnetic
field in the regions $z > 0$ and $z < 0$, respectively, and recalling that the solution for H_y
adjacent to the current sheet is given by

$$H_y = \begin{cases} \dfrac{J_{S0}}{2}\cos\omega t & \text{for } z = 0+ \\[2ex] -\dfrac{J_{S0}}{2}\cos\omega t & \text{for } z = 0- \end{cases} \tag{5.64}$$

we obtain

$$A = \frac{|\overline{\eta}|J_{S0}}{2}, \qquad \theta = \tau \tag{5.65a}$$

$$B = \frac{|\overline{\eta}|J_{S0}}{2}, \qquad \phi = \tau \tag{5.65b}$$

Thus, the electromagnetic field due to the infinite plane current sheet in the
xy-plane having

$$\mathbf{J}_S = -J_{S0}\cos\omega t\,\mathbf{a}_x$$

and with a material medium characterized by σ, ϵ, and μ on either side of it is given by

$$\mathbf{E}(z, t) = \frac{|\overline{\eta}|J_{S0}}{2}e^{\mp\alpha z}\cos(\omega t \mp \beta z + \tau)\,\mathbf{a}_x \qquad \text{for } z \gtrless 0 \tag{5.66a}$$

$$\mathbf{H}(z, t) = \pm\frac{J_{S0}}{2}e^{\mp\alpha z}\cos(\omega t \mp \beta z)\,\mathbf{a}_y \qquad \text{for } z \gtrless 0 \tag{5.66b}$$

We note from (5.66a) and (5.66b) that wave propagation in the material medium is
characterized by phase difference between $\mathbf{E}$ and $\mathbf{H}$ in addition to attenuation. These
properties are illustrated in Figure 5.10, which shows sketches of the current density on
the sheet and the distance-variation of the electric and magnetic fields on either side of
the current sheet for a few values of t.

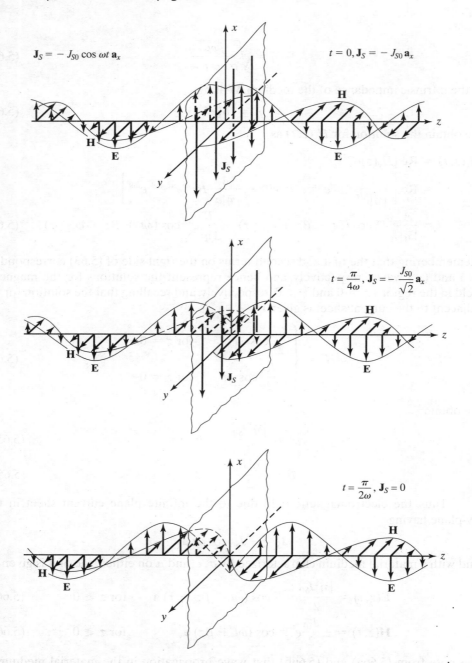

FIGURE 5.10

Time history of uniform plane electromagnetic wave radiating away from an infinite plane current sheet imbedded in a material medium.

Since the fields are attenuated as they progress in their respective directions of propagation, the medium is characterized by power dissipation. In fact, by evaluating the power flow out of a rectangular box lying between z and $z + \Delta z$ and having dimensions Δx and Δy in the x- and y-directions, respectively, as was done in Section 4.6, we obtain

$$\oint_S \mathbf{P} \cdot d\mathbf{S} = \frac{\partial P_z}{\partial z} \Delta x \, \Delta y \, \Delta z = \frac{\partial}{\partial z}(E_x H_y) \, \Delta v$$

$$= \left(E_x \frac{\partial H_y}{\partial z} + H_y \frac{\partial E_x}{\partial z} \right) \Delta v$$

$$= \left[E_x \left(-\sigma E_x - \epsilon \frac{\partial E_x}{\partial t} \right) + H_y \left(-\mu \frac{\partial H_y}{\partial t} \right) \right] \Delta v$$

$$= -\sigma E_x^2 \, \Delta v - \frac{\partial}{\partial t} \left(\frac{1}{2} \epsilon E_x^2 \, \Delta v \right) - \frac{\partial}{\partial t} \left(\frac{1}{2} \mu H_y^2 \, \Delta v \right) \qquad (5.67)$$

The quantity $\sigma E_x^2 \, \Delta v$ is obviously the power dissipated in the volume Δv due to attenuation, and the quantities $\frac{1}{2} \epsilon E_x^2 \, \Delta v$ and $\frac{1}{2} \mu H_y^2 \, \Delta v$ are the energies stored in the electric and magnetic fields, respectively, in the volume Δv. It then follows that the power dissipation density, the stored energy density associated with the electric field, and the stored energy density associated with the magnetic field are given by

$$P_d = \sigma E_x^2 \qquad (5.68)$$

$$w_e = \frac{1}{2} \epsilon E_x^2 \qquad (5.69)$$

and

$$w_m = \frac{1}{2} \mu H_y^2 \qquad (5.70)$$

respectively. Equation (5.67) is the generalization, to the material medium, of the Poynting's theorem given by (4.70) for free space.

5.4 UNIFORM PLANE WAVES IN DIELECTRICS AND CONDUCTORS

In the previous section, we discussed electromagnetic wave propagation for the general case of a material medium characterized by conductivity σ, permittivity ϵ, and permeability μ. We found general expressions for the attenuation constant α, the phase constant β, the phase velocity v_p, the wavelength λ, and the intrinsic impedance $\bar{\eta}$. These are given by (5.56), (5.57), (5.58), (5.59), and (5.61), respectively. For $\sigma = 0$, the medium is a *perfect dielectric*, having the propagation characteristics

$$\alpha = 0 \qquad (5.71a)$$

$$\beta = \omega \sqrt{\mu \epsilon} \qquad (5.71b)$$

$$v_p = \frac{1}{\sqrt{\mu \epsilon}} \qquad (5.71c)$$

$$\lambda = \frac{1}{f\sqrt{\mu\epsilon}} \tag{5.71d}$$

$$\bar{\eta} = \sqrt{\frac{\mu}{\epsilon}} \tag{5.71e}$$

Thus, the waves propagate without attenuation as in free space but with ϵ_0 and μ_0 replaced by ϵ and μ, respectively. For nonzero σ, there are two special cases: (a) imperfect dielectrics or poor conductors and (b) good conductors. The first case is characterized by conduction current small in magnitude compared to the displacement current; the second case is characterized by just the opposite.

Thus, considering the case of *imperfect dielectrics*, we have $|\sigma \bar{E}_x| \ll |j\omega\epsilon\bar{E}_x|$, or $\sigma/\omega\epsilon \ll 1$. We can then obtain approximate expressions for α, β, v_p, λ, and $\bar{\eta}$ as follows:

$$\begin{aligned}
\alpha &= \frac{\omega\sqrt{\mu\epsilon}}{\sqrt{2}} \left[\sqrt{1 + \left(\frac{\sigma}{\omega\epsilon}\right)^2} - 1 \right]^{1/2} \\
&= \frac{\omega\sqrt{\mu\epsilon}}{\sqrt{2}} \left[1 + \frac{\sigma^2}{2\omega^2\epsilon^2} - \frac{\sigma^4}{8\omega^4\epsilon^4} + \cdots - 1 \right]^{1/2} \\
&\approx \frac{\omega\sqrt{\mu\epsilon}}{\sqrt{2}} \frac{\sigma}{\sqrt{2}\omega\epsilon} \left[1 - \frac{\sigma^2}{4\omega^2\epsilon^2} \right]^{1/2} \\
&\approx \frac{\sigma}{2}\sqrt{\frac{\mu}{\epsilon}} \left(1 - \frac{\sigma^2}{8\omega^2\epsilon^2} \right)
\end{aligned} \tag{5.72a}$$

$$\begin{aligned}
\beta &= \frac{\omega\sqrt{\mu\epsilon}}{\sqrt{2}} \left[\sqrt{1 + \left(\frac{\sigma}{\omega\epsilon}\right)^2} + 1 \right]^{1/2} \\
&\approx \frac{\omega\sqrt{\mu\epsilon}}{\sqrt{2}} \left[2 + \frac{\sigma^2}{2\omega^2\epsilon^2} \right]^{1/2} \\
&\approx \omega\sqrt{\mu\epsilon} \left(1 + \frac{\sigma^2}{8\omega^2\epsilon^2} \right)
\end{aligned} \tag{5.72b}$$

$$\begin{aligned}
v_p &= \frac{\sqrt{2}}{\sqrt{\mu\epsilon}} \left[\sqrt{1 + \left(\frac{\sigma}{\omega\epsilon}\right)^2} + 1 \right]^{-1/2} \\
&\approx \frac{\sqrt{2}}{\sqrt{\mu\epsilon}} \left[2 + \frac{\sigma^2}{2\omega^2\epsilon^2} \right]^{-1/2} \\
&\approx \frac{1}{\sqrt{\mu\epsilon}} \left(1 - \frac{\sigma^2}{8\omega^2\epsilon^2} \right)
\end{aligned} \tag{5.72c}$$

$$\begin{aligned}
\lambda &= \frac{\sqrt{2}}{f\sqrt{\mu\epsilon}} \left[\sqrt{1 + \left(\frac{\sigma}{\omega\epsilon}\right)^2} + 1 \right]^{-1/2} \\
&\approx \frac{1}{f\sqrt{\mu\epsilon}} \left(1 - \frac{\sigma^2}{8\omega^2\epsilon^2} \right)
\end{aligned} \tag{5.72d}$$

$$\bar{\eta} = \sqrt{\frac{j\omega\mu}{\sigma + j\omega\epsilon}} = \sqrt{\frac{j\omega\mu}{j\omega\epsilon}}\left(1 - j\frac{\sigma}{\omega\epsilon}\right)^{-1/2}$$

$$= \sqrt{\frac{\mu}{\epsilon}}\left[1 + j\frac{\sigma}{2\omega\epsilon} - \frac{3}{8}\frac{\sigma^2}{\omega^2\epsilon^2} - \cdots\right]$$

$$\approx \sqrt{\frac{\mu}{\epsilon}}\left[\left(1 - \frac{3}{8}\frac{\sigma^2}{\omega^2\epsilon^2}\right) + j\frac{\sigma}{2\omega\epsilon}\right] \qquad (5.72e)$$

In (5.72a)–(5.72e), we have retained all terms up to and including the second power in $\sigma/\omega\epsilon$ and have neglected all higher-order terms. For a value of $\sigma/\omega\epsilon$ equal to 0.1, the quantities β, v_p, and λ are different from those for the corresponding perfect dielectric case by a factor of only 0.01/8, or $\frac{1}{800}$, whereas the intrinsic impedance has a real part differing from the intrinsic impedance of the perfect dielectric medium by a factor of $\frac{3}{800}$ and an imaginary part that is $\frac{1}{20}$ of the intrinsic impedance of the perfect dielectric medium. Thus, the only significant feature different from the perfect dielectric case is the attenuation.

Example 5.2

Let us consider that a material can be classified as a dielectric for $\sigma/\omega\epsilon < 0.1$ and compute the values of the several propagation parameters for three materials: mica, dry earth, and sea water.

Denoting the frequency for which $\sigma/\omega\epsilon = 1$ as f_q, we have $f_q = \sigma/2\pi\epsilon$, assuming that σ and ϵ are independent of frequency. Values of σ, ϵ, and f_q and approximate values of the several propagation parameters for $f > 10f_q$ are listed in Table 5.3, in which c is the velocity of light in free space and β_0 and λ_0 are the phase constant and wavelength in free space for the frequency of operation. It can be seen from Table 5.3 that mica behaves as a dielectric for almost any frequency, but sea water can be classified as a dielectric only for frequencies above approximately 10 GHz. We also note that because of the low value of α, mica is a good dielectric, but the high value of α for sea water makes it a poor dielectric.

TABLE 5.3   Values of Several Propagation Parameters for Three Materials for the Dielectric Range of Frequencies

Material	σ S/m	ϵ_r	f_q Hz	α Np/m	β/β_0	v_p/c	λ/λ_0	$\bar{\eta}$ Ω
Mica	10^{-11}	6	3×10^{-2}	77×10^{-11}	2.45	0.408	0.408	153.9
Dry earth	10^{-5}	5	3.6×10^4	84×10^{-5}	2.24	0.447	0.447	168.6
Sea water	4	80	0.9×10^9	84.3	8.94	0.112	0.112	42.15

Turning now to the case of *good conductors*, we have $|\sigma\bar{E}_x| \gg |j\omega\epsilon\bar{E}_x|$, or $\sigma/\omega\epsilon \gg 1$. We can then obtain approximate expressions for α, β, v_p, λ, and η, as follows:

$$\alpha = \frac{\omega\sqrt{\mu\epsilon}}{\sqrt{2}}\left[\sqrt{1 + \left(\frac{\sigma}{\omega\epsilon}\right)^2} - 1\right]^{1/2}$$

$$\approx \frac{\omega\sqrt{\mu\epsilon}}{\sqrt{2}}\sqrt{\frac{\sigma}{\omega\epsilon}} = \sqrt{\frac{\omega\mu\sigma}{2}}$$

$$= \sqrt{\pi f\mu\sigma} \qquad (5.73a)$$

$$\beta = \frac{\omega\sqrt{\mu\epsilon}}{\sqrt{2}}\left[\sqrt{1+\left(\frac{\sigma}{\omega\epsilon}\right)^2}+1\right]^{1/2}$$

$$\approx \frac{\omega\sqrt{\mu\epsilon}}{\sqrt{2}}\sqrt{\frac{\sigma}{\omega\epsilon}}$$

$$= \sqrt{\pi f\mu\sigma} \tag{5.73b}$$

$$v_p = \frac{\sqrt{2}}{\sqrt{\mu\epsilon}}\left[\sqrt{1+\left(\frac{\sigma}{\omega\epsilon}\right)^2}+1\right]^{-1/2}$$

$$\approx \frac{\sqrt{2}}{\sqrt{\mu\epsilon}}\sqrt{\frac{\omega\epsilon}{\sigma}} = \sqrt{\frac{2\omega}{\mu\sigma}}$$

$$= \sqrt{\frac{4\pi f}{\mu\sigma}} \tag{5.73c}$$

$$\lambda = \frac{\sqrt{2}}{f\sqrt{\mu\epsilon}}\left[\sqrt{1+\left(\frac{\sigma}{\omega\epsilon}\right)^2}+1\right]^{-1/2}$$

$$\approx \sqrt{\frac{4\pi}{f\mu\sigma}} \tag{5.73d}$$

$$\bar{\eta} = \sqrt{\frac{j\omega\mu}{\sigma+j\omega\epsilon}} \approx \sqrt{\frac{j\omega\mu}{\sigma}}$$

$$= (1+j)\sqrt{\frac{\pi f\mu}{\sigma}} \tag{5.73e}$$

We note that α, β, v_p, and $\bar{\eta}$ are proportional to $\sqrt{f}$, provided that σ and μ are constants.

   To discuss the propagation characteristics of a wave inside a good conductor, let us consider the case of copper. The constants for copper are $\sigma = 5.80 \times 10^7$ S/m, $\epsilon = \epsilon_0$, and $\mu = \mu_0$. Hence, the frequency at which α is equal to $\omega\epsilon$ for copper is equal to $5.8 \times 10^7/2\pi\epsilon_0$, or 1.04×10^{18} Hz. Thus, at frequencies of even several gigahertz, copper behaves like an excellent conductor. To obtain an idea of the attenuation of the wave inside the conductor, we note that the attenuation undergone in a distance of one wavelength is equal to $e^{-\alpha\lambda}$ or $e^{-2\pi}$. In terms of decibels, this is equal to $20\log_{10}e^{2\pi} = 54.58$ db. In fact, the field is attenuated by a factor e^{-1}, or 0.368 in a distance equal to $1/\alpha$. This distance is known as the *skin depth* and is denoted by the symbol δ. From (5.73a), we obtain

$$\delta = \frac{1}{\sqrt{\pi f\mu\sigma}} \tag{5.74}$$

The skin depth for copper is equal to

$$\frac{1}{\sqrt{\pi f \times 4\pi \times 10^{-7} \times 5.8 \times 10^7}} = \frac{0.066}{\sqrt{f}}\text{ m.}$$

Thus, in copper the fields are attenuated by a factor e^{-1} in a distance of 0.066 mm even at the low frequency of 1 MHz, thereby resulting in the concentration of the fields near to the skin of the conductor. This phenomenon is known as the *skin effect.* It also explains *shielding* by conductors. This topic is discussed in Section 10.3.

To discuss further the characteristics of wave propagation in a good conductor, we note that the ratio of the wavelength in the conducting medium to the wavelength in a dielectric medium having the same ϵ and μ as those of the conductor is given by

$$\frac{\lambda_{\text{conductor}}}{\lambda_{\text{dielectric}}} \approx \frac{\sqrt{4\pi/f\mu\sigma}}{1/f\sqrt{\mu\epsilon}} = \sqrt{\frac{4\pi f\epsilon}{\sigma}} = \sqrt{\frac{2\omega\epsilon}{\sigma}} \qquad (5.75)$$

Since $\sigma/\omega\epsilon \gg 1$, $\lambda_{\text{conductor}} \ll \lambda_{\text{dielectric}}$. For example, for sea water, $\sigma = 4$ S/m, $\epsilon = 80\epsilon_0$, and $\mu = \mu_0$, so that the ratio of the two wavelengths for $f = 25$ kHz is equal to 0.00745. Thus for $f = 25$ kHz, the wavelength in sea water is $\frac{1}{134}$ of the wavelength in a dielectric having the same ϵ and μ as those of sea water and a still smaller fraction of the wavelength in free space. Furthermore, the lower the frequency, the smaller is this fraction. Since it is the electrical length, that is, the length in terms of the wavelength, instead of the physical length that determines the radiation efficiency of an antenna, this means that antennas of much shorter length can be used in sea water than in free space. Together with the property that $\alpha \propto \sqrt{f}$, this illustrates that low frequencies are more suitable than high frequencies for communication under water, and with underwater objects.

Equation (5.73e) tells us that the intrinsic impedance of a good conductor has a phase angle of 45°. Hence, the electric and magnetic fields in the medium are out of phase by 45°. The magnitude of the intrinsic impedance is given by

$$|\bar{\eta}| = \left|(1+j)\sqrt{\frac{\pi f\mu}{\sigma}}\right| = \sqrt{\frac{2\pi f\mu}{\sigma}} \qquad (5.76)$$

As a numerical example, for copper, this quantity is equal to

$$\sqrt{\frac{2\pi f \times 4\pi \times 10^{-7}}{5.8 \times 10^7}} = 3.69 \times 10^{-7}\sqrt{f}\ \Omega$$

Thus, the intrinsic impedance of copper has as low a magnitude as 0.369 Ω even at a frequency of 10^{12} Hz. In fact, by recognizing that

$$|\bar{\eta}| = \sqrt{\frac{2\pi f\mu}{\sigma}} = \sqrt{\frac{\omega\epsilon}{\sigma}}\sqrt{\frac{\mu}{\epsilon}} \qquad (5.77)$$

we note that the magnitude of the intrinsic impedance of a good conductor medium is a small fraction of the intrinsic impedance of a dielectric medium having the same ϵ and μ. It follows that for the same electric field, the magnetic field inside a good conductor is much larger than the magnetic field inside a dielectric having the same ϵ and μ as those of the conductor.

Finally, for $\sigma = \infty$, the medium is a *perfect conductor*, an idealization of the good conductor. From (5.74), we note that the skin depth is then equal to zero and that there is no penetration of the fields. Thus, no time-varying fields can exist inside a perfect conductor.

5.5 BOUNDARY CONDITIONS

In our study of electromagnetics we will be considering problems involving more than one medium. To solve a problem involving a boundary surface between different media, we need to know the conditions satisfied by the field components at the boundary. These are known as the *boundary conditions*. They are a set of relationships relating the field components at a point adjacent to and on one side of the boundary, to the field components at a corresponding point adjacent to and on the other side of the boundary. These relationships arise from the fact that Maxwell's equations in integral form involve closed paths and surfaces and they must be satisfied for all possible closed paths and surfaces, whether they lie entirely in one medium or encompass a portion of the boundary between two different media. In the latter case, Maxwell's equations in integral form must be satisfied collectively by the fields on either side of the boundary, thereby resulting in the boundary conditions.

We shall derive the boundary conditions by considering the Maxwell's equations in integral form

$$\oint_C \mathbf{E} \cdot d\mathbf{l} = -\frac{d}{dt} \int_S \mathbf{B} \cdot d\mathbf{S} \tag{5.78a}$$

$$\oint_C \mathbf{H} \cdot d\mathbf{l} = \int_S \mathbf{J} \cdot d\mathbf{S} + \frac{d}{dt} \int_S \mathbf{D} \cdot d\mathbf{S} \tag{5.78b}$$

$$\oint_S \mathbf{D} \cdot d\mathbf{S} = \int_V \rho \, dv \tag{5.78c}$$

$$\oint_S \mathbf{B} \cdot d\mathbf{S} = 0 \tag{5.78d}$$

and applying them one at a time to a closed path or a closed surface encompassing the boundary, and in the limit that the area enclosed by the closed path or the volume bounded by the closed surface goes to zero. Thus, let us consider two semi-infinite media separated by a plane boundary, as shown in Figure 5.11. Let us denote the quantities pertinent to medium 1 by subscript 1 and the quantities pertinent to medium 2 by subscript 2. Let $\mathbf{a}_n$ be the unit normal vector to the surface and directed into medium 1, as shown in Figure 5.11, and let all normal components of fields at the boundary in both media denoted by an additional subscript n be directed along $\mathbf{a}_n$.

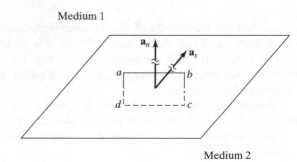

FIGURE 5.11

For deriving the boundary conditions resulting from Faraday's law and Ampere's circuital law.

Let the surface charge density (C/m^2) and the surface current density (A/m) on the boundary be ρ_S and $\mathbf{J}_S$, respectively. Note that, in general, the fields at the boundary in both media and the surface charge and current densities are functions of position on the boundary.

First, we consider a rectangular closed path *abcda* of infinitesimal area in the plane normal to the boundary and with its sides *ab* and *cd* parallel to and on either side of the boundary, as shown in Figure 5.11. Applying Faraday's law (5.78a) to this path in the limit that *ad* and *bc* $\rightarrow 0$ by making the area *abcd* tend to zero, but with *ab* and *cd* remaining on either side of the boundary, we have

$$\lim_{\substack{ad \to 0 \\ bc \to 0}} \oint_{abcda} \mathbf{E} \cdot d\mathbf{l} = -\lim_{\substack{ad \to 0 \\ bc \to 0}} \frac{d}{dt} \int_{\substack{area \\ abcd}} \mathbf{B} \cdot d\mathbf{S} \qquad (5.79)$$

In this limit, the contributions from *ad* and *bc* to the integral on the left side of (5.79) approach zero. Since *ab* and *cd* are infinitesimal, the sum of the contributions from *ab* and *cd* becomes $[E_{ab}(ab) + E_{cd}(cd)]$, where E_{ab} and E_{cd} are the components of $\mathbf{E}_1$ and $\mathbf{E}_2$ along *ab* and *cd*, respectively. The right side of (5.79) is equal to zero, since the magnetic flux crossing the area *abcd* approaches zero as the area *abcd* tends to zero. Thus, (5.79) gives

$$E_{ab}(ab) + E_{cd}(cd) = 0$$

or, since *ab* and *cd* are equal and $E_{dc} = -E_{cd}$,

$$E_{ab} - E_{dc} = 0 \qquad (5.80)$$

Let us now define $\mathbf{a}_s$ to be the unit vector normal to the area *abcd* and in the direction of advance of a right-hand screw as it is turned in the sense of the closed path *abcda*. Noting then that $\mathbf{a}_s \times \mathbf{a}_n$ is the unit vector along *ab*, we can write (5.80) as

$$\mathbf{a}_s \times \mathbf{a}_n \cdot (\mathbf{E}_1 - \mathbf{E}_2) = 0$$

Rearranging the order of the scalar triple product, we obtain

$$\mathbf{a}_s \cdot \mathbf{a}_n \times (\mathbf{E}_1 - \mathbf{E}_2) = 0 \qquad (5.81)$$

Since we can choose the rectangle *abcd* to be in any plane normal to the boundary, (5.81) must be true for all orientations of $\mathbf{a}_s$. It then follows that

$$\mathbf{a}_n \times (\mathbf{E}_1 - \mathbf{E}_2) = 0 \qquad (5.82a)$$

or, in scalar form,

$$E_{t1} - E_{t2} = 0 \qquad (5.82b)$$

where E_{t1} and E_{t2} are the components of $\mathbf{E}_1$ and $\mathbf{E}_2$, respectively, tangential to the boundary. In words, (5.82a) and (5.82b) state that *at any point on the boundary, the components of* $\mathbf{E}_1$ *and* $\mathbf{E}_2$ *tangential to the boundary are equal.*

Similarly, applying Ampere's circuital law (5.78a) to the closed path in the limit that ad and $bc \rightarrow 0$, we have

$$\underset{\substack{ad \to 0 \\ bc \to 0}}{\text{Lim}} \oint_{abcda} \mathbf{H} \cdot d\mathbf{l} = \underset{\substack{ad \to 0 \\ bc \to 0}}{\text{Lim}} \int_{\substack{\text{area} \\ abcd}} \mathbf{J} \cdot d\mathbf{S} + \underset{\substack{ad \to 0 \\ bc \to 0}}{\text{Lim}} \frac{d}{dt} \int_{\substack{\text{area} \\ abcd}} \mathbf{D} \cdot d\mathbf{S} \qquad (5.83)$$

Using the same argument as for the left side of (5.79), we obtain the quantity on the left side of (5.83) to be equal to $[H_{ab}(ab) + H_{cd}(cd)]$, where H_{ab} and H_{cd} are the components of $\mathbf{H}_1$ and $\mathbf{H}_2$ along ab and cd, respectively. The second integral on the right side of (5.83) is zero, since the displacement flux crossing the area $abcd$ approaches zero as the area $abcd$ tends to zero. The first integral on the right side of (5.83) would also be equal to zero but for a contribution from the surface current on the boundary, because letting the area $abcd$ tend to zero with ab and cd on either side of the boundary reduces only the volume current, if any, enclosed by it to zero, keeping the surface current still enclosed by it. This contribution is the surface current flowing normal to the line that $abcd$ approaches as it tends to zero, that is, $[\mathbf{J}_S \cdot \mathbf{a}_s](ab)$. Thus, (5.83) gives

$$H_{ab}(ab) + H_{cd}(cd) = (\mathbf{J}_S \cdot \mathbf{a}_s)(ab)$$

or, since ab and cd are equal and $H_{dc} = -H_{cd}$,

$$H_{ab} - H_{dc} = \mathbf{J}_S \cdot \mathbf{a}_s \qquad (5.84)$$

In terms of $\mathbf{H}_1$ and $\mathbf{H}_2$, we have

$$\mathbf{a}_s \times \mathbf{a}_n \cdot (\mathbf{H}_1 - \mathbf{H}_2) = \mathbf{J}_S \cdot \mathbf{a}_s$$

or

$$\mathbf{a}_s \cdot \mathbf{a}_n \times (\mathbf{H}_1 - \mathbf{H}_2) = \mathbf{a}_s \cdot \mathbf{J}_S \qquad (5.85)$$

Since (5.85) must be true for all orientations of $\mathbf{a}_s$, that is, for a rectangle $abcd$ in any plane normal to the boundary, it follows that

$$\mathbf{a}_n \times (\mathbf{H}_1 - \mathbf{H}_2) = \mathbf{J}_S \qquad (5.86a)$$

or, in scalar form,

$$H_{t1} - H_{t2} = J_S \qquad (5.86b)$$

where H_{t1} and H_{t2} are the components of $\mathbf{H}_1$ and $\mathbf{H}_2$, respectively, tangential to the boundary. In words, (5.86a) and (5.86b) state that *at any point on the boundary, the components of $\mathbf{H}_1$ and $\mathbf{H}_2$ tangential to the boundary are discontinuous by the amount equal to the surface current density at that point.* It should be noted that the information concerning the direction of $\mathbf{J}_S$ relative to that of $(\mathbf{H}_1 - \mathbf{H}_2)$, which is contained in (5.86a), is not present in (5.86b). Thus, in general, (5.86b) is not sufficient, and it is necessary to use (5.86a).

Now, we consider a rectangular box *abcdefgh* of infinitesimal volume enclosing an infinitesimal area of the boundary and parallel to it, as shown in Figure 5.12. Applying Gauss' law for the electric field (5.78d) to this box in the limit that the side surfaces (abbreviated *ss*) tend to zero by making the volume of the box tend to zero but with the sides *abcd* and *efgh* remaining on either side of the boundary, we have

$$\underset{ss \to 0}{\text{Lim}} \oint_{\substack{\text{surface} \\ \text{of the box}}} \mathbf{D} \cdot d\mathbf{S} = \underset{ss \to 0}{\text{Lim}} \int_{\substack{\text{volume} \\ \text{of the box}}} \rho \, dv \qquad (5.87)$$

In this limit, the contributions from the side surfaces to the integral on the left side of (5.87) approach zero. The sum of the contributions from the top and bottom surfaces becomes $[D_{n1}(abcd) - D_{n2}(efgh)]$, since *abcd* and *efgh* are infinitesimal. The quantity on the right side of (5.87) would be zero but for the surface charge on the boundary, since letting the volume of the box tend to zero with the sides *abcd* and *efgh* on either side of it reduces only the volume charge, if any, enclosed by it to zero, keeping the surface charge still enclosed by it. This surface charge is equal to $\rho_S(abcd)$. Thus, (5.87) gives

$$D_{n1}(abcd) - D_{n2}(efgh) = \rho_S(abcd)$$

or, since *abcd* and *efgh* are equal,

$$D_{n1} - D_{n2} = \rho_S \qquad (5.88a)$$

In terms of $\mathbf{D}_1$ and $\mathbf{D}_2$, (5.88a) is given by

$$\mathbf{a}_n \cdot (\mathbf{D}_1 - \mathbf{D}_2) = \rho_S \qquad (5.88b)$$

In words, (5.88a) and (5.88b) state that *at any point on the boundary, the components of* $\mathbf{D}_1$ *and* $\mathbf{D}_2$ *normal to the boundary are discontinuous by the amount of the surface charge density at that point.*

Medium 1

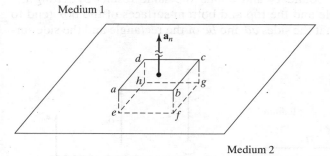

FIGURE 5.12

For deriving the boundary conditions resulting from the two Gauss' laws.

Medium 2

Similarly, applying Gauss' law for the magnetic field (5.78d) to the box *abcdefgh* in the limit that the side surfaces tend to zero, we have

$$\underset{ss \to 0}{\text{Lim}} \oint_{\substack{\text{surface} \\ \text{of the box}}} \mathbf{B} \cdot d\mathbf{S} = 0 \qquad (5.89)$$

Using the same argument as for the left side of (5.87), we obtain the quantity on the left side of (5.89) to be equal to $[B_{n1}(abcd) - B_{n2}(efgh)]$. Thus, (5.89) gives

$$B_{n1}(abcd) - B_{n2}(efgh) = 0$$

or, since $abcd$ and $efgh$ are equal,

$$B_{n1} - B_{n2} = 0 \tag{5.90a}$$

In terms of $\mathbf{B}_1$ and $\mathbf{B}_2$, (5.90a) is given by

$$\mathbf{a}_n \cdot (\mathbf{B}_1 - \mathbf{B}_2) = 0 \tag{5.90b}$$

In words, (5.90a) and (5.90b) state that *at any point on the boundary, the components of* $\mathbf{B}_1$ *and* $\mathbf{B}_2$ *normal to the boundary are equal.*

Summarizing the boundary conditions, we have

$$\mathbf{a}_n \times (\mathbf{E}_1 - \mathbf{E}_2) = 0 \tag{5.91a}$$
$$\mathbf{a}_n \times (\mathbf{H}_1 - \mathbf{H}_2) = \mathbf{J}_S \tag{5.91b}$$
$$\mathbf{a}_n \cdot (\mathbf{D}_1 - \mathbf{D}_2) = \rho_S \tag{5.91c}$$
$$\mathbf{a}_n \cdot (\mathbf{B}_1 - \mathbf{B}_2) = 0 \tag{5.91d}$$

or, in scalar form,

$$E_{t1} - E_{t2} = 0 \tag{5.92a}$$
$$H_{t1} - H_{t2} = J_S \tag{5.92b}$$
$$D_{n1} - D_{n2} = \rho_S \tag{5.92c}$$
$$B_{n1} - B_{n2} = 0 \tag{5.92d}$$

as illustrated in Figure 5.13. Although we have derived these boundary conditions by considering a plane interface between the two media, it should be obvious that we can consider any arbitrary-shaped boundary and obtain the same results by letting the sides ab and cd of the rectangle and the top and bottom surfaces of the box tend to zero, in addition to the limits that the sides ad and bc of the rectangle and the side surfaces of the box tend to zero.

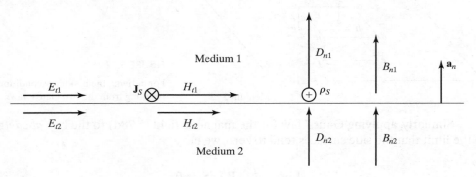

FIGURE 5.13

For illustrating the boundary conditions at an interface between two different media.

The boundary conditions given by (5.91a)–(5.91d) are general. When they are applied to particular cases, the special properties of the pertinent media come into play. Two such cases are important to be considered. They are as follows.

Interface between Two Perfect Dielectric Media

Since for a perfect dielectric, $\sigma = 0$, $\mathbf{J}_c = \sigma\mathbf{E} = 0$. Thus, there cannot be any conduction current in a perfect dielectric, which in turn rules out any accumulation of free charge on the surface of a perfect dielectric. Hence, in applying the boundary conditions (5.91a)–(5.91d) to an interface between two perfect dielectric media, we set ρ_S and $\mathbf{J}_S$ equal to zero, thereby obtaining

$$\mathbf{a}_n \times (\mathbf{E}_1 - \mathbf{E}_2) = 0 \tag{5.93a}$$
$$\mathbf{a}_n \times (\mathbf{H}_1 - \mathbf{H}_2) = 0 \tag{5.93b}$$
$$\mathbf{a}_n \cdot (\mathbf{D}_1 - \mathbf{D}_2) = 0 \tag{5.93c}$$
$$\mathbf{a}_n \cdot (\mathbf{B}_1 - \mathbf{B}_2) = 0 \tag{5.93d}$$

These boundary conditions tell us that the tangential components of $\mathbf{E}$ and $\mathbf{H}$ and the normal components of $\mathbf{D}$ and $\mathbf{B}$ are continuous at the boundary.

Surface of a Perfect Conductor

No time-varying fields can exist in a perfect conductor. In view of this, the boundary conditions on a perfect conductor surface are obtained by setting the fields with subscript 2 in (5.91a)–(5.91d) equal to zero. Thus, we obtain

$$\mathbf{a}_n \times \mathbf{E} = 0 \tag{5.94a}$$
$$\mathbf{a}_n \times \mathbf{H} = \mathbf{J}_S \tag{5.94b}$$
$$\mathbf{a}_n \cdot \mathbf{D} = \rho_S \tag{5.94c}$$
$$\mathbf{a}_n \cdot \mathbf{B} = 0 \tag{5.94d}$$

where we have also omitted subscripts 1, so that $\mathbf{E}$, $\mathbf{H}$, $\mathbf{D}$, and $\mathbf{B}$ are the fields on the perfect conductor surface. The boundary conditions (5.94a) and (5.94d) tell us that on a perfect conductor surface, the tangential component of the electric field intensity and the normal component of the magnetic field intensity are zero. Hence, the electric field must be completely normal, and the magnetic field must be completely tangential to the surface. The remaining two boundary conditions (5.94c) and (5.94b) tell us that the (normal) displacement flux density is equal to the surface charge density and the (tangential) magnetic field intensity is equal in magnitude to the surface current density.

Example 5.3

In Figure 5.14, the region $x < 0$ is a perfect conductor, the region $0 < x < d$ is a perfect dielectric of $\epsilon = 2\epsilon_0$ and $\mu = \mu_0$, and the region $x > d$ is free space. The electric and magnetic fields in the region $0 < x < d$ are given at a particular instant of time by

$$\mathbf{E} = E_1 \cos \pi x \sin 2\pi z \, \mathbf{a}_x + E_2 \sin \pi x \cos 2\pi z \, \mathbf{a}_z$$
$$\mathbf{H} = H_1 \cos \pi x \sin 2\pi z \, \mathbf{a}_y$$

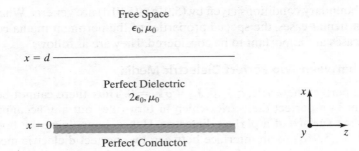

FIGURE 5.14

For illustrating the application of boundary conditions.

We wish to find (a) ρ_S and $\mathbf{J}_S$ on the surface $x = 0$ and (b) $\mathbf{E}$ and $\mathbf{H}$ for $x = d+$, that is, immediately adjacent to the $x = d$ plane and on the free-space side, at that instant of time.

(a) Denoting the perfect dielectric medium ($0 < x < d$) to be medium 1 and the perfect conductor medium ($x < 0$) to be medium 2, we have $\mathbf{a}_n = \mathbf{a}_x$, and all fields with subscript 2 are equal to zero. Then, from (5.91c) and (5.91b), we obtain

$$[\rho_S]_{x=0} = \mathbf{a}_n \cdot [\mathbf{D}_1]_{x=0} = \mathbf{a}_x \cdot 2\epsilon_0 E_1 \sin 2\pi z \, \mathbf{a}_x$$
$$= 2\epsilon_0 E_1 \sin 2\pi z$$
$$[\mathbf{J}_S]_{x=0} = \mathbf{a}_n \times [\mathbf{H}_1]_{x=0} = \mathbf{a}_x \times H_1 \sin 2\pi z \, \mathbf{a}_y$$
$$= H_1 \sin 2\pi z \, \mathbf{a}_z$$

Note that the remaining two boundary conditions (5.91a) and (5.91b) are already satisfied by the given fields, since E_y and B_x do not exist and for $x = 0$, $E_z = 0$. Also note that what we have done here is equivalent to using (5.94a)–(5.94d), since the boundary is the surface of a perfect conductor.

(b) Denoting the perfect dielectric medium ($0 < x < d$) to be medium 1 and the free-space medium ($x > d$) to be medium 2 and setting $\rho_S = 0$, we obtain from (5.91a) and (5.91c)

$$[E_y]_{x=d+} = [E_y]_{x=d-} = 0$$
$$[E_z]_{x=d+} = [E_z]_{x=d-} = E_2 \sin \pi d \cos 2\pi z$$
$$[D_x]_{x=d+} = [D_x]_{x=d-} = 2\epsilon_0[E_x]_{x=d-}$$
$$= 2\epsilon_0 E_1 \cos \pi d \sin 2\pi z$$
$$[E_x]_{x=d+} = \frac{1}{\epsilon_0}[D_x]_{x=d+}$$
$$= 2E_1 \cos \pi d \sin 2\pi z$$

Thus,

$$[\mathbf{E}]_{x=d+} = 2E_1 \cos \pi d \sin 2\pi z \, \mathbf{a}_x + E_2 \sin \pi d \cos 2\pi z \, \mathbf{a}_z$$

Setting $\mathbf{J}_S = 0$ and using (5.91b) and (5.91d), we obtain

$$[H_y]_{x=d+} = [H_y]_{x=d-} = H_1 \cos \pi d \sin 2\pi z$$
$$[H_z]_{x=d+} = [H_z]_{x=d-} = 0$$
$$[B_x]_{x=d+} = [B_x]_{x=d-} = 0$$

Thus,

$$[\mathbf{H}]_{x=d+} = H_1 \cos \pi d \sin 2\pi z \, \mathbf{a}_y$$

Note that what we have done here is equivalent to using (5.93a)–(5.93d), since the boundary is the interface between two perfect dielectrics.

5.6 REFLECTION AND TRANSMISSION OF UNIFORM PLANE WAVES

Thus far, we have considered uniform plane wave propagation in unbounded media. Practical situations are characterized by propagation involving several different media. When a wave is incident on a boundary between two different media, a reflected wave is produced. In addition, if the second medium is not a perfect conductor, a transmitted wave is set up. Together, these waves satisfy the boundary conditions at the interface between the two media. In this section, we shall consider these phenomena for waves incident normally on plane boundaries.

To do this, let us consider the situation shown in Figure 5.15 in which steady-state conditions are established by uniform plane waves of radian frequency ω propagating normal to the plane interface $z = 0$ between two media characterized by two different sets of values of σ, ϵ, and μ, where $\sigma \neq \infty$. We shall assume that a $(+)$ wave is incident from medium 1 ($z < 0$) onto the interface, thereby setting up a reflected $(-)$ wave in that medium, and a transmitted $(+)$ wave in medium 2 ($z > 0$). For convenience, we shall work with the phasor or complex field components. Thus, considering the electric fields to be in the x-direction and the magnetic fields to be in the y-direction, we can write the solution for the complex field components in medium 1 to be

$$\bar{E}_{1x}(z) = \bar{E}_1^+ e^{-\bar{\gamma}_1 z} + \bar{E}_1^- e^{\bar{\gamma}_1 z} \tag{5.95a}$$

$$\bar{H}_{1y}(z) = \bar{H}_1^+ e^{-\bar{\gamma}_1 z} + \bar{H}_1^- e^{\bar{\gamma}_1 z}$$

$$= \frac{1}{\bar{\eta}_1}\left(\bar{E}_1^+ e^{-\bar{\gamma}_1 z} - \bar{E}_1^- e^{\bar{\gamma}_1 z}\right) \tag{5.95b}$$

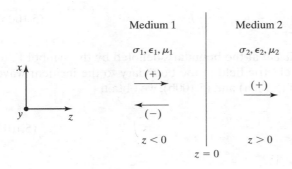

Medium 1

$\sigma_1, \epsilon_1, \mu_1$

$(+)$

$(-)$

$z < 0$

Medium 2

$\sigma_2, \epsilon_2, \mu_2$

$(+)$

$z > 0$

$z = 0$

FIGURE 5.15

Normal incidence of uniform plane waves on a plane interface between two different media.

where $\bar{E}_1^+, \bar{E}_1^-, \bar{H}_1^+$, and $\bar{H}_1^-$ are the incident and reflected wave electric and magnetic field components, respectively, at $z = 0-$ in medium 1 and

$$\bar{\gamma}_1 = \sqrt{j\omega\mu_1(\sigma_1 + j\omega\epsilon_1)} \tag{5.96a}$$

$$\bar{\eta}_1 = \sqrt{\frac{j\omega\mu_1}{\sigma_1 + j\omega\epsilon_1}} \tag{5.96b}$$

Recall that the real field corresponding to a complex field component is obtained by multiplying the complex field component by $e^{j\omega t}$ and taking the real part of the product. The complex field components in medium 2 are given by

$$\bar{E}_{2x}(z) = \bar{E}_2^+ e^{-\bar{\gamma}_2 z} \tag{5.97a}$$
$$\bar{H}_{2y}(z) = \bar{H}_2^+ e^{-\bar{\gamma}_2 z}$$

$$= \frac{\bar{E}_2^+}{\bar{\eta}_2} e^{-\bar{\gamma}_2 z} \tag{5.97b}$$

where $\bar{E}_2^+$ and $\bar{H}_2^+$ are the transmitted wave electric- and magnetic-field components at $z = 0+$ in medium 2 and

$$\bar{\gamma}_2 = \sqrt{j\omega\mu_2(\sigma_2 + j\omega\epsilon_2)} \tag{5.98a}$$

$$\bar{\eta}_2 = \sqrt{\frac{j\omega\mu_2}{\sigma_2 + j\omega\epsilon_2}} \tag{5.98b}$$

To satisfy the boundary conditions at $z = 0$, we note that (1) the components of both electric and magnetic fields are entirely tangential to the interface and (2) in view of the finite conductivities of the media, no surface current exists on the interface (currents flow in the volumes of the media). Hence, from the phasor forms of the boundary conditions (5.92a) and (5.92b), we have

$$[\bar{E}_{1x}]_{z=0} = [\bar{E}_{2x}]_{z=0} \tag{5.99a}$$

$$[\bar{H}_{1y}]_{z=0} = [\bar{H}_{2y}]_{z=0} \tag{5.99b}$$

Applying these to the solution pairs given by (5.95a, b) and (5.97a, b), we have

$$\bar{E}_1^+ + \bar{E}_1^- = \bar{E}_2^+ \tag{5.100a}$$

$$\frac{1}{\bar{\eta}_1}\left(\bar{E}_1^+ - \bar{E}_1^-\right) = \frac{1}{\bar{\eta}_2}\bar{E}_2^+ \tag{5.100b}$$

We now define the *reflection coefficient* at the boundary, denoted by the symbol $\bar{\Gamma}$, to be the ratio of the reflected wave electric field at the boundary to the incident wave electric field at the boundary. From (5.100a) and (5.100b), we obtain

$$\bar{\Gamma} = \frac{\bar{E}_1^-}{\bar{E}_1^+} = \frac{\bar{\eta}_2 - \bar{\eta}_1}{\bar{\eta}_2 + \bar{\eta}_1} \tag{5.101}$$

Note that the ratio of the reflected wave magnetic field at the boundary to the incident wave magnetic field at the boundary is given by

$$\frac{\bar{H}_1^-}{\bar{H}_1^+} = \frac{-\bar{E}_1^-/\bar{\eta}_1}{\bar{E}_1^+/\bar{\eta}_1} = -\frac{\bar{E}_1^-}{\bar{E}_1^+} = -\bar{\Gamma} \qquad (5.102)$$

The ratio of the transmitted wave electric field at the boundary to the incident wave electric field at the boundary, known as the *transmission coefficient* and denoted by the symbol $\bar{\tau}$, is given by

$$\bar{\tau} = \frac{\bar{E}_2^+}{\bar{E}_1^+} = \frac{\bar{E}_1^+ + \bar{E}_1^-}{\bar{E}_1^+} = 1 + \bar{\Gamma} \qquad (5.103)$$

where we have used (5.100a). The ratio of the transmitted wave magnetic field at the boundary to the incident wave magnetic field at the boundary is given by

$$\frac{\bar{H}_2^+}{\bar{H}_1^+} = \frac{\bar{H}_1^+ + \bar{H}_1^-}{\bar{H}_1^+} = 1 - \bar{\Gamma} \qquad (5.104)$$

The reflection and transmission coefficients given by (5.101) and (5.103), respectively, enable us to find the reflected and transmitted wave fields for a given incident wave field. We observe the following properties of $\bar{\Gamma}$ and $\bar{\tau}$:

1. For $\bar{\eta}_2 = \bar{\eta}_1, \bar{\Gamma} = 0$ and $\bar{\tau} = 1$. The incident wave is entirely transmitted. The situation then corresponds to a *matched* condition. A trivial case occurs when the two media have identical values of the material parameters.
2. For $\sigma_1 = \sigma_2 = 0$, that is, when both media are perfect dielectrics, $\bar{\eta}_1$ and $\bar{\eta}_2$ are real. Hence, $\bar{\Gamma}$ and $\bar{\tau}$ are real. In particular, if the two media have the same permeability μ but different permittivities ϵ_1 and ϵ_2, then

$$\bar{\Gamma} = \frac{\sqrt{\mu/\epsilon_2} - \sqrt{\mu/\epsilon_1}}{\sqrt{\mu/\epsilon_2} + \sqrt{\mu/\epsilon_1}}$$

$$= \frac{1 - \sqrt{\epsilon_2/\epsilon_1}}{1 + \sqrt{\epsilon_2/\epsilon_1}} \qquad (5.105)$$

$$\bar{\tau} = \frac{2}{1 + \sqrt{\epsilon_2/\epsilon_1}} \qquad (5.106)$$

3. For $\sigma_2 \rightarrow \infty, \bar{\eta}_2 \rightarrow 0, \bar{\Gamma} \rightarrow -1$, and $\bar{\tau} \rightarrow 0$. Thus, if medium 2 is a perfect conductor, the incident wave is entirely reflected, as it should be, since there cannot be any time-varying fields inside a perfect conductor. The superposition of the reflected and incident waves would then give rise to the so-called complete standing waves in medium 1. Complete standing waves as well as partial standing waves are discussed in Chapter 7.

Example 5.4

Region 1 ($z < 0$) is free space, whereas region 2 ($z > 0$) is a material medium characterized by $\sigma = 10^{-4}$ S/m, $\epsilon = 5\epsilon_0$, and $\mu = \mu_0$. For a uniform plane wave having the electric field

$$\mathbf{E}_i = E_0 \cos(3\pi \times 10^5 t - 10^{-3}\pi z)\, \mathbf{a}_x \text{ V/m}$$

incident on the interface $z = 0$ from region 1, we wish to obtain the expressions for the reflected and transmitted wave electric and magnetic fields.

Substituting $\sigma = 10^{-4}$ S/m, $\epsilon = 5\epsilon_0$, $\mu = \mu_0$, and $f = (3\pi \times 10^5)/2\pi = 1.5 \times 10^5$ Hz, in (5.98a) and (5.98b), we obtain

$$\overline{\gamma}_2 = (6.283 + j9.425) \times 10^{-3}$$

$$\overline{\eta}_2 = 104.559 \underline{/33.69°} = 104.559 \underline{/0.1872\pi}$$

Then, noting that $\overline{\eta}_1 = \eta_0$,

$$\overline{\Gamma} = \frac{\overline{\eta}_2 - \eta_0}{\overline{\eta}_2 + \eta_0} = \frac{104.559\underline{/33.69°} - 377}{104.559\underline{/33.69°} + 377}$$

$$= 0.6325\underline{/161.565°} = 0.6325\underline{/0.8976\pi}$$

$$\overline{\tau} = 1 + \overline{\Gamma} = 1 + 0.6325\underline{/161.565°}$$

$$= 0.4472\underline{/26.565°} = 0.4472\underline{/0.1476\pi}$$

Thus, the reflected and transmitted wave electric and magnetic fields are given by

$$\mathbf{E}_r = 0.6325E_0 \cos(3\pi \times 10^5 t + 10^{-3}\pi z + 0.8976\pi)\, \mathbf{a}_x \text{ V/m}$$

$$\mathbf{H}_r = -\frac{0.6325E_0}{377} \cos(3\pi \times 10^5 t + 10^{-3}\pi z + 0.8976\pi)\, \mathbf{a}_y \text{ A/m}$$

$$= -1.678 \times 10^{-3} E_0 \cos(3\pi \times 10^5 t + 10^{-3}\pi z + 0.8976\pi)\, \mathbf{a}_y \text{ A/m}$$

$$\mathbf{E}_t = 0.4472E_0 e^{-6.283 \times 10^{-3}z}$$
$$\cdot \cos(3\pi \times 10^5 t - 9.425 \times 10^{-3}z + 0.1476\pi)\, \mathbf{a}_x \text{ V/m}$$

$$\mathbf{H}_t = \frac{0.4472E_0}{104.559} e^{-6.283 \times 10^{-3}z}$$
$$\cdot \cos(3\pi \times 10^5 t - 9.425 \times 10^{-3}z + 0.1476\pi - 0.1872\pi)\, \mathbf{a}_y \text{ A/m}$$

$$= 4.277 \times 10^{-3} E_0 e^{-6.283 \times 10^{-3}z}$$
$$\cdot \cos(3\pi \times 10^5 t - 9.425 \times 10^{-3}z - 0.0396\pi)\, \mathbf{a}_y \text{ A/m}$$

Note that at $z = 0$, the boundary conditions of $\mathbf{E}_i + \mathbf{E}_r = \mathbf{E}_t$ and $\mathbf{H}_i + \mathbf{H}_r = \mathbf{H}_t$ are satisfied, since

$$E_0 + 0.6325E_0 \cos 0.8976\pi = 0.4472E_0 \cos 0.1476\pi$$

and

$$\frac{E_0}{377} - 1.678 \times 10^{-3}E_0 \cos 0.8976\pi = 4.277 \times 10^{-3}E_0 \cos(-0.0396\pi)$$

SUMMARY

In this chapter, we studied the principles of uniform plane wave propagation in a material medium. Material media can be classified as (a) conductors, (b) dielectrics, and (c) magnetic materials, depending on the nature of the response of the charged particles in the materials to applied fields. Conductors are characterized by conduction which is the phenomenon of steady drift of free electrons under the influence of an applied electric field. Dielectrics are characterized by polarization which is the phenomenon of the creation and net alignment of electric dipoles, formed by the displacement of the centroids of the electron clouds from the centroids of the nucleii of the atoms, along the direction of an applied electric field. Magnetic materials are characterized by magnetization which is the phenomenon of net alignment of the axes of the magnetic dipoles, formed by the electron orbital and spin motion around the nucleii of the atoms, along the direction of an applied magnetic field.

Under the influence of applied electromagnetic wave fields, all three phenomena described above give rise to currents in the material which in turn influence the wave propagation. These currents are known as the conduction, polarization, and magnetization currents, respectively, for conductors, dielectrics, and magnetic materials. They must be taken into account in the first term on the right side of Ampere's circuital law, that is, $\int_S \mathbf{J} \cdot d\mathbf{S}$ in the case of the integral form and $\mathbf{J}$ in the case of the differential form. The conduction current density is given by

$$\mathbf{J}_c = \sigma \mathbf{E} \tag{5.107}$$

where σ is the conductivity of the material. The conduction current is taken into account explicitly by replacing $\mathbf{J}$ by $\mathbf{J}_c$. The polarization and magnetization currents are taken into account implicitly by revising the definitions of the displacement flux density vector and the magnetic field intensity vector to read as

$$\mathbf{D} = \epsilon_0 \mathbf{E} + \mathbf{P} \tag{5.108}$$

$$\mathbf{H} = \frac{\mathbf{B}}{\mu_0} - \mathbf{M} \tag{5.109}$$

where $\mathbf{P}$ and $\mathbf{M}$ are the polarization and magnetization vectors, respectively. For linear isotropic materials, (5.108) and (5.109) simplify to

$$\mathbf{D} = \epsilon \mathbf{E} \tag{5.110}$$

$$\mathbf{H} = \frac{\mathbf{B}}{\mu} \tag{5.111}$$

where

$$\epsilon = \epsilon_0 \epsilon_r$$
$$\mu = \mu_0 \mu_r$$

are the permittivity and the permeability, respectively, of the material. The quantities ϵ_r and μ_r are the relative permittivity and the relative permeability, respectively, of the material. The parameters σ, ϵ, and μ vary from one material to another and are in general dependent on the frequency of the wave. Equations (5.107), (5.110), and (5.111) are known as the constitutive relations. For anisotropic materials, these relations are expressed in the form of matrix equations with the material parameters represented by tensors.

Together with Maxwell's equations, the constitutive relations govern the behavior of the electromagnetic field in a material medium. Thus, Maxwell's curl equations for a material medium are given by

$$\nabla \times \mathbf{E} = -\frac{\partial \mathbf{B}}{\partial t} = -\mu \frac{\partial \mathbf{H}}{\partial t}$$

$$\nabla \times \mathbf{H} = \mathbf{J}_c + \frac{\partial \mathbf{D}}{\partial t} = \sigma \mathbf{E} + \epsilon \frac{\partial \mathbf{E}}{\partial t}$$

We made use of these equations for the simple case of $\mathbf{E} = E_x(z, t)\mathbf{a}_x$ and $\mathbf{H} = H_y(z, t)\mathbf{a}_y$ to obtain the uniform plane wave solution by considering the infinite plane current sheet in the xy-plane with uniform surface current density

$$\mathbf{J}_S = -\mathbf{J}_{S0} \cos \omega t \, \mathbf{a}_x$$

and with a material medium on either side of it and finding the electromagnetic field due to the current sheet to be given by

$$\mathbf{E} = \frac{|\bar{\eta}|J_{S0}}{2} e^{\mp\alpha z} \cos(\omega t \mp \beta z + \tau) \, \mathbf{a}_x \qquad \text{for } z \gtrless 0 \qquad (5.112a)$$

$$\mathbf{H} = \pm \frac{J_{S0}}{2} e^{\mp\alpha z} \cos(\omega t \mp \beta z) \, \mathbf{a}_y \qquad \text{for } z \gtrless 0 \qquad (5.112b)$$

In (5.112a–b), α and β are the attenuation and phase constants given, respectively, by the real and imaginary parts of the propagation constant, $\bar{\gamma}$. Thus,

$$\bar{\gamma} = \alpha + j\beta = \sqrt{j\omega\mu(\sigma + j\omega\epsilon)}$$

The quantities $|\bar{\eta}|$ and τ are the magnitude and phase angle, respectively, of the intrinsic impedance, $\bar{\eta}$, of the medium. Thus,

$$\bar{\eta} = |\bar{\eta}|e^{j\tau} = \sqrt{\frac{j\omega\mu}{\sigma + j\omega\epsilon}}$$

The uniform plane wave solution given by (5.112a–b) tells us that the wave propagation in the material medium is characterized by attenuation, as indicated by $e^{\mp\alpha z}$, and phase difference between $\mathbf{E}$ and $\mathbf{H}$ by the amount τ. We learned that the attenuation of the wave results from power dissipation due to conduction current flow in the medium. The power dissipation density is given by

$$p_d = \sigma E_x^2$$

The stored energy densities associated with the electric and magnetic fields in the medium are given by

$$w_e = \frac{1}{2}\epsilon E^2$$
$$w_m = \frac{1}{2}\mu H^2$$

Having discussed uniform plane wave propagation for the general case of a medium characterized by σ, ϵ, and μ, we then considered several special cases. These are discussed in the following:

Perfect dielectrics. For these materials, $\sigma = 0$. Wave propagation occurs without attenuation as in free space but with the propagation parameters governed by ϵ and μ instead of ϵ_0 and μ_0, respectively.

Imperfect dielectrics. A material is classified as an imperfect dielectric for $\sigma \ll \omega\epsilon$, that is, conduction current density is small in magnitude compared to the displacement current density. The only significant feature of wave propagation in an imperfect dielectric as compared to that in a perfect dielectric is the attenuation undergone by the wave.

Good conductors. A material is classified as a good conductor for $\sigma \gg \omega\epsilon$, that is, conduction current density is large in magnitude compared to the displacement current density. Wave propagation in a good conductor medium is characterized by attenuation and phase constants both equal to $\sqrt{\pi f \mu \sigma}$. Thus, for large values of f and/or σ, the fields do not penetrate very deeply into the conductor. This phenomenon is known as the skin effect. From considerations of the frequency dependence of the attenuation and wavelength for a fixed σ, we learned that low frequencies are more suitable for communication with underwater objects. We also learned that the intrinsic impedance of a good conductor medium is very low in magnitude compared to that of a dielectric medium having the same ϵ and μ.

Perfect conductors. These are idealizations of good conductors in the limit $\sigma \to \infty$. For $\sigma = \infty$, the skin depth, that is, the distance in which the fields inside a conductor are attenuated by a factor e^{-1}, is zero and hence there can be no penetration of fields into a perfect conductor.

As a prelude to the consideration of problems involving more than one medium, we derived the boundary conditions resulting from the application of Maxwell's equations in integral form to closed paths and closed surfaces encompassing the boundary between two media, and in the limits that the areas enclosed by the closed paths and the volumes bounded by the closed surfaces go to zero. These boundary conditions are given by

$$\mathbf{a}_n \times (\mathbf{E}_1 - \mathbf{E}_2) = 0$$
$$\mathbf{a}_n \times (\mathbf{H}_1 - \mathbf{H}_2) = \mathbf{J}_S$$
$$\mathbf{a}_n \cdot (\mathbf{D}_1 - \mathbf{D}_2) = \rho_S$$
$$\mathbf{a}_n \cdot (\mathbf{B}_1 - \mathbf{B}_2) = 0$$

where the subscripts 1 and 2 refer to media 1 and 2, respectively, and $\mathbf{a}_n$ is unit vector normal to the boundary at the point under consideration and directed into medium 1. In words, the boundary conditions state that at a point on the boundary, the tangential components of $\mathbf{E}$ and the normal components of $\mathbf{B}$ are continuous, whereas the tangential components of $\mathbf{H}$ are discontinuous by the amount equal to J_S at that point, and the normal components of $\mathbf{D}$ are discontinuous by the amount equal to ρ_S at that point.

Two important special cases of boundary conditions are as follows: (a) At the boundary between two perfect dielectrics, the tangential components of $\mathbf{E}$ and $\mathbf{H}$ and the normal components of $\mathbf{D}$ and $\mathbf{B}$ are continuous. (b) On the surface of a perfect conductor, the tangential component of $\mathbf{E}$ and the normal component of $\mathbf{B}$ are zero, whereas the normal component of $\mathbf{D}$ is equal to the surface charge density, and the tangential component of $\mathbf{H}$ is equal in magnitude to the surface current density.

Finally, we considered uniform plane waves incident normally onto a plane boundary between two media, and we learned how to compute the reflected and transmitted wave fields for a given incident wave field.

REVIEW QUESTIONS

5.1. Distinguish between bound electrons and free electrons in an atom.

5.2. Briefly describe the phenomenon of conduction.

5.3. State Ohms' law applicable at a point. How is it taken into account in Maxwell's equations?

5.4. Briefly describe the phenomenon of polarization in a dielectric material.

5.5. What is an electric dipole? How is its strength defined?

5.6. What are the different kinds of polarization in a dielectric?

5.7. What is the polarization vector? How is it related to the electric field intensity?

5.8. Discuss how polarization current arises in a dielectric material.

5.9. State the relationship between polarization current density and electric field intensity. How is it taken into account in Maxwell's equations?

5.10. What is the revised definition of $\mathbf{D}$?

5.11. State the relationship between $\mathbf{D}$ and $\mathbf{E}$ in a dielectric material. How does it simplify the solution of field problems involving dielectrics?

5.12. What is an anisotropic dielectric material?

5.13. When can an effective permittivity be defined for an anisotropic dielectric material?

5.14. Briefly describe the phenomenon of magnetization.

5.15. What is a magnetic dipole? How is its strength defined?

5.16. What are the different kinds of magnetic materials?

5.17. What is the magnetization vector? How is it related to the magnetic flux density?

5.18. Discuss how magnetization current arises in a magnetic material.

5.19. State the relationship between magnetization current density and magnetic flux density. How is it taken into account in Maxwell's equations?

5.20. What is the revised definition of $\mathbf{H}$?

5.21. State the relationship between $\mathbf{H}$ and $\mathbf{B}$ for a magnetic material. How does it simplify the solution of field problems involving magnetic materials?

5.22. What is an anisotropic magnetic material?

5.23. Discuss the relationship between B and H for a ferromagnetic material.

5.24. Summarize the constitutive relations for a material medium.

5.25. What is the propagation constant for a material medium? Discuss the significance of its real and imaginary parts.

5.26. Discuss the consequence of the frequency dependence of the phase velocity of a wave in a material medium.

5.27. What is loss tangent? Discuss its significance.

5.28. What is the intrinsic impedance of a material medium? What is the consequence of its complex nature?

5.29. How do you account for the attenuation undergone by the wave in a material medium?

5.30. What is the power dissipation density in a medium characterized by nonzero conductivity?

5.31. What are the stored energy densities associated with electric and magnetic fields in a material medium?

5.32. What is the condition for a medium to be a perfect dielectric? How do the characteristics of wave propagation in a perfect dielectric medium differ from those of wave propagation in free space?

5.33. What is the criterion for a material to be an imperfect dielectric? What is the significant feature of wave propagation in an imperfect dielectric as compared to that in a perfect dielectric?

5.34. Give two examples of materials that behave as good dielectrics for frequencies down to almost zero.

5.35. What is the criterion for a material to be a good conductor?

5.36. Give two examples of materials that behave as good conductors for frequencies of up to several gigahertz.

5.37. What is skin effect? Discuss skin depth, giving some numerical values.

5.38. Why are low-frequency waves more suitable than high-frequency waves for communication with underwater objects?

5.39. Discuss the consequence of the low intrinsic impedance of a good conductor as compared to that of a dielectric medium having the same ϵ and μ.

5.40. Why can there be no fields inside a perfect conductor?

5.41. What is a boundary condition? How do boundary conditions arise and how are they derived?

5.42. Summarize the boundary conditions for the general case of a boundary between two arbitrary media, indicating correspondingly the Maxwell's equations in integral form from which they are derived.

5.43. Discuss the boundary conditions at the interface between two perfect dielectric media.

5.44. Discuss the boundary conditions on the surface of a perfect conductor.

5.45. Discuss the determination of the reflected and transitted wave fields from the fields of a wave incident normally onto a plane boundary between two material media.

5.46. What is the consequence of a wave incident on a perfect conductor?

PROBLEMS

5.1. Find the electric field intensity required to produce a current of 0.1 A crossing an area of 1 cm^2 normal to the field for the following materials: (a) copper, (b) aluminum, and (c) sea water. Then find the voltage drop along a length of 1 cm parallel to the field, and find the ratio of the voltage drop to the current (resistance) for each material.

5.2. The free electron density in silver is $5.80 \times 10^{28}\,\text{m}^{-3}$. (a) Find the mobility of the electron for silver. (b) Find the drift velocity of the electrons for an applied electric field of intensity 0.1 V/m.

5.3. Use the continuity equation, Ohm's law, and Gauss' law for the electric field to show that the time variation of the charge density at a point inside a conductor is governed by the differential equation

$$\frac{\partial \rho}{\partial t} + \frac{\sigma}{\epsilon_0}\rho = 0$$

Then show that the charge density inside the conductor decays exponentially with a time constant ϵ_0/σ. Compute the value of the time constant for copper.

5.4. Show that the torque acting on an electric dipole of moment **p** due to an applied electric field **E** is $\mathbf{p} \times \mathbf{E}$.

5.5. For an applied electric field $\mathbf{E} = 0.1\cos 2\pi \times 10^9 t\,\mathbf{a}_x$ V/m, find the polarization current crossing an area of 1 cm^2 normal to the field for the following materials: (a) polystyrene, (b) mica, and (c) distilled water.

5.6. For the anisotropic dielectric material having the permittivity tensor given in Example 5.1, find **D** for $\mathbf{E} = E_0(\cos \omega t\,\mathbf{a}_x + \sin \omega t\,\mathbf{a}_y)$. Comment on your result.

5.7. An anisotropic dielectric material is characterized by the permittivity tensor

$$[\epsilon] = \epsilon_0 \begin{bmatrix} 4 & 2 & 2 \\ 2 & 4 & 2 \\ 2 & 2 & 4 \end{bmatrix}$$

(a) Find **D** for $\mathbf{E} = E_0\mathbf{a}_x$. (b) Find **D** for $\mathbf{E} = E_0(\mathbf{a}_x + \mathbf{a}_y + \mathbf{a}_z)$. (c) Find **E**, which produces $\mathbf{D} = 4\epsilon_0 E_0\mathbf{a}_x$.

5.8. An anisotropic dielectric material is characterized by the permittivity tensor

$$[\epsilon] = \begin{bmatrix} \epsilon_{xx} & \epsilon_{xy} & 0 \\ \epsilon_{yx} & \epsilon_{yy} & 0 \\ 0 & 0 & \epsilon_{zz} \end{bmatrix}$$

For $\mathbf{E} = (E_x\mathbf{a}_x + E_y\mathbf{a}_y)\cos \omega t$, find the value(s) of E_y/E_x for which **D** is parallel to **E**. Find the effective permittivity for each case.

5.9. Find the magnetic dipole moment of an electron in circular orbit of radius a normal to a uniform magnetic field of flux density B_0. Compute its value for $a = 10^{-3}\,\text{m}$ and $B_0 = 5 \times 10^{-5}$ Wb/m^2.

5.10. Show that the torque acting on a magnetic dipole of moment **m** due to an applied magnetic field **B** is $\mathbf{m} \times \mathbf{B}$. For simplicity, consider a rectangular loop in the xy-plane and $\mathbf{B} = B_x\mathbf{a}_x + B_y\mathbf{a}_y + B_z\mathbf{a}_z$.

5.11. For an applied magnetic field $\mathbf{B} = 10^{-6}\cos 2\pi z\,\mathbf{a}_y$ Wb/m^2, find the magnetization current crossing an area 1 cm^2 normal to the x-direction for a magnetic material having $\chi_m = 10^{-3}$.

5.12. An anisotropic magnetic material is characterized by the permeability tensor

$$[\mu] = \mu_0 \begin{bmatrix} 7 & 6 & 0 \\ 6 & 12 & 0 \\ 0 & 0 & 3 \end{bmatrix}$$

Find the effective permeability for $\mathbf{H} = H_0(3\mathbf{a}_x - 2\mathbf{a}_y)\cos \omega t$.

5.13. Obtain the wave equation for $\bar{H}_y$ similar to that for $\bar{E}_x$ given by (5.49).

5.14. Obtain the expression for the attenuation per wavelength undergone by a uniform plane wave in a material medium characterized by σ, ϵ, and μ. Using the logarithmic scale for $\sigma/\omega\epsilon$, plot the attenuation per wavelength in decibels versus $\sigma/\omega\epsilon$.

5.15. For dry earth, $\sigma = 10^{-5}$ S/m, $\epsilon = 5\epsilon_0$, and $\mu = \mu_0$. Compute $\alpha, \beta, v_p, \lambda$, and $\bar{\eta}$ for $f = 100$ kHz.

5.16. Obtain the expressions for the real and imaginary parts of the intrinsic impedance of a material medium given by (5.61).

5.17. An infinite plane sheet lying in the xy-plane carries current of uniform density

$$\mathbf{J}_S = -0.1 \cos 2\pi \times 10^6 t\, \mathbf{a}_x \text{ A/m}$$

The medium on either side of the sheet is characterized by $\sigma = 10^{-3}$ S/m, $\epsilon = 18\epsilon_0$, and $\mu = \mu_0$. Find $\mathbf{E}$ and $\mathbf{H}$ on either side of the current sheet.

5.18. Repeat Problem 5.17 for

$$\mathbf{J}_S = -0.1(\cos 2\pi \times 10^6 t\, \mathbf{a}_x + \cos 4\pi \times 10^6 t\, \mathbf{a}_x) \text{ A/m}$$

5.19. For an array of two infinite plane parallel current sheets of uniform densities situated in a medium characterized by $\sigma = 10^{-3}$ S/m, $\epsilon = 18\epsilon_0$, and $\mu = \mu_0$, find the spacing and the relative amplitudes and phase angles of the current densities to obtain an endfire radiation characteristic for $f = 10^6$ Hz.

5.20. Show that energy is not stored equally in the electric and magnetic fields in a material medium for $\sigma \neq 0$.

5.21. The electric field of a uniform plane wave propagating in a perfect dielectric medium having $\mu = \mu_0$ is given by

$$\mathbf{E} = 10 \cos (6\pi \times 10^7 t - 0.4\pi z)\mathbf{a}_x \text{ V/m}$$

Find (a) the frequency, (b) the wavelength, (c) the phase velocity, (d) the permittivity of the medium, and (e) the associated magnetic field vector $\mathbf{H}$.

5.22. The electric and magnetic fields of a uniform plane wave propagating in a perfect dielectric medium are given by

$$\mathbf{E} = 10 \cos (6\pi \times 10^7 t - 0.8\pi z)\mathbf{a}_x \text{V/m}$$
$$\mathbf{H} = \frac{1}{6\pi} \cos (6\pi \times 10^7 t - 0.8\pi z)\mathbf{a}_y \text{ A/m}$$

Find the permittivity and the permeability of the medium.

5.23. Repeat Problem 4.29 for a perfect dielectric medium of $\epsilon = 9\epsilon_0$ and $\mu = \mu_0$ on either side of the current sheet.

5.24. Compute f_q for each of the following materials: (a) fused quartz, (b) Bakelite, and (c) distilled water. Then compute for the imperfect dielectric range of frequencies the values of $\alpha, \beta, v_p, \lambda$, and $\bar{\eta}$ for each material.

5.25. For uniform plane wave propagation in fresh water ($\sigma = 10^{-3}$ S/m, $\epsilon = 80\epsilon_0$, $\mu = \mu_0$), find $\alpha, \beta, v_p, \lambda$, and $\bar{\eta}$ for two frequencies: (a) 100 MHz, and (b) 10 kHz.

5.26. Show that for a given material, the ratio of the attenuation constant for the good conductor range of frequencies to the attenuation constant for the imperfect dielectric range of frequencies is equal to $\sqrt{2\omega\epsilon/\sigma}$ where ω is in the good conductor range of frequencies.

5.27. In Figure 5.16, the points 1 and 2 lie adjacent to each other and on either side of the interface between perfect dielectric media 1 and 2. The fields at point 1 are denoted by subscript 1 and the fields at point 2 are denoted by subscript 2. Assume that medium 1 is characterized by $\epsilon = 12\epsilon_0$ and $\mu = 2\mu_0$ and that medium 2 is characterized by $\epsilon = 9\epsilon_0$ and $\mu = \mu_0$. If $\mathbf{E}_1 = E_0(3\mathbf{a}_x + 2\mathbf{a}_y - 6\mathbf{a}_z)$ and $\mathbf{H}_1 = H_0(2\mathbf{a}_x - 3\mathbf{a}_y)$, find $\mathbf{E}_2$ and $\mathbf{H}_2$.

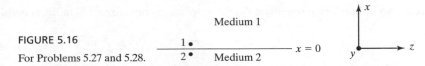

FIGURE 5.16

For Problems 5.27 and 5.28.

5.28. In Figure 5.16, assume that medium 1 is characterized by $\epsilon = 4\epsilon_0$ and $\mu = 3\mu_0$ and that medium 2 is characterized by $\epsilon = 16\epsilon_0$ and $\mu = 9\mu_0$. If $\mathbf{D}_1 = D_0(\mathbf{a}_x - 2\mathbf{a}_y + \mathbf{a}_z)$ and $\mathbf{B}_1 = B_0(\mathbf{a}_x + 2\mathbf{a}_y + 3\mathbf{a}_z)$, find $\mathbf{D}_2$ and $\mathbf{B}_2$.

5.29. A boundary separates free space from a perfect dielectric medium. At a point on the boundary, the electric field intensity on the free space side is $\mathbf{E}_1 = E_0(4\mathbf{a}_x + 2\mathbf{a}_y + 5\mathbf{a}_z)$, whereas on the dielectric side, it is $\mathbf{E}_2 = 3E_0(\mathbf{a}_x + \mathbf{a}_z)$, where E_0 is a constant. Find the permittivity of the dielectric medium.

5.30. The plane $x + 2y + 3z = 5$ defines the surface of a perfect conductor. Find the possible direction(s) of the electric field intensity at a point on the conductor surface.

5.31. Given $\mathbf{E} = y\mathbf{a}_x + x\mathbf{a}_y$, determine if a perfect conductor can be placed in the surface $xy = 2$ without disturbing the field.

5.32. A perfect conductor occupies the region $x + 2y \leq 2$. Find the surface current density at a point on the conductor at which $\mathbf{H} = H_0\mathbf{a}_z$.

5.33. The displacement flux density at a point on the surface of a perfect conductor is given by $\mathbf{D} = D_0(\mathbf{a}_x + \sqrt{3}\mathbf{a}_y + 2\sqrt{3}\mathbf{a}_z)$. Find the magnitude of the surface charge density at that point.

5.34. It is known that at a point on the surface of a perfect conductor $\mathbf{D} = D_0(\mathbf{a}_x + 2\mathbf{a}_y + 2\mathbf{a}_z)$, $\mathbf{H} = H_0(2\mathbf{a}_x - 2\mathbf{a}_y + \mathbf{a}_z)$, and ρ_S is positive. Find ρ_S and $\mathbf{J}_S$ at that point.

5.35. Two infinite plane conducting sheets occupy the planes $x = 0$ and $x = 0.1$ m. An electric field given by

$$\mathbf{E} = E_0 \sin 10\pi x \cos 3\pi \times 10^9 t \, \mathbf{a}_z$$

where E_0 is a constant, exists in the region between the plates, which is free space. (a) Show that $\mathbf{E}$ satisfies the boundary condition on the sheets. (b) Obtain $\mathbf{H}$ associated with the given $\mathbf{E}$. (c) Find the surface current densities on the two sheets.

5.36. Region $1 (z < 0)$ is free space, whereas region $2 (z > 0)$ is a material medium characterized by $\sigma = 10^{-3}$ S/m, $\epsilon = 12\epsilon_0$, and $\mu = \mu_0$. For a uniform plane wave having the electric field

$$\mathbf{E}_i = E_0 \cos (3\pi \times 10^6 t - 0.01\pi z)\mathbf{a}_x \text{ V/m}$$

incident on the interface $z = 0$ from region 1, obtain the expression for the reflected and transmitted wave electric and magnetic fields.

5.37. The regions $z < 0$ and $z > 0$ are nonmagnetic ($\mu = \mu_0$) perfect dielectrics of permittivities ϵ_1 and ϵ_2, respectively. For a uniform plane wave incident from the region $z < 0$ normally onto the boundary $z = 0$, find ϵ_2/ϵ_1 for each of the following to hold at $z = 0$: (a) the electric field of the reflected wave is $-1/3$ times the electric field of the incident wave; (b) the electric field of the transmitted wave is 0.4 times the electric field of the incident wave; and (c) the electric field of the transmitted wave is six times the electric field of the reflected wave.

5.38. A uniform plane wave propagating in the $+z$-direction and having the electric field $\mathbf{E}_i = E_{xi}(t)\mathbf{a}_x$, where $E_{xi}(t)$ in the $z = 0$ plane is as shown in Figure 5.17, is incident normally from free space ($z < 0$) onto a nonmagnetic ($\mu = \mu_0$), perfect dielectric ($z > 0$) of permittivity $4\epsilon_0$. Find and sketch the following: (a) E_x versus z for $t = 1\,\mu$s and (b) H_y versus z for $t = 1\,\mu$s.

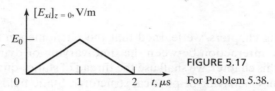

FIGURE 5.17

For Problem 5.38.

5.39. The region $z < 0$ is a perfect dielectric, whereas the region $z > 0$ is a perfect conductor. For a uniform plane wave having the electric and magnetic fields

$$\mathbf{E}_i = E_0 \cos(\omega t - \beta z)\,\mathbf{a}_x$$
$$\mathbf{H}_i = \frac{E_0}{\eta} \cos(\omega t - \beta z)\,\mathbf{a}_y$$

where $\beta = \omega\sqrt{\mu\epsilon}$ and $\eta = \sqrt{\mu/\epsilon}$, obtain the expressions for the reflected wave electric and magnetic fields and hence the expressions for the total (incident + reflected) electric and magnetic fields in the dielectric, and the current density on the surface of the perfect conductor.

5.40. In Figure 5.18, medium 3 extends to infinity so that no reflected ($-$) wave exists in that medium. For a uniform plane wave having the electric field

$$\mathbf{E}_i = E_0 \cos(3 \times 10^8\pi t - \pi z)\,\mathbf{a}_x \text{ V/m}$$

incident from medium 1 onto the interface $z = 0$, obtain the expressions for the phasor electric- and magnetic-field components in all three media.

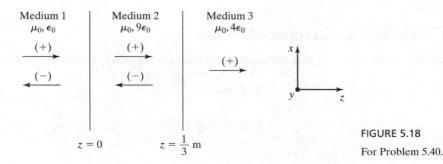

FIGURE 5.18

For Problem 5.40.

CHAPTER
6
Statics, Quasistatics, and Transmission Lines

In the preceding chapters, we learned that the phenomenon of wave propagation is based upon the interaction between the time-varying or dynamic electric and magnetic fields. In this chapter, we shall use the thread of statics-quasistatics-waves to bring out the frequency behavior of physical structures. Static fields are studied by setting the time derivatives in Maxwell's equations equal to zero. We will introduce the *lumped* circuit elements familiar in circuit theory, through the different classifications of static fields. For a nonzero frequency, the fields are dynamic. The exact solutions are solutions to the complete Maxwell's equations for time-varying fields. However, a class of fields, known as quasistatic fields, can be studied as low-frequency extensions of static fields. They are approximations to the exact solutions. We will learn that for quasistatic fields, the circuit equivalent for the input behavior of a physical structure is essentially same as the lumped circuit equivalent for the corresponding static case. As the frequency is increased beyond the quasistatic approximation, the lumped circuit equivalent is no longer valid and the *distributed* circuit equivalent comes into play, leading to the transmission line.

We begin the chapter with electric potential, a scalar that is related to the static electric field intensity through a vector operation known as the *gradient*. We shall introduce the gradient and the electric potential and then consider two important differential equations involving the potential, known as *Poisson's equation* and *Laplace's equation.* Beginning with static field involving the solution of the Laplace's equation, we shall then embark on the study based on the thread of statics-quasistatics-waves.

6.1 GRADIENT AND ELECTRIC POTENTIAL

For static fields, $\partial/\partial t = 0$, and Maxwell's curl equations given for time-varying fields by

$$\nabla \times \mathbf{E} = -\frac{\partial \mathbf{B}}{\partial t} \tag{6.1}$$

$$\nabla \times \mathbf{H} = \mathbf{J} + \frac{\partial \mathbf{D}}{\partial t} \tag{6.2}$$

reduce to

$$\nabla \times \mathbf{E} = 0 \tag{6.3}$$

$$\nabla \times \mathbf{H} = \mathbf{J} \tag{6.4}$$

respectively. Equation (6.3) states that the curl of the static electric field is equal to zero. If the curl of a vector is zero, then that vector can be expressed as the *gradient* of a scalar, since the curl of the gradient of a scalar is identically equal to zero. The gradient of a scalar, say Φ, denoted $\nabla\Phi$ (del Φ) is given in Cartesian coordinates by

$$\nabla\Phi = \left(\mathbf{a}_x \frac{\partial}{\partial x} + \mathbf{a}_y \frac{\partial}{\partial y} + \mathbf{a}_z \frac{\partial}{\partial z} \right) \Phi$$

$$= \frac{\partial \Phi}{\partial x} \mathbf{a}_x + \frac{\partial \Phi}{\partial y} \mathbf{a}_y + \frac{\partial \Phi}{\partial z} \mathbf{a}_z \tag{6.5}$$

The curl of $\nabla\Phi$ is then given by

$$\nabla \times \nabla\Phi = \begin{vmatrix} \mathbf{a}_x & \mathbf{a}_y & \mathbf{a}_z \\ \dfrac{\partial}{\partial x} & \dfrac{\partial}{\partial y} & \dfrac{\partial}{\partial z} \\ (\nabla\Phi)_x & (\nabla\Phi)_y & (\nabla\Phi)_z \end{vmatrix}$$

$$= \begin{vmatrix} \mathbf{a}_x & \mathbf{a}_y & \mathbf{a}_z \\ \dfrac{\partial}{\partial x} & \dfrac{\partial}{\partial y} & \dfrac{\partial}{\partial z} \\ \dfrac{\partial \Phi}{\partial x} & \dfrac{\partial \Phi}{\partial y} & \dfrac{\partial \Phi}{\partial z} \end{vmatrix}$$

$$= 0 \tag{6.6}$$

To discuss the physical interpretation of the gradient, we note that

$$\nabla\Phi \cdot d\mathbf{l} = \left(\frac{\partial \Phi}{\partial x} \mathbf{a}_x + \frac{\partial \Phi}{\partial y} \mathbf{a}_y + \frac{\partial \Phi}{\partial z} \mathbf{a}_z \right) \cdot (dx\, \mathbf{a}_x + dy\, \mathbf{a}_y + dz\, \mathbf{a}_z)$$

$$= \frac{\partial \Phi}{\partial x} dx + \frac{\partial \Phi}{\partial y} dy + \frac{\partial \Phi}{\partial z} dz$$

$$= d\Phi \tag{6.7}$$

Let us consider a surface on which Φ is equal to a constant, say Φ_0, and a point P on that surface, as shown in Figure 6.1(a). If we now consider another point Q_1 on the same surface and an infinitesimal distance away from P, $d\Phi$ between these two points is zero since Φ is constant on the surface. Thus, for the vector $d\mathbf{l}_1$ drawn from P to Q_1, $[\nabla\Phi]_P \cdot d\mathbf{l}_1 = 0$ and hence $[\nabla\Phi]_P$ is perpendicular to $d\mathbf{l}_1$. Since this is true for all points $Q_1, Q_2, Q_3, \ldots$ on the constant Φ surface, it follows that $[\nabla\Phi]_P$ must be normal to all possible infinitesimal displacement vectors $d\mathbf{l}_1, d\mathbf{l}_2, d\mathbf{l}_3, \ldots$ drawn at P and hence

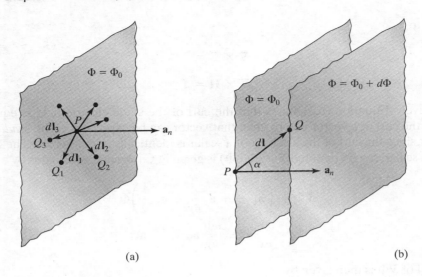

(a)   (b)

FIGURE 6.1

For discussing the physical interpretation of the gradient of a scalar function.

is normal to the surface. Denoting $\mathbf{a}_n$ to be the unit normal vector to the surface at P, we then have

$$[\nabla \Phi]_P = |\nabla \Phi|_P \, \mathbf{a}_n \tag{6.8}$$

Let us now consider two surfaces on which Φ is constant, having values Φ_0 and $\Phi_0 + d\Phi$, as shown in Figure 6.1(b). Let P and Q be points on the $\Phi = \Phi_0$ and $\Phi = \Phi_0 + d\Phi$ surfaces, respectively, and $d\mathbf{l}$ be the vector drawn from P to Q. Then from (6.7) and (6.8),

$$\begin{aligned}
d\Phi &= [\nabla \Phi]_P \cdot d\mathbf{l} \\
&= |\nabla \Phi|_P \, \mathbf{a}_n \cdot d\mathbf{l} \\
&= |\nabla \Phi|_P \, dl \cos \alpha
\end{aligned} \tag{6.9}$$

where α is the angle between $\mathbf{a}_n$ at P and $d\mathbf{l}$. Thus,

$$|\nabla \Phi|_P = \frac{d\Phi}{dl \cos \alpha} \tag{6.10}$$

Since $dl \cos \alpha$ is the distance between the two surfaces along $\mathbf{a}_n$ and hence is the shortest distance between them, it follows that $|\nabla \Phi|_P$ is the maximum rate of increase of Φ at the point P. Thus, the gradient of a scalar function Φ at a point is a vector having magnitude equal to the maximum rate of increase of Φ at that point and is directed along the direction of the maximum rate of increase, which is normal to the constant Φ surface passing through that point. This concept of the gradient of a scalar function is often utilized to find a unit vector normal to a given surface. We shall illustrate this by means of an example.

Example 6.1

Let us find the unit vector normal to the surface $y = x^2$ at the point $(2, 4, 1)$ by using the concept of the gradient of a scalar.

Writing the equation for the surface as

$$x^2 - y = 0$$

we note that the scalar function that is constant on the surface is given by

$$\Phi(x, y, z) = x^2 - y$$

The gradient of the scalar function is then given by

$$\nabla\Phi = \nabla(x^2 - y)$$
$$= \frac{\partial(x^2 - y)}{\partial x}\mathbf{a}_x + \frac{\partial(x^2 - y)}{\partial y}\mathbf{a}_y + \frac{\partial(x^2 - y)}{\partial z}\mathbf{a}_z$$
$$= 2x\mathbf{a}_x - \mathbf{a}_y$$

The value of the gradient at the point $(2, 4, 1)$ is $2(2)\mathbf{a}_x - \mathbf{a}_y = 4\mathbf{a}_x - \mathbf{a}_y$. Thus, the required unit vector is

$$\mathbf{a}_n = \pm\frac{4\mathbf{a}_x - \mathbf{a}_y}{|4\mathbf{a}_x - \mathbf{a}_y|} = \pm\left(\frac{4}{\sqrt{17}}\mathbf{a}_x - \frac{1}{\sqrt{17}}\mathbf{a}_y\right)$$

Returning to Maxwell's curl equation for the static electric field given by (6.3), we can now express **E** as the gradient of a scalar function, say, Φ. The question then arises as to what this scalar function is. To obtain the answer, let us consider a region of static electric field. Then we can draw a set of surfaces orthogonal everywhere to the field lines, as shown in Figure 6.2. These surfaces correspond to the constant Φ

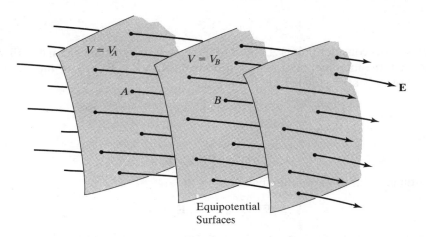

Equipotential
Surfaces

FIGURE 6.2

A set of equipotential surfaces in a region of static electric field.

surfaces. Since on any such surface $\mathbf{E} \cdot d\mathbf{l} = 0$, no work is involved in the movement of a test charge from one point to another on the surface. Such surfaces are known as the *equipotential surfaces*. Since they are orthogonal to the field lines, they may physically be occupied by conductors without affecting the field distribution.

Movement of a test charge from a point, say A, on one equipotential surface to a point, say B, on another equipotential surface involves an amount of work per unit charge equal to $\int_A^B \mathbf{E} \cdot d\mathbf{l}$ to be done by the field. This quantity is known as the *electric potential difference* between the points A and B and is denoted by the symbol $[V]_A^B$. It has the units of volts. There is a potential drop from A to B if work is done by the field and a potential rise if work is done against the field by an external agent. The situation is similar to that in the earth's gravitational field for which there is a potential drop associated with the movement of a mass from a point of higher elevation to a point of lower elevation and a potential rise for just the opposite case.

It is convenient to define an *electric potential* associated with each point. The potential at point A, denoted V_A, is simply the potential difference between point A and a reference point, say O. It is the amount of work per unit charge done by the field in connection with the movement of a test charge from A to O, or the amount of work per unit charge done against the field by an external agent in moving the test charge from O to A. Thus,

$$V_A = \int_A^O \mathbf{E} \cdot d\mathbf{l} = -\int_O^A \mathbf{E} \cdot d\mathbf{l} \tag{6.11}$$

and

$$
\begin{aligned}
[V]_A^B &= \int_A^B \mathbf{E} \cdot d\mathbf{l} = \int_A^O \mathbf{E} \cdot d\mathbf{l} + \int_O^B \mathbf{E} \cdot d\mathbf{l} \\
&= \int_A^O \mathbf{E} \cdot d\mathbf{l} - \int_B^O \mathbf{E} \cdot d\mathbf{l} \\
&= V_A - V_B
\end{aligned}
\tag{6.12}
$$

If we now consider points A and B to be separated by infinitesimal length $d\mathbf{l}$ from A to B, then the incremental potential drop from A to B is $\mathbf{E}_A \cdot d\mathbf{l}$, or the incremental potential rise dV along the length $d\mathbf{l}$ is given by

$$dV = -\mathbf{E}_A \cdot d\mathbf{l} \tag{6.13}$$

Writing

$$dV = [\nabla V]_A \cdot d\mathbf{l} \tag{6.14}$$

in accordance with (6.7), we then have

$$[\nabla V]_A \cdot d\mathbf{l} = -\mathbf{E}_A \cdot d\mathbf{l} \tag{6.15}$$

Since (6.15) is true at any point A in the static electric field, it follows that

$$\mathbf{E} = -\nabla V \tag{6.16}$$

Thus, we have obtained the result that the static electric field is the negative of the gradient of the electric potential.

Before proceeding further, we note that the potential difference we have defined here has the same meaning as the voltage between two points, defined in Section 2.1. We, however, recall that the voltage between two points A and B in a time-varying field is in general dependent on the path followed from A to B to evaluate $\int_A^B \mathbf{E} \cdot d\mathbf{l}$, since, according to Faraday's law,

$$\oint_C \mathbf{E} \cdot d\mathbf{l} = -\frac{d}{dt} \int_S \mathbf{B} \cdot d\mathbf{S} \tag{6.17}$$

is not in general equal to zero. On the other hand, the potential difference (or voltage) between two points A and B in a static electric field is independent of the path followed from A to B to evaluate $\int_A^B \mathbf{E} \cdot d\mathbf{l}$, since, for static fields, $\partial/\partial t = 0$, and (6.17) reduces to

$$\oint_C \mathbf{E} \cdot d\mathbf{l} = 0 \tag{6.18}$$

Thus, the potential difference between two points in a static electric field has a unique value. Fields for which the line integral around a closed path is zero are known as *conservative* fields. The static electric field is a conservative field. The earth's gravitational field is another example of a conservative field, since the work done in moving a mass around a closed path is equal to zero.

Returning now to the discussion of electric potential, let us consider the electric field of a point charge and investigate the electric potential due to the point charge. To do this, we recall from Section 1.5 that the electric field intensity due to a point charge Q is directed radially away from the point charge and its magnitude is $Q/4\pi\epsilon_0 R^2$, where R is the radial distance from the point charge. Since the equipotential surfaces are everywhere orthogonal to the field lines, it then follows that they are spherical surfaces centered at the point charge, as shown by the cross-sectional view in Figure 6.3. If we now consider two equipotential surfaces of radii R and $R + dR$, the potential drop from the surface of radius R to the surface of radius $R + dR$ is $(Q/4\pi\epsilon_0 R^2)\,dR$, or, the incremental potential rise dV is given by

$$dV = -\frac{Q}{4\pi\epsilon_0 R^2}\,dR$$

$$= d\left(\frac{Q}{4\pi\epsilon_0 R} + C\right) \tag{6.19}$$

where C is a constant. Thus,

$$V(R) = \frac{Q}{4\pi\epsilon_0 R} + C \tag{6.20}$$

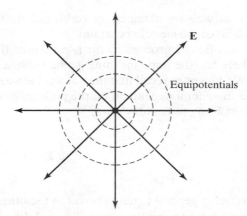

FIGURE 6.3

Cross-sectional view of equipotential surfaces and electric field lines for a point charge.

We can conveniently set C equal to zero by noting that it is equal to $V(\infty)$ and by choosing $R = \infty$ for the reference point. Thus, we obtain the electric potential due to a point charge Q to be

$$V = \frac{Q}{4\pi\epsilon_0 R} \qquad (6.21)$$

We note that the potential drops off inversely with the radial distance away from the point charge. Equation (6.21) is often the starting point for the computation of the potential field due to static charge distributions and the subsequent determination of the electric field by using (6.16).

6.2 POISSON'S AND LAPLACE'S EQUATIONS

In the previous section, we learned that for the static electric field, $\nabla \times \mathbf{E}$ is equal to zero, and hence

$$\mathbf{E} = -\nabla V$$

Substituting this result into Maxwell's divergence equation for $\mathbf{D}$, and assuming ϵ to be uniform, we obtain

$$\nabla \cdot \mathbf{D} = \nabla \cdot \epsilon\mathbf{E} = \epsilon\nabla \cdot \mathbf{E}$$
$$= \epsilon\nabla \cdot (-\nabla V) = \rho$$

or

$$\nabla \cdot \nabla V = -\frac{\rho}{\epsilon}$$

The quantity $\nabla \cdot \nabla V$ is known as the *Laplacian* of V, denoted $\nabla^2 V$ (del squared V). Thus, we have

$$\nabla^2 V = -\frac{\rho}{\epsilon} \qquad (6.22)$$

This equation is known as the *Poisson's equation*. It governs the relationship between the volume charge density ρ in a region and the potential in that region. In Cartesian coordinates,

$$\nabla^2 V = \nabla \cdot \nabla V$$

$$= \left(\mathbf{a}_x \frac{\partial}{\partial x} + \mathbf{a}_y \frac{\partial}{\partial y} + \mathbf{a}_z \frac{\partial}{\partial z} \right) \cdot \left(\frac{\partial V}{\partial x} \mathbf{a}_x + \frac{\partial V}{\partial y} \mathbf{a}_y + \frac{\partial V}{\partial z} \mathbf{a}_z \right)$$

$$= \frac{\partial^2 V}{\partial x^2} + \frac{\partial^2 V}{\partial y^2} + \frac{\partial^2 V}{\partial z^2} \tag{6.23}$$

and Poisson's equation becomes

$$\frac{\partial^2 V}{\partial x^2} + \frac{\partial^2 V}{\partial y^2} + \frac{\partial^2 V}{\partial z^2} = -\frac{\rho}{\epsilon} \tag{6.24}$$

For the one-dimensional case in which V varies with x only, $\partial^2 V/\partial y^2$ and $\partial^2 V/\partial z^2$ are both equal to zero, and (6.24) reduces to

$$\frac{\partial^2 V}{\partial x^2} = \frac{d^2 V}{dx^2} = -\frac{\rho}{\epsilon} \tag{6.25}$$

We shall illustrate the application of (6.25) by means of an example.

Example 6.2

Let us consider the space charge layer in a *p-n* junction semiconductor with zero bias, as shown in Figure 6.4(a), in which the region $x < 0$ is doped *p*-type and the region $x > 0$ is doped *n*-type. To review briefly the formation of the space charge layer, we note that since the density of the holes on the *p* side is larger than that on the *n* side, there is a tendency for the holes to diffuse to the *n* side and recombine with the electrons. Similarly, there is a tendency for the electrons on the *n* side to diffuse to the *p* side and recombine with the holes. The diffusion of holes leaves behind negatively charged acceptor atoms, and the diffusion of electrons leaves behind positively charged donor atoms. Since these acceptor and donor atoms are immobile, a space charge layer, also known as the *depletion layer*, is formed in the region of the junction, with negative charges on the *p* side and positive charges on the *n* side. This space charge gives rise to an electric field directed from the *n* side of the junction to the *p* side so that it opposes diffusion of the mobile carriers across the junction, thereby resulting in an equilibrium. For simplicity, let us consider an abrupt junction, that is, a junction in which the impurity concentration is constant on either side of the junction. Let N_A and N_D be the acceptor and donor ion concentrations, respectively, and d_p and d_n be the widths in the *p* and *n* regions, respectively, of the depletion layer. The space charge density ρ is then given by

$$\rho = \begin{cases} -|e|N_A & \text{for } -d_p < x < 0 \\ |e|N_D & \text{for } 0 < x < d_n \end{cases} \tag{6.26}$$

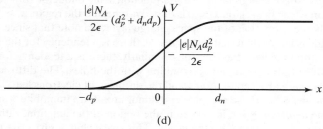

FIGURE 6.4

For illustrating the application of Poisson's equation for the determination of the potential distribution for a *p-n* junction semiconductor.

as shown in Figure 6.4(b), where $|e|$ is the magnitude of the electronic charge. Since the semiconductor is electrically neutral, the total acceptor charge must be equal to the total donor charge; that is,

$$|e|N_A d_p = |e|N_D d_n \qquad (6.27)$$

We wish to find the potential distribution in the depletion layer and the depletion layer width in terms of the potential difference across the depletion layer and the acceptor and donor ion concentrations.

Substituting (6.26) into (6.25), we obtain the equation governing the potential distribution to be

$$
\frac{d^2V}{dx^2} =
\begin{cases}
\dfrac{|e|N_A}{\epsilon} & \text{for } -d_p < x < 0 \\[2ex]
-\dfrac{|e|N_D}{\epsilon} & \text{for } 0 < x < d_n
\end{cases}
\tag{6.28}
$$

To solve (6.28) for V, we integrate it once and obtain

$$
\frac{dV}{dx} =
\begin{cases}
\dfrac{|e|N_A}{\epsilon} x + C_1 & \text{for } -d_p < x < 0 \\[2ex]
-\dfrac{|e|N_D}{\epsilon} x + C_2 & \text{for } 0 < x < d_n
\end{cases}
$$

where C_1 and C_2 are constants of integration. To evaluate C_1 and C_2, we note that since $\mathbf{E} = -\nabla V = -(\partial V/\partial x)\mathbf{a}_x$, $\partial V/\partial x$ is simply equal to $-E_x$. Since the electric field lines begin on the positive charges and end on the negative charges, and in view of (6.27), the field and, hence, $\partial V/\partial x$ must vanish at $x = -d_p$ and $x = d_n$, giving us

$$
\frac{dV}{dx} =
\begin{cases}
\dfrac{|e|N_A}{\epsilon}(x + d_p) & \text{for } -d_p < x < 0 \\[2ex]
-\dfrac{|e|N_D}{\epsilon}(x - d_n) & \text{for } 0 < x < d_n
\end{cases}
\tag{6.29}
$$

The field intensity, that is, $-dV/dx$, may now be sketched as a function of x, as shown in Figure 6.4(c).

Proceeding further, we integrate (6.29) and obtain

$$
V =
\begin{cases}
\dfrac{|e|N_A}{2\epsilon}(x + d_p)^2 + C_3 & \text{for } -d_p < x < 0 \\[2ex]
-\dfrac{|e|N_D}{2\epsilon}(x - d_n)^2 + C_4 & \text{for } 0 < x < d_n
\end{cases}
$$

where C_3 and C_4 are constants of integration. To evaluate C_3 and C_4, we first set the potential at $x = -d_p$ arbitrarily equal to zero to obtain C_3 equal to zero. Then we make use of the condition that the potential be continuous at $x = 0$, since the discontinuity in dV/dx at $x = 0$ is finite, to obtain

$$
\frac{|e|N_A}{2\epsilon} d_p^2 = -\frac{|e|N_D}{2\epsilon} d_n^2 + C_4
$$

or

$$
C_4 = \frac{|e|}{2\epsilon}(N_A d_p^2 + N_D d_n^2)
$$

Substituting this value for C_4 and setting C_3 equal to zero in the expression for V, we get the required solution

$$V = \begin{cases} \dfrac{|e|N_A}{2\epsilon}(x + d_p)^2 & \text{for } -d_p < x < 0 \\[3mm] -\dfrac{|e|N_D}{2\epsilon}(x^2 - 2xd_n) + \dfrac{|e|N_A}{2\epsilon}d_p^2 & \text{for } 0 < x < d_n \end{cases} \tag{6.30}$$

The variation of potential with x as given by (6.30) is shown in Figure 6.4(d).

We can proceed further and find the width $d = d_p + d_n$ of the depletion layer by setting $V(d_n)$ equal to the contact potential, V_0, that is, the potential difference across the depletion layer resulting from the electric field in the layer. Thus,

$$\begin{aligned} V_0 = V(d_n) &= \frac{|e|N_D}{2\epsilon}d_n^2 + \frac{|e|N_A}{2\epsilon}d_p^2 \\[2mm] &= \frac{|e|}{2\epsilon}\frac{N_D(N_A + N_D)}{N_A + N_D}d_n^2 + \frac{|e|}{2\epsilon}\frac{N_A(N_A + N_D)}{N_A + N_D}d_p^2 \\[2mm] &= \frac{|e|}{2\epsilon}\frac{N_A N_D}{N_A + N_D}(d_n^2 + d_p^2 + 2d_n d_p) \\[2mm] &= \frac{|e|}{2\epsilon}\frac{N_A N_D}{N_A + N_D}d^2 \end{aligned}$$

where we have made use of (6.27). Finally, we obtain the result that

$$d = \sqrt{\frac{2\epsilon V_0}{|e|}\left(\frac{1}{N_A} + \frac{1}{N_D}\right)}$$

which tells us that the depletion layer width is smaller, the heavier the doping is. This property is used in tunnel diodes to achieve layer widths on the order of 10^{-6} cm by heavy doping as compared to widths on the order of 10^{-4} cm in ordinary p-n junctions.

We have just illustrated an example of the application of Poisson's equation involving the solution for the potential distribution for a given charge distribution. Poisson's equation is even more useful for the solution of problems in which the charge distribution is the quantity to be determined, given the functional dependence of the charge density on the potential. We shall, however, not pursue this topic any further.

If the charge density in a region is zero, then Poisson's equation reduces to

$$\nabla^2 V = 0 \tag{6.31}$$

This equation is known as *Laplace's equation*. It governs the behavior of the potential in a charge-free region. In Cartesian coordinates, it is given by

$$\frac{\partial^2 V}{\partial x^2} + \frac{\partial^2 V}{\partial y^2} + \frac{\partial^2 V}{\partial z^2} = 0 \tag{6.32}$$

The problems for which Laplace's equation is applicable consist of finding the potential distribution in the region between two conductors given the charge distribution on the surfaces of the conductors or the potentials of the conductors or a combination of the two. The procedure involves the solving of Laplace's equation subject to the boundary conditions on the surfaces of the conductors. We shall do this in the following section.

6.3 STATIC FIELDS AND CIRCUIT ELEMENTS

In the previous two sections, we considered static fields with reference to electric field alone. In this section, we shall expand the treatment to all types of static fields, for the purpose of introducing circuit elements. Thus, for static fields, $\partial/\partial t = 0$. Maxwell's equations in integral form and the law of conservation of charge become

$$\oint_C \mathbf{E} \cdot d\mathbf{l} = 0 \tag{6.33a}$$

$$\oint_C \mathbf{H} \cdot d\mathbf{l} = \int_S \mathbf{J} \cdot d\mathbf{S} \tag{6.33b}$$

$$\oint_S \mathbf{D} \cdot d\mathbf{S} = \int_V \rho \, dv \tag{6.33c}$$

$$\oint_S \mathbf{B} \cdot d\mathbf{S} = 0 \tag{6.33d}$$

$$\oint_S \mathbf{J} \cdot d\mathbf{S} = 0 \tag{6.33e}$$

whereas Maxwell's equations in differential form and the continuity equation reduce to

$$\nabla \times \mathbf{E} = 0 \tag{6.34a}$$
$$\nabla \times \mathbf{H} = \mathbf{J} \tag{6.34b}$$
$$\nabla \cdot \mathbf{D} = \rho \tag{6.34c}$$
$$\nabla \cdot \mathbf{B} = 0 \tag{6.34d}$$
$$\nabla \cdot \mathbf{J} = 0 \tag{6.34e}$$

Immediately, one can see that, unless $\mathbf{J}$ includes a component due to conduction current, the equations involving the electric field are completely independent of those involving the magnetic field. Thus, the fields can be subdivided into *static electric fields*, or *electrostatic fields*, governed by (6.33a) and (6.33c), or (6.34a) and (6.34c), and *static magnetic fields*, or *magnetostatic fields*, governed by (6.33b) and (6.33d), or (6.34b) and (6.34d). The source of a static electric field is ρ, whereas the source of a static magnetic field is $\mathbf{J}$. One can also see from (6.33e) or (6.34e) that no relationship exists between $\mathbf{J}$ and ρ. If $\mathbf{J}$ includes a component due to conduction current, then, since $\mathbf{J}_c = \sigma \mathbf{E}$, a coupling between the electric and magnetic fields exists for that part of the total field associated with $\mathbf{J}_c$. However, the coupling is only one way, since the right side of (6.33a) or (6.34a) is still zero. The field is then referred to as *electromagnetostatic field*. It can also be seen, then, that for consistency, the right sides of (6.33c) and (6.34c) must be zero, since the right sides of (6.33e) and (6.34e) are zero. We shall now consider each of the three types of static fields separately and discuss some fundamental aspects.

Electrostatic Fields and Capacitance

The equations of interest are (6.33a) and (6.33c), or (6.34a) and (6.34c). The first of each pair of these equations simply tells us that the electrostatic field is a conservative field, and the second of each pair of these equations enables us, in principle, to determine the electrostatic field for a given charge distribution. Alternatively, the Poisson's equation, equation (6.22), can be used to find the electric scalar potential, V, from which the electrostatic field can be determined by using (6.16).

In a charge-free region, the Poisson's equation reduces to the Laplace's equation, (6.31). The field is then due to charges outside the region, such as surface charge on conductors bounding the region. The situation is then one of solving a boundary value problem, as we shall illustrate by means of an example.

Example 6.3

Figure 6.5(a) is that of a parallel-plate arrangement in which two parallel, perfectly conducting plates ($\sigma = \infty$, $\mathbf{E} = 0$) of dimensions w along the y-direction and l along the z-direction lie in the $x = 0$ and $x = d$ planes. The region between the plates is a perfect dielectric ($\sigma = 0$) of material parameters ϵ and μ. The thickness of the plates is shown exaggerated for convenience in illustration. A potential difference of V_0 is maintained between the plates by connecting a direct voltage source at the end $z = -l$. If fringing of the field due to the finite dimensions of the structure normal to the x-direction is neglected, or, if it is assumed that the structure is part of one which is infinite in extent normal to the x-direction, then the problem can be treated as one-dimensional with x as the variable, and (6.31) reduces to

$$\frac{d^2V}{dx^2} = 0 \tag{6.35}$$

We wish to carry out the electrostatic field analysis for this arrangement.

The solution for the potential in the charge-free region between the plates is given by

$$V(x) = \frac{V_0}{d}(d - x) \tag{6.36}$$

which satisfies (6.35), as well as the boundary conditions of $V = 0$ at $x = d$ and $V = V_0$ at $x = 0$. The electric field intensity between the plates is then given by

$$\mathbf{E} = -\nabla V = \frac{V_0}{d}\mathbf{a}_x \tag{6.37}$$

as depicted in the cross-sectional view in Figure 6.5(b), and resulting in displacement flux density

$$\mathbf{D} = \frac{\epsilon V_0}{d}\mathbf{a}_x \tag{6.38}$$

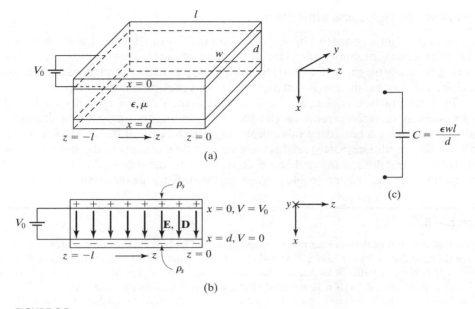

FIGURE 6.5

Electrostatic field analysis for a parallel-plate arrangement.

Then, using the boundary condition for the normal component of **D** given by (5.94c), we obtain the magnitude of the charge on either plate to be

$$Q = \left(\frac{\epsilon V_0}{d}\right)(wl) = \frac{\epsilon wl}{d} V_0 \tag{6.39}$$

We can now find the familiar circuit parameter, the capacitance, C, of the parallel-plate arrangement, which is defined as the ratio of the magnitude of the charge on either plate to the potential difference V_0. Thus,

$$C = \frac{Q}{V_0} = \frac{\epsilon wl}{d} \tag{6.40}$$

Note that the units of C are the units of ϵ times meter, that is, farads. The phenomenon associated with the arrangement is that energy is stored in the capacitor in the form of electric field energy between the plates, as given by

$$W_e = \left(\frac{1}{2}\epsilon E_x^2\right)(wld)$$

$$= \frac{1}{2}\left(\frac{\epsilon wl}{d}\right)V_0^2$$

$$= \frac{1}{2}CV_0^2 \tag{6.41}$$

the familiar expression for energy stored in a capacitor.

Magnetostatic Fields and Inductance

The equations of interest are (6.33b) and (6.33d), or (6.34b) and (6.34d). The second of each pair of these equations simply tells us that the magnetostatic field is solenoidal, which as we know holds for any magnetic field, and the first of each pair of these equations enables us, in principle, to determine the magnetostatic field for a given current distribution.

In a current-free region, $\mathbf{J} = 0$. The field is then due to currents outside the region, such as surface currents on conductors bounding the region. The situation is then one of solving a boundary value problem as in the case of (6.31). However, since the boundary condition (5.94b) relates the magnetic field directly to the surface current density, it is straightforward and more convenient to determine the magnetic field directly by using (6.34b) and (6.34d). We shall illustrate by means of an example.

Example 6.4

Figure 6.6(a) is that of the parallel-plate arrangement of Figure 6.5(a) with the plates connected by another conductor at the end $z = 0$ and driven by a source of direct current I_0 at the end $z = -l$. If fringing of the field due to the finite dimensions of the structure normal to the x-direction is neglected, or, if it is assumed that the structure is part of one which is infinite in extent normal to the x-direction, then the problem can be treated as one-dimensional with x as the variable and we can write the current density on the plates to be

$$\mathbf{J}_S = \begin{cases} (I_0/w)\mathbf{a}_z & \text{on the plate } x = 0 \\ (I_0/w)\mathbf{a}_x & \text{on the plate } z = 0 \\ -(I_0/w)\mathbf{a}_z & \text{on the plate } x = d \end{cases} \tag{6.42}$$

We wish to carry out the magnetostatic field analysis for this arrangement.

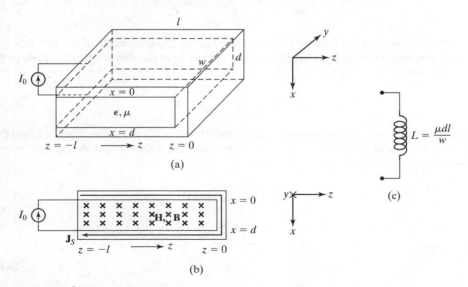

(a)

(b)

(c)

FIGURE 6.6

Magnetostatic field analysis for a parallel-plate arrangement.

In the current-free region between the plates, (6.34b) reduces to

$$\begin{vmatrix} \mathbf{a}_x & \mathbf{a}_y & \mathbf{a}_z \\ \dfrac{\partial}{\partial x} & 0 & 0 \\ H_x & H_y & H_z \end{vmatrix} = 0 \tag{6.43}$$

and (6.34d) reduces to

$$\frac{\partial B_x}{\partial x} = 0 \tag{6.44}$$

so that each component of the field, if it exists, has to be uniform. This automatically forces H_x and H_z to be zero, since nonzero value of these components do not satisfy the boundary conditions (5.94b) and (5.94d) on the plates, keeping in mind that the field is entirely in the region between the conductors. Thus, as depicted in the cross-sectional view in Figure 6.6(b),

$$\mathbf{H} = \frac{I_0}{w} \mathbf{a}_y \tag{6.45}$$

which satisfies the boundary condition (5.94b) on all three plates, and results in magnetic flux density

$$\mathbf{B} = \frac{\mu I_0}{w} \mathbf{a}_y \tag{6.46}$$

The magnetic flux, ψ, linking the current I_0, is then given by

$$\psi = \left(\frac{\mu I_0}{w}\right)(dl) = \left(\frac{\mu dl}{w}\right) I_0 \tag{6.47}$$

We can now find the familiar circuit parameter, the inductance, L, of the parallel-plate arrangement, which is defined as the ratio of the magnetic flux linking the current to the current. Thus,

$$L = \frac{\psi}{I_0} = \frac{\mu dl}{w} \tag{6.48}$$

Note that the units of L are the units of μ times meter, that is, henrys. The phenomenon associated with the arrangement is that energy is stored in the inductor in the form of magnetic field energy between the plates, as given by

$$\begin{aligned} W_m &= \left(\frac{1}{2} \mu H^2\right)(wld) \\ &= \frac{1}{2}\left(\frac{\mu dl}{w}\right) I_0^2 \\ &= \frac{1}{2} L I_0^2 \end{aligned} \tag{6.49}$$

the familiar expression for energy stored in an inductor.

Electromagnetostatic Fields and Conductance

The equations of interest are

$$\oint_C \mathbf{E} \cdot d\mathbf{l} = 0 \tag{6.50a}$$

$$\oint_C \mathbf{H} \cdot d\mathbf{l} = \oint_S \mathbf{J}_c \cdot d\mathbf{S} = \sigma \oint_S \mathbf{E} \cdot d\mathbf{S} \tag{6.50b}$$

$$\oint_S \mathbf{D} \cdot d\mathbf{S} = 0 \tag{6.50c}$$

$$\oint_S \mathbf{B} \cdot d\mathbf{S} = 0 \tag{6.50d}$$

or, in differential form,

$$\nabla \times \mathbf{E} = 0 \tag{6.51a}$$
$$\nabla \times \mathbf{H} = \mathbf{J}_c = \sigma\mathbf{E} \tag{6.51b}$$
$$\nabla \cdot \mathbf{D} = 0 \tag{6.51c}$$
$$\nabla \cdot \mathbf{B} = 0 \tag{6.51d}$$

From (6.51a) and (6.51c), we note that Laplace's equation for the electrostatic potential, (6.31), is satisfied, so that, for a given problem, the electric field can be found in the same manner as in the case of the example of Figure 6.6. The magnetic field is then found by using (6.51b), and making sure that (6.51d) is also satisfied. We shall illustrate by means of an example.

Example 6.5

Figure 6.7(a) is that of the parallel-plate arrangement of Figure 6.5(a) but with an imperfect dielectric material of parameters σ, ϵ, and μ, between the plates. We wish to carry out the electromagnetostatic field analysis of the arrangement.

The electric field between the plates is the same as that given by (6.37), that is,

$$\mathbf{E} = \frac{V_0}{d} \mathbf{a}_x \tag{6.52}$$

resulting in a conduction current of density

$$\mathbf{J}_c = \frac{\sigma V_0}{d} \mathbf{a}_x \tag{6.53}$$

from the top plate to the bottom plate, as depicted in the cross-sectional view of Figure 6.7(b). Since $\partial\rho/\partial t = 0$ at the boundaries between the plates and the slab, continuity of current is satisfied by the flow of surface current on the plates. At the input $z = -l$, this surface current, which is the current drawn from the source, must be equal to the total current flowing from the top to the bottom plate. It is given by

$$I_c = \left(\frac{\sigma V_0}{d}\right)(wl) = \frac{\sigma wl}{d} V_0 \tag{6.54}$$

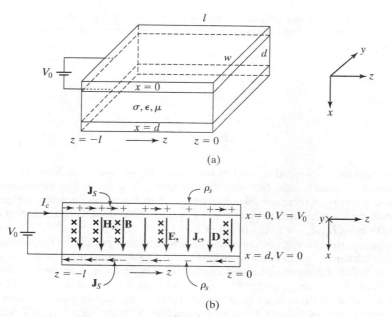

FIGURE 6.7

Electromagnetostatic field analysis for a parallel-plate arrangement.

We can now find the familiar circuit parameter, the conductance, G, of the parallel-plate arrangement, which is defined as the ratio of the current drawn from the source to the source voltage V_0. Thus,

$$G = \frac{I_c}{V_0} = \frac{\sigma wl}{d} \tag{6.55}$$

Note that the units of G are the units of σ times meter, that is, siemens (S). The reciprocal quantity, R, the resistance of the parallel-plate arrangement, is given by

$$R = \frac{V_0}{I_c} = \frac{d}{\sigma wl} \tag{6.56}$$

The unit of R is ohms. The phenomenon associated with the arrangement is that power is dissipated in the material between the plates, as given by

$$
\begin{aligned}
P_d &= (\sigma E^2)(wld) \\
&= \left(\frac{\sigma wl}{d}\right)V_0^2 \\
&= GV_0^2 \\
&= \frac{V_0^2}{R}
\end{aligned}
\tag{6.57}
$$

the familiar expression for power dissipated in a resistor.

Proceeding further, we find the magnetic field between the plates by using (6.51b), and noting that the geometry of the situation requires a y-component of $\mathbf{H}$, dependent on z, to satisfy the equation. Thus,

$$\mathbf{H} = H_y(z)\,\mathbf{a}_y \tag{6.58a}$$

$$\frac{\partial H_y}{\partial z} = -\frac{\sigma V_0}{d} \tag{6.58b}$$

$$\mathbf{H} = -\frac{\sigma V_0}{d}\, z\, \mathbf{a}_y \tag{6.58c}$$

where the constant of integration is set to zero, since the boundary condition at $z = 0$ requires H_y to be zero for z equal to zero. Note that the magnetic field is directed in the positive y-direction (since z is negative) and increases linearly from $z = 0$ to $z = -l$, as depicted in Figure 6.7(b). It also satisfies the boundary condition at $z = -l$ by being consistent with the current drawn from the source to be $w[H_y]_{z=-l} = (\sigma V_0/d)(wl) = I_c$.

Because of the existence of the magnetic field, the arrangement is characterized by an inductance, which can be found either by using the flux linkage concept or by the energy method. To use the flux linkage concept, we recognize that a differential amount of magnetic flux $d\psi' = \mu H_y d(dz')$ between z equal to $(z' - dz')$ and z equal to z', where $-l < z' < 0$, links only that part of the current that flows from the top plate to the bottom plate between $z = z'$ and $z = 0$, thereby giving a value of $(-z'/l)$ for the fraction, N, of the total current linked. Thus, the inductance, familiarly known as the internal inductance, denoted L_i, since it is due to magnetic field internal to the current distribution, as compared to that in (6.48) for which the magnetic field is external to the current distribution, is given by

$$L_i = \frac{1}{I_c} \int_{z'=-1}^{0} N\, d\psi'$$

$$= \frac{1}{3}\frac{\mu dl}{w} \tag{6.59}$$

or, $1/3$ times the inductance of the structure if $\sigma = 0$ and the plates are joined at $z = 0$, as in Figure 6.6(b).

Alternatively, if the energy method is used by computing the energy stored in the magnetic field and setting it equal to $\frac{1}{2}L_i I_c^2$, then we have

$$L_i = \frac{1}{I_c^2}(dw) \int_{z=-l}^{0} \mu H_y^2\, dz$$

$$= \frac{1}{3}\frac{\mu dl}{w} \tag{6.60}$$

same as in (6.59).

Finally, recognizing that there is energy storage associated with the electric field between the plates, we note that the arrangement has also associated with it a capacitance C, equal to $\epsilon wl/d$. Thus, all three properties of conductance, capacitance, and inductance are associated with the structure. Since for $\sigma = 0$ the situation reduces to that of Figure 6.5, we can represent the arrangement of Figure 6.7 to be equivalent to the circuit shown in Figure 6.8. Note that the capacitor is charged to the voltage V_0 and the current through it is zero (open circuit condition).

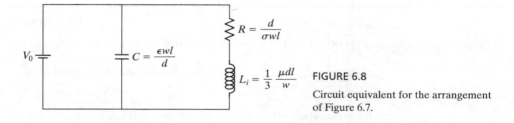

FIGURE 6.8

Circuit equivalent for the arrangement of Figure 6.7.

The voltage across the inductor is zero (short circuit condition) and the current through it is V_0/R. Thus, the current drawn from the voltage source is V_0/R and the voltage source views a single resistor R, as far as the current drawn from it is concerned.

6.4 LOW-FREQUENCY BEHAVIOR VIA QUASISTATICS

In the preceding section, we introduced circuit elements via static fields. A class of dynamic fields for which certain features can be analyzed as though the fields were static are known as *quasistatic fields*. In terms of behavior in the frequency domain, they are low-frequency extensions of static fields present in a physical structure, when the frequency of the source driving the structure is zero, or low-frequency approximations of time-varying fields in the structure that are complete solutions to Maxwell's equations. In this section, we consider the approach of low-frequency extensions of static fields. Thus, for a given structure, we begin with a time-varying field having the same spatial characteristics as that of the static field solution for the structure, and obtain field solutions containing terms up to and including the first power (which is the lowest power) in ω for their amplitudes. Depending on whether the predominant static field is electric or magnetic, quasistatic fields are called *electroquasistatic fields* or *magneto-quasistatic fields*. We shall now consider these separately.

Electroquasistatic Fields

For electroquasistatic fields, we begin with the electric field having the spatial dependence of the static field solution for the given arrangement. We shall illustrate by means of an example.

Example 6.6

Figure 6.9 shows the cross-sectional view of the arrangement of Figure 6.5(a) excited by a sinusoidally time-varying voltage source $V_g(t) = V_0 \cos \omega t$ instead of a direct voltage source. We wish to carry out the electroquasistatic field analysis for the arrangement.

From (6.37), we write

$$\mathbf{E}_0 = \frac{V_0}{d} \cos \omega t \, \mathbf{a}_x \tag{6.61}$$

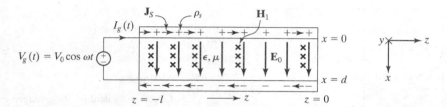

FIGURE 6.9

Electroquasistatic field analysis for the parallel-plate structure of Figure 6.5.

where the subscript 0 denotes that the amplitude of the field is of the zeroth power in ω. This results in a magnetic field in accordance with Maxwell's equation for the curl of **H**, given by (3.28). Thus, noting that $\mathbf{J} = 0$ in view of the perfect dielectric medium, we have for the geometry of the arrangement,

$$\frac{\partial H_{y1}}{\partial z} = -\frac{\partial D_{x0}}{\partial t} = \frac{\omega \epsilon V_0}{d} \sin \omega t$$

$$\mathbf{H}_1 = \frac{\omega \epsilon V_0 z}{d} \sin \omega t \, \mathbf{a}_y \qquad (6.62)$$

where we have also satisfied the boundary condition at $z = 0$ by choosing the constant of integration such that $[H_{y1}]_{z=0}$ is zero, and the subscript 1 denotes that the amplitude of the field is of the first power in ω. Note that the amplitude of H_{y1} varies linearly with z, from zero at $z = 0$ to a maximum at $z = -l$.

We stop the solution here, because continuing the process by substituting (6.62) into Maxwell's curl equation for **E**, (3.17), to obtain the resulting electric field will yield a term having amplitude proportional to the second power in ω. This simply means that the fields given as a pair by (6.61) and (6.62) do not satisfy (3.17), and hence are not complete solutions to Maxwell's equations. They are the quasistatic fields. The complete solutions are obtained by solving Maxwell's equations simultaneously and subject to the boundary conditions for the given problem.

Proceeding further, we obtain the current drawn from the voltage source to be

$$I_g(t) = w[H_{y1}]_{z=-l}$$

$$= -\omega \left(\frac{\epsilon w l}{d} \right) V_0 \sin \omega t$$

$$= C \frac{dV_g(t)}{dt} \qquad (6.63a)$$

or,

$$\bar{I}_g = j\omega C \bar{V}_g \qquad (6.63b)$$

where $C = (\epsilon w l / d)$ is the capacitance of the arrangement obtained from static field considerations. Thus, the input admittance of the structure is $j\omega C$, such that its low frequency input behavior is essentially that of a single capacitor of value same as that found from static field analysis of the structure. Indeed, from considerations of power flow, using Poynting's theorem, we obtain the power flowing into the structure to be

$$P_{\text{in}} = wd[E_{x0}H_{y1}]_{z=0}$$

$$= -\left(\frac{\epsilon wl}{d}\right)\omega V_0^2 \sin \omega t \cos \omega t$$

$$= \frac{d}{dt}\left(\frac{1}{2}CV_g^2\right) \tag{6.64}$$

which is consistent with the electric energy stored in the structure for the static case, as given by (6.41).

Magnetoquasistatic Fields

For magnetoquasistatic fields, we begin with the magnetic field having the spatial dependence of the static field solution for the given arrangement. We shall illustrate by means of an example.

Example 6.7

Figure 6.10 shows the cross-sectional view of the arrangement of Figure 6.6(a), excited by a sinusoidally time-varying current source $I_g(t) = I_0 \cos \omega t$ instead of a direct current source. We wish to carry out the magnetoquasistatic field analysis for the arrangement.
From (6.45) we write

$$\mathbf{H}_0 = \frac{I_0}{w} \cos \omega t\, \mathbf{a}_y \tag{6.65}$$

where the subscript 0 again denotes that the amplitude of the field is of the zeroth power in ω. This results in an electric field in accordance with Maxwell's curl equation for $\mathbf{E}$, given by (3.17). Thus, we have for the geometry of the arrangement,

$$\frac{\partial E_{x1}}{\partial z} = -\frac{\partial B_{y0}}{\partial t} = \frac{\omega \mu I_0}{w} \sin \omega t$$

$$\mathbf{E}_1 = \frac{\omega \mu I_0 z}{w} \sin \omega t\, \mathbf{a}_x \tag{6.66}$$

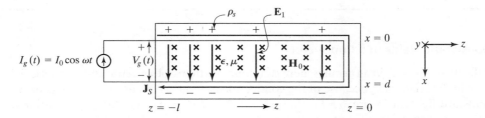

FIGURE 6.10

Magnetoquasistatic field analysis for the parallel-plate structure of Figure 6.6.

where we have also satisfied the boundary condition at $z = 0$ by choosing the constant of integration such that $[E_{x1}]_{z=0}$ is equal to zero, and again the subscript 1 denotes that the amplitude of the field is of the first power in ω. Note that the amplitude of E_{x1} varies linearly with z, from zero at $z = 0$ to a maximum at $z = -l$.

As in the case of electroquasistatic fields, we stop the process here, because continuing it by substituting (6.66) into Maxwell's curl equation for $\mathbf{H}$, (3.28), to obtain the resulting magnetic field will yield a term having amplitude proportional to the second power in ω. This simply means that the fields given as a pair by (6.65) and (6.66) do not satisfy (3.28), and hence are not complete solutions to Maxwell's equations. They are the quasistatic fields. The complete solutions are obtained by solving Maxwell's equations simultaneously and subject to the boundary conditions for the given problem.

Proceeding further, we obtain the voltage across the current source to be

$$V_g(t) = d[E_{x1}]_{z=-l}$$

$$= -\omega\left(\frac{\mu dl}{w}\right)I_0 \sin \omega t$$

$$= L\frac{dI_g(t)}{dt} \tag{6.67a}$$

or

$$\bar{V}_g = j\omega L\bar{I}_g \tag{6.67b}$$

where $L = (\mu dl/w)$ is the inductance of the arrangement obtained from static field considerations. Thus, the input impedance of the structure is $j\omega L$, such that its low frequency input behavior is essentially that of a single inductor of value same as that found from static field analysis of the structure. Indeed, from considerations of power flow, using Poynting's theorem, we obtain the power flowing into the structure to be

$$P_{\text{in}} = wd[E_{x1}H_{y0}]_{z=-l}$$

$$= -\left(\frac{\mu dl}{w}\right)\omega I_0^2 \sin \omega t \cos \omega t$$

$$= \frac{d}{dt}\left(\frac{1}{2}LI_g^2\right) \tag{6.68}$$

which is consistent with the magnetic energy stored in the structure for the static case, as given by (6.49).

Quasistatic Fields in a Conductor

If the dielectric slab in an arrangement is conductive, then both electric and magnetic fields exist in the static case, because of the conduction current, as discussed under electromagnetostatic fields in Section 6.3. Furthermore, the electric field of amplitude proportional to the first power in ω contributes to the creation of magnetic field of amplitude proportional to the first power in ω, in addition to that from electric field of amplitude proportional to the zeroth power in ω. We shall illustrate by means of an example.

Example 6.8

Let us consider that the dielectric slab in the arrangement of Figure 6.9 is conductive, as shown in Figure 6.11(a), and carry out the quasistatic field analysis for the arrangement.

Using the results from the static field analysis from the arrangement of Figure 6.7, we have for the arrangement of Figure 6.11(a),

$$\mathbf{E}_0 = \frac{V_0}{d} \cos \omega t \, \mathbf{a}_x \tag{6.69}$$

$$\mathbf{J}_{c0} = \sigma \mathbf{E}_0 = \frac{\sigma V_0}{d} \cos \omega t \, \mathbf{a}_x \tag{6.70}$$

$$\mathbf{H}_0 = -\frac{\sigma V_0 z}{d} \cos \omega t \, \mathbf{a}_y \tag{6.71}$$

as depicted in the figure. Also, the variations with z of the amplitudes of E_{x0} and H_{y0} are shown in Figure 6.11(b).

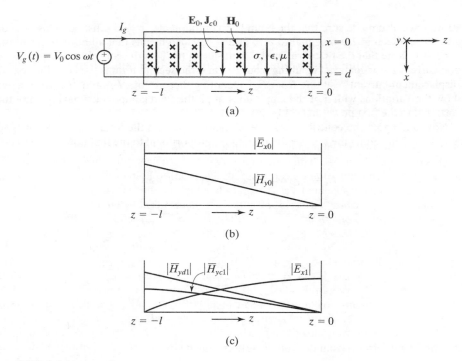

FIGURE 6.11

(a) Zero-order fields for the parallel-plate structure of Figure 6.7. (b) Variations of amplitudes of the zero-order fields along the structure. (c) Variations of amplitdes of the first-order fields along the structure.

The magnetic field given by (6.69) gives rise to an electric field having amplitude proportional to the first power in ω, in accordance with Maxwell's curl equation for **E**, (3.17). Thus,

$$\frac{\partial E_{x1}}{\partial z} = -\frac{\partial B_{y0}}{\partial t} = -\frac{\omega\mu\sigma V_0 z}{d}\sin\omega t$$

$$E_{x1} = -\frac{\omega\mu\sigma V_0}{2d}(z^2 - l^2)\sin\omega t \tag{6.72}$$

where we have also made sure that the boundary condition at $z = -l$ is satisfied. This boundary condition requires that E_x be equal to V_g/d at $z = -l$. Since this is satisfied by E_{x0} alone, it follows that E_{x1} must be zero at $z = -l$.

The electric field given by (6.69) and that given by (6.72) together give rise to a magnetic field having terms with amplitudes proportional to the first power in ω, in accordance with Maxwell's curl equation for **H**, (3.28). Thus,

$$\frac{\partial H_{y1}}{\partial z} = -\sigma E_{x1} - \epsilon\frac{\partial E_{x0}}{\partial t}$$

$$= \frac{\omega\mu\sigma^2 V_0}{2d}(z^2 - l^2)\sin\omega t + \frac{\omega\epsilon V_0}{d}\sin\omega t$$

$$H_{y1} = \frac{\omega\mu\sigma^2 V_0(z^3 - 3zl^2)}{6d}\sin\omega t + \frac{\omega\epsilon V_0 z}{d}\sin\omega t \tag{6.73}$$

where we have also made sure that the boundary condition at $z = 0$ is satisfied. This boundary condition requires that H_y be equal to zero at $z = 0$, which means that all of its terms must be zero at $z = 0$. Note that the first term on the right side of (6.73) is the contribution from the conduction current in the material resulting from E_{x1}, and the second term is the contribution from the displacement current resulting from E_{x0}. Denoting these to be H_{yc1} and H_{yd1}, respectively, we show the variations with z of the amplitudes of all the field components having amplitudes proportional to the first power in ω in Figure 6.11(c).

Now, adding up the contributions to each field, we obtain the total electric and magnetic fields up to and including the terms with amplitudes proportional to the first power in ω to be

$$E_x = \frac{V_0}{d}\cos\omega t - \frac{\omega\mu\sigma V_0}{2d}(z^2 - l^2)\sin\omega t \tag{6.74a}$$

$$H_y = -\frac{\sigma V_0 z}{d}\cos\omega t + \frac{\omega\epsilon V_0 z}{d}\sin\omega t + \frac{\omega\mu\sigma^2 V_0(z^3 - 3zl^2)}{6d}\sin\omega t \tag{6.74b}$$

or

$$\bar{E}_x = \frac{\bar{V}_g}{d} + j\omega\frac{\mu\sigma}{2d}(z^2 - l^2)\bar{V}_g \tag{6.75a}$$

$$\bar{H}_y = -\frac{\sigma z}{d}\bar{V}_g - j\omega\frac{\epsilon z}{d}\bar{V}_g - j\omega\frac{\mu\sigma^2(z^3 - 3zl^2)}{6d}\bar{V}_g \tag{6.75b}$$

Finally, the current drawn from the voltage source is given by

$$\bar{I}_g = w[\bar{H}_y]_{z=-l}$$

$$= \left(\frac{\sigma wl}{d} + j\omega\frac{\epsilon wl}{d} - j\omega\frac{\mu\sigma^2 wl^3}{3d}\right)\bar{V}_g \tag{6.76}$$

The input admittance of the structure is given by

$$\bar{Y}_{in} = \frac{\bar{I}_g}{\bar{V}_g} = j\omega\frac{\epsilon wl}{d} + \frac{\sigma wl}{d}\left(1 - j\omega\frac{\mu\sigma l^2}{3}\right)$$

$$\approx j\omega\frac{\epsilon wl}{d} + \frac{1}{\dfrac{d}{\sigma wl}\left(1 + j\omega\dfrac{\mu\sigma l^2}{3}\right)} \tag{6.77}$$

where we have used the approximation $[1 + j\omega(\mu\sigma l^2/3)]^{-1} \approx [1 - j\omega(\mu\sigma l^2/3)]$. Proceeding further, we have

$$\bar{Y}_{in} = j\omega\frac{\epsilon wl}{d} + \frac{1}{\dfrac{d}{\sigma wl} + j\omega\dfrac{\mu dl}{3w}}$$

$$= j\omega C + \frac{1}{R + j\omega L_i} \tag{6.78}$$

where $C = \epsilon wl/d$ is the capacitance of the structure if the material is a perfect dielectric, $R = d/\sigma wl$ is the resistance of the structure, and $L_i = \mu dl/3w$ is the internal inductance of the structure, all computed from static field analysis of the structure.

The equivalent circuit corresponding to (6.78) consists of capacitance C in parallel with the series combination of resistance R and internal inductance L_i, the same as in Figure 6.8. Thus, the low-frequency input behavior of the structure is essentially the same as that of the equivalent circuit of Figure 6.8, with the understanding that its input admittance must also be approximated to first-order terms. Note that for $\sigma = 0$, the input admittance of the structure is purely capacitive. For nonzero σ, a critical value of σ equal to $\sqrt{3\epsilon/\mu l^2}$ exists for which the input admittance is purely conductive. For values of σ smaller than the critical value, the input admittance is complex and capacitive, and for values of σ larger than the critical value, the input admittance is complex and inductive.

6.5   THE DISTRIBUTED CIRCUIT CONCEPT AND THE PARALLEL-PLATE TRANSMISSION LINE

In the preceding section, we have seen that, from the circuit point of view, the parallel-plate structure of Figure 6.5 behaves like a capacitor for the static case and the capacitive character is essentially retained for its input behavior for sinusoidally time-varying excitation at frequencies low enough to be within the range of validity of the quasistatic approximation. Likewise, we have seen that, from a circuit point of view, the parallel-plate structure of Figure 6.6 behaves like an inductor for the static case and the inductive character is essentially retained for its input behavior for sinusoidally time-varying excitation at frequencies low enough to be within the range of validity of the quasistatic approximation. For both structures, at an arbitrarily high enough frequency, the input behavior can be obtained only by obtaining complete (wave) solutions to Maxwell's equations, subject to the appropriate boundary conditions.

Two questions to ask at this point are (1) whether there is a circuit equivalent for the structure itself, independent of the termination, that is representative of the phenomenon taking place along the structure and valid at any arbitrary frequency, to the extent that the material parameters themselves are independent of frequency, and (2) what the limit on frequency is beyond which the quasistatic approximation is not valid. The answer to the first question is, yes, under a certain condition, giving rise to the concept of the *distributed circuit*, which we shall develop in this section by considering the parallel-plate structure, to be then known as the *parallel-plate transmission line*. The condition is that the waves propagating along the structure be the so-called *transverse electromagnetic* or TEM waves, meaning that the directions of the electric and magnetic fields are entirely traverse to the direction of propagation of the waves. The answer to the second question is that for the quasistatic approximation to hold, the length of the physical structure along the direction of propagation of the waves must be very small compared to the wavelength corresponding to the frequency of the source, in the dielectric region between the plates. While this can be obtained by extending the solution for the quasistatic case beyond the terms of the first power in ω by successive solution of Maxwell's equations (as in Section 4.3) and finding the condition under which the term of the first power in ω is predominant, it is more straightforward to obtain the exact solution by resorting to simultaneous solution of Maxwell's equations and finding the condition for which it approximates to the quasistatic solution. We shall do this in Section 7.1 by considering the structure of Figure 6.10 as a short-circuited transmission line and finding its input impedance.

Now, to develop and discuss the concept of the distributed circuit, we consider the parallel-plate arrangement of Figure 6.7(a) excited by a sinusoidally time-varying source of arbitrary frequency, as shown in Figure 6.12(a). Then, for an exact solution, the equations to be solved are

$$\nabla \times \mathbf{E} = -\frac{\partial \mathbf{B}}{\partial t} = -\mu \frac{\partial \mathbf{H}}{\partial t} \tag{6.79a}$$

$$\nabla \times \mathbf{H} = \mathbf{J}_c + \frac{\partial \mathbf{D}}{\partial t} = \sigma \mathbf{E} + \epsilon \frac{\partial \mathbf{E}}{\partial t} \tag{6.79b}$$

For the geometry of the arrangement, neglecting fringing of the fields at the edges or assuming that the structure is part of a much larger-sized configuration, $\mathbf{E} = E_x(z, t)\mathbf{a}_x$ and $\mathbf{H} = H_y(z, t)\mathbf{a}_y$, so that (6.79a) and (6.79b) simplify to

$$\frac{\partial E_x}{\partial z} = -\mu \frac{\partial H_y}{\partial t} \tag{6.80a}$$

$$\frac{\partial H_y}{\partial z} = -\sigma E_x - \epsilon \frac{\partial E_x}{\partial t} \tag{6.80b}$$

The situation is one of uniform plane electromagnetic waves propagating in the z-direction as though the conductors are not present, being guided by them, since all the boundary conditions are satisfied. We then have the simple case of a *parallel-plate transmission line*. Now, since E_z and H_z are zero in a given constant-z plane, that is, a plane *transverse* to the direction of propagation of the wave, as shown in Figure 6.12(b), we can uniquely define a voltage between the plates in terms of the electric field intensity in that plane, and a current crossing that plane in one direction on the top plate and in the opposite direction on the bottom plate in terms of the magnetic

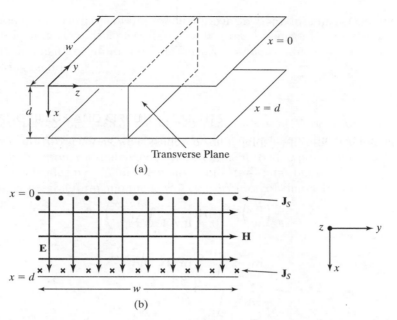

FIGURE 6.12

(a) Parallel-plate transmission line. (b) A transverse plane of the parallel-plate transmission line.

field intensity in that plane. These are given by

$$V(z, t) = \int_{x=0}^{d} E_x(z, t)\, dx = E_x(z, t) \int_{x=0}^{d} dx = dE_x(z, t) \tag{6.81a}$$

$$I(z, t) = \int_{y=0}^{w} J_S(z, t)\, dy = \int_{y=0}^{w} H_y(z, t)\, dy = H_y(z, t) \int_{y=0}^{w} dy$$

$$= wH_y(z, t) \tag{6.81b}$$

Proceeding further, we can find the power flow down the line by evaluating the surface integral of the Poynting vector over a given transverse plane. Thus,

$$P(z, t) = \int_{\text{transverse plane}} (\mathbf{E} \times \mathbf{H}) \cdot d\mathbf{S}$$

$$= \int_{x=0}^{d} \int_{y=0}^{w} E_x(z, t) H_y(z, t)\, \mathbf{a}_z \cdot dx\, dy\, \mathbf{a}_z$$

$$= \int_{x=0}^{d} \int_{y=0}^{w} \frac{V(z, t)}{d} \frac{I(z, t)}{w}\, dx\, dy$$

$$= V(z, t) I(z, t) \tag{6.82}$$

which is the familiar relationship employed in circuit theory.

From (6.81a) and (6.81b), we have

$$E_x = \frac{V}{d} \tag{6.83a}$$

$$H_y = \frac{I}{w} \tag{6.83b}$$

Substituting for E_x and H_y in (6.80a) and (6.80b) from (6.83a) and (6.83b), respectively, we now obtain two differential equations for voltage and current along the line as

$$\frac{\partial}{\partial z}\left(\frac{V}{d}\right) = -\mu\frac{\partial}{\partial t}\left(\frac{I}{w}\right) \tag{6.84a}$$

$$\frac{\partial}{\partial z}\left(\frac{I}{w}\right) = -\sigma\left(\frac{V}{d}\right) - \epsilon\frac{\partial}{\partial t}\left(\frac{V}{d}\right) \tag{6.84b}$$

or

$$\frac{\partial V}{\partial z} = -\left(\frac{\mu d}{w}\right)\frac{\partial I}{\partial t} \tag{6.85a}$$

$$\frac{\partial I}{\partial z} = -\left(\frac{\sigma w}{d}\right)V - \left(\frac{\epsilon w}{d}\right)\frac{\partial V}{\partial t} \tag{6.85b}$$

We now recognize the quantities in parentheses in (6.85a) and (6.85b) to be the circuit parameters $L, G,$ and $C,$ divided by the length l of the structure in the z-direction. Thus, these are the inductance per unit length, capacitance per unit length, and conductance per unit length, of the line, denoted to be $\mathscr{L}, \mathscr{G},$ and $\mathscr{C},$ respectively, and we can write the equations in terms of these parameters as

$$\frac{\partial V}{\partial z} = -\mathscr{L}\frac{\partial I}{\partial t} \tag{6.86a}$$

$$\frac{\partial I}{\partial z} = -\mathscr{G}V - \mathscr{C}\frac{\partial V}{\partial t} \tag{6.86b}$$

where

$$\mathscr{L} = \frac{\mu d}{w} \tag{6.87a}$$

$$\mathscr{C} = \frac{\epsilon w}{d} \tag{6.87b}$$

$$\mathscr{G} = \frac{\sigma w}{d} \tag{6.87c}$$

We note that $\mathscr{L}, \mathscr{C},$ and $\mathscr{G}$ are purely dependent on the dimensions of the line and

$$\mathscr{L}\mathscr{C} = \mu\epsilon \tag{6.88a}$$

$$\frac{\mathscr{G}}{\mathscr{C}} = \frac{\sigma}{\epsilon} \tag{6.88b}$$

Equations (6.86a) and (6.86b) are known as the *transmission line equations*. They characterize the wave propagation along the line in terms of the circuit quantities instead of in terms of the field quantities. It should, however, not be forgotten that the actual phenomenon is one of electromagnetic waves guided by the conductors of the line.

It is customary to represent a transmission line by means of its circuit equivalent, derived from the transmission-line equations (6.86a) and (6.86b). To do this, let us consider a section of infinitesimal length Δz along the line between z and $z + \Delta z$. From (6.86a), we then have

$$\underset{\Delta z \to 0}{\text{Lim}} \frac{V(z + \Delta z, t) - V(z, t)}{\Delta z} = -\mathscr{L}\frac{\partial I(z, t)}{\partial t}$$

or, for $\Delta z \to 0$,

$$V(z + \Delta z, t) - V(z, t) = -\mathscr{L}\,\Delta z\,\frac{\partial I(z, t)}{\partial t} \qquad (6.89a)$$

This equation can be represented by the circuit equivalent shown in Figure 6.13(a), since it satisfies Kirchhoff's voltage law written around the loop *abcda*. Similarly, from (6.86b), we have

$$\underset{\Delta z \to 0}{\text{Lim}} \frac{I(z + \Delta z, t) - I(z, t)}{\Delta z} = \underset{\Delta z \to 0}{\text{Lim}}\left[-\mathscr{G}V(z + \Delta z, t) - \mathscr{C}\frac{\partial V(z + \Delta z, t)}{\partial t}\right]$$

or, for $\Delta z \to 0$,

$$I(z + \Delta z, t) - I(z, t) = -\mathscr{G}\,\Delta z\,V(z + \Delta z, t) - \mathscr{C}\,\Delta z\,\frac{\partial V(z + \Delta z, t)}{\partial t} \qquad (6.89b)$$

This equation can be represented by the circuit equivalent shown in Figure 6.13(b), since it satisfies Kirchhoff's current law written for node *c*. Combining the two equations, we then obtain the equivalent circuit shown in Figure 6.13(c) for a section Δz of the line. It

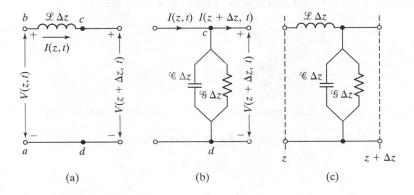

(a) (b) (c)

FIGURE 6.13

Development of circuit equivalent for an infinitesimal length Δz of a transmission line.

then follows that the circuit representation for a portion of length l of the line consists of an infinite number of such sections in cascade, as shown in Figure 6.14. Such a circuit is known as a *distributed circuit* as opposed to the *lumped circuits* that are familiar in circuit theory. The distributed circuit notion arises from the fact that the inductance, capacitance, and conductance are distributed uniformly and overlappingly along the line.

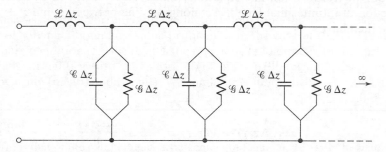

FIGURE 6.14

Distributed circuit representation of a transmission line.

A more physical interpretation of the distributed circuit concept follows from energy considerations. We know that the uniform plane wave propagation between the conductors of the line is characterized by energy storage in the electric and magnetic fields and power dissipation due to the conduction current flow. If we consider a section Δz of the line, the energy stored in the electric field in this section is given by

$$W_e = \frac{1}{2}\,\epsilon E_x^2 (\text{volume}) = \frac{1}{2}\,\epsilon E_x^2 (dw\,\Delta z)$$

$$= \frac{1}{2}\frac{\epsilon w}{d}\,(E_x d)^2\,\Delta z = \frac{1}{2}\,\mathcal{C}\,\Delta z\,V^2 \tag{6.90a}$$

The energy stored in the magnetic field in that section is given by

$$W_m = \frac{1}{2}\,\mu H_y^2\,(\text{volume}) = \frac{1}{2}\,\mu H_y^2 (dw\,\Delta z)$$

$$= \frac{1}{2}\frac{\mu d}{w}\,(H_y w)^2\,\Delta z = \frac{1}{2}\,\mathcal{L}\,\Delta z\,I^2 \tag{6.90b}$$

The power dissipated due to conduction current flow in that section is given by

$$P_d = \sigma E_x^2 (\text{volume}) = \sigma E_x^2 (dw\,\Delta z)$$

$$= \frac{\sigma w}{d}\,(E_x d)^2\,\Delta z = \mathcal{G}\,\Delta z\,V^2 \tag{6.90c}$$

Thus, $\mathcal{L}$, $\mathcal{C}$, and $\mathcal{G}$ are elements associated with energy storage in the magnetic field, energy storage in the electric field, and power dissipation due to the conduction current flow in the dielectric, respectively, for a given infinitesimal section of the line. Since these phenomena occur continuously and since they overlap, the inductance, capacitance, and conductance must be distributed uniformly and overlappingly along the line. In actual practice, the conductors of the transmission line are imperfect, resulting in slight penetration of the fields into the conductors, in accordance with the skin effect phenomenon. This gives rise to power dissipation and magnetic field energy storage in the conductors, which are taken into account by including a resistance and additional inductance in the series branch of the transmission-line equivalent circuit.

6.6   TRANSMISSION LINE WITH AN ARBITRARY CROSS SECTION

In the previous section, we considered the parallel-plate transmission line made up of perfectly conducting sheets lying in the planes $x = 0$ and $x = $ d so that the boundary conditions of zero tangential component of the electric field and zero normal component of the magnetic field are satisfied by the uniform plane wave characterized by the fields

$$\mathbf{E} = E_x(z, t)\mathbf{a}_x$$
$$\mathbf{H} = H_y(z, t)\mathbf{a}_y$$

thereby leading to the situation in which the uniform plane wave is guided by the conductors of the transmission line. In the general case, however, the conductors of the transmission line have arbitrary cross sections and the fields consist of both x- and y-components and are dependent on x- and y-coordinates in addition to the z-coordinate. Thus, the fields between the conductors are given by

$$\mathbf{E} = E_x(x, y, z, t)\mathbf{a}_x + E_y(x, y, z, t)\mathbf{a}_y$$
$$\mathbf{H} = H_x(x, y, z, t)\mathbf{a}_x + H_y(x, y, z, t)\mathbf{a}_y$$

These fields are no longer uniform in x and y but are directed entirely transverse to the direction of propagation, that is, the z-axis, which is the axis of the transmission line. Hence, they are known as *transverse electromagnetic waves*, or *TEM waves*. The uniform plane waves are simply a special case of the transverse electromagnetic waves.

To extend the computation of the transmission line parameters $\mathcal{L}, \mathcal{C}$, and $\mathcal{G}$ to the general case, let us consider a transmission line made up of parallel, perfect conductors of arbitrary cross sections, as shown by the cross-sectional view in Figure 6.15(a). Let us assume that the inner conductor is positive with respect to the outer conductor and that the current flows along the positive z-direction (into the page) on the inner conductor and along the negative z-direction (out of the page) on the outer conductor. We can then draw a *field map*, that is, a graphical sketch of the direction lines of the fields between the conductors, from the following considerations: (a) The electric field lines must originate on the inner conductor and be normal to it and must terminate on the outer conductor and be normal to it, since the tangential component of the electric field on a perfect conductor surface must be zero. (b) The magnetic field lines must be everywhere perpendicular to the electric field lines; although this can be shown by a rigorous mathematical proof, it is intuitively obvious, since, first, the magnetic field lines must be tangential near the conductor surfaces and, second, at any arbitrary point the fields correspond to those of a locally uniform plane wave. Thus, suppose that we

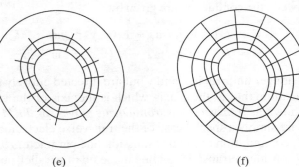

FIGURE 6.15

Construction of a transmission line field map consisting of
curvilinear rectangles.

start with the inner conductor and draw several lines normal to it at several points on
the surface, as shown in Figure 6.15(b). We can then draw a curved line displaced from
the conductor surface and such that it is perpendicular everywhere to the electric field
lines of Figure 6.15(b), as shown in Figure 6.15(c). This contour represents a magnetic
field line and forms the basis for further extension of the electric field lines, as shown in
Figure 6.15(d). A second magnetic field line can then be drawn so that it is everywhere
perpendicular to the extended electric field lines, as shown in Figure 6.15(e). This

procedure is continued until the entire cross section between the conductors is filled with two sets of orthogonal contours, as shown in Figure 6.15(f), thereby resulting in a field map made up of curvilinear rectangles.

By drawing the field lines with very small spacings, we can make the rectangles so small that each of them can be considered to be the cross section of a parallel-plate line. In fact, by choosing the spacings appropriately, we can even make them a set of squares. If we now replace the magnetic field lines by perfect conductors, since it does not violate any boundary condition, it can be seen that the arrangement can be viewed as the parallel combination, in the angular direction, of m number of series combinations of n number of parallel-plate lines in the radial direction, where m is the number of squares in the angular direction, that is, along a magnetic field line, and n is the number of squares in the radial direction, that is, along an electric field line. We can then find simple expressions for $\mathcal{L}$, $\mathcal{C}$, and $\mathcal{G}$ of the line in the following manner.

Let us for simplicity consider the field map of Figure 6.16, consisting of eight segments $1, 2, \ldots, 8$ in the angular direction and two segments a and b in the radial direction. The arrangement is then a parallel combination, in the angular direction, of eight series combinations of two lines in the radial direction, each having a curvilinear rectangular cross section. Let $I_1, I_2, \ldots, I_8$ be the currents associated with the segments $1, 2, \ldots, 8$, respectively, and let ψ_a and ψ_b be the magnetic fluxes per unit length in the z-direction associated with the segments a and b, respectively. Then the inductance per

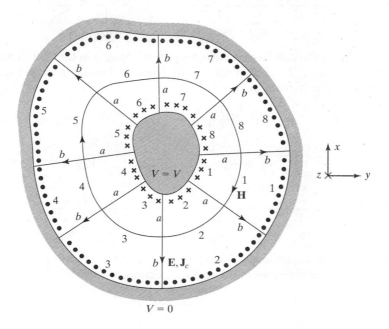

FIGURE 6.16

For deriving the expressions for the transmission-line parameters from the field map.

unit length of the transmission line is given by

$$
\begin{aligned}
\mathscr{L} &= \frac{\psi}{I} = \frac{\psi_a + \psi_b}{I_1 + I_2 + \cdots + I_8} \\
&= \frac{1}{\dfrac{I_1}{\psi_a} + \dfrac{I_2}{\psi_a} + \cdots + \dfrac{I_8}{\psi_a}} + \frac{1}{\dfrac{I_1}{\psi_b} + \dfrac{I_2}{\psi_b} \cdots + \dfrac{I_8}{\psi_b}} \\
&= \frac{1}{\dfrac{1}{\mathscr{L}_{1a}} + \dfrac{1}{\mathscr{L}_{2a}} + \cdots + \dfrac{1}{\mathscr{L}_{8a}}} + \frac{1}{\dfrac{1}{\mathscr{L}_{1b}} + \dfrac{1}{\mathscr{L}_{2b}} + \cdots + \dfrac{1}{\mathscr{L}_{8b}}}
\end{aligned}
\tag{6.91a}
$$

Let $Q_1, Q_2, \ldots, Q_8$ be the charges per unit length in the z-direction associated with the segments $1, 2, \ldots, 8$, respectively, and let V_a and V_b be the voltages associated with the segments a and b, respectively. Then the capacitance per unit length of the transmission line is given by

$$
\begin{aligned}
\mathscr{C} &= \frac{Q}{V} = \frac{Q_1 + Q_2 + \cdots + Q_8}{V_a + V_b} \\
&= \frac{1}{\dfrac{V_a}{Q_1} + \dfrac{V_b}{Q_1}} + \frac{1}{\dfrac{V_a}{Q_2} + \dfrac{V_b}{Q_2}} + \cdots + \frac{1}{\dfrac{V_a}{Q_8} + \dfrac{V_b}{Q_8}} \\
&= \frac{1}{\dfrac{1}{\mathscr{C}_{1a}} + \dfrac{1}{\mathscr{C}_{1b}}} + \frac{1}{\dfrac{1}{\mathscr{C}_{2a}} + \dfrac{1}{\mathscr{C}_{2b}}} + \cdots + \frac{1}{\dfrac{1}{\mathscr{C}_{8a}} + \dfrac{1}{\mathscr{C}_{8b}}}
\end{aligned}
\tag{6.91b}
$$

Let $I_{c1}, I_{c2}, \ldots, I_{c8}$ be the conduction currents per unit length in the z-direction associated with the segments $1, 2, \ldots, 8$, respectively. Then the conductance per unit length of the transmission line is given by

$$
\begin{aligned}
\mathscr{G} &= \frac{I_c}{V} = \frac{I_{c1} + I_{c2} + \cdots + I_{c8}}{V_a + V_b} \\
&= \frac{1}{\dfrac{V_a}{I_{c1}} + \dfrac{V_b}{I_{c1}}} + \frac{1}{\dfrac{V_a}{I_{c2}} + \dfrac{V_b}{I_{c2}}} + \cdots + \frac{1}{\dfrac{V_a}{I_{c8}} + \dfrac{V_b}{I_{c8}}} \\
&= \frac{1}{\dfrac{1}{\mathscr{G}_{1a}} + \dfrac{1}{\mathscr{G}_{1b}}} + \frac{1}{\dfrac{1}{\mathscr{G}_{2a}} + \dfrac{1}{\mathscr{G}_{2b}}} + \cdots + \frac{1}{\dfrac{1}{\mathscr{G}_{8a}} + \dfrac{1}{\mathscr{G}_{8b}}}
\end{aligned}
\tag{6.91c}
$$

Generalizing the expressions (6.91a), (6.91b), and (6.91c) to m segments in the angular direction and n segments in the radial direction, we obtain

$$
\mathscr{L} = \sum_{j=1}^{n} \frac{1}{\displaystyle\sum_{i=1}^{m} \frac{1}{\mathscr{L}_{ij}}}
\tag{6.92a}
$$

$$
\mathscr{C} = \sum_{i=1}^{m} \frac{1}{\displaystyle\sum_{j=1}^{n} \frac{1}{\mathscr{C}_{ij}}}
\tag{6.92b}
$$

$$\mathcal{G} = \sum_{i=1}^{m} \frac{1}{\displaystyle\sum_{j=1}^{n} \frac{1}{\mathcal{G}_{ij}}} \tag{6.92c}$$

where $\mathcal{L}_{ij}$, $\mathcal{C}_{ij}$, and $\mathcal{G}_{ij}$ are the inductance, capacitance, and conductance per unit length corresponding to the rectangle ij. If the map consists of curvilinear squares, then $\mathcal{L}_{ij}$, $\mathcal{C}_{ij}$, and $\mathcal{G}_{ij}$ are equal to μ, ϵ, and σ, respectively, according to (6.87a), (6.87b), and (6.87c), respectively, since the width w of the plates is equal to the spacing d of the plates for each square. Thus, we obtain simple expressions for $\mathcal{L}$, $\mathcal{C}$, and $\mathcal{G}$ as given by

$$\mathcal{L} = \mu \frac{n}{m} \tag{6.93a}$$

$$\mathcal{C} = \epsilon \frac{m}{n} \tag{6.93b}$$

$$\mathcal{G} = \sigma \frac{m}{n} \tag{6.93c}$$

The computation of $\mathcal{L}$, $\mathcal{C}$, and $\mathcal{G}$ then consists of sketching a field map consisting of curvilinear squares, counting the number of squares in each direction, and substituting these values in (6.93a), (6.93b), and (6.93c). Note that once again

$$\mathcal{L}\mathcal{C} = \mu\epsilon \tag{6.94a}$$

$$\frac{\mathcal{G}}{\mathcal{C}} = \frac{\sigma}{\epsilon} \tag{6.94b}$$

We shall now consider an example of the application of the curvilinear squares technique.

Example 6.9

The coaxial cable is a transmission line made up of parallel, coaxial, cylindrical conductors. Let the radius of the inner conductor be a and that of the outer conductor be b. We wish to find expressions for $\mathcal{L}$, $\mathcal{C}$, and $\mathcal{G}$ of the coaxial cable by using the curvilinear squares technique.

Figure 6.17 shows the cross-sectional view of the coaxial cable and the field map. In view of the symmetry associated with the conductor configuration, the construction of the field map is simplified in this case. The electric field lines are radial lines from one conductor to the other, and the magnetic field lines are circles concentric with the conductors, as shown in the figure. Let the number of curvilinear squares in the angular direction be m. Then to find the number of curvilinear squares in the radial direction, we note that the angle subtended at the center of the conductors by adjacent pairs of electric field lines is equal to $2\pi/m$. Hence, at any arbitrary radius r between the two conductors, the side of the curvilinear square is equal to $r(2\pi/m)$. The number of squares in an infinitesimal distance dr in the radial direction is then equal to $\dfrac{dr}{r(2\pi/m)}$, or $\dfrac{m}{2\pi}\dfrac{dr}{r}$. The total number of squares in the radial direction from the inner to the outer conductor is given by

$$n = \int_{r=a}^{b} \frac{m}{2\pi} \frac{dr}{r} = \frac{m}{2\pi} \ln \frac{b}{a}$$

The required expressions for $\mathcal{L}$, $\mathcal{C}$, and $\mathcal{G}$ are then given by

$$\mathcal{L} = \mu \frac{n}{m} = \frac{\mu}{2\pi} \ln \frac{b}{a} \tag{6.95a}$$

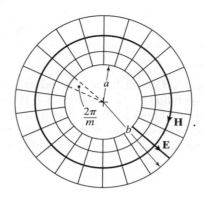

FIGURE 6.17

Field map consisting of curvilinear squares for a coaxial cable.

$$\mathscr{C} = \epsilon \frac{m}{n} = \frac{2\pi\epsilon}{\ln (b/a)} \tag{6.95b}$$

$$\mathscr{G} = \sigma \frac{m}{n} = \frac{2\pi\sigma}{\ln (b/a)} \tag{6.95c}$$

These expressions are exact. We have been able to obtain exact expressions in this case because of the geometry involved. When the geometry is not so simple, we can only obtain approximate values for $\mathscr{L}$, $\mathscr{C}$, and $\mathscr{G}$.

We have just discussed an example of the determination of the transmission-line parameters $\mathscr{L}$, $\mathscr{C}$, and $\mathscr{G}$ for a coaxial cable. There are other configurations having different cross sections for which one can obtain the parameters either by the curvilinear squares technique or by other analytical or experimental techniques. The parameters for some cases for which exact expressions are available are listed in Table 6.1, along with those for the parallel-plate line and coaxial cable.

TABLE 6.1 Conductance, Capacitance, and Inductance per Unit Length for Some Structures Consisting of Infinitely Long Conductors Having the Cross Sections Shown in Figure 6.18

Description	Capacitance per unit length, $\mathscr{C}$	Conductance per unit length, $\mathscr{G}$	Inductance per unit length, $\mathscr{L}$
Parallel-plane conductors, Figure 6.18(a)	$\epsilon \dfrac{w}{d}$	$\sigma \dfrac{w}{d}$	$\mu \dfrac{d}{w}$
Coaxial cylindrical conductors, Figure 6.18(b)	$\dfrac{2\pi\epsilon}{\ln (b/a)}$	$\dfrac{2\pi\sigma}{\ln (b/a)}$	$\dfrac{\mu}{2\pi} \ln \dfrac{b}{a}$
Parallel cylindrical wires, Figure 6.18(c)	$\dfrac{\pi\epsilon}{\cosh^{-1} (d/a)}$	$\dfrac{\pi\sigma}{\cosh^{-1} (d/a)}$	$\dfrac{\mu}{\pi} \cosh^{-1} \dfrac{d}{a}$
Eccentric inner conductor, Figure 6.18(d)	$\dfrac{2\pi\epsilon}{\cosh^{-1}\left(\dfrac{a^2 + b^2 - d^2}{2ab}\right)}$	$\dfrac{2\pi\sigma}{\cosh^{-1}\left(\dfrac{a^2 + b^2 - d^2}{2ab}\right)}$	$\dfrac{\mu}{2\pi} \cosh^{-1}\left(\dfrac{a^2 + b^2 + d^2}{2ab}\right)$
Shielded parallel cylindrical wires, Figure 6.18(e)	$\dfrac{\pi\epsilon}{\ln \dfrac{d(b^2 - d^2/4)}{a(b^2 + d^2/4)}}$	$\dfrac{\pi\sigma}{\ln \dfrac{d(b^2 - d^2/4)}{a(b^2 + d^2/4)}}$	$\dfrac{\mu}{\pi} \ln \dfrac{d(b^2 - d^2/4)}{a(b^2 + d^2/4)}$

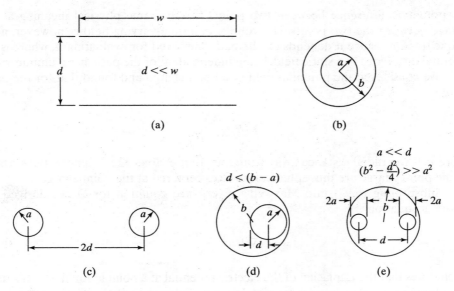

FIGURE 6.18

Cross sections of some common configurations of parallel, infinitely long conductors.

SUMMARY

In this chapter, we first introduced the electric potential from the fact that for the static case,

$$\nabla \times \mathbf{E} = 0 \tag{6.96}$$

and, since the curl of the gradient of a scalar function is identically zero, $\mathbf{E}$ can be expressed as the gradient of a scalar function. The gradient of a scalar function Φ is given in Cartesian coordinates by

$$\nabla \Phi = \frac{\partial \Phi}{\partial x}\, \mathbf{a}_x + \frac{\partial \Phi}{\partial y}\, \mathbf{a}_y + \frac{\partial \Phi}{\partial z}\, \mathbf{a}_z$$

The magnitude of $\nabla \Phi$ at a given point is the maximum rate of increase of Φ at that point, and its direction is the direction in which the maximum rate of increase occurs, that is, normal to the constant Φ surface passing through that point.

From considerations of work associated with the movement of a test charge in the static electric field, we found that for the case of the static electric field, the scalar function is $-V$, so that

$$\mathbf{E} = -\nabla V \tag{6.97}$$

where V is the electric potential. The electric potential V_A at a point A is the amount of work per unit charge done by the field in the movement of a test charge from the point A to a reference point O. It is the potential difference between A and O. Thus,

$$V_A = [V]_A^O = \int_A^O \mathbf{E} \cdot d\mathbf{l} = -\int_O^A \mathbf{E} \cdot d\mathbf{l}$$

The potential difference between two points has the same physical meaning as the voltage between the two points. The voltage in a time-varying field is, however, not a unique quantity, since it depends on the path employed for evaluating it, whereas the potential difference in a static field, being independent of the path, has a unique value.

We considered the potential field of a point charge and found that for the point charge

$$V = \frac{Q}{4\pi\epsilon R}$$

where R is the radial distance away from the point charge. The equipotential surfaces for the point charge are thus spherical surfaces centered at the point charge.

Substituting (6.97) into Maxwell's divergence equation for $\mathbf{D}$, we derived the Poisson's equation

$$\nabla^2 V = -\frac{\rho}{\epsilon} \tag{6.98}$$

which states that the Laplacian of the electric potential at a point is equal to $-1/\epsilon$ times the volume charge density at that point. In Cartesian coordinates,

$$\nabla^2 V = \frac{\partial^2 V}{\partial x^2} + \frac{\partial^2 V}{\partial y^2} + \frac{\partial^2 V}{\partial z^2}$$

For the one-dimensional case in which the charge density is a function of x only, (6.98) reduces to

$$\frac{\partial^2 V}{\partial x^2} = \frac{\partial^2 V}{\partial x^2} = -\frac{\rho}{\epsilon}$$

We illustrated the solution of this equation by considering the example of a p–n junction diode.

If $\rho = 0$, Poisson's equation reduces to Laplace's equation

$$\nabla^2 V = 0 \tag{6.99}$$

This equation is applicable for a charge-free dielectric region as well as for a conducting medium.

To introduce circuit elements, we next began with Maxwell's equations in differential form and the continuity equation for static fields, given by

$$\nabla \times \mathbf{E} = 0 \tag{6.100a}$$

$$\nabla \times \mathbf{H} = \mathbf{J} \tag{6.100b}$$

$$\nabla \cdot \mathbf{D} = \rho \tag{6.100c}$$

$$\nabla \cdot \mathbf{B} = 0 \tag{6.100d}$$

$$\nabla \cdot \mathbf{J} = 0 \tag{6.100e}$$

and considered three cases of static fields: (a) electrostatic fields, (b) magnetostatic fields, and (c) electromagnetostatic fields. From these three cases, we introduced the circuit elements, capacitance (C), inductance (L) and conductance (G), respectively, by considering a parallel-plate arrangement.

We then turned to the quasistatic extension of the static field solution as a means of obtaining the low-frequency behavior of a physical structure. The quasistatic field approach involves starting with a time-varying field having the same spatial character-istics as the static field in the physical structure and then obtaining field solutions con-taining terms up to and including the first power in frequency by using Maxwell's curl equations for time-varying fields. We applied this approach for the same three cases as for the static fields, and found that the input behavior of the structure remains essen-tially the same as for the corresponding static case.

The quasistatic approximation holds for frequencies for which the wavelength corresponding to the frequency of the source is large compared to the length of the structure along the direction of propagation of the waves, which is to be derived in Section 7.1. Beyond the range of validity of the quasistatic approximation, the input behavior can be obtained only by obtaining complete solutions to Maxwell's equations, subject to the boundary conditions, which led us to the concept of the distributed cir-cuit and the parallel-plate structure becoming a parallel-plate transmission line. We derived the *transmission-line equations*,

$$\frac{\partial V}{\partial z} = -\mathcal{L}\frac{\partial I}{\partial t} \tag{6.101a}$$

$$\frac{\partial I}{\partial z} = -\mathcal{G}V - \mathcal{C}\frac{\partial V}{\partial t} \tag{6.101b}$$

These equations are applicable to all transmission lines, characterized by transverse electromagnetic wave propagation. They govern the wave propagation along the line in terms of circuit quantities instead of in terms of field quantities.

The parameters $\mathcal{L}$, $\mathcal{C}$, and $\mathcal{G}$ in (6.101a) and (6.101b) are the inductance, capaci-tance, and conductance per unit length of line, which differ from one line to another. For the parallel-plate line having width w of the plates and spacing d between the plates, they are given by

$$\mathcal{L} = \frac{\mu d}{w}$$

$$\mathcal{C} = \frac{\epsilon w}{d}$$

$$\mathcal{G} = \frac{\sigma w}{d}$$

where μ, ϵ, and σ are the material parameters of the medium between the plates and fringing of the fields is neglected. We learned how to compute $\mathcal{L}$, $\mathcal{C}$, and $\mathcal{G}$ for a line of arbitrary cross section by constructing a field map of the transverse electromagnetic wave fields, consisting of curvilinear squares in the cross-sectional plane of the line.

If m is the number of squares tangential to the conductors and n is the number of squares normal to the conductors, then

$$\mathcal{L} = \mu \frac{n}{m}$$

$$\mathcal{C} = \epsilon \frac{m}{n}$$

$$\mathcal{G} = \sigma \frac{m}{n}$$

By applying this technique to the coaxial cable, we found that for a cable of inner radius a and outer radius b,

$$\mathcal{L} = \frac{\mu}{2\pi} \ln \frac{b}{a}$$

$$\mathcal{C} = \frac{2\pi\epsilon}{\ln (b/a)}$$

$$\mathcal{G} = \frac{2\pi\sigma}{\ln (b/a)}$$

REVIEW QUESTIONS

6.1. State Maxwell's curl equations for static fields.

6.2. What is the expansion for the gradient of a scalar in Cartesian coordinates? When can a vector be expressed as the gradient of a scalar?

6.3. Discuss the physical interpretation for the gradient of a scalar function.

6.4. Discuss the application of the gradient concept for the determination of unit vector normal to a surface.

6.5. How would you find the rate of increase of a scalar function along a specified direction by using the gradient concept?

6.6. Define electric potential. What is its relationship to the static electric field intensity?

6.7. Distinguish between voltage, as applied to time-varying fields, and potential difference.

6.8. What is a conservative field? Give two examples of conservative fields.

6.9. Describe the equipotential surfaces for a point charge.

6.10. Discuss the determination of the electric field intensity due to a charge distribution by using the potential concept.

6.11. What is the Laplacian of a scalar? What is its expansion in Cartesian coordinates?

6.12. State Poisson's equation.

6.13. Outline the solution of Poisson's equation for the potential in a region of known charge density varying in one dimension.

6.14. State Laplace's equation. In what regions is it valid?

6.15. State Maxwell's equations for static fields in (a) integral form, and (b) differential form.

6.16. Discuss the classification of static fields with reference to subsets of Maxwell's equations.

6.17. Outline the steps involved in the electrostatic field analysis of a parallel plate structure and the determination of its capacitance.

6.18. Outline the steps involved in the magnetostatic field analysis of a parallel plate structure and the determination of its inductance.

6.19. Outline the steps involved in the electromagnetostatic field analysis of a parallel plate structure and the determination of its circuit equivalent.

6.20. Explain the term *internal inductance*.

6.21. What is meant by the quasistatic extension of the static field in a physical structure?

6.22. Outline the steps involved in the electroquasistatic field analysis of a parallel plate structure and the determination of its input behavior. Compare the input behavior with the electrostatic case.

6.23. Outline the steps involved in the magnetoquasistatic field analysis of a parallel plate structure and the determination of its input behavior. Compare the input behavior with the magnetostatic case.

6.24. Outline the steps involved in the quasistatic field analysis of a parallel plate structure with a conducting slab between the plates and the determination of its input behavior. Compare the input behavior with the electromagnetostatic case.

6.25. Discuss the phenomenon taking place along a parallel-plate structure at any arbitrary frequency and the need for the concept of the *distributed circuit*.

6.26. What is the limit on the frequency beyond which the quasistatic approximation for the input behavior of a physical structure is not valid?

6.27. How is the voltage between the two conductors in a given cross-sectional plane of a parallel-plate transmission line related to the electric field in that plane?

6.28. How is the current flowing on the plates across a given cross-sectional plane of a parallel-plate transmission line related to the magnetic field in that plane?

6.29. What are transmission-line equations? How are they obtained from Maxwell's equations?

6.30. What are the expressions for $\mathscr{L}$, the inductance per unit length, $\mathscr{C}$, the capacitance per unit length, and $\mathscr{G}$, the conductance per unit length, for a parallel-plate transmission line?

6.31. Are the three quantities $\mathscr{L}$, $\mathscr{C}$, and $\mathscr{G}$ independent? If not, how are they dependent on each other?

6.32. Draw the transmission-line equivalent circuit. How is it derived from the transmission-line equations?

6.33. Discuss the concept of the distributed circuit and compare it to a lumped circuit.

6.34. Discuss the physical phenomena associated with each of the elements in the transmission-line equivalent circuit.

6.35. What is a transverse electromagnetic wave?

6.36. What is a field map? Describe the procedure for drawing the field map for a transmission line of arbitrary cross section.

6.37. Draw a rough sketch of the field map for a line made up of two identical parallel cylindrical conductors with their axes separated by four times their radii.

6.38. Describe the procedure for computing the transmission line parameters $\mathscr{L}$, $\mathscr{C}$, and $\mathscr{G}$ from the field map.

6.39. How does a field map consisting of curvilinear squares simplify the computation of the line parameters?

6.40. Discuss the determination of $\mathscr{L}$, $\mathscr{C}$, and $\mathscr{G}$ for a coaxial cable by using the curvilinear squares technique.

PROBLEMS

6.1. Find the gradients of the following scalar functions: (a) $\sqrt{x^2 + y^2 + z^2}$; (b) xyz.

6.2. Determine which of the following vectors can be expressed as the gradient of a scalar function: (a) $y\mathbf{a}_x - x\mathbf{a}_y$; (b) $x\mathbf{a}_x + y\mathbf{a}_y + z\mathbf{a}_z$; (c) $2xy^3 z\mathbf{a}_x + 3x^2y^2 z\mathbf{a}_y + x^2y^3\mathbf{a}_z$.

6.3. Find the unit vector normal to the plane surface $5x + 2y + 4z = 20$.

6.4. Find the unit vector normal to the surface $x^2 - y^2 = 5$ at the point $(3, 2, 1)$.

6.5. Find the rate of increase of the scalar function $x^2 y$ at the point $(1, 2, 1)$ in the direction of the vector $\mathbf{a}_x - \mathbf{a}_y$.

6.6. For the static electric field given by $\mathbf{E} = y\mathbf{a}_x + x\mathbf{a}_y$, find the potential difference between points $A(1, 1, 1)$ and $B(2, 2, 2)$.

6.7. For a point charge Q situated at the point $(1, 2, 0)$, find the potential difference between the point $A(3, 4, 1)$ and the point $B(5, 5, 0)$.

6.8. For the arrangement of a linear electric dipole consisting of point charges Q and $-Q$ at the points $(0, 0, d/2)$ and $(0, 0, -d/2)$, respectively, obtain the expression for the electric potential and hence for the electric field intensity at distances from the dipole large compared to d.

6.9. For a line charge of uniform density 10^{-3} C/m situated along the z-axis between $(0, 0, -1)$ and $(0, 0, 1)$, obtain the series expression for the electric potential at the point $(0, y, 0)$ by dividing the line charge into 100 equal segments and considering the charge in each segment to be a point charge located at the center of the segment. Then find the series expression for the electric field intensity at the point $(0, 1, 0)$.

6.10. Repeat Problem 6.9, assuming the line charge density to be $10^{-3}|z|$ C/m.

6.11. The potential distribution in a simplified model of a vacuum diode consisting of cathode in the plane $x = 0$ and anode in the plane $x = d$ and held at a potential V_0 relative to the cathode is given by

$$V = V_0 \left(\frac{x}{d} \right)^{4/3} \quad \text{for } 0 < x < d$$

(a) Find the space charge density distribution in the region $0 < x < d$.

(b) Find the surface charge densities on the cathode and the anode.

6.12. Show that for the p–n junction diode of Figure 6.4(a), the boundary condition of the continuity of the normal component of displacement flux density at $x = 0$ is automatically satisfied by equation (6.29).

6.13. Assume that the impurity concentration for the p–n junction diode of Figure 6.4(a) is a linear function of distance across the junction. The space charge density distribution is then given by

$$\rho = kx \quad \text{for } -d/2 < x < d/2$$

where d is the width of the space charge region and k is the proportionality constant. Find the solution for the potential in the space charge region.

6.14. A space-charge density distribution is given by

$$\rho = \begin{cases} \rho_0 \sin x & \text{for } -\pi < x < \pi \\ 0 & \text{otherwise} \end{cases}$$

where ρ_0 is a constant. Find the sketch the potential V versus x for all x. Assume $V = 0$ for $x = 0$.

6.15. The region between the two plates of Figure 6.5 is filled with two perfect dielectric media having permittivities ϵ_1 for $0 < x < t$ and ϵ_2 for $t < x < d$. (a) Find the solutions for the potentials in the two regions $0 < x < t$ and $t < x < d$. (b) Find the potential at the interface $x = t$. (c) Find the capacitance of the arrangement.

6.16. For a dielectric medium of nonuniform permittivity, show that the Poisson's equation is given by

$$\epsilon \nabla^2 V + \nabla \epsilon \cdot \nabla V = -\rho$$

Assume that the region between the two plates of Figure 6.5 is filled with a perfect dielectric of nonuniform permittivity

$$\epsilon = \frac{\epsilon_0}{1 - (x/2d)}$$

Find the solution for the potential between the plates and obtain the expression for the capacitance per unit area of the plates.

6.17. The region between the plates of Figure 6.6 is divided into half in the y-direction. Assume that one half is filled with a material of permeability μ_1 and the other half with a material of permeability μ_2. Find the inductance of the arrangement.

6.18. The region between the two plates of Figure 6.7 is filled with two imperfect dielectric media having conductivities σ_1 for $0 < x < t$ and σ_2 for $t < x < d$. (a) Find the solutions for the potentials in the two regions $0 < x < t$ and $t < x < d$. (b) Find the potential at the interface $x = t$.

6.19. For the structure of Figure 6.9, continue the analysis beyond the quasistatic extension and obtain the input admittance correct to the third power in ω. Determine the equivalent circuit.

6.20. For the structure of Figure 6.10, continue the analysis beyond the quasistatic extension to obtain the input impedance correct to the third power in ω. Determine the equivalent circuit.

6.21. For the structure of Figure 6.10, assume that the medium between the plates is an imperfect dielectric of conductivity σ. (a) Show that the input impedance correct to the first power in ω is the same as if σ were zero. (b) Obtain the input impedance correct to the second power in ω and determine the equivalent circuit.

6.22. Find the condition(s) under which the quasistatic input behavior of the structure of Figure 6.11 is essentially equivalent to (a) a capacitor in parallel with a resistor and (b) a resistor in series with an inductor.

6.23. A parallel-plate transmission line is made up of perfect conductors of width $w = 0.1$ m and lying in the planes $x = 0$ and $x = 0.02$ m. The medium between the conductors is a perfect dielectric of $\mu = \mu_0$. For a uniform plane wave having the electric field

$$\mathbf{E} = 100\pi \cos(2\pi \times 10^6 t - 0.02\pi z) \, \mathbf{a}_x \text{ V/m}$$

propagating between the conductors, find (a) the voltage between the conductors, (b) the current along the conductors, and (c) the power flow along the line.

6.24. A parallel-plate transmission line made up of perfect conductors has $\mathscr{L}$ equal to 10^{-7} H/m. If the medium between the plates is characterized by $\sigma = 10^{-11}$ S/m, $\epsilon = 6\epsilon_0$, and $\mu = \mu_0$, find $\mathscr{C}$ and $\mathscr{G}$ of the line.

6.25. If the conductors of a transmission line are imperfect, then the transmission-line equivalent circuit contains a resistance and additional inductance in the series branch. Assuming that the thickness of the (imperfect) conductors of a parallel-plate line is several skin depths at the frequency of interest, show from considerations of skin effect phenomenon in a good conductor medium that the resistance and inductance per unit length along the conductors are $2/\sigma_c \delta w$ and $2/\omega \sigma_c \delta w$, respectively, where σ_c is the conductivity of the (imperfect) conductors, w is the width, and δ is the skin depth. The factor 2 arises because of two conductors.

6.26. Show that two alternative representations of the circuit equivalent of the transmission-line equations (6.86a) and (6.86b) are as shown in Figures 6.19(a) and (b).

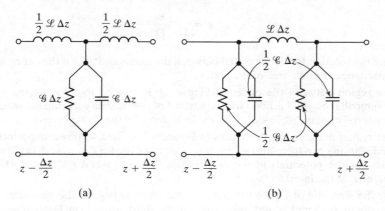

(a)   (b)

FIGURE 6.19

For Problem 6.26.

6.27. Show that for a transverse electromagnetic wave, the voltage between the conductors and the current along the conductors in a given transverse plane are uniquely defined in terms of the electric and magnetic fields, respectively, in that plane.

6.28. By constructing a field map consisting of curvilinear squares for a coaxial cable having $b/a = 3.5$, obtain the approximate values of the line parameters $\mathscr{L}, \mathscr{C}$, and $\mathscr{G}$ in terms of μ, ϵ, and σ of the dielectric. Compare the approximate values with the exact values given by expressions derived in Example 6.9.

6.29. For $d/a = 2$ for the parallel-wire line [see Figure 6.18(c)], construct a field map consisting of curvilinear squares and obtain approximate values for the line parameters $\mathscr{L}, \mathscr{C}$, and $\mathscr{G}$. Compare approximate values with the exact values given by the expressions in Table 6.1.

6.30. The shielded strip line, employed in microwave integrated circuits, consists of a center conductor photoetched on the inner faces of two substrates sandwiched between two conductors, as shown by the cross-sectional view in Figure 6.20. For the dimensions shown in the figure, construct a field map consisting of curvilinear squares and compute $\mathscr{L}$ and $\mathscr{C}$, considering the substrate to be a perfect dielectric having $\epsilon = 9\epsilon_0$ and $\mu = \mu_0$. Assume for simplicity that the field is confined to the substrate region.

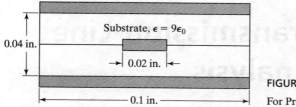

Substrate, $\epsilon = 9\epsilon_0$

0.04 in.

0.02 in.

0.1 in.

FIGURE 6.20

For Problem 6.30.

6.31. The cross section of an eccentric coaxial cable [see Figure 6.18(d)] consists of an outer circle of radius $a = 5$ cm and an inner circle of radius $b = 2$ cm, with their centers separated by $d = 2$ cm. By constructing a field map consisting of curvilinear squares, obtain the approximate values of the line parameters $\mathscr{L}$, $\mathscr{C}$, and $\mathscr{G}$ in terms of μ, ϵ, and σ of the dielectric.

6.32. Consider a transmission line having the cross section shown in Figure 6.21. The inner conductor is a circle of radius a and the outer conductor is a square of sides $2a$. Find the approximate values of $\mathscr{L}$, $\mathscr{C}$, and $\mathscr{G}$, by using the method of curvilinear squares.

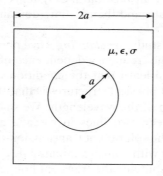

2a

μ, ϵ, σ

a

FIGURE 6.21

For Problem 6.32.

7 Transmission-Line Analysis

In the previous chapter, we introduced the transmission line and the transmission-line equations. The transmission-line equations enable us to discuss the wave propagation phenomena along an arrangement of two parallel conductors having uniform cross section in terms of circuit quantities instead of in terms of field quantities. This chapter is devoted to the analysis of lossless transmission-line systems first in frequency domain, that is, for sinusoidal steady state, and then in time domain, that is, for arbitrary variation with time.

In the frequency domain, we shall study the *standing wave* phenomenon by considering the short-circuited line. From the frequency dependence of the input impedance of the short-circuited line, we shall learn that the condition for the quasistatic approximation for the input behavior of physical structures is that the physical length of the structure must be a small fraction of the wavelength. We shall study reflection and transmission at the junction between two lines in cascade and introduce the Smith®. Chart, a useful graphical aid in the solution of transmission-line problems.

In the time domain, we shall begin with a line terminated by a resistive load and learn the *bounce diagram* technique of studying the transient bouncing of waves back and forth on the line for a constant voltage source as well as for a pulse voltage source. We shall apply the bounce diagram technique for an initially charged line. Finally, we shall introduce the *load-line* technique of analysis of a line terminated by a nonlinear element and apply it for the analysis of interconnections between logic gates.

A. FREQUENCY DOMAIN

In Chapter 6, we introduced transmission lines, and learned that the voltage and current on a line are governed by the transmission-line equations

$$\frac{\partial V}{\partial z} = -\mathcal{L}\frac{\partial I}{\partial t} \tag{7.1a}$$

Smith® Chart is a registered trademark of Analog Instrument Co., P.O. Box 950, New Providence, NJ 07974, USA.

$$\frac{\partial I}{\partial z} = -\mathscr{G}V - \mathscr{C}\frac{\partial V}{\partial t} \tag{7.1b}$$

For the sinusoidally time-varying case, the corresponding differential equations for the phasor voltage $\bar{V}$ and phasor current $\bar{I}$ are given by

$$\frac{\partial \bar{V}}{\partial z} = -j\omega\mathscr{L}\bar{I} \tag{7.2a}$$

$$\frac{\partial \bar{I}}{\partial z} = -\mathscr{G}\bar{V} - j\omega\mathscr{C}\bar{V} = -(\mathscr{G} + j\omega\mathscr{C})\bar{V} \tag{7.2b}$$

Combining (7.2a) and (7.2b) by eliminating $\bar{I}$, we obtain the wave equation for $\bar{V}$ as

$$\frac{\partial^2 \bar{V}}{\partial z^2} = -j\omega\mathscr{L}\frac{\partial \bar{I}}{\partial z} = j\omega\mathscr{L}(\mathscr{G} + j\omega\mathscr{C})\bar{V}$$

$$= \bar{\gamma}^2\bar{V} \tag{7.3}$$

where

$$\bar{\gamma} = \sqrt{j\omega\mathscr{L}(\mathscr{G} + j\omega\mathscr{C})} \tag{7.4}$$

is the propagation constant associated with the wave propagation on the line. The solution for $\bar{V}$ is given by

$$\bar{V}(z) = \bar{A}e^{-\bar{\gamma}z} + \bar{B}e^{\bar{\gamma}z} \tag{7.5}$$

where $\bar{A}$ and $\bar{B}$ are arbitrary constants to be determined from the boundary conditions. The corresponding solution for $\bar{I}$ is then given by

$$\bar{I}(z) = -\frac{1}{j\omega\mathscr{L}}\frac{\partial \bar{V}}{\partial z} = -\frac{1}{j\omega\mathscr{L}}(-\bar{\gamma}\bar{A}e^{-\bar{\gamma}z} + \bar{\gamma}\bar{B}e^{\bar{\gamma}z})$$

$$= \sqrt{\frac{\mathscr{G} + j\omega\mathscr{C}}{j\omega\mathscr{L}}}(\bar{A}e^{-\bar{\gamma}z} - \bar{B}e^{\bar{\gamma}z})$$

$$= \frac{1}{\bar{Z}_0}(\bar{A}e^{-\bar{\gamma}z} - \bar{B}e^{\bar{\gamma}z}) \tag{7.6}$$

where

$$\bar{Z}_0 = \sqrt{\frac{j\omega\mathscr{L}}{\mathscr{G} + j\omega\mathscr{C}}} \tag{7.7}$$

is known as the *characteristic impedance* of the transmission line.

The solutions for the line voltage and line current given by (7.5) and (7.6), respectively, represent the superposition of $(+)$ and $(-)$ waves, that is, waves propagating in the positive z- and negative z-directions, respectively. They are completely analogous to the solutions for the electric and magnetic fields in the medium between the conductors of the line. In fact, the propagation constant given by (7.4) is the same as the propagation constant $\sqrt{j\omega\mu(\sigma + j\omega\epsilon)}$, as it should be. The characteristic impedance of the line is analogous to (but not equal to) the intrinsic impedance of the material medium between the conductors of the line.

For a *lossless line*, that is, for a line consisting of a perfect dielectric medium between the conductors, $\mathcal{G} = 0$, and

$$\bar{\gamma} = \alpha + j\beta = \sqrt{j\omega\mathcal{L} \cdot j\omega\mathcal{C}} = j\omega\sqrt{\mathcal{L}\mathcal{C}} \tag{7.8}$$

Thus, the attenuation constant α is equal to zero, which is to be expected, and the phase constant β is equal to $\omega\sqrt{\mathcal{L}\mathcal{C}}$. We can then write the solutions for $\bar{V}$ and $\bar{I}$ as

$$\bar{V}(z) = \bar{A}e^{-j\beta z} + \bar{B}e^{j\beta z} \tag{7.9a}$$

$$\bar{I}(z) = \frac{1}{Z_0}(\bar{A}e^{-j\beta z} - \bar{B}e^{j\beta z}) \tag{7.9b}$$

where

$$Z_0 = \sqrt{\frac{\mathcal{L}}{\mathcal{C}}} \tag{7.10}$$

is purely real and independent of frequency. Note also that

$$v_p = \frac{\omega}{\beta} = \frac{1}{\sqrt{\mathcal{L}\mathcal{C}}} = \frac{1}{\sqrt{\mu\epsilon}} \tag{7.11}$$

as it should be, and independent of frequency.

Thus, provided that $\mathcal{L}$ and $\mathcal{C}$ are independent of frequency, which is the case if μ and ϵ are independent of frequency and the transmission line is uniform, that is, its dimensions remain constant transverse to the direction of propagation of the waves, the lossless line is characterized by no dispersion, a phenomenon discussed in Section 8.3. We shall be concerned with such lines only in this book.

7.1 SHORT-CIRCUITED LINE AND FREQUENCY BEHAVIOR

Let us now consider a lossless line short-circuited at the far end $z = 0$, as shown in Figure 7.1(a), in which the double-ruled lines represent the conductors of the transmission line. Note that the line is characterized by Z_0 and β, which is equivalent to specifying $\mathcal{L}$, $\mathcal{C}$, and ω. In actuality, the arrangement may consist, for example, of a perfectly conducting rectangular sheet joining the two conductors of a parallel-plate line as in Figure 7.1(b) or a perfectly conducting ring-shaped sheet joining the two conductors of a coaxial cable as in Figure 7.1(c). We shall assume that the line is driven by a voltage generator of frequency ω at the left end $z = -l$ so that waves are set up on the line. The short circuit at $z = 0$ requires that the tangential electric field on the surface of the conductor comprising the short circuit be zero. Since the voltage between the conductors of the line is proportional to this electric field, which is transverse to them, it follows that the voltage across the short circuit has to be zero. Thus, we have

$$\bar{V}(0) = 0 \tag{7.12}$$

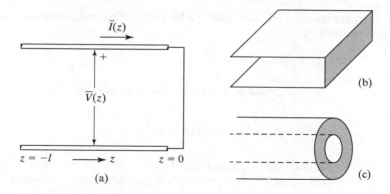

FIGURE 7.1

Transmission line short-circuited at the far end.

Applying the boundary condition given by (7.12) to the general solution for $\bar{V}$ given by (7.9a), we have

$$\bar{V}(0) = \bar{A}e^{-j\beta(0)} + \bar{B}e^{j\beta(0)} = 0$$

or

$$\bar{B} = -\bar{A} \qquad (7.13)$$

Thus, we find that the short circuit gives rise to a $(-)$ or reflected wave whose voltage is exactly the negative of the $(+)$ or incident wave voltage, at the short circuit. Substituting this result in (7.9a) and (7.9b), we get the particular solutions for the complex voltage and current on the short-circuited line to be

$$\bar{V}(z) = \bar{A}e^{-j\beta z} - \bar{A}e^{j\beta z} = -2j\bar{A}\sin\beta z \qquad (7.14a)$$

$$\bar{I}(z) = \frac{1}{Z_0}(\bar{A}e^{-j\beta z} + \bar{A}e^{j\beta z}) = \frac{2\bar{A}}{Z_0}\cos\beta z \qquad (7.14b)$$

The real voltage and current are then given by

$$V(z, t) = \text{Re}[\bar{V}(z)e^{j\omega t}] = \text{Re}(2e^{-j\pi/2}Ae^{j\theta}\sin\beta z\, e^{j\omega t})$$

$$= 2A\sin\beta z\sin(\omega t + \theta) \qquad (7.15a)$$

$$I(z, t) = \text{Re}[\bar{I}(z)e^{j\omega t}] = \text{Re}\left[\frac{2}{Z_0}Ae^{j\theta}\cos\beta z\, e^{j\omega t}\right]$$

$$= \frac{2A}{Z_0}\cos\beta z\cos(\omega t + \theta) \qquad (7.15b)$$

where we have replaced $\bar{A}$ by $Ae^{j\theta}$ and $-j$ by $e^{-j\pi/2}$. The instantaneous power flow down the line is given by

$$P(z, t) = V(z, t)I(z, t)$$

$$= \frac{4A^2}{Z_0} \sin \beta z \cos \beta z \sin (\omega t + \theta) \cos (\omega t + \theta)$$

$$= \frac{A^2}{Z_0} \sin 2\beta z \sin 2(\omega t + \theta) \qquad (7.15c)$$

These results for the voltage, current, and power flow on the short-circuited line given by (7.15a), (7.15b), and (7.15c), respectively, are illustrated in Figure 7.2, which shows the variation of each of these quantities with distance from the short circuit for several values of time. The numbers 1, 2, 3, . . . , 9 beside the curves in Figure 7.2 represent the order of the curves corresponding to values of $(\omega t + \theta)$ equal to 0, $\pi/4$, $\pi/2$, . . . , 2π. It can be seen that the phenomenon is one in which the voltage, current, and power flow oscillate sinusoidally with time with different amplitudes at different locations on the line, unlike in the case of traveling waves, in which a given point on the waveform progresses in distance with time. These waves are therefore known as *standing waves*. In particular, they represent *complete standing waves*, in view of the zero amplitudes of the voltage, current, and power flow at certain locations on the line, as shown by Figure 7.2.

The line voltage amplitude is zero for values of z given by $\sin \beta z = 0$ or $\beta z = -m\pi, m = 1, 2, 3, \ldots$, or $z = -m\lambda/2, m = 1, 2, 3, \ldots$, that is, at multiples of $\lambda/2$ from the short circuit. The line current amplitude is zero for values of z given by $\cos \beta z = 0$ or $\beta z = -(2m + 1)\pi/2$, $m = 0, 1, 2, 3, \ldots$, or $z = -(2m + 1)\lambda/4$, $m = 0, 1, 2, 3, \ldots$, that is, at odd multiples of $\lambda/4$ from the short circuit. The power flow amplitude is zero for values of z given by $\sin 2\beta z = 0$ or $\beta z = -m\pi/2, m = 1, 2, 3,$ . . . , or $z = -m\lambda/4, m = 1, 2, 3, \ldots$, that is, at multiples of $\lambda/4$ from the short circuit. Proceeding further, we find that the time-average power flow down the line, that is, power flow averaged over one period of the source voltage, is

$$\langle P \rangle = \frac{1}{T} \int_{t=0}^{T} P(z, t)\, dt = \frac{\omega}{2\pi} \int_{t=0}^{2\pi/\omega} P(z, t)\, dt$$

$$= \frac{\omega}{2\pi} \frac{A^2}{Z_0} \sin 2\beta z \int_{t=0}^{2\pi/\omega} \sin 2(\omega t + \theta)\, dt = 0$$

Thus, the time average power flow down the line is zero at all points on the line. This is characteristic of complete standing waves.

From (7.14a) and (7.14b) or (7.15a) and (7.15b), or from Figures 7.2(a) and 7.2(b), we find that the amplitudes of the sinusoidal time-variations of the line voltage and line current as functions of distance along the line are

$$|\bar{V}(z)| = 2A|\sin \beta z| = 2A\left|\sin \frac{2\pi}{\lambda} z\right| \qquad (7.16a)$$

$$|\bar{I}(z)| = \frac{2A}{Z_0}|\cos \beta z| = \frac{2A}{Z_0}\left|\cos \frac{2\pi}{\lambda} z\right| \qquad (7.16b)$$

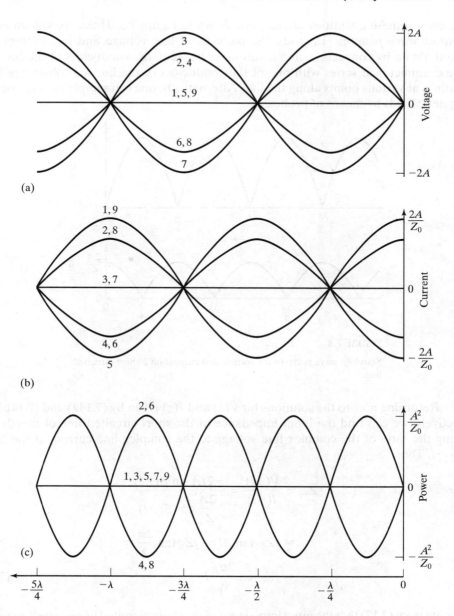

FIGURE 7.2

Time variations of voltage, current, and power flow associated with standing waves on a short-circuited transmission line.

Sketches of these quantities versus z are shown in Figure 7.3. These are known as the *standing wave patterns*. They are the patterns of line voltage and line current one would obtain by connecting an a.c. voltmeter between the conductors of the line and an a.c. ammeter in series with one of the conductors of the line and observing their readings at various points along the line. Alternatively, one can sample the electric and magnetic fields by means of probes.

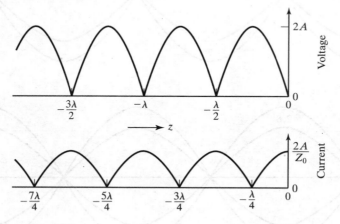

FIGURE 7.3

Standing wave patterns for voltage and current on a short-circuited line.

Returning now to the solutions for $\bar{V}(z)$ and $\bar{I}(z)$ given by (7.14a) and (7.14b), respectively, we can find the input impedance of the short-circuited line of length l by taking the ratio of the complex line voltage to the complex line current at the input $z = -l$. Thus,

$$\bar{Z}_{in} = \frac{\bar{V}(-l)}{\bar{I}(-l)} = \frac{-2j\bar{A}\,\sin\beta(-l)}{\dfrac{2\bar{A}}{Z_0}\cos\beta(-l)}$$

$$= jZ_0\,\tan\beta l = jZ_0\,\tan\frac{2\pi}{\lambda}l$$

$$= jZ_0\,\tan\frac{2\pi f}{v_p}l \qquad (7.17)$$

We note from (7.17) that the input impedance of the short-circuited line is purely reactive. As the frequency is varied from a low value upward, the input reactance changes from inductive to capacitive and back to inductive, and so on, as illustrated in Figure 7.4. The input reactance is zero for values of frequency equal to multiples of $v_p/2l$. These are the frequencies for which l is equal to multiples of $\lambda/2$ so that the line voltage is zero at the input and hence the input sees a short circuit. The input reactance is infinity for values of frequency equal to odd multiples of $v_p/4l$. These are the frequencies for which l is equal to odd multiples of $\lambda/4$ so that the line current is zero at the input and hence the input sees an open circuit.

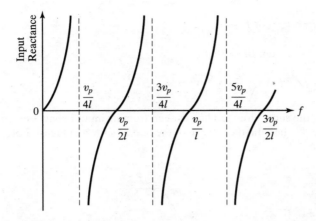

FIGURE 7.4

Variation of the input reactance of a short-circuited transmission line with frequency.

Example 7.1

From the foregoing discussion of the input reactance of the short-circuited line, we note that as the frequency of the generator is varied continuously upward, the current drawn from it undergoes alternatively maxima and minima corresponding to zero input reactance and infinite input reactance conditions, respectively. This behavior can be utilized for determining the location of a short circuit in the line.

Since the difference between a pair of consecutive frequencies for which the input reactance values are zero and infinity is $v_p/4l$, as can be seen from Figure 7.4, it follows that the difference between successive frequencies for which the currents drawn from the generator are maxima and minima is $v_p/4l$. As a numerical example, if for an air dielectric line it is found that as the frequency is varied from 50 MHz upward, the current reaches a minimum for 50.01 MHz and then a maximum for 50.04 MHz, then the distance l of the short circuit from the generator is given by

$$\frac{v_p}{4l} = (50.04 - 50.01) \times 10^6 = 0.03 \times 10^6 = 3 \times 10^4$$

Since $v_p = 3 \times 10^8$ m/s, it follows that

$$l = \frac{3 \times 10^8}{4 \times 3 \times 10^4} = 2500 \text{ m} = 2.5 \text{ km}$$

Example 7.2

We found that the input impedance of a short-circuited line of length l is given by

$$\bar{Z}_{\text{in}} = jZ_0 \tan \beta l$$

Let us investigate the low-frequency behavior of this input impedance.

First, we note that for any arbitrary value of βl,

$$\tan \beta l = \beta l + \frac{1}{3}(\beta l)^3 + \frac{2}{15}(\beta l)^5 + \cdots$$

For $\beta l \ll 1$, that is, $\dfrac{2\pi}{\lambda} l \ll 1$ or $l \ll \dfrac{\lambda}{2\pi}$ or $f \ll \dfrac{v_p}{2\pi l}$,

$$\tan \beta l \approx \beta l$$

$$\bar{Z}_{in} \approx j Z_0 \beta l = j\sqrt{\frac{\mathscr{L}}{\mathscr{C}}}\, \omega \sqrt{\mathscr{L}\mathscr{C}}\, l = j\omega \mathscr{L} l$$

Thus, for frequencies $f \ll v_p/2\pi l$, the short-circuited line as seen from its input behaves essentially like a single inductor of value $\mathscr{L}l$, the total inductance of the line, as shown in Figure 7.5(a).

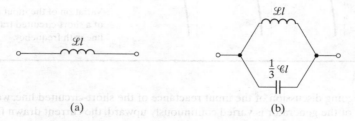

(a)   (b)

FIGURE 7.5

Equivalent circuits for the input behavior of a short-circuited transmission line.

Proceeding further, we observe that if the frequency is slightly beyond the range for which the above approximation is valid, then

$$\tan \beta l \approx \beta l + \frac{1}{3}(\beta l)^3$$

$$\bar{Z}_{in} \approx j Z_0 \left(\beta l + \frac{1}{3}\beta^3 l^3 \right)$$

$$= j\sqrt{\frac{\mathscr{L}}{\mathscr{C}}} \left(\omega \sqrt{\mathscr{L}\mathscr{C}}\, l + \frac{1}{3}\omega^3 \mathscr{L}^{3/2}\mathscr{C}^{3/2} l^3 \right)$$

$$= j\omega \mathscr{L} l \left(1 + \frac{1}{3}\omega^2 \mathscr{L}\mathscr{C} l^2 \right)$$

$$\bar{Y}_{in} = \frac{1}{\bar{Z}_{in}} = \frac{1}{j\omega \mathscr{L} l} \left(1 + \frac{1}{3}\omega^2 \mathscr{L}\mathscr{C} l^2 \right)^{-1}$$

$$\approx \frac{1}{j\omega \mathscr{L} l} \left(1 - \frac{1}{3}\omega^2 \mathscr{L}\mathscr{C} l^2 \right)$$

$$= \frac{1}{j\omega \mathscr{L} l} + j\frac{1}{3}\omega \mathscr{C} l$$

Thus, for frequencies somewhat above those for which the approximation $f \ll v_p/2\pi l$ is valid, the short-circuited line as seen from its input behaves like an inductor of value $\mathscr{L}l$ in parallel with a capacitance of value $\dfrac{1}{3}\mathscr{C}l$, as shown in Figure 7.5(b).

These findings illustrate that a physical structure that can be considered as an inductor at low frequencies $f \ll v_p/2\pi l$ no longer behaves like an inductor, if the frequency is increased beyond that range. In fact, it has a *stray* capacitance associated with it. As the frequency is still increased, the equivalent circuit becomes further complicated. With reference to the question posed in Section 6.5 as to the limit on the frequency beyond which the quasistatic approximation for the input behavior of a physical structure is not valid, it can now be seen that the condition $\beta l \ll 1$ dictates the range of validity for the quasistatic approximation. In terms of the frequency f of the source, this condition means that $f \ll v_p/2\pi l$, or in terms of the period $T = 1/f$, it means that $T \gg 2\pi(l/v_p)$. Thus, quasistatic fields are low-frequency approximations of time-varying fields that are complete solutions to Maxwell's equations, which represent wave propagation phenomena and can be approximated to the quasistatic character only when the times of interest are much greater than the propagation time, l/v_p, corresponding to the length of the structure. In terms of space variations of the fields at a fixed time, the wavelength $\lambda(= 2\pi/\beta)$, must be such that $l \ll \lambda/2\pi$; thus, the physical length of the structure must be a small fraction of the wavelength. In terms of the line voltage and current amplitudes, what this means is that over the length of the structure, these amplitudes are fractional portions of the first one-quarter sinusoidal variations at the $z = 0$ end in Figure 7.3, with the boundary conditions at the two ends of the structure always satisfied. Thus, because of the $\sin \beta z$ dependence of V on z, the line voltage amplitude varies linearly with z, whereas because of the $\cos \beta z$ dependence of I on z, the line current amplitude is essentially a constant. These are exactly the nature of the variations of the zero-order electric field and the first-order magnetic field, as discussed under magnetoquasistatic fields in Example 6.7.

7.2 TRANSMISSION-LINE DISCONTINUITY

Let us now consider the case of two transmission lines, 1 and 2, having different characteristic impedances Z_{01} and Z_{02}, respectively, and phase constants β_1 and β_2, respectively, connected in cascade and driven by a generator at the left end of line 1, as shown in Figure 7.6(a). Physically, the arrangement may, for example, consist of two parallel-plate lines or two coaxial cables of different dielectrics in cascade, as shown in Figures 7.6(b) and 7.6(c), respectively. In view of the discontinuity at the junction $z = 0$

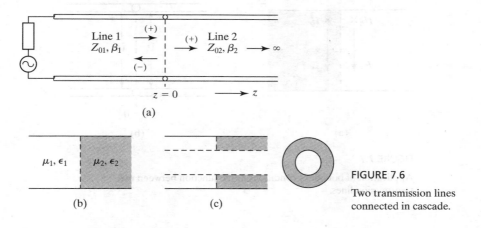

(a)

(b)   (c)

FIGURE 7.6

Two transmission lines connected in cascade.

between the two lines, the incident $(+)$ wave on the junction sets up a reflected $(-)$ wave in line 1 and a transmitted $(+)$ wave in line 2. We shall assume that line 2 is infinitely long so that there is no $(-)$ wave in that line.

We can now write the solutions for the complex voltage and complex current in line 1 as

$$\bar{V}_1(z) = \bar{V}_1^+ e^{-j\beta_1 z} + \bar{V}_1^- e^{j\beta_1 z} \tag{7.18a}$$

$$\bar{I}_1(z) = \bar{I}_1^+ e^{-j\beta_1 z} + \bar{I}_1^- e^{j\beta_1 z}$$

$$= \frac{1}{Z_{01}} (\bar{V}_1^+ e^{-j\beta_1 z} - \bar{V}_1^- e^{j\beta_1 z}) \tag{7.18b}$$

where $\bar{V}_1^+, \bar{V}_1^-, \bar{I}_1^+,$ and $\bar{I}_1^-$ are the $(+)$ and $(-)$ wave voltages and currents at $z = 0-$ in line 1, that is, just to the left of the junction. The solutions for the complex voltage and complex current in line 2 are

$$\bar{V}_2(z) = \bar{V}_2^+ e^{-j\beta_2 z} \tag{7.19a}$$

$$\bar{I}_2(z) = \bar{I}_2^+ e^{-j\beta_2 z} = \frac{1}{Z_{02}} \bar{V}_2^+ e^{-j\beta_2 z} \tag{7.19b}$$

where $\bar{V}_2^+$ and $\bar{I}_2^+$ are the $(+)$ wave voltage and current at $z = 0+$ in line 2, that is, just to the right of the junction.

At the junction, the boundary conditions require that the components of **E** and **H** tangential to the dielectric interface be continuous, as shown, for example, for the parallel-plate arrangement in Figure 7.7(a). These are, in fact, the only components present, since the transmission line fields are entirely transverse to the direction of propagation. Now, since the line voltage and current are related to these electric and magnetic fields, respectively, it then follows that the line voltage and line current be

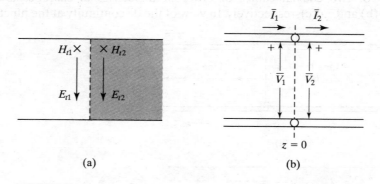

(a)   (b)

FIGURE 7.7

Application of boundary conditions at the junction between two transmission lines.

continuous at the junction, as shown in Figure 7.7(b). Thus, we obtain the boundary conditions at the junction in terms of line voltage and line current as

$$[\bar{V}_1]_{z=0-} = [\bar{V}_2]_{z=0+} \tag{7.20a}$$

$$[\bar{I}_1]_{z=0-} = [\bar{I}_2]_{z=0+} \tag{7.20b}$$

Applying these boundary conditions to the solutions given by (7.18a) and (7.18b), we obtain

$$\bar{V}_1^+ + \bar{V}_1^- = \bar{V}_2^+ \tag{7.21a}$$

$$\frac{1}{Z_{01}}(\bar{V}_1^+ - \bar{V}_1^-) = \frac{1}{Z_{02}}\bar{V}_2^+ \tag{7.21b}$$

Eliminating $\bar{V}_2^+$ from (7.21a) and (7.21b), we get

$$\bar{V}_1^+\left(\frac{1}{Z_{02}} - \frac{1}{Z_{01}}\right) + \bar{V}_1^-\left(\frac{1}{Z_{02}} + \frac{1}{Z_{01}}\right) = 0$$

or

$$\bar{V}_1^- = \bar{V}_1^+\frac{Z_{02} - Z_{01}}{Z_{02} + Z_{01}} \tag{7.22}$$

We now define the voltage reflection coefficient at the junction, Γ_V, as the ratio of the reflected wave voltage ($\bar{V}_1^-$) at the junction to the incident wave voltage ($\bar{V}_1^+$) at the junction. Thus,

$$\Gamma_V = \frac{\bar{V}_1^-}{\bar{V}_1^+} = \frac{Z_{02} - Z_{01}}{Z_{02} + Z_{01}} \tag{7.23}$$

The current reflection coefficient at the junction, Γ_I, which is the ratio of the reflected wave current ($\bar{I}_1^-$) at the junction to the incident wave current ($\bar{I}_1^+$) at the junction is then given by

$$\Gamma_I = \frac{\bar{I}_1^-}{\bar{I}_1^+} = \frac{-\bar{V}_1^-/Z_{01}}{\bar{V}_1^+/Z_{01}} = -\frac{\bar{V}_1^-}{\bar{V}_1^+} = -\Gamma_V \tag{7.24}$$

We also define the voltage transmission coefficient at the junction, τ_V, as the ratio of the transmitted wave voltage ($\bar{V}_2^+$) at the junction to the incident wave voltage ($\bar{V}_1^+$) at the junction. Thus,

$$\tau_V = \frac{\bar{V}_2^+}{\bar{V}_1^+} = \frac{\bar{V}_1^+ + \bar{V}_1^-}{\bar{V}_1^+} = 1 + \frac{\bar{V}_1^-}{\bar{V}_1^+} = 1 + \Gamma_V \tag{7.25}$$

The current transmission coefficient at the junction, τ_I, which is the ratio of the transmitted wave current ($\bar{I}_2^+$) at the junction to the incident wave current ($\bar{I}_1^+$) at the junction, is given by

$$\tau_I = \frac{\bar{I}_2^+}{\bar{I}_1^+} = \frac{\bar{I}_1^+ + \bar{I}_1^-}{\bar{I}_1^+} = 1 + \frac{\bar{I}_1^-}{\bar{I}_1^+} = 1 - \Gamma_V \tag{7.26}$$

We note that for $Z_{02} = Z_{01}, \Gamma_V = 0, \Gamma_I = 0, \tau_V = 1$, and $\tau_I = 1$. Thus, the incident wave is entirely transmitted, as we may expect since there is no discontinuity at the junction.

Example 7.3

Let us consider the junction of two lines having characteristic impedances $Z_{01} = 50\ \Omega$ and $Z_{02} = 75\ \Omega$, as shown in Figure 7.8, and compute the various quantities.

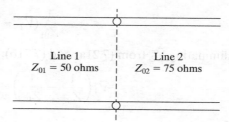

FIGURE 7.8

For the computation of several quantities pertinent to reflection and transmission at the junction between two transmission lines.

From (7.23)–(7.26), we have

$$\Gamma_V = \frac{75 - 50}{75 + 50} = \frac{25}{125} = \frac{1}{5}; \qquad \bar{V}_1^- = \frac{1}{5}\bar{V}_1^+$$

$$\Gamma_I = -\Gamma_V = -\frac{1}{5}; \qquad \bar{I}_1^- = -\frac{1}{5}\bar{I}_1^+$$

$$\tau_V = 1 + \Gamma_V = 1 + \frac{1}{5} = \frac{6}{5}; \qquad \bar{V}_2^+ = \frac{6}{5}\bar{V}_1^+$$

$$\tau_I = 1 - \Gamma_V = 1 - \frac{1}{5} = \frac{4}{5}; \qquad \bar{I}_2^+ = \frac{4}{5}\bar{I}_1^+$$

The fact that the transmitted wave voltage is greater than the incident wave voltage should not be of concern, since it is the power balance that must be satisfied at the junction. We can verify this by noting that if the incident power on the junction is P_i, then

$$\text{reflected power, } P_r = \Gamma_V \Gamma_I P_i = -\frac{1}{25}P_i$$

$$\text{transmitted power, } P_t = \tau_V \tau_I P_i = \frac{24}{25}P_i$$

Recognizing that the minus sign for P_r signifies power flow in the negative z-direction, we find that power balance is indeed satisfied at the junction.

Returning now to the solutions for the voltage and current in line 1 given by (7.18a) and (7.18b), respectively, we obtain, by replacing $\bar{V}_1^-$ by $\Gamma_V \bar{V}_1^+$,

$$\bar{V}_1(z) = \bar{V}_1^+ e^{-j\beta_1 z} + \Gamma_V \bar{V}_1^+ e^{j\beta_1 z}$$

$$= \bar{V}_1^+ e^{-j\beta_1 z}(1 + \Gamma_V e^{j2\beta_1 z}) \tag{7.27a}$$

$$\bar{I}_1(z) = \frac{1}{Z_{01}}(\bar{V}_1^+ e^{-j\beta_1 z} - \Gamma_V \bar{V}_1^+ e^{j\beta_1 z})$$

$$= \frac{\bar{V}_1^+}{Z_{01}} e^{-j\beta_1 z}(1 - \Gamma_V e^{j2\beta_1 z}) \tag{7.27b}$$

The amplitudes of the sinusoidal time-variations of the line voltage and line current as functions of distance along the line are then given by

$$|\bar{V}_1(z)| = |\bar{V}_1^+||e^{-j\beta_1 z}||1 + \Gamma_V e^{j2\beta_1 z}|$$

$$= |\bar{V}_1^+||1 + \Gamma_V \cos 2\beta_1 z + j\Gamma_V \sin 2\beta_1 z|$$

$$= |\bar{V}_1^+|\sqrt{1 + \Gamma_V^2 + 2\Gamma_V \cos 2\beta_1 z} \tag{7.28a}$$

$$|\bar{I}_1(z)| = \frac{|\bar{V}_1^+|}{Z_{01}}|e^{-j\beta_1 z}||1 - \Gamma_V e^{j2\beta_1 z}|$$

$$= \frac{|\bar{V}_1^+|}{Z_{01}}|1 - \Gamma_V \cos 2\beta_1 z - j\Gamma_V \sin 2\beta_1 z|$$

$$= \frac{|\bar{V}_1^+|}{Z_{01}}\sqrt{1 + \Gamma_V^2 - 2\Gamma_V \cos 2\beta_1 z} \tag{7.28b}$$

From (7.28a) and (7.28b), we note the following:

1. The line voltage amplitude undergoes alternate maxima and minima equal to $|\bar{V}_1^+|(1 + |\Gamma_V|)$ and $|\bar{V}_1^+|(1 - |\Gamma_V|)$, respectively. The line voltage amplitude at $z = 0$ is a maximum on minimum depending on whether Γ_V is positive or negative. The distance between a voltage maximum and the adjacent voltage minimum is $\pi/2\beta_1$ or $\lambda_1/4$.

2. The line current amplitude undergoes alternate maxima and minima equal to $(\bar{V}_1^+/Z_{01})(1 + |\Gamma_V|)$ and $(\bar{V}_1^+/Z_{01})(1 - |\Gamma_V|)$, respectively. The line current amplitude at $z = 0$ is a minimum or maximum depending on whether Γ_V is positive or negative. The distance between a current maximum and the adjacent current minimum is $\pi/2\beta_1$ or $\lambda_1/4$.

Knowing these properties of the line voltage and current amplitudes, we now sketch the voltage and current standing wave patterns, as shown in Figure 7.9, assuming $\Gamma_V > 0$. Since these standing wave patterns do not contain perfect nulls, as in the case of the short-circuited line of Section 7.1, these are said to correspond to *partial standing waves*.

We now define a quantity known as the *standing wave ratio* (SWR) as the ratio of the maximum voltage, V_{max}, to the minimum voltage, V_{min}, of the standing wave pattern. Thus, we find that

$$\text{SWR} = \frac{V_{max}}{V_{min}} = \frac{|\bar{V}_1^+|(1 + |\Gamma_V|)}{|\bar{V}_1^+|(1 - |\Gamma_V|)} = \frac{1 + |\Gamma_V|}{1 - |\Gamma_V|} \tag{7.29}$$

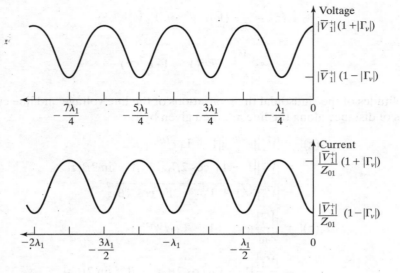

FIGURE 7.9

Standing wave patterns for voltage and current on a transmission line
terminated by another transmission line.

The SWR is an important parameter in transmission-line matching. It is an indicator
of the degree of the existence of standing waves on the line. We shall, however, not
pursue the topic here any further. Finally, we note that for the case of Example 7.3, the
SWR in line 1 is $(1 + \frac{1}{5})/(1 - \frac{1}{5})$, or 1.5. The SWR in line 2 is, of course, equal to 1 since
there is no reflected wave in that line.

7.3   THE SMITH CHART

In the previous section, we studied reflection and transmission at the junction of two
transmission lines, shown in Figure 7.10. In this section, we shall introduce the Smith
Chart, which is a useful graphical aid in the solution of transmission-line and many
other problems.

First we define the line impedance $\bar{Z}(z)$ at a given value of z on the line as the
ratio of the complex line voltage to the complex line current at that value of z, that is,

$$\bar{Z}(z) = \frac{\bar{V}(z)}{\bar{I}(z)} \tag{7.30}$$

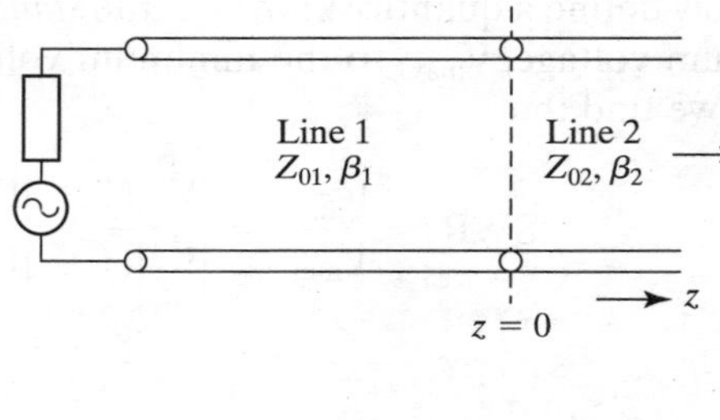

FIGURE 7.10

A transmission line terminated
by another infinitely long
transmission line.

From the solutions for the line voltage and line current on line 2 given by (7.19a) and (7.19b), respectively, the line impedance in line 2 is given by

$$\bar{Z}_2(z) = \frac{\bar{V}_2(z)}{\bar{I}_2(z)} = Z_{02}$$

Thus, the line impedance at all points on line 2 is simply equal to the characteristic impedance of that line. This is because the line is infinitely long and hence there is only a (+) wave on the line. From the solutions for the line voltage and line current in line 1 given by (7.18a) and (7.18b), respectively, the line impedance for that line is given by

$$\bar{Z}_1(z) = \frac{\bar{V}_1(z)}{\bar{I}_1(z)} = Z_{01}\frac{\bar{V}_1^+e^{-j\beta_1 z} + \bar{V}_1^-e^{j\beta_1 z}}{\bar{V}_1^+e^{-j\beta_1 z} - \bar{V}_1^-e^{j\beta_1 z}}$$

$$= Z_{01}\frac{1 + \bar{\Gamma}_V(z)}{1 - \bar{\Gamma}_V(z)} \tag{7.31}$$

where

$$\bar{\Gamma}_V(z) = \frac{\bar{V}_1^-e^{j\beta_1 z}}{\bar{V}_1^+e^{-j\beta_1 z}} = \bar{\Gamma}_V(0)e^{j2\beta_1 z} \tag{7.32}$$

$$\bar{\Gamma}_V(0) = \frac{\bar{V}_1^-}{\bar{V}_1^+} = \frac{Z_{02} - Z_{01}}{Z_{02} + Z_{01}} \tag{7.33}$$

The quantity $\bar{\Gamma}_V(0)$ is the voltage reflection coefficient at the junction $z = 0$, and $\bar{\Gamma}_V(z)$ is the voltage reflection coefficient at any value of z.

To compute the line impedance at a particular value of z, we first compute $\bar{\Gamma}_V(0)$ from a knowledge of Z_{02}, which is the terminating impedance to line 1. We then compute $\bar{\Gamma}_V(z) = \bar{\Gamma}_V(0)e^{j2\beta_1 z}$, which is a complex number having the same magnitude as that of $\bar{\Gamma}_V(0)$ but a phase angle equal to $2\beta_1 z$ plus the phase angle of $\bar{\Gamma}_V(0)$. The computed value of $\bar{\Gamma}_V(z)$ is then substituted in (7.31) to find $\bar{Z}_1(z)$. All of this complex algebra is eliminated through the use of the Smith Chart.

The Smith Chart is a mapping of the values of normalized line impedance onto the reflection coefficient $(\bar{\Gamma}_V)$ plane. The normalized line impedance $\bar{Z}_n(z)$ is the ratio of the line impedance to the characteristic impedance of the line. From (7.31), and omitting the subscript 1 for the sake of generality, we have

$$\bar{Z}_n(z) = \frac{\bar{Z}(z)}{Z_0} = \frac{1 + \bar{\Gamma}_V(z)}{1 - \bar{\Gamma}_V(z)} \tag{7.34}$$

Conversely,

$$\bar{\Gamma}_V(z) = \frac{\bar{Z}_n(z) - 1}{\bar{Z}_n(z) + 1} \tag{7.35}$$

Writing $\bar{Z}_n = r + jx$ and substituting into (10.35), we find that

$$|\bar{\Gamma}_V| = \left|\frac{r + jx - 1}{r + jx + 1}\right| = \frac{\sqrt{(r - 1)^2 + x^2}}{\sqrt{(r + 1)^2 + x^2}} \le 1 \qquad \text{for } r \ge 0$$

Thus, we note that all passive values of normalized line impedances, that is, points in the right half of the complex $\overline{Z}_n$-plane shown in Figure 7.11(a) are mapped onto the region within the circle of radius unity in the complex $\overline{\Gamma}_V$-plane shown in Figure 7.11(b).

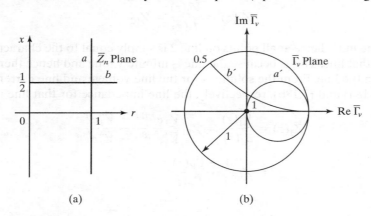

FIGURE 7.11

For illustrating the development of the Smith Chart.

We can now assign values for $\overline{Z}_n$, compute the corresponding values of $\overline{\Gamma}_V$, and plot them on the $\overline{\Gamma}_V$-plane but indicating the values of $\overline{Z}_n$ instead of the values of $\overline{\Gamma}_V$. To do this in a systematic manner, we consider contours in the $\overline{Z}_n$-plane corresponding to constant values of r, as shown for example by the line marked a for $r = 1$, and corresponding to constant values of x, as shown for example by the line marked b for $x = \frac{1}{2}$ in Figure 7.11(a).

By considering several points along line a, computing the corresponding values of $\overline{\Gamma}_V$, plotting them on the $\overline{\Gamma}_V$-plane, and joining them, we obtain the contour marked a' in Figure 7.11(b). Although it can be shown analytically that this contour is a circle of radius $\frac{1}{2}$ and centered at $(1/2, 0)$, it is a simple task to write a computer program to perform this operation, including the plotting. Similarly, by considering several points along line b and following the same procedure, we obtain the contour marked b' in Figure 7.11(b). Again, it can be shown analytically that this contour is a portion of a circle of radius 2 and centered at $(1, 2)$. We can now identify the points on contour a' as corresponding to $r = 1$ by placing the number 1 beside it and the points on contour b' as corresponding to $x = \frac{1}{2}$ by placing the number 0.5 beside it. The point of intersection of contours a' and b' then corresponds to $\overline{Z}_n = 1 + j0.5$.

When the procedure discussed above is applied to many lines of constant r and constant x covering the entire right half of the $\overline{Z}_n$-plane, we obtain the Smith Chart. In a commercially available form shown in Figure 7.12, the Smith Chart contains contours of constant r and constant x at appropriate increments of r and x in the range $0 < r < \infty$ and $-\infty < x < \infty$ so that interpolation between the contours can be carried out to a good degree of accuracy.

Let us now consider the transmission line system shown in Figure 7.13, which is the same as that in Figure 7.10 except that a reactive element having susceptance (reciprocal of reactance) B is connected in parallel with line 1 at a distance l from the junction.

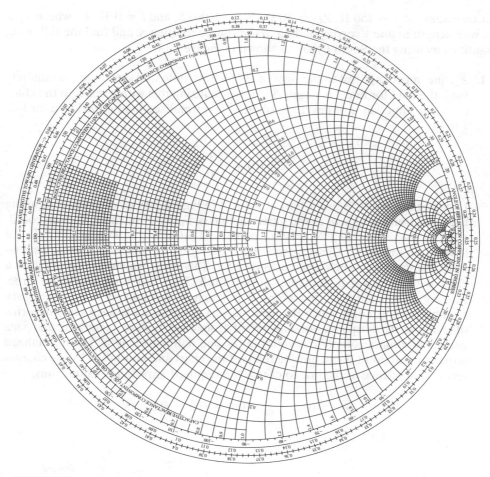

FIGURE 7.12

A commercially available form of the Smith Chart (reproduced with the courtesy of Analog Instrument Co., P.O. Box 950, New Providence, NJ 07974, USA).

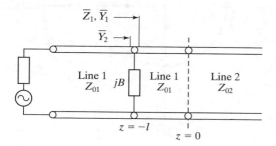

FIGURE 7.13

A transmission-line system for illustrating the computation of several quantities by using the Smith Chart.

Let us assume $Z_{01} = 150 \ \Omega$, $Z_{02} = 50 \ \Omega$, $B = -0.003$ S, and $l = 0.375\lambda_1$, where λ_1 is the wavelength in line 1 corresponding to the source frequency, and find the following quantities by using the Smith Chart, as shown in Figure 7.14:

1. $\bar{Z}_1$, line impedance just to the right of jB: First we note that since line 2 is infinitely long, the load for line 1 is simply 50 Ω. Normalizing this with respect to the characteristic impedance of line 1, we obtain the normalized load impedance for line 1 to be

$$\bar{Z}_n(0) = \frac{50}{150} = \frac{1}{3}$$

Locating this on the Smith Chart at point A in Figure 7.14 amounts to computing the reflection coefficient at the junction, that is, $\bar{\Gamma}_V(0)$. Now the reflection coefficient at $z = -l = -0.375\lambda_1$, being equal to $\bar{\Gamma}_V(0)e^{-j2\beta_1 l} = \bar{\Gamma}_V(0)e^{-j1.5\pi}$, can be located on the Smith Chart by moving A such that the magnitude remains constant but the phase angle decreases by 1.5π. This is equivalent to moving it on a circle with its center at the center of the Smith Chart and in the clockwise direction by 1.5π or $270°$ so that point B is reached. Actually, it is not necessary to compute this angle, since the Smith Chart contains a distance scale in terms of λ along its periphery for movement from load toward generator and vice versa, based on a complete revolution for one-half wavelength. The normalized impedance at point B can now be read off the chart and multiplied by the characteristic impedance of the line to obtain the required impedance value. Thus,

$$\bar{Z}_1 = (0.6 - j0.8)150 = (90 - j120) \ \Omega.$$

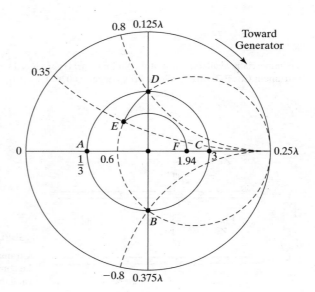

FIGURE 7.14

For illustrating the use of the Smith Chart in the computation of several quantities for the transmission-line system of Figure 7.13.

2. SWR on line 1 to the right of jB: From (7.29)

$$\text{SWR} = \frac{1 + |\Gamma_V|}{1 - |\Gamma_V|} = \frac{1 + |\bar{\Gamma}_V|e^{j0}}{1 - |\bar{\Gamma}_V|e^{j0}} \tag{7.36}$$

Comparing the right side of (7.36) with the expression for $\bar{Z}_n$ given by (7.34), we note that it is simply equal to $\bar{Z}_n$ corresponding to phase angle of $\bar{\Gamma}_V$ equal to zero. Thus, to find the SWR, we locate the point on the Smith Chart having the same $|\bar{\Gamma}_V|$ as that for $z = 0$, but having a phase angle equal to zero, that is, the point C in Figure 7.14, and then read off the normalized resistance value at that point. Here, it is equal to 3 and hence the required SWR is equal to 3. In fact, the circle passing through C and having its center at the center of the Smith Chart is known as the *constant SWR* (=3) *circle*, since for any normalized load impedance to line 1 lying on that circle, the SWR is the same (and equal to 3).

3. $\bar{Y}_1$, line admittance just to the right of jB: To find this, we note that the normalized line admittance $\bar{Y}_n$ at any value of z, that is, the line admittance normalized with respect to the line characteristic admittance Y_0 (reciprocal of Z_0) is given by

$$\begin{aligned}
\bar{Y}_n(z) &= \frac{\bar{Y}(z)}{Y_0} = \frac{Z_0}{\bar{Z}(z)} = \frac{1}{\bar{Z}_n(z)} \\
&= \frac{1 - \bar{\Gamma}_V(z)}{1 + \bar{\Gamma}_V(z)} = \frac{1 + \bar{\Gamma}_V(z)e^{\pm j\pi}}{1 - \bar{\Gamma}_V(z)e^{\pm j\pi}} \\
&= \frac{1 + \bar{\Gamma}_V(z)e^{\pm j2\beta\lambda/4}}{1 - \bar{\Gamma}_V(z)e^{\pm j2\beta\lambda/4}} = \frac{1 + \bar{\Gamma}_V(z \pm \lambda/4)}{1 - \bar{\Gamma}_V(z \pm \lambda/4)} \\
&= \bar{Z}_n\left(z \pm \frac{\lambda}{4}\right) \tag{7.37}
\end{aligned}$$

Thus, $\bar{Y}_n$ at a given value of z is equal to $\bar{Z}_n$ at a value of z located $\lambda/4$ from it. On the Smith Chart, this corresponds to the point on the constant SWR circle passing through B and diametrically opposite to it, that is, the point D. Thus,

$$\bar{Y}_{n1} = 0.6 + j0.8$$

and

$$\begin{aligned}
\bar{Y}_1 &= Y_{01}\bar{Y}_{n1} = \frac{1}{150}(0.6 + j0.8) \\
&= (0.004 + j0.0053) \text{ S}
\end{aligned}$$

In fact, the Smith Chart can be used as an admittance chart instead of as an impedance chart, that is, by knowing the line admittance at one point on the line, the line admittance at another point on the line can be found by proceeding in the same manner as for impedances. As an example, to find $\bar{Y}_1$, we can first find the normalized line admittance at $z = 0$ by locating the point C diametrically opposite to point A on the constant SWR circle. Then we find $\bar{Y}_{n1}$ by simply going on the constant SWR circle by the distance $l(= 0.375\lambda_1)$ toward the generator. This leads to point D, thereby giving us the same result for $\bar{Y}_1$ as found above.

4. SWR on line 1 to the left of jB: To find this, we first locate the normalized line admittance just to the left of jB, which then determines the constant SWR circle corresponding to the portion of line 1 to the left of jB. Thus, noting that $\bar{Y}_2 = \bar{Y}_1 + jB$, or $\bar{Y}_{n2} = \bar{Y}_{n1} + jB/Y_{01}$, and hence

$$\text{Re}[\bar{Y}_{n2}] = \text{Re}[\bar{Y}_{n1}] \tag{7.38a}$$

$$\text{Im}[\bar{Y}_{n2}] = \text{Im}[\bar{Y}_{n1}] + \frac{B}{Y_{01}} \tag{7.38b}$$

we start at point D and go along the constant real part (conductance) circle to reach point E for which the imaginary part differs from the imaginary part at D by the amount B/Y_{01}, that is, $-0.003/(1/150)$, or -0.45. We then draw the constant SWR circle passing through E and then read off the required SWR value at point F. This value is equal to 1.94.

The steps outlined above in part 4 can be applied is reverse to determine the location and the value of the susceptance required to achieve an SWR of unity to the left of it, that is, a condition of no standing waves. This procedure is known as transmission-line *matching*. It is important from the point of view of eliminating or minimizing certain undesirable effects of standing waves in electromagnetic energy transmission.

To illustrate the solution to the matching problem, we first recognize that an SWR of unity is represented by the center point of the Smith Chart. Hence, matching is achieved if $\bar{Y}_{n2}$ falls at the center of the Smith Chart. Now since the difference between $\bar{Y}_{n1}$ and $\bar{Y}_{n2}$ is only in the imaginary part as indicated by (7.38a) and (7.38b), $\bar{Y}_{n1}$ must lie on the constant conductance circle passing through the center of the Smith Chart (this circle is known as the *unit conductance circle*, since it corresponds to normalized real part equal to unity). $\bar{Y}_{n1}$ must also lie on the constant SWR circle corresponding to the portion of the line to the right of jB. Hence, it is given by the point(s) of intersection of this constant SWR circle and the unit conductance circle. There are two such points, G and H, as shown in Figure 7.15, in which the points A and C are repeated from Figure 7.14. There are thus two solutions to the matching problem. If we choose G to correspond to $\bar{Y}_{n1}$, then, since the distance from C to G is $(0.333 - 0.250)\lambda_1$, or $0.083\lambda_1$, jB must be located at $z = -0.083\lambda_1$. To find the value of jB, we note that the normalized susceptance value corresponding to G is -1.16 and hence $B/Y_{01} = 1.16$, or $jB = j1.16\, Y_{01} = j0.00773$ S. If, however, we choose the point H to correspond to $\bar{Y}_{n1}$, then we find in a similar manner that jB must be located at $z = (0.250 + 0.167)\lambda_1$ or $0.417\lambda_1$ and its value must be $-j0.00773$ S.

The reactive element jB used to achieve the matching is commonly realized by means of a short-circuited section of line, known as a *stub*. This is based on the fact that the input impedance of a short-circuited line is purely reactive, as shown in Section 7.1.

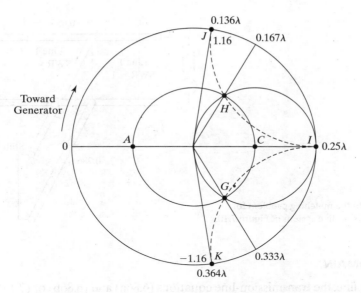

FIGURE 7.15

Solution of transmission-line matching problem by using the Smith Chart.

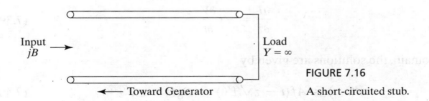

FIGURE 7.16

A short-circuited stub.

The length of the stub for a required input susceptance can be found by considering the short circuit as the load, as shown in Figure 7.16, and using the Smith Chart. The admittance corresponding to a short circuit is infinity, and hence the load admittance normalized with respect to the characteristic admittance of the stub is also equal to infinity. This is located on the Smith Chart at point I in Figure 7.15. We then go along the constant SWR circle passing through I (the outermost circle) toward the generator (input) until we reach the point corresponding to the required input susceptance of the stub normalized with respect to the characteristic admittance of the stub. Assuming the characteristic impedance of the stub to be the same as that of the line, this quantity is here equal to $j1.16$ or $-j1.16$, depending on whether point G or point H is chosen for the location of the stub. This leads us to point J or point K, and hence the stub length is $(0.25 + 0.136)\lambda_1$, or $0.386\lambda_1$, for $jB = j1.16$, and $(0.364 - 0.25)\lambda_1$, or $0.114\lambda_1$, for $jB = -j1.16$. The arrangement of the stub corresponding to the solution for which the stub location is at $z = -0.083\lambda_1$, and the stub length is $0.386\lambda_1$, is shown in Figure 7.17.

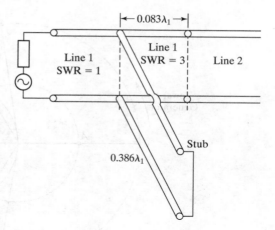

FIGURE 7.17

A solution to the matching problem for
the transmission-line system of Figure 7.10.

B. TIME DOMAIN

For a lossless line, the transmission-line equations (6.86a) and (6.86b) or (7.1a) and (7.1b)
reduce to

$$\frac{\partial V}{\partial z} = -\mathscr{L}\frac{\partial I}{\partial t} \tag{7.39a}$$

$$\frac{\partial I}{\partial z} = -\mathscr{C}\frac{\partial V}{\partial t} \tag{7.39b}$$

In time domain, the solutions are given by

$$V(z, t) = Af(t - z\sqrt{\mathscr{L}\mathscr{C}}) + Bg(t + z\sqrt{\mathscr{L}\mathscr{C}}) \tag{7.40a}$$

$$I(z, t) = \frac{1}{\sqrt{\mathscr{L}/\mathscr{C}}}[Af(t - z\sqrt{\mathscr{L}\mathscr{C}}) - Bg(t + z\sqrt{\mathscr{L}\mathscr{C}})] \tag{7.40b}$$

which can be verified by substituting them into (7.39a) and (7.39b). These solutions
represent voltage and current traveling waves propagating with velocity

$$v_p = \frac{1}{\sqrt{\mathscr{L}\mathscr{C}}} \tag{7.41}$$

in view of the arguments $(t \mp z\sqrt{\mathscr{L}\mathscr{C}})$ for the functions f and g, and characteristic
impedance

$$Z_0 = \sqrt{\frac{\mathscr{L}}{\mathscr{C}}} \tag{7.42}$$

They can also be inferred from the fact that v_p and Z_0 are independent of frequency.

We now rewrite (7.40a) and (7.40b) as

$$V(z,t) = V^+\left(t - \frac{z}{v_p}\right) + V^-\left(t + \frac{z}{v_p}\right) \tag{7.43a}$$

$$I(z,t) = \frac{1}{Z_0}\left[V^+\left(t - \frac{z}{v_p}\right) - V^-\left(t + \frac{z}{v_p}\right)\right] \tag{7.43b}$$

or, more concisely,

$$V = V^+ + V^- \tag{7.44a}$$

$$I = \frac{1}{Z_0}(V^+ - V^-) \tag{7.44b}$$

with the understanding that V^+ is a function of $(t - z/v_p)$ and V^- is a function of $(t + z/v_p)$. In terms of $(+)$ and $(-)$ wave currents, the solution for the current may also be written as

$$I = I^+ + I^- \tag{7.45}$$

Comparing (7.44b) and (7.45), we see that

$$I^+ = \frac{V^+}{Z_0} \tag{7.46a}$$

$$I^- = -\frac{V^-}{Z_0} \tag{7.46b}$$

The minus sign in (7.46b) can be understood if we recognize that in writing (7.44a) and (7.45), we follow the notation that both V^+ and V^- have the same polarities with one conductor (say, a) positive with respect to the other conductor (say, b) and that both I^+ and I^- flow in the positive z-direction along conductor a and return in the negative z-direction along conductor b, as shown in Figure 7.18. The power flow associated with either wave, as given by the product of the corresponding voltage and current, is then directed in the positive z-direction, as shown in Figure 7.18. Thus,

$$P^+ = V^+I^+ = V^+\left(\frac{V^+}{Z_0}\right) = \frac{(V^+)^2}{Z_0} \tag{7.47a}$$

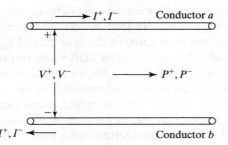

FIGURE 7.18

Polarities for voltages and currents associated with $(+)$ and $(-)$ waves.

Since $(V^+)^2$ is always positive, regardless of whether V^+ is numerically positive or negative, (7.47a) indicates that the $(+)$ wave power does actually flow in the positive z-direction, as it should. On the other hand,

$$P^- = V^- I^- = V^- \left(-\frac{V^-}{Z_0} \right) = -\frac{(V^-)^2}{Z_0} \tag{7.47b}$$

Since $(V^-)^2$ is always positive, regardless of whether V^- is numerically positive or negative, the minus sign in (7.47b) indicates that P^- is negative, and, hence, the $(-)$ wave power actually flows in the negative z-direction, as it should.

7.4   LINE TERMINATED BY RESISTIVE LOAD

Let us now consider a line of length l terminated by a load resistance R_L and driven by a constant voltage source V_0 in series with internal resistance R_g, as shown in Figure 7.19. Note again that the conductors of the transmission line are represented by double-ruled lines, whereas the connections to the conductors are single-ruled lines, to be treated as lumped circuits. We assume that no voltage and current exist on the line for $t < 0$ and the switch S is closed at $t = 0$. We wish to discuss the transient wave phenomena on the line for $t > 0$. The characteristic impedance of the line and the velocity of propagation are Z_0 and v_p, respectively.

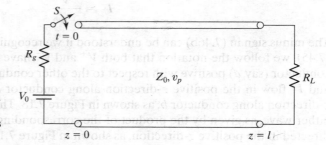

FIGURE 7.19

Transmission line terminated by a load resistance R_L and driven by a constant voltage source in series with an internal resistance R_g.

When the switch S is closed at $t = 0$, a $(+)$ wave originates at $z = 0$ and travels toward the load. Let the voltage and current of this $(+)$ wave be V^+ and I^+, respectively. Then we have the situation at $z = 0$, as shown in Figure 7.20(a). Note that the load resistor does not come into play here since the phenomenon is one of wave propagation; hence, until the $(+)$ wave goes to the load, sets up a reflection, and the reflected wave arrives back at the source, the source does not even know of the existence of R_L. This is a fundamental distinction between ordinary (lumped-) circuit theory and transmission-line (distributed-circuit) theory. In ordinary circuit theory, no time delay is involved; the effect of a transient in one part of the circuit is felt in all branches of the circuit instantaneously. In a transmission-line system, the effect of a transient at one location on the line is felt at a different location on the line only after an interval of time that the wave takes to travel from the first location to the second. Returning now to the circuit

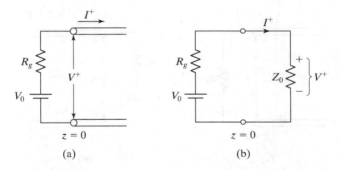

FIGURE 7.20

(a) For obtaining the (+) wave voltage and current at $z = 0$ for the line of Figure 7.19. (b) Equivalent circuit for (a).

in Figure 7.20(a), the various quantities must satisfy the boundary condition, that is, Kirchhoff's voltage law around the loop. Thus, we have

$$V_0 - I^+ R_g - V^+ = 0 \qquad (7.48a)$$

We, however, know from (6.31a) that $I^+ = V^+/Z_0$. Hence, we get

$$V_0 - \frac{V^+}{Z_0} R_g - V^+ = 0 \qquad (7.48b)$$

or

$$V^+ = V_0 \frac{Z_0}{R_g + Z_0} \qquad (7.49a)$$

$$I^+ = \frac{V^+}{Z_0} = \frac{V_0}{R_g + Z_0} \qquad (7.49b)$$

Thus, we note that the situation in Figure 7.20(a) is equivalent to the circuit shown in Figure 7.20(b); that is, the voltage source views a resistance equal to the characteristic impedance of the line, across $z = 0$. This is to be expected, since only a (+) wave exists at $z = 0$ and the ratio of the voltage to current in the (+) wave is equal to Z_0.

The (+) wave travels toward the load and reaches the termination at $t = l/v_p$. It does not, however, satisfy the boundary condition there, since this condition requires the voltage across the load resistance to be equal to the current through it times its value, R_L. But the voltage-to-current ratio in the (+) wave is equal to Z_0. To resolve this inconsistency, there is only one possibility, which is the setting up of a (−) wave, or a reflected wave. Let the voltage and current in this reflected wave be V^- and I^-, respectively. Then the total voltage across R_L is $V^+ + V^-$, and the total current through it is $I^+ + I^-$, as shown in Figure 7.21(a). To satisfy the boundary condition, we have

$$V^+ - V^- = R_L(I^+ + I^-) \qquad (7.50a)$$

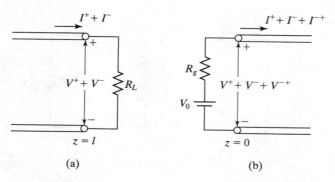

FIGURE 7.21

For obtaining the voltages and currents associated with (a) the $(-)$
wave and (b) the $(-+)$ wave, for the line of Figure 7.19.

But from (7.46a) and (7.46b), we know that $I^+ = V^+/Z_0$ and $I^- = -V^-/Z_0$, respectively.
Hence,

$$V^+ - V^- = R_L\left(\frac{V^+}{Z_0} - \frac{V^-}{Z_0}\right) \tag{7.50b}$$

or

$$V^- = V^+\frac{R_L - Z_0}{R_L + Z_0} \tag{7.51}$$

We now denote the *voltage reflection coefficient*, that is, the ratio of the reflected voltage to the incident voltage, by the symbol Γ (previously Γ_V). Thus,

$$\Gamma = \frac{V^-}{V^+} = \frac{R_L - Z_0}{R_L + Z_0} \tag{7.52}$$

We then note that the *current reflection coefficient* is

$$\frac{I^-}{I^+} = \frac{-V^-/Z_0}{V^+/Z_0} = -\frac{V^-}{V^+} = -\Gamma \tag{7.53}$$

Now, returning to the reflected wave, we observe that this wave travels back
toward the source and that it reaches there at $t = 2l/v_p$. Since the boundary condition
at $z = 0$, which was satisfied by the original $(+)$ wave alone, is then violated, a reflection of the reflection, or a re-reflection, will be set up and it travels toward the load. Let
us assume the voltage and current in this re-reflected wave, which is a $(+)$ wave, to be
V^{-+} and I^{-+}, respectively, with the superscripts denoting that the $(+)$ wave is a consequence of the $(-)$ wave. Then the total line voltage and the line current at $z = 0$ are
$V^+ + V^- + V^{-+}$ and $I^+ + I^- + I^{-+}$, respectively, as shown in Figure 7.21(b). To satisfy the boundary condition, we have

$$V^+ + V^- + V^{-+} = V_0 - R_g(I^+ + I^- + I^{-+}) \tag{7.54a}$$

But we know that $I^+ = V^+/Z_0$, $I^- = -V^-/Z_0$, and $I^{-+} = V^{-+}/Z_0$. Hence,

$$V^+ + V^- + V^{-+} = V_0 - \frac{R_g}{Z_0}(V^+ - V^- + V^{-+}) \tag{7.54b}$$

Furthermore, substituting for V^+ from (7.49a), simplifying, and rearranging, we get

$$V^{-+}\left(1 + \frac{R_g}{Z_0}\right) = V^-\left(\frac{R_g}{Z_0} - 1\right)$$

or

$$V^{-+} = V^-\frac{R_g - Z_0}{R_g + Z_0} \tag{7.55}$$

Comparing (7.55) with (7.51), we note that the reflected wave views the source with internal resistance as the internal resistance alone; that is, the voltage source is equivalent to a short circuit insofar as the $(-)$ wave is concerned. A moment's thought will reveal that superposition is at work here. The effect of the voltage source is taken into account by the constant outflow of the original $(+)$ wave from the source. Hence, for the reflection of the reflection, that is, for the $(-+)$ wave, we need only consider the internal resistance R_g. Thus, the voltage reflection coefficient formula (7.52) is a general formula and will be used repeatedly. In view of its importance, a brief discussion of the values of Γ for some special cases is in order, as follows:

1. $R_L = 0$, or short-circuited line.

$$\Gamma = \frac{0 - Z_0}{0 + Z_0} = -1$$

The reflected voltage is exactly the negative of the incident voltage, thereby keeping the voltage across R_L (short circuit) always zero.

2. $R_L = \infty$, or open-circuited line.

$$\Gamma = \frac{\infty - Z_0}{\infty + Z_0} = 1$$

and the current reflection coefficient $= -\Gamma = -1$. Thus, the reflected current is exactly the negative of the incident current, thereby keeping the current through R_L (open circuit) always zero.

3. $R_L = Z_0$, or line terminated by its characteristic impedance.

$$\Gamma = \frac{Z_0 - Z_0}{Z_0 + Z_0} = 0$$

This corresponds to no reflection, which is to be expected since $R_L (= Z_0)$ is consistent with the voltage-to-current ratio in the $(+)$ wave alone, and, hence, there is no violation of boundary condition and no need for the setting up of a reflected wave. Thus, a line terminated by its characteristic impedance is equivalent to an infinitely long line insofar as the source is concerned.

Returning to the discussion of the re-reflected wave, we note that this wave reaches the load at time $t = 3l/v_p$ and sets up another reflected wave. This process of bouncing back and forth of waves goes on indefinitely until the steady state is reached. To keep track of this transient phenomenon, we make use of the *bounce-diagram* technique. Some other names given for this diagram are *reflection diagram* and *space-time diagram*. We shall introduce the bounce diagram through a numerical example.

Example 7.4

Let us consider the system shown in Figure 7.22. Note that we have introduced a new quantity T, which is the one-way travel time along the line from $z = 0$ to $z = l$; that is, instead of specifying two quantities l and v_p, we specify $T(= l/v_p)$. Using the bounce-diagram technique, we wish to obtain and plot line voltage and current versus t for fixed values z and line voltage and current versus z for fixed values t.

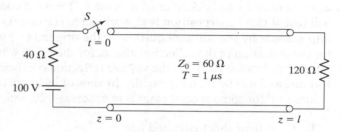

FIGURE 7.22

Transmission-line system for illustrating the bounce-diagram technique of keeping track of the transient phenomenon.

Before we construct the bounce diagram, we need to compute the following quantities:

$$\text{Voltage carried by the initial } (+) \text{ wave} = 100 \, \frac{60}{40 + 60} = 60 \text{ V}$$

$$\text{Current carried by the initial } (+) \text{ wave} = \frac{60}{60} = 1 \text{ A}$$

$$\text{Voltage reflection coefficient at load, } \Gamma_R = \frac{120 - 60}{120 + 60} = \frac{1}{3}$$

$$\text{Voltage reflection coefficient at source, } \Gamma_S = \frac{40 - 60}{40 + 60} = -\frac{1}{5}$$

The bounce diagram is essentially a two-dimensional representation of the transient waves bouncing back and forth on the line. Separate bounce diagrams are drawn for voltage and current, as shown in Figure 7.23(a) and (b), respectively. Position (z) on the line is represented horizontally and the time (t) vertically. Reflection coefficient values for the two ends are shown at the top of the diagrams for quick reference. Note that current reflection coefficients are $-\Gamma_R = -\frac{1}{3}$ and $-\Gamma_S = \frac{1}{5}$, respectively, at the load and at the source. Crisscross lines are drawn as shown in the figures to indicate the progress of the wave as a function of both z and t, with the numerical value for each leg of travel shown beside the line corresponding to that leg and approximately at the center of the line. The arrows indicate the directions of travel. Thus, for example, the first line on the voltage bounce diagram indicates that the initial $(+)$ wave of 60 V takes a time of 1 μs to reach the load end of the line. It sets up a reflected wave of 20 V, which travels back to the source, reaching there at a time of 2 μs, which then gives rise to a $(+)$ wave of voltage -4 V, and so on, with the process continuing indefinitely.

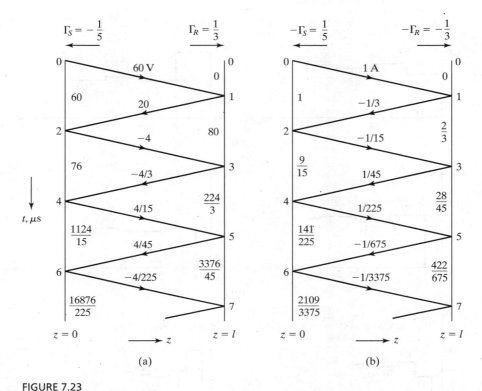

FIGURE 7.23

(a) Voltage and (b) current bounce diagrams, depicting the bouncing back and forth of the transient waves for the system of Figure 7.22.

Now, to sketch the line voltage and/or current versus time at any value of z, we note that since the voltage source is a constant voltage source, each individual wave voltage and current, once the wave is set up at that value of z, continues to exist there forever. Thus, at any particular time, the voltage (or current) at that value of z is a superposition of all the voltages (or currents) corresponding to the crisscross lines preceding that value of time. These values are marked on the bounce diagrams for $z = 0$ and $z = l$. Noting that each value corresponds to the 2-μs time interval between adjacent crisscross lines, we now sketch the time variations of line voltage and current at $z = 0$ and $z = l$, as shown in Figures 7.24(a) and (b), respectively. Similarly, by observing that the numbers written along the time axis for $z = 0$ are actually valid for any pair of z and t within the triangle ($\triangleright$) inside which they lie and that the numbers written along the time axis for $z = l$ are actually valid for any pair of z and t within the triangle ($\triangleleft$) inside which they lie, we can draw the sketches of line voltage and current versus time for any other value of z. This is done for $z = l/2$ in Figure 7.24(c).

It can be seen from the sketches of Figure 7.24 that as time progresses, the line voltage and current tend to converge to certain values, which we can expect to be the steady-state values. In the steady state, the situation consists of a single (+) wave, which is actually a superposition of the infinite number of transient (+) waves, and a single (−) wave, which is actually a superposition of the infinite number of transient (−) waves. Denoting the steady-state (+) wave voltage and

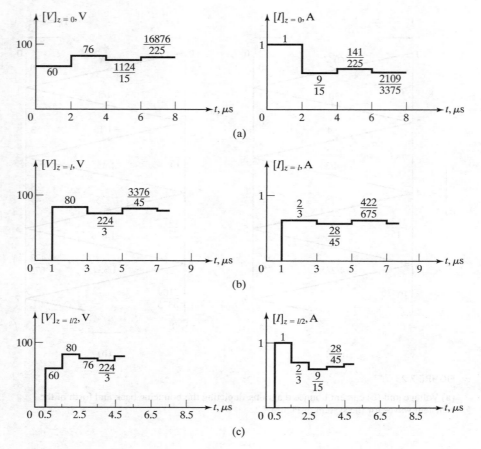

FIGURE 7.24

Time variations of line voltage and line current at (a) $z = 0$, (b) $z = l$, and (c) $z = l/2$ for the system of Figure 7.22.

current to be V_{SS}^+ and I_{SS}^+, respectively, and the steady-state $(-)$ wave voltage and current to be V_{SS}^- and I_{SS}^-, respectively, we obtain from the bounce diagrams

$$V_{SS}^+ = 60 - 4 + \frac{4}{15} - \cdots = 60\left(1 - \frac{1}{15} + \frac{1}{15^2} - \cdots\right) = 56.25 \text{ V}$$

$$I_{SS}^+ = 1 - \frac{1}{15} + \frac{1}{225} - \cdots = 1 - \frac{1}{15} + \frac{1}{15^2} - \cdots = 0.9375 \text{ A}$$

$$V_{SS}^- = 20 - \frac{4}{3} + \frac{4}{45} - \cdots = 20\left(1 - \frac{1}{15} + \frac{1}{15^2} - \cdots\right) = 18.75 \text{ V}$$

$$I_{SS}^- = -\frac{1}{3} + \frac{1}{45} - \frac{1}{675} + \cdots = -\frac{1}{3}\left(1 - \frac{1}{15} + \frac{1}{15^2} - \cdots\right) = -0.3125 \text{ A}$$

Note that $I_{SS}^+ = V_{SS}^+/Z_0$ and $I_{SS}^- = -V_{SS}^-/Z_0$, as they should be. The steady-state line voltage and current can now be obtained to be

$$V_{SS} = V_{SS}^+ + V_{SS}^- = 75\text{ V}$$

$$I_{SS} = I_{SS}^+ + I_{SS}^- = 0.625\text{ A}$$

These are the same as the voltage across R_L and current through R_L if the source and its internal resistance were connected directly to R_L, as shown in Figure 7.25. This is to be expected since the series inductors and shunt capacitors of the distributed equivalent circuit behave like short circuits and open circuits, respectively, for the constant voltage source in the steady state.

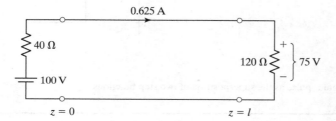

FIGURE 7.25

Steady-state equivalent for the system of Figure 7.22.

Sketches of line voltage and current as functions of distance (z) along the line for any particular time can also be drawn from considerations similar to those employed for the sketches of Figure 7.24. For example, suppose we wish to draw the sketch of line voltage versus z for $t = 2.5\ \mu$s. Then we note from the voltage bounce diagram that for $t = 2.5\ \mu$s, the line voltage is 76 V from $z = 0$ to $z = l/2$ and 80 V from $z = l/2$ to $z = l$. This is shown in Figure 7.26(a). Similarly, Figure 7.26(b) shows the variation of line current versus z for $t = 1\frac{1}{3}\ \mu$s.

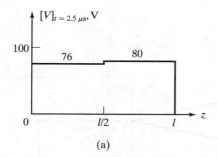

(a)

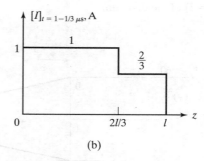

(b)

FIGURE 7.26

Variations with z of (a) line voltage for $t = 2.5\ \mu$s and (b) line current for $t = 1\frac{1}{3}\ \mu$s, for the system of Figure 7.22.

In Example 7.4, we introduced the bounce-diagram technique for a constant-voltage source. The technique can also be applied if the voltage source is a pulse. In the case of a rectangular pulse, this can be done by representing the pulse as the superposition of two step functions, as shown in Figure 7.27, and superimposing the bounce diagrams for the two sources one on another. In doing so, we should note that the bounce diagram for one source begins at a value of time greater than zero. Alternatively, the time variation for each wave can be drawn alongside the time axes beginning at the time of start of the wave. These can then be used to plot the required sketches. An example is in order, to illustrate this technique, which can also be used for a pulse of arbitrary shape.

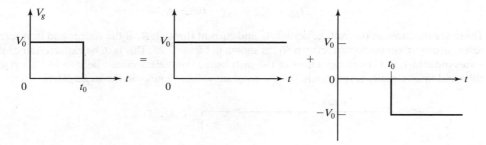

FIGURE 7.27

Representation of a rectangular pulse as the superposition of two step functions.

Example 7.5

Let us assume that the voltage source in the system of Figure 7.22 is a 100-V rectangular pulse extending from $t = 0$ to $t = 1 \mu s$ and extend the bounce-diagram technique.

Considering, for example, the voltage bounce diagram, we reproduce in Figure 7.28 part of the voltage bounce diagram of Figure 7.23(a) and draw the time variations of the individual pulses alongside the time axes, as shown in the figure. Note that voltage axes are chosen such that positive values are to the left at the left end ($z = 0$) of the diagram, but to the right at the right end ($z = l$) of the diagram.

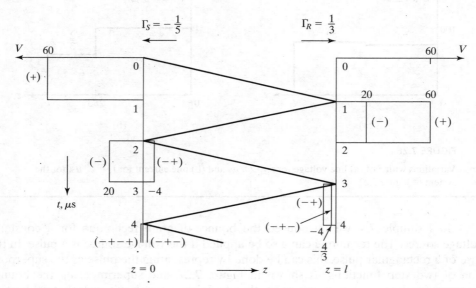

FIGURE 7.28

Voltage bounce diagram for the system of Figure 7.22 except that the voltage source is a rectangular pulse of 1-μs duration from $t = 0$ to $t = 1 \mu s$.

From the voltage bounce diagram, sketches of line voltage versus time at $z = 0$ and $z = l$ can be drawn, as shown in Figures 7.29(a) and (b), respectively. To draw the sketch of line voltage versus time for any other value of z, we note that as time progresses, the $(+)$ wave pulses slide down the crisscross lines from left to right, whereas the $(-)$ wave pulses slide down from right to

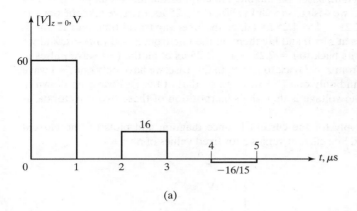

(a)

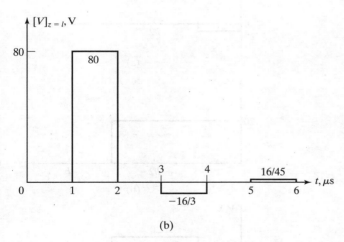

(b)

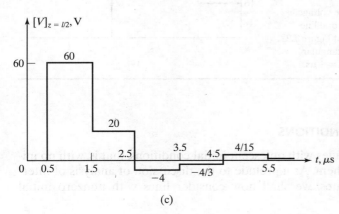

(c)

FIGURE 7.29

Time variations of line voltage at (a) $z = 0$, (b) $z = l$, and (c) $z = l/2$ for the system of Figure 7.22, except that the voltage source is a rectangular pulse of 1-μs duration from $t = 0$ to $t = 1$ μs.

left. Thus, to draw the sketch for $z = l/2$, we displace the time plots of the $(+)$ waves at $z = 0$ and of the $(-)$ waves at $z = l$ forward in time by 0.5 μs, that is, delay them by 0.5 μs, and add them to obtain the plot shown in Figure 7.29(c).

Sketches of line voltage versus distance (z) along the line for fixed values of time can also be drawn from the bounce diagram, based on the phenomenon of the individual pulses sliding down the crisscross lines. Thus, if we wish to sketch $V(z)$ for $t = 2.25$ μs, then we take the portion from $t = 2.25$ μs back to $t = 2.25 - 1 = 1.25$ μs (since the one-way travel time on the line is 1 μs) of all the $(+)$ wave pulses at $z = 0$ and lay them on the line from $z = 0$ to $z = l$, and we take the portion from $t = 2.25$ μs back to $t = 2.25 - 1 = 1.25$ μs of all the $(-)$ wave pulses at $z = l$ and lay them on the line from $z = l$ back to $z = 0$. In this case, we have only one $(+)$ wave pulse, that of the $(-+)$ wave, and only one $(-)$ wave pulse, that of the $(-)$ wave, as shown in Figures 7.30(a) and (b). The line voltage is then the superposition of these two waveforms, as shown in Figure 7.30(c).

Similar considerations apply for the current bounce diagram and plots of line current versus t for fixed values of z and line current versus z for fixed values of t.

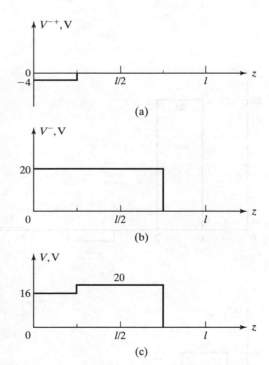

FIGURE 7.30

Variations with z of (a) the $(-+)$ wave voltage, (b) the $(-)$ wave voltage, and (c) the total line voltage, at $t = 2.25$ μs for the system of Figure 7.22, except that the voltage source is a rectangular pulse of 1-μs duration from $t = 0$ to $t = 1$ μs.

7.5 LINES WITH INITIAL CONDITIONS

Thus far, we have considered lines with quiescent initial conditions, that is, with no initial voltages and currents on them. As a prelude to the discussion of analysis of interconnections between logic gates, we shall now consider lines with nonzero initial conditions. We discuss first the general case of arbitrary initial voltage and current distributions by decomposing them into $(+)$ and $(-)$ wave voltages and currents. To do this, we consider the example shown in Figure 7.31, in which a line open-circuited at both ends is charged initially, say, at $t = 0$, to the voltage and current distributions shown in the figure.

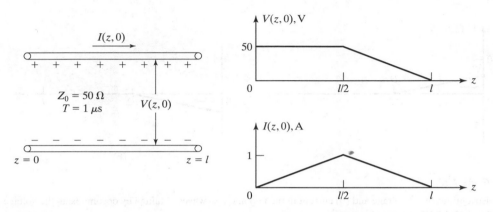

FIGURE 7.31

Line open-circuited at both ends and initially charged to the voltage and current distributions $V(z, 0)$ and $I(z, 0)$, respectively.

Writing the line voltage and current distributions as sums of $(+)$ and $(-)$ wave voltages and currents, we have

$$V^+(z, 0) + V^-(z, 0) = V(z, 0) \tag{7.56a}$$

$$I^+(z, 0) + I^-(z, 0) = I(z, 0) \tag{7.56b}$$

But we know that $I^+ = V^+/Z_0$ and $I^- = -V^-/Z_0$. Substituting these into (7.56b) and multiplying by Z_0, we get

$$V^+(z, 0) - V^-(z, 0) = Z_0 I(z, 0) \tag{7.57}$$

Solving (7.56a) and (7.57), we obtain

$$V^+(z, 0) = \tfrac{1}{2}[V(z, 0) + Z_0 I(z, 0)] \tag{7.58a}$$

$$V^-(z, 0) = \tfrac{1}{2}[V(z, 0) - Z_0 I(z, 0)] \tag{7.58b}$$

Thus, for the distributions $V(z, 0)$ and $I(z, 0)$ given in Figure 7.31, we obtain the distributions of $V^+(z, 0)$ and $V^-(z, 0)$, as shown by Figure 7.32(a), and hence of $I^+(z, 0)$ and $I^-(z, 0)$, as shown by Figure 7.32(b).

Suppose that we wish to find the voltage and current distributions at some later value of time, say, $t = 0.5 \mu s$. Then, we note that as the $(+)$ and $(-)$ waves propagate

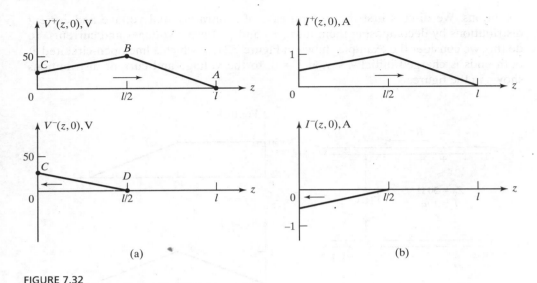

FIGURE 7.32

Distributions of (a) voltage and (b) current in the $(+)$ and $(-)$ waves obtained by decomposing the voltage and current distributions of Figure 7.31.

and impinge on the open circuits at $z = l$ and $z = 0$, respectively, they produce the $(-)$ and $(+)$ waves, respectively, consistent with a voltage reflection coefficient of 1 and current reflection coefficient of -1 at both ends. Hence, at $t = 0.5\ \mu s$, the $(+)$ and $(-)$ wave voltage and current distributions and their sum distributions are as shown in Figure 7.33, in which the points A, B, C, and D correspond to the points A, B, C, and D, respectively, in Figure 7.32. Proceeding in this manner, one can obtain the voltage and current distributions for any value of time.

Suppose that we connect a resistor of value Z_0 at the end $z = l$ at $t = 0$ instead of keeping it open-circuited. Then the reflection coefficient at that end becomes zero thereafter, and the $(+)$ wave, as it impinges on the resistor, gets absorbed in it instead of producing the $(-)$ wave. The line therefore completely discharges into the resistor by the time $t = 1.5\ \mu s$, with the resulting time variation of voltage across R_L, as shown in Figure 7.34, where the points A, B, C, and D correspond to the points A, B, C, and D, respectively, in Figure 7.32.

For a line with uniform initial voltage and current distributions, the analysis can be performed in the same manner as for arbitrary initial voltage and current distributions. Alternatively, and more conveniently, the analysis can be carried out with the aid of superposition and bounce diagrams. The basis behind this method lies in the fact that the uniform distribution corresponds to a situation in which the line voltage and current remain constant with time at all points on the line until a change is made at some point on the line. The boundary condition is then violated at that point, and a transient wave of constant voltage and current is set up, to be superimposed on the initial distribution. We shall illustrate this technique of analysis by means of an example.

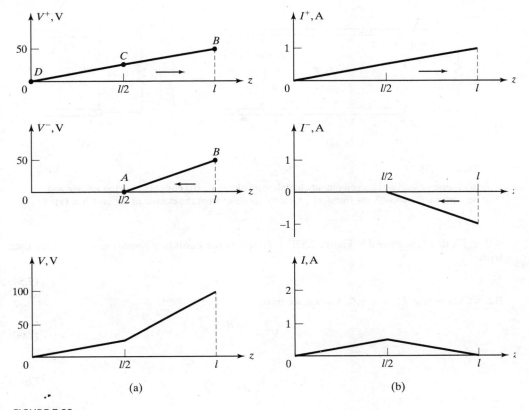

FIGURE 7.33

Distributions of (a) voltage and (b) current in the (+) and (−) waves and their sum for $t = 0.5 \ \mu s$ for the initially charged line of Figure 7.31.

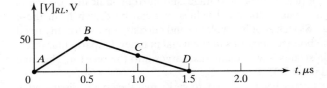

FIGURE 7.34

Voltage across $R_L (= Z_0 = 50 \ \Omega)$ resulting from connecting it at $t = 0$ to the end $z = l$ of the line of Figure 7.31.

Example 7.6

Let us consider a line of $Z_0 = 50 \ \Omega$ and $T = 1 \ \mu s$ initially charged to uniform voltage $V_0 = 100$ V and zero current. A resistor $R_L = 150 \ \Omega$ is connected at $t = 0$ to the end $z = 0$ of the line, as shown in Figure 7.35(a). We wish to obtain the time variation of the voltage across R_L for $t > 0$.

Since the change is made at $z = 0$ by connecting R_L to the line, a (+) wave originates at $z = 0$, so that the total line voltage at that point is $V_0 + V^+$ and the total line current

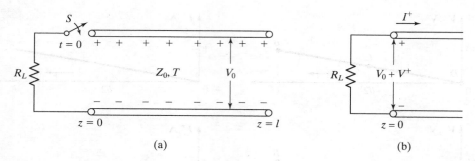

FIGURE 7.35

(a) Transmission line charged initially to uniform voltage V_0. (b) For obtaining the voltage and current associated with the transient $(+)$ wave resulting from the closure of the switch in (a).

is $0 + I^+$, or I^+, as shown in Figure 7.35(b). To satisfy the boundary condition at $z = 0$, we then write

$$V_0 + V^+ = -R_L I^+ \tag{7.59}$$

But we know that $I^+ = V^+/Z_0$. Hence, we have

$$V_0 + V^+ = -\frac{R_L}{Z_0} V^+ \tag{7.60}$$

or

$$V^+ = -V_0 \frac{Z_0}{R_L + Z_0} \tag{7.61a}$$

$$I^+ = -V_0 \frac{1}{R_L + Z_0} \tag{7.61b}$$

For $V_0 = 100$ V, $Z_0 = 50\ \Omega$, and $R_L = 150\ \Omega$, we obtain $V^+ = -25$ V and $I^+ = -0.5$ A.

We may now draw the voltage and current bounce diagrams, as shown in Figure 7.36. We note that in these bounce diagrams, the initial conditions are accounted for by the horizontal lines drawn at the top, with the numerical values of voltage and current indicated on them. Sketches of line voltage and current versus z for fixed values of t can be drawn from these bounce diagrams in the usual manner. Sketches of line voltage and current versus t for any fixed value of z also can be drawn from the bounce diagrams in the usual manner. Of particular interest is the voltage across R_L, which illustrates how the line discharges into the resistor. The time variation of this voltage is shown in Figure 7.37.

It is also instructive to check the energy balance, that is, to verify that the energy dissipated in the 150-Ω resistor for $t > 0$ is indeed equal to the energy stored in the line at $t = 0-$, since the line is lossless. To do this, we note that, in general, energy is stored in both electric and magnetic fields in the line, with energy densities $\frac{1}{2}\mathscr{C}V^2$ and $\frac{1}{2}\mathscr{L}I^2$, respectively. Thus, for a line charged uniformly to voltage V_0 and current I_0, the total electric and magnetic stored energies are given by

$$
\begin{aligned}
W_e &= \frac{1}{2}\mathscr{C}V_0^2 l = \frac{1}{2}\mathscr{C}V_0^2 v_p T \\
&= \frac{1}{2}\mathscr{C}V_0^2 \frac{1}{\sqrt{\mathscr{L}\mathscr{C}}} T = \frac{1}{2}\frac{V_0^2}{Z_0}T
\end{aligned}
\tag{7.62a}
$$

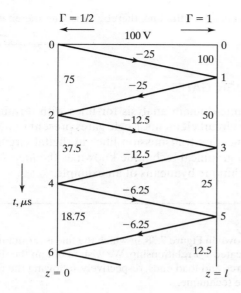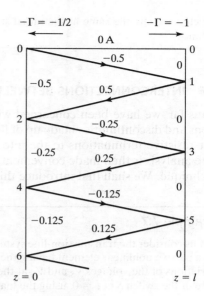

FIGURE 7.36

Voltage and current bounce diagrams depicting the transient phenomenon for $t > 0$ for the line of Figure 7.35(a), for $V_0 = 100$ V, $Z_0 = 50 \, \Omega$, $R_L = 150 \, \Omega$, and $T = 1 \, \mu s$.

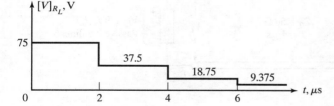

FIGURE 7.37

Time variation of voltage across R_L for $t > 0$ in Figure 7.35(a) for $V_0 = 100$ V, $Z_0 = 50 \, \Omega$, $R_L = 150 \, \Omega$, and $T = 1 \, \mu s$.

$$W_m = \frac{1}{2} \mathcal{L} I_0^2 l = \frac{1}{2} \mathcal{L} I_0^2 v_p T$$

$$= \frac{1}{2} \mathcal{L} I_0^2 \frac{1}{\sqrt{\mathcal{L}\mathcal{C}}} T = \frac{1}{2} I_0^2 Z_0 T \qquad (7.62b)$$

Since, for the example under consideration, $V_0 = 100$ V, $I_0 = 0$, and $T = 1 \, \mu s$, $W_e = 10^{-4}$ J and $W_m = 0$. Thus, the total initial stored energy in the line is 10^{-4} J. Now, denoting the power dissipated in the resistor to be P_d, we obtain the energy dissipated in the resistor to be

$$W_d = \int_{t=0}^{\infty} P_d \, dt$$

$$= \int_0^{2\times10^{-6}} \frac{75^2}{150} \, dt + \int_{2\times10^{-6}}^{4\times10^{-6}} \frac{37.5^2}{150} \, dt + \int_{4\times10^{-6}}^{6\times10^{-6}} \frac{18.75^2}{150} \, dt + \cdots$$

$$= \frac{2 \times 10^{-6}}{150} \times 75^2 \left(1 + \frac{1}{4} + \frac{1}{16} + \cdots \right) = 10^{-4} \, J$$

which is exactly the same as the initial stored energy in the line, thereby satisfying the energy balance.

7.6 INTERCONNECTIONS BETWEEN LOGIC GATES

Thus far, we have been concerned with time-domain analysis for lines with terminations and discontinuities made up of linear circuit elements. Logic gates present nonlinear resistive terminations to the interconnecting transmission lines in digital circuits. The analysis is then made convenient by a graphical technique known as the *load-line* technique. We shall first introduce this technique by means of an example.

Example 7.7

Let us consider the transmission-line system shown in Figure 7.38, in which the line is terminated by a passive nonlinear element having the indicated *V-I* relationship. We wish to obtain the time variations of the voltages V_S and V_L at the source and load ends, respectively, following the closure of the switch S at $t = 0$, using the load-line technique.

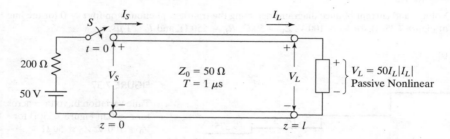

FIGURE 7.38

Line terminated by a passive nonlinear element and driven by a constant-voltage source in series with internal resistance.

With reference to the notation shown in Figure 7.38, we can write the following equations pertinent to $t = 0+$ at $z = 0$:

$$50 = 200I_S + V_S \tag{7.63a}$$

$$V_S = V^+$$

$$I_S = I^+ = \frac{V^+}{Z_0} = \frac{V_S}{50} \tag{7.63b}$$

where V^+ and I^+ are the voltage and current, respectively, of the $(+)$ wave set up immediately after closure of the switch. The two equations (7.63a) and (7.63b) can be solved graphically by constructing the straight lines representing them, as shown in Figure 7.39, and obtaining the point of intersection A, which gives the values of V_S and I_S. Note in particular that (7.63b) is a straight line of slope 1/50 and passing through the origin.

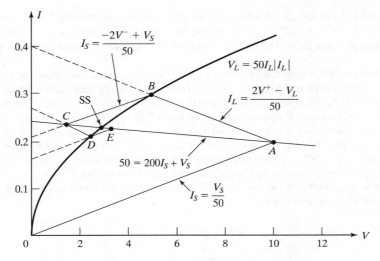

FIGURE 7.39

Graphical solution for obtaining time variations of V_S and V_L for $t > 0$ in the transmission-line system of Figure 7.38.

When the $(+)$ wave reaches the load end $z = l$ at $t = T$, a $(-)$ wave is set up. We can then write the following equations pertinent to $t = T+$ at $z = l$:

$$V_L = 50I_L|I_L| \tag{7.64a}$$

$$V_L = V^+ + V^-$$

$$I_L = I^+ + I^- = \frac{V^+ - V^-}{Z_0}$$

$$= \frac{V^+ - (V_L - V^+)}{50} = \frac{2V^+ - V_L}{50} \tag{7.64b}$$

where V^- and I^- are the $(-)$ wave voltage and current, respectively. The solution for V_L and I_L is then given by the intersection of the nonlinear curve representing (7.64a) and the straight line of slope $-1/50$ corresponding to (7.64b). Noting from (7.64b) that for $V_L = V^+$, $I_L = V^+/50$, we see that the straight line passes through point A. Thus, the solution of (7.64a) and (7.64b) is given by point B in Figure 7.39.

When the $(-)$ wave reaches the source end $z = 0$ at $t = 2T$, it sets up a reflection. Denoting this to be the $(-+)$ wave, we can then write the following equations pertinent to $t = 2T+$ at $z = 0$:

$$50 = 200I_S + V_S \tag{7.65a}$$

$$V_S = V^+ + V^- + V^{-+}$$

$$I_S = I^+ + I^- + I^{-+} = \frac{V^+ - V^- + V^{-+}}{Z_0}$$

$$= \frac{V^+ - V^- + (V_S - V^+ - V^-)}{50} = \frac{-2V^- + V_S}{50} \tag{7.65b}$$

where V^{-+} and I^{-+} are the $(-+)$ wave voltage and current, respectively. Noting from (7.65a) that for $V_S = V^+ + V^-$, $I_S = (V^+ - V^-)/50$, we see that (7.65b) represents a straight line of slope $1/50$ passing through B. Thus, the solution of (7.65a) and (7.65b) is given by point C in Figure 7.39.

Continuing in this manner, we observe that the solution consists of obtaining the points of intersection on the source and load V-I characteristics by drawing successively straight lines of slope $1/Z_0$ and $-1/Z_0$, beginning at the origin (the initial state) and with each straight line originating at the previous point of intersection, as shown in Figure 7.39. The points $A, C, E, \ldots,$ give the voltage and current at the source end for $0 < t < 2T, 2T < t < 4T, 4T < t < 6T, \ldots,$ whereas the points $B, D, \ldots,$ give the voltage and current at the load end for $T < t < 3T, 3T < t < 5T, \ldots$. Thus, for example, the time variations of V_S and V_L are shown in Figures 7.40(a) and (b), respectively. Finally, it can be seen from Figure 7.39 that the steady-state values of line voltage and current are reached at the point of intersection (denoted SS) of the source and load V-I characteristics.

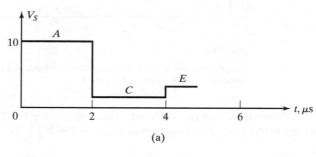

(a)

FIGURE 7.40

Time variations of (a) V_S and (b) V_L, for the transmission-line system of Figure 7.38. The voltage levels $A, B,$ $C, \ldots$ correspond to those in Figure 7.39.

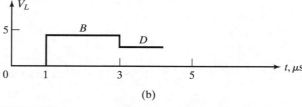

(b)

Now, going back to Example 7.6, the behavior of the system for the uniformly charged line can be analyzed by using the load-line technique, as an alternative to the solution using the bounce-diagram technique. Thus, noting that the terminal voltage-current characteristics at the ends $z = 0$ and $z = l$ of the system in Figure 7.35 are given by $V = -IR_L = -150I$ and $I = 0$, respectively, and that the characteristic impedance of the line is 50 Ω, we can carry out the load-line construction, as shown in Figure 7.41, beginning at the point A $(100 \text{ V}, 0 \text{ A})$, and drawing alternately straight lines of slope $1/50$ and $-1/50$ to obtain the points of intersection $B, C, D, \ldots$. The points $B,$ $D, F, \ldots$ give the line voltage and current values at the end $z = 0$ for intervals of 2 μs beginning at $t = 0 \ \mu\text{s}, 2 \ \mu\text{s}, 4 \ \mu\text{s}, \ldots,$ whereas the points $C, E, \ldots$ give the line voltage and current values at the end $z = l$ for intervals of 2 μs beginning at $t = 1 \ \mu\text{s}, 3 \ \mu\text{s}, \ldots$. For example, the time variation of the line voltage at $z = 0$ provided by the load-line construction is the same as in Figure 7.37.

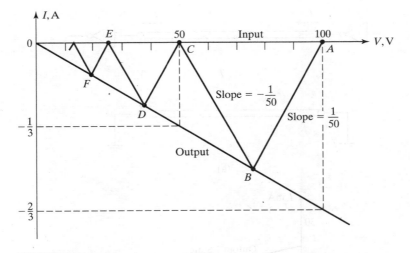

FIGURE 7.41

Load-line construction for the analysis of the system of Figure 7.35(a).

We shall now apply the procedure for the use of the load-line technique for a line with uniform initial distribution, just illustrated, to the analysis of the system in Figure 7.42(a) in which two transistor-transistor logic (TTL) inverters are interconnected by using a transmission line of characteristic impedance Z_0 and one-way travel time T. As the name inverter implies, the gate has an output that is the inverse of the input. Thus, if the input is in the HIGH (logic 1) range, the output will be in the LOW (logic 0) range, and vice versa. Typical V-I characteristics for a TTL inverter are shown in Figure 7.42(b). As shown in this figure, when the system is in the steady state with the output of the first inverter in the 0 state, the voltage and current along the line are given by the intersection of the output 0 characteristic and the input characteristic; when the system is in the steady state with the output of the first inverter in the 1 state, the voltage and current along the line are given by the intersection of the output 1 characteristic and the input characteristic. Thus, the line is charged to 0.2 V for the steady-state 0 condition and to 4 V for the steady-state 1 condition. We wish to study the transient phenomena corresponding to the transition when the output of the first gate switches from the 0 to the 1 state, and vice versa, assuming Z_0 of the line to be 30 Ω.

Considering first the transition from the 0 state to the 1 state, and following the line of argument in Example 7.7, we carry out the construction shown in Figure 7.43(a). This construction consists of beginning at the point corresponding to the steady-state 0 (the initial state) and drawing a straight line of slope 1/30 to intersect with the output 1 characteristic at point A, then drawing from point A a straight line of slope $-1/30$ to intersect the input characteristic at point B, and so on. From this construction, the variation of the voltage V_i at the input of the second gate can be sketched as shown in Figure 7.43(b), in which the voltage levels correspond to the

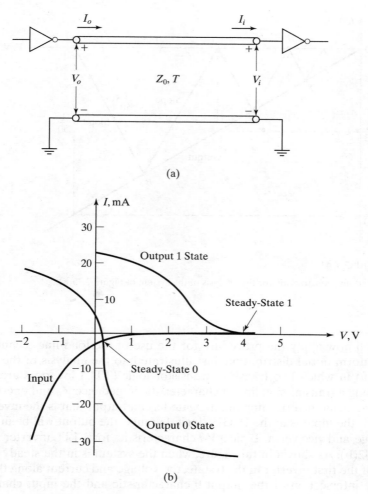

(a)

(b)

(a) Transmission-line interconnection between two logic gates. (b) Typical V-I characteristics for the logic gates.

points $0, B, D, \ldots$, in Figure 7.43(a). The effect of the transients on the performance of the system may now be seen by noting from Figure 7.43(b) that depending on the value of the minimum gate voltage that will reliably be recognized as logic 1, a time delay in excess of T may be involved in the transition from 0 to 1. Thus, if this minimum voltage is 2 V, the interconnecting line will result in an extra time delay of $2T$ for the input of the second gate to switch from 0 to 1, since V_i does not exceed 2 V until $t = 3T+$.

Considering next the transition from the 1 state to the 0 state, we carry out the construction shown in Figure 7.44(a), with the crisscross lines beginning at the point

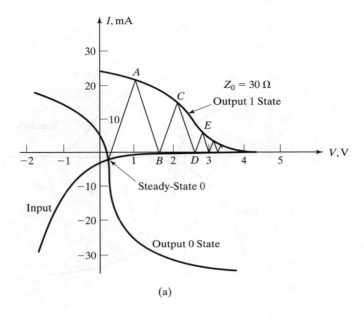

(a)

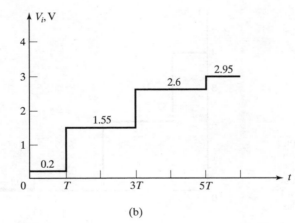

(b)

FIGURE 7.43

(a) Construction based on the load-line technique for analysis of the 0-to-1 transition for the system of Figure 7.42(a). (b) Plot of V_i versus t obtained from the construction in (a).

corresponding to the steady-state 1. From this construction, we obtain the plot of V_i versus t, as shown in Figure 7.44(b), in which the voltage levels correspond to the points $1, B, D, \ldots$, in Figure 7.44(a). If we assume a maximum gate input voltage that can be readily recognized as logic 0 to be 1 V, it can once again be seen that an extra time delay of $2T$ is involved in the switching of the input of the second gate from 1 to 0, since V_i does not drop below 1 V until $t = 3T+$.

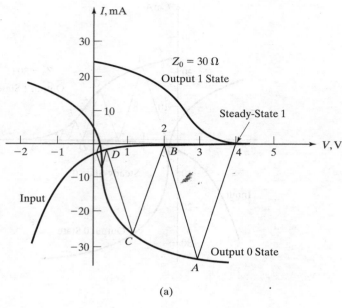

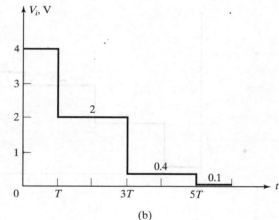

FIGURE 7.44

(a) Construction based on the load-line technique for analysis of the 1-to-0 transition for the system of Figure 7.42(a). (b) Plot of V_i versus t obtained from the construction in (a).

SUMMARY

In this chapter, we first studied frequency domain analysis of transmission lines. The general solutions to the transmission-line equations, expressed in phasor form, that is,

$$\frac{\partial \bar{V}}{\partial z} = -j\omega \mathcal{L} \bar{I} \tag{7.66a}$$

$$\frac{\partial \bar{I}}{\partial z} = -\mathcal{G} \bar{V} - j\omega \mathcal{C} \bar{V} \tag{7.66b}$$

are given by

$$\bar{V}(z) = \bar{A}e^{-\bar{\gamma}z} + \bar{B}e^{\bar{\gamma}z} \qquad (7.67a)$$

$$\bar{I}(z) = \frac{1}{\bar{Z}_0}(\bar{A}e^{-\bar{\gamma}z} - \bar{B}e^{\bar{\gamma}z}) \qquad (7.67b)$$

where

$$\bar{\gamma} = \sqrt{j\omega\mathscr{L}(\mathscr{G} + j\omega\mathscr{C})} \qquad [=\sqrt{j\omega\mu(\sigma + j\omega\epsilon)}]$$

$$\bar{Z}_0 = \sqrt{\frac{j\omega\mathscr{L}}{\mathscr{G} + j\omega\mathscr{C}}} \qquad \left[\neq\sqrt{\frac{j\omega\mu}{\sigma + j\omega\epsilon}}\right]$$

are the propagation constant and the characteristic impedance, respectively, of the line. For a lossless line ($\mathscr{G} = 0$), these reduce to

$$\bar{\gamma} = j\beta = j\omega\sqrt{\mathscr{L}\mathscr{C}} \qquad (=j\omega\sqrt{\mu\epsilon})$$

$$\bar{Z}_0 = Z_0 = \sqrt{\frac{\mathscr{L}}{\mathscr{C}}} \qquad (\neq\sqrt{\mu/\epsilon})$$

so that for a lossless line,

$$\bar{V}(z) = \bar{A}e^{-j\beta z} + \bar{B}e^{j\beta z} \qquad (7.68a)$$

$$\bar{I}(z) = \frac{1}{Z_0}(\bar{A}e^{-j\beta z} - \bar{B}e^{j\beta z}) \qquad (7.68b)$$

The solutions given by (7.67a) and (7.67b) or (7.68a) and (7.68b) represent the superposition of $(+)$ and $(-)$ waves propagating in the medium between the conductors of the line, expressed in terms of the line voltage and current instead of in terms of the electric and magnetic fields.

By applying these general solutions to the case of a lossless line short circuited at the far end and obtaining the particular solutions for that case, we discussed the standing wave phenomenon and the standing wave patterns resulting from the complete reflection of waves by the short circuit. We also examined the frequency behavior of the input impedance of a short-circuited line of length l, given by

$$\bar{Z}_{in} = jZ_0 \tan \beta l$$

and (a) illustrated its application in a technique for the location of short circuit in a line, and (b) learned that for a circuit element to behave as assumed by conventional (lumped) circuit theory, its dimensions must be a small fraction of the wavelength corresponding to the frequency of operation.

Next, we studied reflection and transmission of waves at a junction between two lossless lines. By applying them to the general solutions for the line voltage and current on either side of the junction, we deduced the ratio of the reflected wave voltage to the incident wave voltage, that is, the voltage reflection coefficient, to be

$$\Gamma_V = \frac{Z_{02} - Z_{01}}{Z_{02} + Z_{01}}$$

where Z_{01} is the characteristic impedance of the line from which the wave is incident and Z_{02} is the characteristic impedance of the line on which the wave is incident. The ratio of the transmitted wave voltage to the incident wave voltage, that is, the voltage transmission coefficient, is given by

$$\tau_V = 1 + \Gamma_V$$

The current reflection and transmission coefficients are given by

$$\Gamma_I = -\Gamma_V$$
$$\tau_I = 1 - \Gamma_V$$

We discussed the standing wave pattern resulting from the partial reflection of the wave at the junction and defined a quantity known as the standing wave ratio (SWR), which is a measure of the reflection phenomenon. In terms of Γ_V, it is given by

$$\text{SWR} = \frac{1 + |\Gamma_V|}{1 - |\Gamma_V|}$$

We then introduced the Smith Chart, which is a graphical aid in the solution of transmission-line problems. After first discussing the basis behind the construction of the Smith Chart, we illustrated its use by considering a transmission-line system and computing several quantities of interest. We concluded the section on Smith Chart with the solution of a transmission-line matching problem.

We devoted the remainder of the chapter to time-domain analysis of transmission lines. For a lossless line, the transmission-line equations in time domain are given by

$$\frac{\partial V}{\partial z} = -\mathcal{L}\frac{\partial I}{\partial t} \tag{7.69a}$$

$$\frac{\partial I}{\partial z} = -\mathcal{C}\frac{\partial V}{\partial t} \tag{7.69b}$$

The solutions to these equations are

$$V(z, t) = Af\left(t - \frac{z}{v_p}\right) + Bg\left(t + \frac{z}{v_p}\right) \tag{7.70a}$$

$$I(z, t) = \frac{1}{Z_0}\left[Af\left(t - \frac{z}{v_p}\right) - Bg\left(t + \frac{z}{v_p}\right)\right] \tag{7.70b}$$

where $Z_0 = \sqrt{\mathcal{L}/\mathcal{C}}$ is the characteristic impedance of the line, and $v_p = 1/\sqrt{\mathcal{L}\mathcal{C}}$ is the velocity of propagation on the line.

We then discussed time-domain analysis of a transmission line terminated by a load resistance R_L and excited by a constant voltage source V_0 in series with internal resistance R_g. Writing the general solutions (7.70a) and (7.70b) concisely in the manner

$$V = V^+ + V^-$$
$$I = I^+ + I^-$$

where

$$I^+ = \frac{V^+}{Z_0}$$

$$I^- = -\frac{V^-}{Z_0}$$

we found that the situation consists of the bouncing back and forth of transient $(+)$ and $(-)$ waves between the two ends of the line. The initial $(+)$ wave voltage is $V^+ Z_0/(R_g + Z_0)$. All other waves are governed by the reflection coefficients at the two ends of the line, given for the voltage by

$$\Gamma_R = \frac{R_L - Z_0}{R_L + Z_0}$$

and

$$\Gamma_S = \frac{R_g - Z_0}{R_g + Z_0}$$

for the load and source ends, respectively. In the steady state, the situation is the superposition of all the transient waves, equivalent to the sum of a single $(+)$ wave and a single $(-)$ wave. We discussed the bounce-diagram technique of keeping track of the transient phenomenon and extended it to a pulse voltage source.

As a prelude to the consideration of interconnections between logic gates, we discussed time-domain analysis of lines with nonzero initial conditions. For the general case, the initial voltage and current distributions $V(z, 0)$ and $I(z, 0)$ are decomposed into $(+)$ and $(-)$ wave voltages and currents as given by

$$V^+(z, 0) = \frac{1}{2}[V(z, 0) + Z_0 I(z, 0)]$$

$$V^-(z, 0) = \frac{1}{2}[V(z, 0) - Z_0 I(z, 0)]$$

$$I^+(z, 0) = \frac{1}{Z_0}V^+(z, 0)$$

$$I^-(z, 0) = -\frac{1}{Z_0}V^-(z, 0)$$

The voltage and current distributions for $t > 0$ are then obtained by keeping track of the bouncing of these waves at the two ends of the line. For the special case of uniform distribution, the analysis can be performed more conveniently by considering the situation as one in which a transient wave is superimposed on the initial distribution and using the bounce-diagram technique. We then introduced the load-line technique of time-domain analysis, and applied it to the analysis of transmission-line interconnection between logic gates.

REVIEW QUESTIONS

7.1. Discuss the solutions for the transmission-line equations in frequency domain.

7.2. Discuss the propagation constant and characteristic impedance associated with wave propagation on transmission lines.

7.3. What is the boundary condition to be satisfied at a short circuit on a line?

7.4. For an open-circuited line, what would be the boundary condition to be satisfied at the open circuit?

7.5. What is a standing wave? How do complete standing waves arise? Discuss their characteristics and give an example in mechanics.

7.6. What is a standing wave pattern? Discuss the voltage and current standing wave patterns for the short-circuited line.

7.7. What would be the voltage and current standing wave patterns for an open-circuited line?

7.8. Discuss the variation with frequency of the input reactance of a short-circuited line and its application in the determination of the location of a short circuit.

7.9. Can you suggest an alternative procedure to that described in Example 7.1 to locate a short circuit in a transmission line?

7.10. Discuss the condition for the validity of the quasistatic approximation for the input behavior of a physical structure.

7.11. Discuss the input behavior of a short-circuited line for frequencies slightly beyond those for which the quasistatic approximation is valid.

7.12. What are the boundary conditions for the voltage and current at the junction between two transmission lines?

7.13. What is the voltage reflection coefficient at the junction between two transmission lines? How are the current reflection coefficient and the voltage and current transmission coefficients related to the voltage reflection coefficient?

7.14. What is the voltage reflection coefficient at the short circuit for a short-circuited line?

7.15. Can the transmitted wave current at the junction between two transmission lines be greater than the incident wave current? Explain.

7.16. What is a partial standing wave? Discuss the standing wave patterns corresponding to partial standing waves.

7.17. Define standing wave ratio (SWR). What are the standing wave ratios for (a) an infinitely long line, (b) a short-circuited line, (c) an open-circuited line, and (d) a line terminated by its characteristic impedance?

7.18. Define line impedance. What is its value for an infinitely long line?

7.19. What is the basis behind the construction of the Smith Chart? How does the Smith Chart simplify the solution of transmission-line problems?

7.20. Briefly discuss the mapping of the normalized line impedances from the complex $\bar{Z}_n$-plane onto the Smith Chart.

7.21. Why is a circle with its center at the center of the Smith Chart known as a constant SWR circle? Where on the circle is the corresponding SWR value marked?

7.22. Using the Smith Chart, how do you find the normalized line admittance at a point on the line given the normalized line impedance at that point?

7.23. Briefly discuss the solution of the transmission-line matching problem.

7.24. How is the length of a short-circuited stub for a required input susceptance determined by using the Smith Chart?

7.25. Discuss the general solutions for the line voltage and current in time-domain and the notation associated with their representation in concise form.

7.26. What is the fundamental distinction between the occurrence of the response in one branch of a lumped circuit to the application of an excitation in a different branch of the circuit and the occurrence of the response at one location on a transmission line to the application of an excitation at a different location on the line?

7.27. Describe the phenomenon of the bouncing back and forth of transient waves on a transmission line excited by a constant voltage source in series with internal resistance and terminated by a resistance.

7.28. Discuss the values of the voltage reflection coefficient for some special cases.

7.29. What is the steady-state equivalent of a line excited by a constant voltage source? What is the actual situation in the steady state?

7.30. Discuss the bounce-diagram technique of keeping track of the bouncing back and forth of the transient waves on a transmission line for a constant voltage source.

7.31. Discuss the bounce-diagram technique of keeping track of the bouncing back and forth of the transient waves on a transmission line for a pulse voltage source.

7.32. Discuss the determination of the voltage and current distributions on an initially charged line for any given time from the knowledge of the initial voltage and current distributions.

7.33. Discuss with the aid of an example the discharging of an initially charged line into a resistor.

7.34. Discuss the bounce-diagram technique of transient analysis of a line with uniform initial voltage and current distributions.

7.35. Discuss the load-line technique of obtaining the time variations of the voltages and currents at the source and load ends of a line from a knowledge of the terminal $V\text{-}I$ characteristics.

7.36. Discuss the analysis of transmission-line interconnection between two logic gates.

PROBLEMS

7.1. For a transmission line of arbitrary cross section and with the medium between the conductors characterized by $\sigma = 10^{-16}$ S/m, $\epsilon = 2.5\epsilon_0$, and $\mu = \mu_0$, it is known that $\mathscr{C} = 10^{-10}$ F/m. (a) Find $\mathscr{L}$ and $\mathscr{G}$. (b) Find $\bar{Z}_0$ for $f = 10^6$ Hz.

7.2. For the coaxial cable of Example 6.9 employing air dielectric, find the ratio of the outer to the inner radii for which the characteristic impedance of the cable is 75 Ω.

7.3. Using the general solutions for the complex line voltage and current on a lossless line given by (7.9a) and (7.9b), respectively, obtain the particular solutions for the complex voltage and current on an open-circuited line. Then find the input impedance of an open-circuited line of length l.

7.4. Solve Example 7.1 by considering the standing wave patterns between the short circuit and the generator for the two frequencies of interest and by deducing the number of wavelengths at one of the two frequencies.

7.5. For an air dielectric short-circuited line of characteristic impedance 50 Ω, find the minimum values of the length for which its input impedance is equivalent to that of (a) an inductor of value 0.25×10^{-6} H at 100 MHz and (b) a capacitor of value 10^{-10} F at 100 MHz.

7.6. A transmission line of length 2 m having a nonmagnetic ($\mu = \mu_0$) perfect dielectric is short-circuited at the far end. A variable-frequency generator is connected at its input and the current drawn is monitored. It is found that the current reaches a maximum for $f = 500$ MHz and then a minimum for $f = 525$ MHz. Find the permittivity of the dielectric.

7.7. A voltage generator is connected to the input of a lossless line short-circuited at the far end. The frequency of the generator is varied and the line voltage and line current at the input terminals are monitored. It is found that the voltage reaches a maximum value of 10 V at 405 MHz and the current reaches a maximum value of 0.2 A at 410 MHz. (a) Find the characteristic impedance of the line. (b) Find the voltage and current values at 407 MHz.

7.8. Assuming that the criterion $f \ll v_p/2\pi l$ is satisfied for frequencies less than $0.1 v_p/2\pi l$, compute the maximum length of an air dielectric short-circuited line for which the input impedance is approximately that of an inductor of value equal to the total inductance of the line for $f = 100$ MHz.

7.9. A lossless transmission line of length 2 m and having $\mathcal{L} = 0.5\mu_0$ and $\mathcal{C} = 18\epsilon_0$ is short circuited at the far end. (a) Find the phase velocity, v_p. (b) Find the wavelength, the length of the line in terms of the number of wavelengths, and the input impedance of the line for each of the following frequencies: 100 Hz; 100 MHz; and 12.5 MHz.

7.10. Repeat Example 7.3 with the values of Z_{01} and Z_{02} interchanged.

7.11. In the transmission-line system shown in Figure 7.45, a power P_i is incident on the junction from line 1. Find (a) the power reflected into line 1, (b) the power transmitted into line 2, and (c) the power transmitted into line 3.

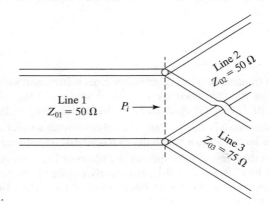

Line 2
$Z_{02} = 50\ \Omega$

Line 1
$Z_{01} = 50\ \Omega$ $P_i \longrightarrow$

Line 3
$Z_{03} = 75\ \Omega$

FIGURE 7.45

For Problem 7.11.

7.12. Show that the voltage minima of the standing wave pattern of Figure 7.9 are sharper than the voltage maxima by computing the voltage amplitude halfway between the locations of voltage maxima and minima.

7.13. A line assumed to be infinitely long and of unknown characteristic impedance is connected to a line of characteristic impedance 50 Ω on which standing wave measurements are made. It is found that the standing wave ratio is 3 and that two consecutive voltage minima exist at 15 cm and 25 cm from the junction of the two lines. Find the unknown characteristic impedance.

7.14. A line assumed to be infinitely long and of unknown characteristic impedance when connected to a line of characteristic impedance 50 Ω produces a standing wave ratio of value 2 in the 50-Ω line. The same line when connected to a line of characteristic impedance 150 Ω produces a standing wave ratio of value 1.5 in the 150-Ω line. Find the unknown characteristic impedance.

7.15. Compute values of $\bar{\Gamma}_V$ corresponding to several points along line a in Figure 7.11(a) and show that the contour a' in Figure 7.11(b) is a circle of radius $\frac{1}{2}$ and centered at $(1/2, 0)$.

7.16. Compute values of $\bar{\Gamma}_V$ corresponding to several points along line b in Figure 7.11(a) and show that the contour b' in Figure 7.11(b) is a portion of a circle of radius 2 and centered at $(1, 2)$.

7.17. For the transmission-line system of Figure 7.13, and for the values of Z_{01}, Z_{02}, and l specified in the text, find the value of B that minimizes the SWR to the left of jB. What is the minimum value of SWR?

7.18. In Figure 7.13, assume $Z_{01} = 300$ Ω, $Z_{02} = 75$ Ω, $B = 0.002$ S, and $l = 0.145\lambda_1$, and find (a) $\bar{Z}_1$, (b) SWR on line 1 to the right of jB, (c) $\bar{Y}_1$, and (d) SWR on line 1 to the left of jB.

7.19. A transmission line of characteristic impedance 50 Ω is terminated by a load impedance of $(73 + j0)$ Ω. Find the location and the length of a short-circuited stub of characteristic impedance 50 Ω for achieving a match between the line and the load.

7.20. Show that (7.40a) and (7.40b) satisfy the transmission-line equations (7.39a) and (7.39b).

7.21. In the system shown in Figure 7.46, assume that V_g is a constant voltage source of 100 V and the switch S is closed at $t = 0$. Find and sketch: (a) the line voltage versus z for $t = 0.2$ μs; (b) the line current versus z for $t = 0.4$ μs; (c) the line voltage versus t for $z = 30$ m; and (d) the line current versus t for $z = -40$ m.

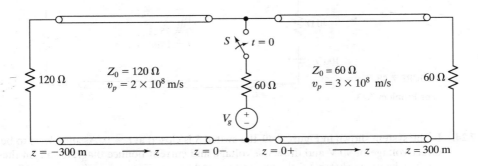

FIGURE 7.46

For Problem 7.21.

7.22. In the system shown in Figure 7.47(a), the switch S is closed at $t = 0$. The line voltage variations with time at $z = 0$ and $z = l$ for the first 5 μs are observed to be as shown in Figure 7.47(b) and (c), respectively. Find the values of V_0, R_g, R_L, and T.

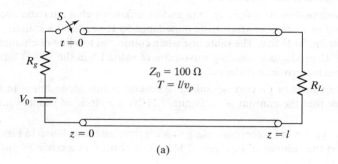

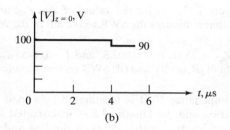

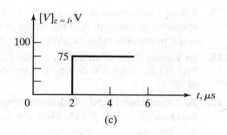

FIGURE 7.47

For Problem 7.22.

7.23. The system shown in Figure 7.48 is in steady state. Find (a) the line voltage and current, (b) the (+) wave voltage and current, and (c) the (−) wave voltage and current.

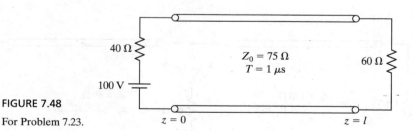

FIGURE 7.48

For Problem 7.23.

7.24. In the system shown in Figure 7.49, the switch S is closed at $t = 0$. Assume $V_g(t)$ to be a direct voltage of 90 V and draw the voltage and current bounce diagrams. From these bounce diagrams, sketch: (a) the line voltage and line current versus t (up to $t = 7.25$ μs) at $z = 0$, $z = l$, and $z = l/2$; and (b) the line voltage and line current versus z for $t = 1.2$ μs and $t = 3.5$ μs.

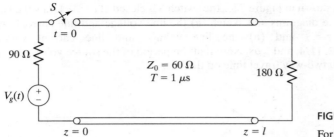

FIGURE 7.49

For Problem 7.24.

7.25. Repeat Problem 7.21 assuming V_g to be a triangular pulse, as shown in Figure 7.50.

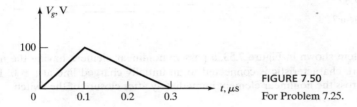

FIGURE 7.50

For Problem 7.25.

7.26. For the system of Problem 7.24, assume that the voltage source is of 0.3-μs duration instead of being of infinite duration. Find and sketch the line voltage and line current versus z for $t = 1.2\ \mu$s and $t = 3.5\ \mu$s.

7.27. In the system shown in Figure 7.51, the switch S is closed at $t = 0$. Find and sketch: (a) the line voltage versus z for $t = 2\frac{1}{2}\ \mu$s; (b) the line current versus z for $t = 2\frac{1}{2}\ \mu$s; and (c) the line voltage at $z = l$ versus t up to $t = 4\ \mu$s.

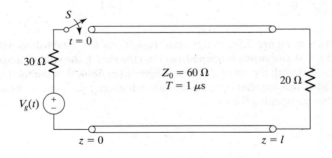

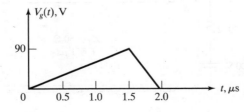

FIGURE 7.51

For Problem 7.27.

7.28. In the system shown in Figure 7.52, the switch S is closed at $t = 0$. Draw the voltage and current-bounce diagrams and sketch (a) the line voltage and line current versus t for $z = 0$ and $z = l$ and (b) the line voltage and line current versus z for $t = 2, 9/4, 5/2, 11/4$, and 3 μs. Note that the period of the source voltage is 2 μs, which is equal to the two-way travel time on the line.

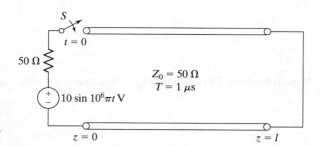

FIGURE 7.52

For Problem 7.28.

7.29. In the system shown in Figure 7.53, a passive nonlinear element having the indicated volt-ampere characteristic is connected to an initially charged line at $t = 0$. Find the voltage across the nonlinear element immediately after closure of the switch.

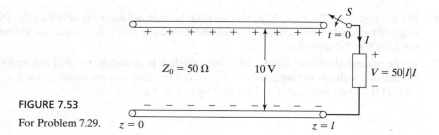

FIGURE 7.53

For Problem 7.29.

7.30. In the system shown in Figure 7.54, steady-state conditions are established with the switch S closed. At $t = 0$, the switch is opened. (a) Find the sketch the voltage across the 150-Ω resistor for $t \geq 0$, with the aid of a bounce diagram. (b) Show that the total energy dissipated in the 150-Ω resistor after opening the switch is exactly the same as the energy stored in the line before opening the switch.

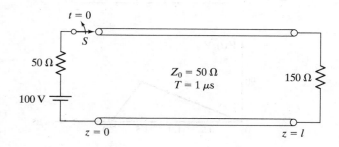

FIGURE 7.54

For Problem 7.30.

7.31. In the system shown in Figure 7.55, steady-state conditions are established with the switch S closed. At $t = 0$, the switch is opened. (a) Sketch the voltage and current along the system for $t = 0-$. (b) Find the total energy stored in the lines for $t = 0-$. (c) Find and sketch the voltages across the two resistors for $t > 0$. (d) From your sketches of part (c), find the total energy dissipated in the resistors for $t > 0$.

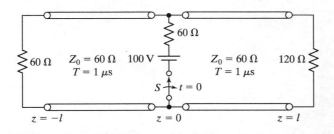

FIGURE 7.55

For Problem 7.31.

7.32. For the system of Problem 7.24, use the load-line technique to obtain and plot line voltage and line current versus t (up to $t = 5.25\ \mu s$) at $z = 0$ and $z = l$. Also obtain the steady-state values of line voltage and current from the load-line construction.

7.33. For the system of Problem 7.29, use the load-line technique to obtain and plot line voltage versus t from $t = 0$ up to $t = 7l/v_p$ at $z = 0$ and $z = l$.

7.34. For the example of interconnection between logic gates of Figure 7.42(a), repeat the load-line constructions for $Z_0 = 50\ \Omega$ and draw graphs of V_i versus t for both 0-to-1 and 1-to-0 transitions.

7.35. For the example of interconnection between logic gates of Figure 7.42(a), find (a) the minimum value of Z_0 such that for the transition form 0 to 1, the voltage V_i reaches 2V at $t = T+$ and (b) the minimum value of Z_0 such that for the transition from 1 to 0, the voltage V_i reaches 1 V at $t = T+$.

CHAPTER

Waveguide Principles

In Chapter 6, we introduced transmission lines, and in Chapter 7, we studied their analysis. We learned that transmission lines are made up of two (or more) parallel conductors. In this chapter, we shall learn the principles of waveguides in which guiding of waves is accomplished by the bouncing of waves obliquely within the guide, as compared to the case of a transmission line in which the waves slide parallel to the conductors of the line.

We shall introduce waveguides by first considering a parallel-plate waveguide, that is, a waveguide consisting of two parallel, plane conductors and then extend it to the rectangular waveguide, which is a hollow metallic pipe of rectangular cross section, a common form of waveguide. We shall learn that waveguides are characterized by cut-off, which is the phenomenon of no propagation in a certain range of frequencies, and dispersion, which is the phenomenon of propagating waves of different frequencies possessing different phase velocities along the waveguide. In connection with the latter characteristic, we shall introduce the concept of group velocity. We shall also discuss the principles of cavity resonators, the microwave counterparts of resonant circuits, and of optical waveguides.

We shall study the topic of reflection and refraction of plane waves at an interface between two dielectrics, and finally introduce the dielectric slab waveguide, based on the phenomenon of total internal reflection at the interface, when the angle of incidence of the wave on the interface is greater than a certain critical value.

8.1 UNIFORM PLANE WAVE PROPAGATION IN AN ARBITRARY DIRECTION

In Chapter 4, we introduced the uniform plane wave propagating in the z-direction by considering an infinite plane current sheet lying in the xy-plane. If the current sheet lies in a plane making an angle to the xy-plane, the uniform plane wave would then propagate in a direction different from the z-direction. Thus, let us consider a uniform plane wave propagating in the z'-direction making an angle θ with the negative x-axis, as shown in Figure 8.1. Let the electric field of the wave be entirely in the y-direction. The magnetic field would then be directed as shown in the figure so that $\mathbf{E} \times \mathbf{H}$ points in the z'-direction.

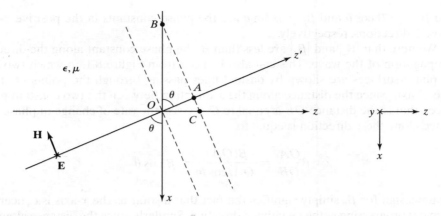

FIGURE 8.1

Uniform plane wave propagating in the z'-direction lying in the xz-plane and making an angle θ with the negative x-axis.

We can write the expression for the electric field of the wave as

$$\mathbf{E} = E_0 \cos{(\omega t - \beta z')}\, \mathbf{a}_y \tag{8.1}$$

where $\beta = \omega\sqrt{\mu\epsilon}$ is the phase constant, that is, the rate of change of phase with distance along the z'-direction for a fixed value of time. From the construction of Figure 8.2(a), we, however, have

$$z' = -x \cos\theta + z \sin\theta \tag{8.2}$$

so that

$$
\begin{aligned}
\mathbf{E} &= E_0 \cos{[\omega t - \beta(-x \cos\theta + z \sin\theta)]}\, \mathbf{a}_y \\
&= E_0 \cos{[\omega t - (-\beta \cos\theta)x - (\beta \sin\theta)z]}\, \mathbf{a}_y \\
&= E_0 \cos{(\omega t - \beta_x x - \beta_z z)}\, \mathbf{a}_y
\end{aligned}
\tag{8.3}
$$

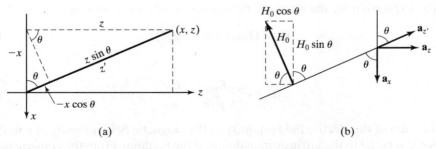

(a)   (b)

FIGURE 8.2

Constructions pertinent to the formulation of the expressions for the fields of the uniform plane wave of Figure 8.1.

where $\beta_x = -\beta \cos \theta$ and $\beta_z = \beta \sin \theta$ are the phase constants in the positive x- and positive z-directions, respectively.

We note that $|\beta_x|$ and $|\beta_z|$ are less than β, the phase constant along the direction of propagation of the wave. This can also be seen from Figure 8.1, in which two constant phase surfaces are shown by dashed lines passing through the points O and A on the z'-axis. Since the distance along the x-direction between the two constant phase surfaces, that is, the distance OB is equal to $OA/\cos \theta$, the rate of change of phase with distance along the x-direction is equal to

$$\beta \frac{OA}{OB} = \frac{\beta(OA)}{OA/\cos \theta} = \beta \cos \theta$$

The minus sign for β_x simply signifies the fact that insofar as the x-axis is concerned, the wave is progressing in the negative x-direction. Similarly, since the distance along the z-direction between the two constant phase surfaces, that is, the distance OC is equal to $OA/\sin \theta$, the rate of change of phase with distance along the z-direction is equal to

$$\beta \frac{OA}{OC} = \frac{\beta(OA)}{OA/\sin \theta} = \beta \sin \theta$$

Since the wave is progressing along the positive z-direction, β_z is positive. We further note that

$$\beta_x^2 + \beta_z^2 = (-\beta \cos \theta)^2 + (\beta \sin \theta)^2 = \beta^2 \tag{8.4}$$

and that

$$-\cos \theta \, \mathbf{a}_x + \sin \theta \, \mathbf{a}_z = \mathbf{a}_{z'} \tag{8.5}$$

where $\mathbf{a}_{z'}$ is the unit vector directed along z'-direction, as shown in Figure 8.2(b). Thus, the vector

$$\boldsymbol{\beta} = (-\beta \cos \theta)\mathbf{a}_x + (\beta \sin \theta)\mathbf{a}_z = \beta_x \mathbf{a}_x + \beta_z \mathbf{a}_z \tag{8.6}$$

defines completely the direction of propagation and the phase constant along the direction of propagation. Hence, the vector $\boldsymbol{\beta}$ is known as the *propagation vector.*

The expression for the magnetic field of the wave can be written as

$$\mathbf{H} = \mathbf{H}_0 \cos (\omega t - \beta z') \tag{8.7}$$

where

$$|\mathbf{H}_0| = \frac{E_0}{\sqrt{\mu/\epsilon}} = \frac{E_0}{\eta} \tag{8.8}$$

since the ratio of the electric field intensity to the magnetic field intensity of a uniform plane wave is equal to the intrinsic impedance of the medium. From the construction in Figure 8.2(b), we observe that

$$\mathbf{H}_0 = H_0(-\sin \theta \, \mathbf{a}_x - \cos \theta \, \mathbf{a}_z) \tag{8.9}$$

Thus, using (8.9) and substituting for z' from (8.2), we obtain

$$\mathbf{H} = H_0(-\sin\theta\,\mathbf{a}_x - \cos\theta\,\mathbf{a}_z)\cos\left[\omega t - \beta(-x\cos\theta + z\sin\theta)\right]$$

$$= -\frac{E_0}{\eta}(\sin\theta\,\mathbf{a}_x + \cos\theta\,\mathbf{a}_z)\cos\left[\omega t - \beta_x x - \beta_z z\right] \qquad (8.10)$$

Generalizing the foregoing treatment to the case of a uniform plane wave propagating in a completely arbitrary direction in three dimensions, as shown in Figure 8.3, and characterized by phase constants β_x, β_y, and β_z in the x-, y-, and z-directions, respectively, we can write the expression for the electric field as

$$\begin{aligned}\mathbf{E} &= \mathbf{E}_0\cos(\omega t - \beta_x x - \beta_y y - \beta_z z + \phi_0)\\ &= \mathbf{E}_0\cos\left[\omega t - (\beta_x\mathbf{a}_x + \beta_y\mathbf{a}_y + \beta_z\mathbf{a}_z)\cdot(x\mathbf{a}_x + y\mathbf{a}_y + z\mathbf{a}_z) + \phi_0\right]\\ &= \mathbf{E}_0\cos(\omega t - \boldsymbol{\beta}\cdot\mathbf{r} + \phi_0)\end{aligned} \qquad (8.11)$$

where

$$\boldsymbol{\beta} = \beta_x\mathbf{a}_x + \beta_y\mathbf{a}_y + \beta_z\mathbf{a}_z \qquad (8.12)$$

is the propagation vector,

$$\mathbf{r} = x\mathbf{a}_x + y\mathbf{a}_y + z\mathbf{a}_z \qquad (8.13)$$

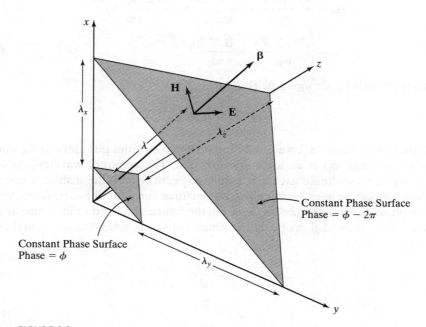

FIGURE 8.3

The various quantities associated with a uniform plane wave propagating in an arbitrary direction.

is the position vector, and ϕ_0 is the phase at the origin at $t = 0$. The position vector is the vector drawn from the origin to the point (x, y, z) and hence has components $x, y,$ and z along the x-, y-, and z-axes, respectively. The expression for the magnetic field of the wave is then given by

$$\mathbf{H} = \mathbf{H}_0 \cos(\omega t - \boldsymbol{\beta} \cdot \mathbf{r} + \phi_0) \qquad (8.14)$$

where

$$|\mathbf{H}_0| = \frac{|\mathbf{E}_0|}{\eta} \qquad (8.15)$$

Since $\mathbf{E}, \mathbf{H}$, and the direction of propagation are mutually perpendicular to each other, it follows that

$$\mathbf{E}_0 \cdot \boldsymbol{\beta} = 0 \qquad (8.16a)$$

$$\mathbf{H}_0 \cdot \boldsymbol{\beta} = 0 \qquad (8.16b)$$

$$\mathbf{E}_0 \cdot \mathbf{H}_0 = 0 \qquad (8.16c)$$

In particular, $\mathbf{E} \times \mathbf{H}$ should be directed along the propagation vector $\boldsymbol{\beta}$ as illustrated in Figure 8.3, so that $\boldsymbol{\beta} \times \mathbf{E}_0$ is directed along $\mathbf{H}_0$. We can therefore combine the facts (8.16) and (8.15) to obtain

$$\mathbf{H}_0 = \frac{\mathbf{a}_\beta \times \mathbf{E}_0}{\eta} = \frac{\mathbf{a}_\beta \times \mathbf{E}_0}{\sqrt{\mu/\epsilon}} = \frac{\omega\sqrt{\mu\epsilon}\,\mathbf{a}_\beta \times \mathbf{E}_0}{\omega\mu}$$

$$= \frac{\beta\mathbf{a}_\beta \times \mathbf{E}_0}{\omega\mu} = \frac{\boldsymbol{\beta} \times \mathbf{E}_0}{\omega\mu} \qquad (8.17)$$

where $\mathbf{a}_\beta$ is the unit vector along $\boldsymbol{\beta}$. Thus,

$$\mathbf{H} = \frac{1}{\omega\mu}\boldsymbol{\beta} \times \mathbf{E} \qquad (8.18)$$

Returning to Figure 8.3, we can define several quantities pertinent to the uniform plane wave propagation in an arbitrary direction. The apparent wavelengths λ_x, λ_y, and λ_z along the coordinate axes x, y, and z, respectively, are the distances measured along those respective axes between two consecutive constant phase surfaces between which the phase difference is 2π, as shown in the figure, at a fixed time. From the interpretations of β_x, β_y, and β_z as being the phase constants along the x-, y-, and z-axes, respectively, we have

$$\lambda_x = \frac{2\pi}{\beta_x} \qquad (8.19a)$$

$$\lambda_y = \frac{2\pi}{\beta_y} \qquad (8.19b)$$

$$\lambda_z = \frac{2\pi}{\beta_z} \qquad (8.19c)$$

We note that the wavelength λ along the direction of propagation is related to λ_x, λ_y, and λ_z in the manner

$$\frac{1}{\lambda^2} = \frac{1}{(2\pi/\beta)^2} = \frac{\beta^2}{4\pi^2} = \frac{\beta_x^2 + \beta_y^2 + \beta_z^2}{4\pi^2}$$

$$= \frac{1}{\lambda_x^2} + \frac{1}{\lambda_y^2} + \frac{1}{\lambda_z^2} \tag{8.20}$$

The apparent phase velocities v_{px}, v_{py}, and v_{pz} along the x-, y-, and z-axes, respectively, are the velocities with which the phase of the wave progresses with time along the respective axes. Thus,

$$v_{px} = \frac{\omega}{\beta_x} \tag{8.21a}$$

$$v_{py} = \frac{\omega}{\beta_y} \tag{8.21b}$$

$$v_{pz} = \frac{\omega}{\beta_z} \tag{8.21c}$$

The phase velocity v_p along the direction of propagation is related to v_{px}, v_{py}, and v_{pz} in the manner

$$\frac{1}{v_p^2} = \frac{1}{(\omega/\beta)^2} = \frac{\beta^2}{\omega^2} = \frac{\beta_x^2 + \beta_y^2 + \beta_z^2}{\omega^2}$$

$$= \frac{1}{v_{px}^2} + \frac{1}{v_{py}^2} + \frac{1}{v_{pz}^2} \tag{8.22}$$

The apparent wavelengths and phase velocities along the coordinate axes are greater than the actual wavelength and phase velocity, respectively, along the direction of propagation of the wave. This fact can be understood physically by considering, for example, water waves in an ocean striking the shore at an angle. The distance along the shoreline between two successive crests is greater than the distance between the same two crests measured along a line normal to the orientation of the crests. Also, an observer has to run faster along the shoreline in order to keep pace with a particular crest than he has to do in a direction normal to the orientation of the crests. We shall now consider an example.

Example 8.1

Let us consider a 30-MHz uniform plane wave propagating in free space and given by the electric field vector

$$\mathbf{E} = 5(\mathbf{a}_x + \sqrt{3}\mathbf{a}_y) \cos [6\pi \times 10^7 t - 0.05\pi(3x - \sqrt{3}y + 2z)] \text{ V/m}$$

Then comparing with the general expression for **E** given by (8.11), we have

$$\mathbf{E}_0 = 5(\mathbf{a}_x + \sqrt{3}\mathbf{a}_y)$$

$$\boldsymbol{\beta} \cdot \mathbf{r} = 0.05\pi(3x - \sqrt{3}y + 2z)$$

$$= 0.05\pi(3\mathbf{a}_x - \sqrt{3}\mathbf{a}_y + 2\mathbf{a}_z) \cdot (x\mathbf{a}_x + y\mathbf{a}_y + z\mathbf{a}_z)$$

$$\boldsymbol{\beta} = 0.05\pi(3\mathbf{a}_x - \sqrt{3}\mathbf{a}_y + 2\mathbf{a}_z)$$

$$\boldsymbol{\beta} \cdot \mathbf{E}_0 = 0.05\pi(3\mathbf{a}_x - \sqrt{3}\mathbf{a}_y + 2\mathbf{a}_z) \cdot 5(\mathbf{a}_x + \sqrt{3}\mathbf{a}_y)$$

$$= 0.25\pi(3 - 3) = 0$$

Hence, (8.16a) is satisfied; $\mathbf{E}_0$ is perpendicular to $\boldsymbol{\beta}$.

$$\beta = |\boldsymbol{\beta}| = 0.05\pi|3\mathbf{a}_x - \sqrt{3}\mathbf{a}_y + 2\mathbf{a}_z| = 0.05\pi\sqrt{9 + 3 + 4} = 0.2\pi$$

$$\lambda = \frac{2\pi}{\beta} = \frac{2\pi}{0.2\pi} = 10 \text{ m}$$

This does correspond to a frequency of $(3 \times 10^8)/10$ Hz or 30 MHz in free space. The direction of propagation is along the unit vector

$$\mathbf{a}_\beta = \frac{\boldsymbol{\beta}}{|\boldsymbol{\beta}|} = \frac{3\mathbf{a}_x - \sqrt{3}\mathbf{a}_y + 2\mathbf{a}_z}{\sqrt{9 + 3 + 4}} = \frac{3}{4}\mathbf{a}_x - \frac{\sqrt{3}}{4}\mathbf{a}_y + \frac{1}{2}\mathbf{a}_z$$

From (8.17),

$$\mathbf{H}_0 = \frac{1}{\omega\mu_0}\boldsymbol{\beta} \times \mathbf{E}_0$$

$$= \frac{0.05\pi \times 5}{6\pi \times 10^7 \times 4\pi \times 10^{-7}}(3\mathbf{a}_x - \sqrt{3}\mathbf{a}_y + 2\mathbf{a}_z) \times (\mathbf{a}_x + \sqrt{3}\mathbf{a}_y)$$

$$= \frac{1}{96\pi}\begin{vmatrix} \mathbf{a}_x & \mathbf{a}_y & \mathbf{a}_z \\ 3 & -\sqrt{3} & 2 \\ 1 & \sqrt{3} & 0 \end{vmatrix}$$

$$= \frac{1}{48\pi}(-\sqrt{3}\mathbf{a}_x + \mathbf{a}_y + 2\sqrt{3}\mathbf{a}_z)$$

Thus,

$$\mathbf{H} = \frac{1}{48\pi}(-\sqrt{3}\mathbf{a}_x + \mathbf{a}_y + 2\sqrt{3}\mathbf{a}_z)\cos[6\pi \times 10^7 t - 0.05\pi(3x - \sqrt{3}y + 2z)] \text{ A/m}$$

To verify the expression for **H** just derived, we note that

$$\mathbf{H}_0 \cdot \boldsymbol{\beta} = \left[\frac{1}{48\pi}(-\sqrt{3}\mathbf{a}_x + \mathbf{a}_y + 2\sqrt{3}\mathbf{a}_z)\right] \cdot [0.05\pi(3\mathbf{a}_x - \sqrt{3}\mathbf{a}_y + 2\mathbf{a}_z)]$$

$$= \frac{0.05}{48}(-3\sqrt{3} - \sqrt{3} + 4\sqrt{3}) = 0$$

$$\mathbf{E}_0 \cdot \mathbf{H}_0 = 5(\mathbf{a}_x + \sqrt{3}\mathbf{a}_y) \cdot \frac{1}{48\pi}(-\sqrt{3}\mathbf{a}_x + \mathbf{a}_y + 2\sqrt{3}\mathbf{a}_z)$$

$$= \frac{5}{48\pi}(-\sqrt{3} + \sqrt{3}) = 0$$

$$\frac{|\mathbf{E}_0|}{|\mathbf{H}_0|} = \frac{5|\mathbf{a}_x + \sqrt{3}\mathbf{a}_y|}{(1/48\pi)|-\sqrt{3}\mathbf{a}_x + \mathbf{a}_y + 2\sqrt{3}\mathbf{a}_z|} = \frac{5\sqrt{1 + 3}}{(1/48\pi)\sqrt{3 + 1 + 12}}$$

$$= \frac{10}{1/12\pi} = 120\pi = \eta_0$$

Hence, (8.16b), (8.16c), and (8.15) are satisfied.
   Proceeding further, we find that

$$\beta_x = 0.05\pi \times 3 = 0.15\pi$$
$$\beta_y = -0.05\pi \times \sqrt{3} = -0.05\sqrt{3}\pi$$
$$\beta_z = 0.05\pi \times 2 = 0.1\pi$$

We then obtain

$$\lambda_x = \frac{2\pi}{\beta_x} = \frac{2\pi}{0.15\pi} = \frac{40}{3}\text{ m} = 13.333\text{ m}$$

$$\lambda_y = \frac{2\pi}{|\beta_y|} = \frac{2\pi}{0.05\sqrt{3}\pi} = \frac{40}{\sqrt{3}}\text{ m} = 23.094\text{ m}$$

$$\lambda_z = \frac{2\pi}{\beta_z} = \frac{2\pi}{0.1\pi} = 20\text{ m}$$

$$v_{px} = \frac{\omega}{\beta_x} = \frac{6\pi \times 10^7}{0.15\pi} = 4 \times 10^8\text{ m/s}$$

$$v_{py} = \frac{\omega}{|\beta_y|} = \frac{6\pi \times 10^7}{0.05\sqrt{3}\pi} = 4\sqrt{3} \times 10^8\text{ m/s} = 6.928 \times 10^8\text{ m/s}$$

$$v_{pz} = \frac{\omega}{\beta_z} = \frac{6\pi \times 10^7}{0.1\pi} = 6 \times 10^8\text{ m/s}$$

Finally, to verify (8.20) and (8.22), we note that

$$\frac{1}{\lambda_x^2} + \frac{1}{\lambda_y^2} + \frac{1}{\lambda_z^2} = \frac{1}{(40/3)^2} + \frac{1}{(40/\sqrt{3})^2} + \frac{1}{20^2}$$

$$= \frac{9}{1600} + \frac{3}{1600} + \frac{4}{1600} = \frac{1}{100} = \frac{1}{10^2} = \frac{1}{\lambda^2}$$

and

$$\frac{1}{v_{px}^2} + \frac{1}{v_{py}^2} + \frac{1}{v_{pz}^2} = \frac{1}{(4 \times 10^8)^2} + \frac{1}{(4\sqrt{3} \times 10^8)^2} + \frac{1}{(6 \times 10^8)^2}$$

$$= \frac{1}{16 \times 10^{16}} + \frac{1}{48 \times 10^{16}} + \frac{1}{36 \times 10^{16}}$$

$$= \frac{1}{9 \times 10^{16}} = \frac{1}{(3 \times 10^8)^2} = \frac{1}{v_p^2}$$

8.2　TRANSVERSE ELECTRIC WAVES IN A PARALLEL-PLATE WAVEGUIDE

Let us now consider the superposition of two uniform plane waves propagating symmetrically with respect to the z-axis, as shown in Figure 8.4, and having the electric fields

$$\mathbf{E}_1 = E_0 \cos(\omega t - \boldsymbol{\beta}_1 \cdot \mathbf{r})\,\mathbf{a}_y$$
$$= E_0 \cos(\omega t + \beta x \cos\theta - \beta z \sin\theta)\,\mathbf{a}_y \tag{8.23a}$$

$$\mathbf{E}_2 = -E_0 \cos(\omega t - \boldsymbol{\beta}_2 \cdot \mathbf{r})\,\mathbf{a}_y$$
$$= -E_0 \cos(\omega t - \beta x \cos\theta - \beta z \sin\theta)\,\mathbf{a}_y \tag{8.23b}$$

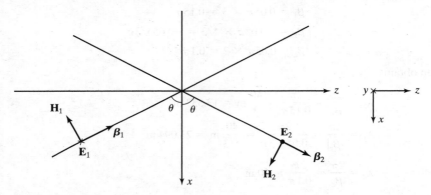

FIGURE 8.4

Superposition of two uniform plane waves propagating symmetrically with respect to the z-axis.

where $\beta = \omega\sqrt{\mu\epsilon}$, with ϵ and μ being the permittivity and the permeability, respectively, of the medium. The corresponding magnetic fields are given by

$$\mathbf{H}_1 = \frac{E_0}{\eta}(-\sin\theta\,\mathbf{a}_x - \cos\theta\,\mathbf{a}_z)\cos(\omega t + \beta x \cos\theta - \beta z \sin\theta) \tag{8.24a}$$

$$\mathbf{H}_2 = \frac{E_0}{\eta}(\sin\theta\,\mathbf{a}_x - \cos\theta\,\mathbf{a}_z)\cos(\omega t - \beta x \cos\theta - \beta z \sin\theta) \tag{8.24b}$$

where $\eta = \sqrt{\mu/\epsilon}$. The electric and magnetic fields of the superposition of the two waves are given by

$$\mathbf{E} = \mathbf{E}_1 + \mathbf{E}_2$$
$$= E_0[\cos(\omega t - \beta z \sin\theta + \beta x \cos\theta)$$
$$- \cos(\omega t - \beta z \sin\theta - \beta x \cos\theta)]\,\mathbf{a}_y$$
$$= -2E_0 \sin(\beta x \cos\theta)\sin(\omega t - \beta z \sin\theta)\,\mathbf{a}_y \tag{8.25a}$$

$$\mathbf{H} = \mathbf{H}_1 + \mathbf{H}_2$$

$$= -\frac{E_0}{\eta} \sin\theta \, [\cos(\omega t - \beta z \sin\theta + \beta x \cos\theta)$$

$$- \cos(\omega t - \beta z \sin\theta - \beta x \cos\theta)]\mathbf{a}_x$$

$$- \frac{E_0}{\eta} \cos\theta \, [\cos(\omega t - \beta z \sin\theta + \beta x \cos\theta)$$

$$+ \cos(\omega t - \beta z \sin\theta - \beta x \cos\theta)]\mathbf{a}_z$$

$$= \frac{2E_0}{\eta} \sin\theta \sin(\beta x \cos\theta) \sin(\omega t - \beta z \sin\theta)\,\mathbf{a}_x$$

$$- \frac{2E_0}{\eta} \cos\theta \cos(\beta x \cos\theta) \cos(\omega t - \beta z \sin\theta)\,\mathbf{a}_z \qquad (8.25\text{b})$$

In view of the factors $\sin(\beta x \cos\theta)$ and $\cos(\beta x \cos\theta)$ for the x-dependence and the factors $\sin(\omega t - \beta z \sin\theta)$ and $\cos(\omega t - \beta z \sin\theta)$ for the z-dependence, the composite fields have standing wave character in the x-direction and traveling wave character in the z-direction. Thus, we have standing waves in the x-direction moving bodily in the z-direction, as illustrated in Figure 8.5, by considering the electric field for two different times. In fact, we find that the Poynting vector is given by

$$\mathbf{P} = \mathbf{E} \times \mathbf{H} = E_y\mathbf{a}_y \times (H_x\mathbf{a}_x + H_z\mathbf{a}_z)$$

$$= -E_yH_x\mathbf{a}_z + E_yH_z\mathbf{a}_x$$

$$= \frac{4E_0^2}{\eta} \sin\theta \sin^2(\beta x \cos\theta) \sin^2(\omega t - \beta z \sin\theta)\,\mathbf{a}_z$$

$$+ \frac{E_0^2}{\eta} \cos\theta \sin(2\beta x \cos\theta) \sin 2(\omega t - \beta z \sin\theta)\,\mathbf{a}_x \qquad (8.26)$$

The time-average Poynting vector is given by

$$\langle \mathbf{P} \rangle = \frac{4E_0^2}{\eta} \sin\theta \sin^2(\beta x \cos\theta) \langle \sin^2(\omega t - \beta z \sin\theta)\rangle\mathbf{a}_z$$

$$+ \frac{E_0^2}{\eta} \cos\theta \sin(2\beta x \cos\theta) \langle \sin 2(\omega t - \beta z \sin\theta)\rangle\mathbf{a}_x$$

$$= \frac{2E_0^2}{\eta} \sin\theta \sin^2(\beta x \cos\theta)\,\mathbf{a}_z \qquad (8.27)$$

Thus, the time-average power flow is entirely in the z-direction, thereby verifying our interpretation of the field expressions. Since the composite electric field is directed entirely transverse to the z-direction, that is, the direction of time-average power flow, whereas the composite magnetic field is not, the composite wave is known as the *transverse electric*, or TE wave.

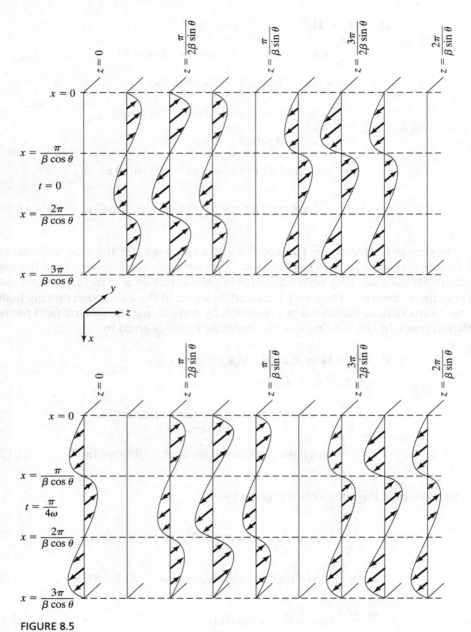

FIGURE 8.5

Standing waves in the *x*-direction moving bodily in the *z*-direction.

From the expressions for the fields for the TE wave given by (8.25a) and (8.25b), we note that the electric field is zero for sin ($\beta x \cos \theta$) equal to zero, or

$$\beta x \cos \theta = \pm m\pi, \qquad m = 0, 1, 2, 3, \ldots$$

$$x = \pm \frac{m\pi}{\beta \cos \theta} = \pm \frac{m\lambda}{2 \cos \theta}, \qquad m = 0, 1, 2, 3, \ldots \tag{8.28}$$

where

$$\lambda = \frac{2\pi}{\beta} = \frac{2\pi}{\omega\sqrt{\mu\epsilon}} = \frac{1}{f\sqrt{\mu\epsilon}}$$

Thus, if we place perfectly conducting sheets in these planes, the waves will propagate undisturbed, that is, as though the sheets were not present, since the boundary condition that the tangential component of the electric field be zero on the surface of a perfect conductor is satisfied in these planes. The boundary condition that the normal component of the magnetic field be zero on the surface of a perfect conductor is also satisfied since H_x is zero in these planes.

If we consider any two adjacent sheets, the situation is actually one of uniform plane waves bouncing obliquely between the sheets, as illustrated in Figure 8.6 for two sheets in the planes $x = 0$ and $x = \lambda/(2 \cos \theta)$, thereby guiding the wave and hence the energy in the z-direction, parallel to the plates. Thus, we have a *parallel-plate waveguide*, as compared to the parallel-plate transmission line in which the uniform plane wave slides parallel to the plates. We note from the constant phase surfaces of the obliquely bouncing wave shown in Figure 8.6 that $\lambda/(2 \cos \theta)$ is simply one-half of the apparent wavelength of that wave in the x-direction, that is, normal to the plates. Thus, the fields have one-half apparent wavelength in the x-direction. If we place the perfectly conducting sheets in the planes $x = 0$ and $x = m\lambda/(2 \cos \theta)$, the fields will then have m number of one-half apparent wavelengths in the x-direction between the plates. The fields have no variations in the y-direction. Thus, the fields are said to correspond to $TE_{m,0}$ *modes*, where the subscript m refers to the x-direction, denoting m number of one-half apparent wavelengths in that direction and the subscript 0 refers to the y-direction, denoting zero number of one-half apparent wavelengths in that direction.

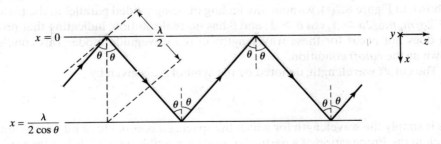

FIGURE 8.6

Uniform plane waves bouncing obliquely between two parallel plane perfectly conducting sheets.

Let us now consider a parallel-plate waveguide with perfectly conducting plates situated in the planes $x = 0$ and $x = a$, that is, having a fixed spacing a between them, as shown in Figure 8.7(a). Then, for $TE_{m,0}$ waves guided by the plates, we have from (8.28),

$$a = \frac{m\lambda}{2 \cos \theta}$$

or

$$\cos \theta = \frac{m\lambda}{2a} = \frac{m}{2a} \frac{1}{f\sqrt{\mu\epsilon}} \tag{8.29}$$

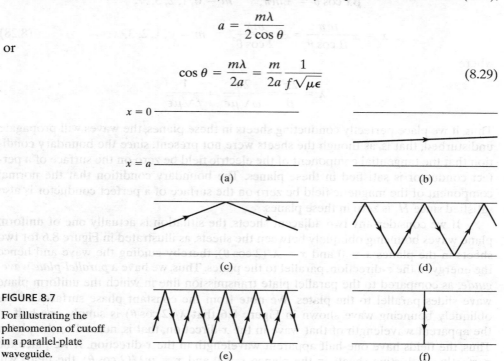

FIGURE 8.7

For illustrating the phenomenon of cutoff in a parallel-plate waveguide.

Thus, waves of different wavelengths (or frequencies) bounce obliquely between the plates at different values of the angle θ. For very small wavelengths (very high frequencies), $m\lambda/2a$ is small, $\cos \theta \approx 0, \theta \approx 90°$, and the waves simply slide between the plates as in the case of the transmission line, as shown in Figure 8.7(b). As λ increases (f decreases), $m\lambda/2a$ increases, θ decreases, and the waves bounce more and more obliquely, as shown in Figure 8.7(c)–(e), until λ becomes equal to $2a/m$, for which $\cos \theta = 1, \theta = 0°$, and the waves simply bounce back and forth normally to the plates, as shown in Figure 8.7(f), without any feeling of being guided parallel to the plates. For $\lambda > 2a/m$, $m\lambda/2a > 1$, $\cos \theta > 1$, and θ has no real solution, indicating that propagation does not occur for these wavelengths in the waveguide mode. This condition is known as the *cutoff* condition.

The cutoff wavelength, denoted by the symbol λ_c, is given by

$$\lambda_c = \frac{2a}{m} \tag{8.30}$$

This is simply the wavelength for which the spacing a is equal to m number of one-half wavelengths. Propagation of a particular mode is possible only if λ is less than the value of λ_c for that mode. The cutoff frequency is given by

$$f_c = \frac{m}{2a\sqrt{\mu\epsilon}} \tag{8.31}$$

Propagation of a particular mode is possible only if f is greater than the value of f_c for that mode. Consequently, waves of a given frequency f can propagate in all modes for which the cutoff wavelengths are greater than the wavelength or the cutoff frequencies are less than the frequency.

Substituting λ_c for $2a/m$ in (8.29), we have

$$\cos \theta = \frac{\lambda}{\lambda_c} = \frac{f_c}{f} \tag{8.32a}$$

$$\sin \theta = \sqrt{1 - \cos^2 \theta} = \sqrt{1 - \left(\frac{\lambda}{\lambda_c}\right)^2} = \sqrt{1 - \left(\frac{f_c}{f}\right)^2} \tag{8.32b}$$

$$\beta \cos \theta = \frac{2\pi}{\lambda} \frac{\lambda}{\lambda_c} = \frac{2\pi}{\lambda_c} = \frac{m\pi}{a} \tag{8.32c}$$

$$\beta \sin \theta = \frac{2\pi}{\lambda} \sqrt{1 - \left(\frac{\lambda}{\lambda_c}\right)^2} \tag{8.32d}$$

We see from (8.32d) that the phase constant along the z-direction, that is, $\beta \sin \theta$, is real for $\lambda < \lambda_c$ and imaginary for $\lambda > \lambda_c$, thereby explaining once again the cutoff phenomenon. We now define the guide wavelength, λ_g, to be the wavelength in the z-direction, that is, along the guide. This is given by

$$\lambda_g = \frac{2\pi}{\beta \sin \theta} = \frac{\lambda}{\sqrt{1 - (\lambda/\lambda_c)^2}} = \frac{\lambda}{\sqrt{1 - (f_c/f)^2}} \tag{8.33}$$

This is simply the apparent wavelength, in the z-direction, of the obliquely bouncing uniform plane waves. The phase velocity along the guide axis, which is simply the apparent phase velocity, in the z-direction, of the obliquely bouncing uniform plane waves, is

$$v_{pz} = \frac{\omega}{\beta \sin \theta} = \frac{v_p}{\sin \theta} = \frac{v_p}{\sqrt{1 - (\lambda/\lambda_c)^2}} = \frac{v_p}{\sqrt{1 - (f_c/f)^2}} \tag{8.34}$$

We note that the phase velocity along the guide axis is a function of frequency and hence the propagation along the guide axis is characterized by *dispersion*. The topic of dispersion is discussed in the next section.

Finally, substituting (8.32a)–(8.32d) in the field expressions (8.25a) and (8.25b), we obtain

$$\mathbf{E} = -2E_0 \sin\left(\frac{m\pi x}{a}\right) \sin\left(\omega t - \frac{2\pi}{\lambda_g}z\right)\mathbf{a}_y \tag{8.35a}$$

$$\mathbf{H} = \frac{2E_0}{\eta} \frac{\lambda}{\lambda_g} \sin\left(\frac{m\pi x}{a}\right) \sin\left(\omega t - \frac{2\pi}{\lambda_g}z\right)\mathbf{a}_x$$

$$-\frac{2E_0}{\eta} \frac{\lambda}{\lambda_c} \cos\left(\frac{m\pi x}{a}\right) \cos\left(\omega t - \frac{2\pi}{\lambda_g}z\right)\mathbf{a}_z \tag{8.35b}$$

These expressions for the TE$_{m,0}$ mode fields in the parallel-plate waveguide do not contain the angle θ. They clearly indicate the standing wave character of the fields in the x-direction, having m one-half sinusoidal variations between the plates. We shall now consider an example.

Example 8.2

Let us assume the spacing a between the plates of a parallel-plate waveguide to be 5 cm and investigate the propagating $TE_{m,0}$ modes for $f = 10,000$ MHz.

From (8.30), the cutoff wavelengths for $TE_{m,0}$ modes are given by

$$\lambda_c = \frac{2a}{m} = \frac{10}{m} \text{ cm} = \frac{0.1}{m} \text{ m}$$

This result is independent of the dielectric between the plates. If the medium between the plates is free space, then the cutoff frequencies for the $TE_{m,0}$ modes are

$$f_c = \frac{3 \times 10^8}{\lambda_c} = \frac{3 \times 10^8}{0.1/m} = 3m \times 10^9 \text{ Hz}$$

For $f = 10,000$ MHz $= 10^{10}$ Hz, the propagating modes are $TE_{1,0}(f_c = 3 \times 10^9 \text{ Hz})$, $TE_{2,0}(f_c = 6 \times 10^9 \text{ Hz})$, and $TE_{3,0}(f_c = 9 \times 10^9 \text{ Hz})$.

For each propagating mode, we can find θ, λ_g, and v_{pz} by using (8.32a), (8.33), and (8.34), respectively. Values of these quantities are listed in the following:

Mode	λ_c, cm	f_c, MHz	θ, deg	λ_g, cm	v_{pz}, m/s
$TE_{1,0}$	10	3000	72.54	3.145	3.145×10^8
$TE_{2,0}$	5	6000	53.13	3.75	3.75×10^8
$TE_{3,0}$	3.33	9000	25.84	6.882	6.882×10^8

8.3 DISPERSION AND GROUP VELOCITY

In Section 8.2, we learned that for the propagating range of frequencies, the phase velocity and the wavelength along the axis of the parallel-plate waveguide are given by

$$v_{pz} = \frac{v_p}{\sqrt{1 - (f_c/f)^2}} \tag{8.36}$$

and

$$\lambda_g = \frac{\lambda}{\sqrt{1 - (f_c/f)^2}} \tag{8.37}$$

where $v_p = 1/\sqrt{\mu\epsilon}$, $\lambda = v_p/f = 1/f\sqrt{\mu\epsilon}$, and f_c is the cutoff frequency. We note that for a particular mode, the phase velocity of propagation along the guide axis varies with the frequency. As a consequence of this characteristic of the guided wave propagation, the field patterns of the different frequency components of a signal comprising a band of frequencies do not maintain the same phase relationships as they propagate down the guide. This phenomenon is known as *dispersion*, so termed after the phenomenon of dispersion of colors by a prism.

To discuss dispersion, let us consider a simple example of two infinitely long trains A and B traveling in parallel, one below the other, with each train made up of boxcars of identical size and having wavy tops, as shown in Figure 8.8. Let the spacings between the peaks (centers) of successive boxcars be 50 m and 90 m, and let the speeds

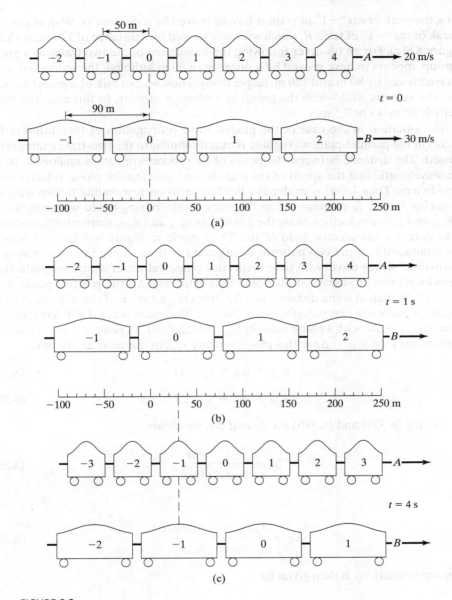

FIGURE 8.8

For illustrating the concept of group velocity.

of the trains be 20 m/s and 30 m/s, for trains A and B, respectively. Let the peaks of the cars numbered 0 for the two trains be aligned at time $t = 0$, as shown in Figure 8.8(a). Now, as time progresses, the two peaks get out of alignment as shown, for example, for $t = 1$ s in Figure 8.8(b), since train B is traveling faster than train A. But at the same time, the gap between the peaks of cars numbered -1 decreases. This continues until at

$t = 4$ s, the peak of car "−1" of train A having moved by a distance of 80 m aligns with the peak of car "−1" of train B, which will have moved by a distance of 120 m, as shown in Figure 8.8(c). For an observer following the movement of the two trains as a group, the group appears to have moved by a distance of 30 m although the individual trains will have moved by 80 m and 120 m, respectively. Thus, we can talk of a *group velocity*, that is, the velocity with which the group as a whole is moving. In this case, the group velocity is 30 m/4 s or 7.5 m/s.

The situation in the case of the guided wave propagation of two different frequencies in the parallel-plate waveguide is exactly similar to the two-train example just discussed. The distance between the peaks of two successive cars is analogous to the guide wavelength, and the speed of the train is analogous to the phase velocity along the guide axis. Thus, let us consider the field patterns corresponding to two waves of frequencies f_A and f_B propagating in the same mode, having guide wavelengths λ_{gA} and λ_{gB}, and phase velocities along the guide axis v_{pzA} and v_{pzB}, respectively, as shown, for example, for the electric field of the $TE_{1,0}$ mode in Figure 8.9. Let the positive peaks numbered 0 of the two patterns be aligned $t = 0$, as shown in Figure 8.9(a). As the individual waves travel with their respective phase velocities along the guide, these two peaks get out of alignment but some time later, say Δt, the positive peaks numbered −1 will align at some distance, say Δz, from the location of the alignment of the "0" peaks, as shown in Figure 8.9(b). Since the "−1"th peak of wave A will have traveled a distance $\lambda_{gA} + \Delta z$ with a phase velocity v_{pzA} and the "−1"th peak of wave B will have traveled a distance $\lambda_{gB} + \Delta z$ with a phase velocity v_{pzB} in this time Δt, we have

$$\lambda_{gA} + \Delta z = v_{pzA} \Delta t \tag{8.38a}$$

$$\lambda_{gB} + \Delta z = v_{pzB} \Delta t \tag{8.38b}$$

Solving (8.38a) and (8.38b) for Δt and Δz, we obtain

$$\Delta t = \frac{\lambda_{gA} - \lambda_{gB}}{v_{pzA} - v_{pzB}} \tag{8.39a}$$

and

$$\Delta z = \frac{\lambda_{gA} v_{pzB} - \lambda_{gB} v_{pzA}}{v_{pzA} - v_{pzB}} \tag{8.39b}$$

The group velocity, v_g, is then given by

$$v_g = \frac{\Delta z}{\Delta t} = \frac{\lambda_{gA} v_{pzB} - \lambda_{gB} v_{pzA}}{\lambda_{gA} - \lambda_{gB}} = \frac{\lambda_{gA} \lambda_{gB} f_B - \lambda_{gB} \lambda_{gA} f_A}{\lambda_{gA} \lambda_{gB} \left(\dfrac{1}{\lambda_{gB}} - \dfrac{1}{\lambda_{gA}} \right)}$$

$$= \frac{f_B - f_A}{\dfrac{1}{\lambda_{gB}} - \dfrac{1}{\lambda_{gA}}} = \frac{\omega_B - \omega_A}{\beta_{zB} - \beta_{zA}} \tag{8.40}$$

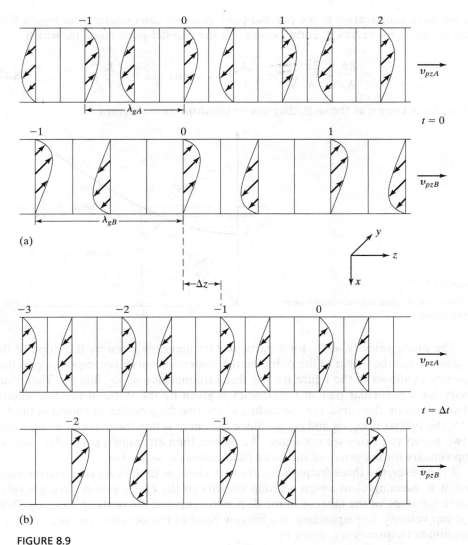

FIGURE 8.9

For illustrating the concept of group velocity for guided wave propagation.

where β_{zA} and β_{zB} are the phase constants along the guide axis, corresponding to f_A and f_B, respectively. Thus, the group velocity of a signal comprised of two frequencies is the ratio of the difference between the two radian frequencies to the difference between the corresponding phase constants along the guide axis.

If we now have a signal comprised of a number of frequencies, then a value of group velocity can be obtained for each pair of these frequencies in accordance with (8.40). In general, these values of group velocity will all be different. In fact, this is the case for wave propagation in the parallel-plate guide, as can be seen from Figure 8.10, which is a plot of ω versus β_z corresponding to the parallel-plate guide for which

$$\beta_z = \frac{2\pi}{\lambda_g} = \frac{2\pi}{\lambda}\sqrt{1 - \left(\frac{\lambda}{\lambda_c}\right)^2} = \omega\sqrt{\mu\epsilon}\sqrt{1 - \left(\frac{f_c}{f}\right)^2} \qquad (8.41)$$

Such a plot is known as the ω–β_z *diagram* or the *dispersion diagram.*

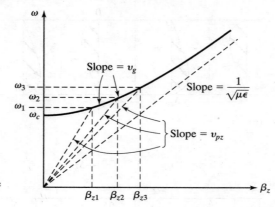

FIGURE 8.10

Dispersion diagram for the parallel-plate waveguide.

The phase velocity, ω/β_z, for a particular frequency is given by the slope of the line drawn from the origin to the point, on the dispersion curve, corresponding to that frequency, as shown in the figure for the three frequencies ω_1, ω_2, and ω_3. The group velocity for a particular pair of frequencies is given by the slope of the line joining the two points, on the curve, corresponding to the two frequencies, as shown in the figure for the two pairs ω_1, ω_2 and ω_2, ω_3. Since the curve is nonlinear, it can be seen that the two group velocities are not equal. We cannot then attribute a particular value of group velocity for the group of the three frequencies ω_1, ω_2, and ω_3.

If, however, the three frequencies are very close, as in the case of a narrow-band signal, it is meaningful to assign a group velocity to the entire group having a value equal to the slope of the tangent to the dispersion curve at the center frequency. Thus, the group velocity corresponding to a narrow band of frequencies centered around a predominant frequency ω is given by

$$v_g = \frac{d\omega}{d\beta_z} \qquad (8.42)$$

For the parallel-plate waveguide under consideration, we have from (8.41),

$$\frac{d\beta_z}{d\omega} = \sqrt{\mu\epsilon}\sqrt{1 - \left(\frac{f_c}{f}\right)^2} + \omega\sqrt{\mu\epsilon} \cdot \frac{1}{2}\left(1 - \frac{f_c^2}{f^2}\right)^{-1/2}\frac{f_c^2}{\pi f^3}$$

$$= \sqrt{\mu\epsilon}\left(1 - \frac{f_c^2}{f^2} + \frac{\omega}{2\pi}\frac{f_c^2}{f^3}\right)\left(1 - \frac{f_c^2}{f^2}\right)^{-1/2}$$

$$= \sqrt{\mu\epsilon}\left(1 - \frac{f_c^2}{f^2}\right)^{-1/2}$$

and

$$v_g = \frac{d\omega}{d\beta_z} = \frac{1}{\sqrt{\mu\epsilon}}\sqrt{1 - \frac{f_c^2}{f^2}} = v_p\sqrt{1 - \left(\frac{f_c}{f}\right)^2} \qquad (8.43)$$

As a numerical example, for the case of Example 8.2, the group velocities for $f = 10{,}000$ MHz for the three propagating modes $TE_{1,0}$, $TE_{2,0}$, and $TE_{3,0}$ are 2.862×10^8 m/s, 2.40×10^8 m/s, and 1.308×10^8 m/s, respectively. From (8.36) and (8.43), we note that

$$v_{pz}v_g = v_p^2 \qquad (8.44)$$

An example of a narrow-band signal is an amplitude modulated signal, having a carrier frequency ω modulated by a low frequency $\Delta\omega \ll \omega$, as given by

$$E_x(t) = E_{x0}(1 + m \cos \Delta\omega \cdot t) \cos \omega t \qquad (8.45)$$

where m is the percentage modulation. Such a signal is actually equivalent to a superposition of unmodulated signals of three frequencies $\omega - \Delta\omega$, ω, and $\omega + \Delta\omega$, as can be seen by expanding the right side of (8.45). Thus

$$E_x(t) = E_{x0} \cos \omega t + m E_{x0} \cos \omega t \cos \Delta\omega \cdot t$$

$$= E_{x0} \cos \omega t + \frac{m E_{x0}}{2}[\cos(\omega - \Delta\omega)t + \cos(\omega + \Delta\omega)t] \qquad (8.46)$$

The frequencies $\omega - \Delta\omega$ and $\omega + \Delta\omega$ are the side frequencies. When the amplitude modulated signal propagates in a dispersive channel such as the parallel-plate waveguide under consideration, the different frequency components undergo phase changes in accordance with their respective phase constants. Thus, if $\beta_z - \Delta\beta_z$, β_z, and $\beta_z + \Delta\beta_z$ are the phase constants corresponding to $\omega - \Delta\omega$, ω, and $\omega + \Delta\omega$, respectively, assuming linearity of the dispersion curve within the narrow band, the amplitude modulated wave is given by

$$E_x(z, t) = E_{x0} \cos(\omega t - \beta_z z)$$

$$+ \frac{m E_{x0}}{2}\{\cos[(\omega - \Delta\omega)t - (\beta_z - \Delta\beta_z)z]$$

$$+ \cos[(\omega + \Delta\omega)t - (\beta_z + \Delta\beta_z)z]\}$$

$$= E_{x0} \cos(\omega t - \beta_z z)$$

$$+ \frac{m E_{x0}}{2}\{\cos[(\omega t - \beta_z z) - (\Delta\omega \cdot t - \Delta\beta_z \cdot z)]$$

$$+ \cos[(\omega t - \beta_z z) + (\Delta\omega \cdot t - \Delta\beta_z \cdot z)]\}$$

$$= E_{x0} \cos(\omega t - \beta_z z) + m E_{x0} \cos(\omega t - \beta_z z) \cos(\Delta\omega \cdot t - \Delta\beta_z \cdot z)$$

$$= E_{x0}[1 + m \cos(\Delta\omega \cdot t - \Delta\beta_z \cdot z)] \cos(\omega t - \beta_z z) \qquad (8.47)$$

This indicates that although the carrier frequency phase changes in accordance with the phase constant β_z, the modulation envelope and hence the information travels with the group velocity $\Delta\omega/\Delta\beta_z$, as shown in Figure 8.11. In view of this and since v_g is less than v_p, the fact that v_{pz} is greater than v_p is not a violation of the theory of relativity. Since it is always necessary to use some modulation technique to convey information from one point to another, the information always takes more time to reach from one point to another in a dispersive channel than in the corresponding nondispersive medium.

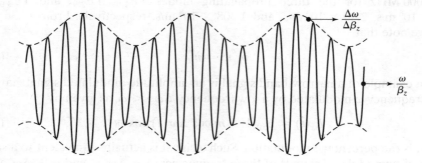

FIGURE 8.11

For illustrating that the modulation envelope travels with the group velocity.

8.4   RECTANGULAR WAVEGUIDE AND CAVITY RESONATOR

Thus far, we have restricted our discussion to $TE_{m,0}$ wave propagation in a parallel-plate waveguide. From Section 8.2, we recall that the parallel-plate waveguide is made up of two perfectly conducting sheets in the planes $x = 0$ and $x = a$ and that the electric field of the $TE_{m,0}$ mode has only a y-component with m number of one-half sinusoidal variations in the x-direction and no variations in the y-direction. If we now introduce two perfectly conducting sheets in two constant y-planes, say, $y = 0$ and $y = b$, the field distribution will remain unaltered, since the electric field is entirely normal to the plates, and hence the boundary condition of zero tangential electric field is satisfied for both sheets. We then have a metallic pipe with rectangular cross section in the xy-plane, as shown in Figure 8.12. Such a structure is known as the *rectangular waveguide* and is, in fact, a common form of waveguide.

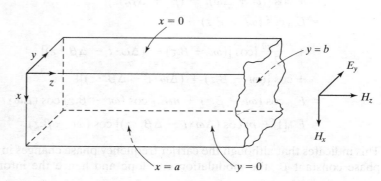

FIGURE 8.12

A rectangular waveguide.

Since the $\text{TE}_{m,0}$ mode field expressions derived for the parallel-plate waveguide satisfy the boundary conditions for the rectangular waveguide, those expressions as well as the entire discussion of the parallel-plate waveguide case hold also for $\text{TE}_{m,0}$ mode propagation in the rectangular waveguide case. We learned that the $\text{TE}_{m,0}$ modes can be interpreted as due to uniform plane waves having electric field in the y-direction and bouncing obliquely between the conducting walls $x = 0$ and $x = a$, and with the associated cutoff condition characterized by bouncing of the waves back and forth normally to these walls, as shown in Figure 8.13(a). For the cutoff condition, the dimension a is equal to m number of one-half wavelengths such that

$$[\lambda_c]_{\text{TE}_{m,0}} = \frac{2a}{m} \tag{8.48}$$

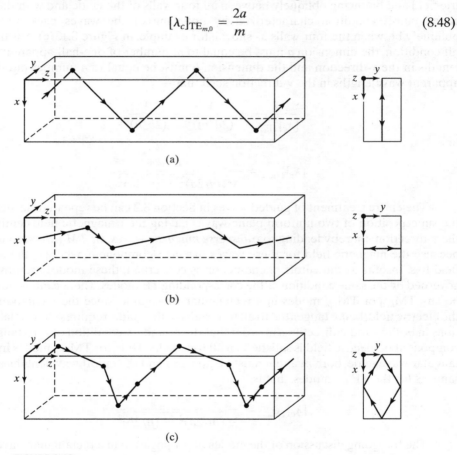

(a)

(b)

(c)

FIGURE 8.13

Propagation and cutoff of (a) $\text{TE}_{m,0}$, (b) $\text{TE}_{0,n}$, and (c) $\text{TE}_{m,n}$ modes in a rectangular waveguide.

In a similar manner, we can have uniform plane waves having electric field in the x-direction and bouncing obliquely between the walls $y = 0$ and $y = b$, and with the associated cutoff condition characterized by bouncing of the waves back and forth normally to these walls, as shown in Figure 8.13(b), thereby resulting in $\text{TE}_{0,n}$ modes having no variations in the x-direction and n number of one-half sinusoidal variations in the y-direction. For the cutoff condition, the dimension b is equal to n number of one-half wavelengths such that

$$[\lambda_c]_{TE_{0,n}} = \frac{2b}{n} \tag{8.49}$$

We can even have $TE_{m,n}$ modes having m number of one-half sinusoidal variations in the x-direction and n number of one-half sinusoidal variations in the y-direction due to uniform plane waves having both x- and y-components of the electric field and bouncing obliquely between all four walls of the guide and with the associated cutoff condition characterized by bouncing of the waves back and forth obliquely between the four walls as shown, for example, in Figure 8.13(c). For the cutoff condition, the dimension a must be equal to m number of one-half apparent wavelengths in the x-direction and the dimension b must be equal to n number of one-half apparent wavelengths in the y-direction such that

$$\frac{1}{[\lambda_c]_{TE_{m,n}}^2} = \frac{1}{(2a/m)^2} + \frac{1}{(2b/n)^2} \tag{8.50}$$

or

$$[\lambda_c]_{TE_{m,n}} = \frac{1}{\sqrt{(m/2a)^2 + (n/2b)^2}} \tag{8.51}$$

The entire treatment of guided waves in Section 8.2 can be repeated starting with the superposition of two uniform plane waves having their magnetic fields entirely in the y-direction, thereby leading to *transverse magnetic waves*, or *TM waves*, so termed because the magnetic field for these waves has no z-component, whereas the electric field has. Insofar as the cutoff phenomenon is concerned, these modes are obviously governed by the same condition as the corresponding TE modes. There cannot, however, be any $TM_{m,0}$ or $TM_{0,n}$ modes in a rectangular waveguide, since the z-component of the electric field, being tangential to all four walls of the guide, requires sinusoidal variations in both x- and y-directions in order that the boundary condition of zero tangential component of electric field is satisfied on all four walls. Thus, for $TM_{m,n}$ modes in a rectangular waveguide, both m and n must be nonzero and the cutoff wavelengths are the same as for the $TE_{m,n}$ modes, that is,

$$[\lambda_c]_{TM_{m,n}} = \frac{1}{\sqrt{(m/2a)^2 + (n/2b)^2}} \tag{8.52}$$

The foregoing discussion of the modes of propagation in a rectangular waveguide points out that a signal of given frequency can propagate in several modes, namely, all modes for which the cutoff frequencies are less than the signal frequency or the cutoff wavelengths are greater than the signal wavelength. Waveguides are, however, designed so that only one mode, the mode with the lowest cutoff frequency (or the largest cutoff wavelength), propagates. This is known as the *dominant mode*. From (8.48), (8.49), (8.51), and (8.52), we can see that the dominant mode is the $TE_{1,0}$ mode or the $TE_{0,1}$ mode, depending on whether the dimension a or the dimension b is the larger of the two. By convention, the larger dimension is designated to be a, and hence the $TE_{1,0}$ mode is the dominant mode. We shall now consider an example.

Example 8.3

It is desired to determine the lowest four cutoff frequencies referred to the cutoff frequency of the dominant mode for three cases of rectangular waveguide dimensions: (i) $b/a = 1$, (ii) $b/a = 1/2$, and (iii) $b/a = 1/3$. Given $a = 3$ cm, it is then desired to find the propagating mode(s) for $f = 9000$ MHz for each of the three cases.

From (8.51) and (8.52), the expression for the cutoff wavelength for a $TE_{m,n}$ mode where $m = 0, 1, 2, 3, \ldots$ and $n = 0, 1, 2, 3, \ldots$ but not both m and n equal to zero and for a $TM_{m,n}$ mode where $m = 1, 2, 3, \ldots$ and $n = 1, 2, 3, \ldots$ is given by

$$\lambda_c = \frac{1}{\sqrt{(m/2a)^2 + (n/2b)^2}}$$

The corresponding expression for the cutoff frequency is

$$f_c = \frac{v_p}{\lambda_c} = \frac{1}{\sqrt{\mu\epsilon}} \sqrt{\left(\frac{m}{2a}\right)^2 + \left(\frac{n}{2b}\right)^2}$$

$$= \frac{1}{2a\sqrt{\mu\epsilon}} \sqrt{m^2 + \left(n\frac{a}{b}\right)^2}$$

The cutoff frequency of the dominant mode $TE_{1,0}$ is $1/2a\sqrt{\mu\epsilon}$. Hence,

$$\frac{f_c}{[f_c]_{TE_{1,0}}} = \sqrt{m^2 + \left(n\frac{a}{b}\right)^2}$$

By assigning different pairs of values for m and n, the lowest four values of $f_c/[f_c]_{TE_{1,0}}$ can be computed for each of the three specified values of b/a. These computed values and the corresponding modes are shown in Figure 8.14.

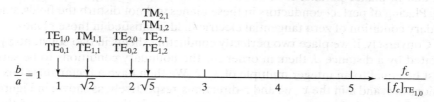

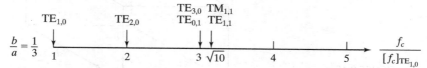

FIGURE 8.14

Lowest four cutoff frequencies referred to the cutoff frequency of the dominant mode for three cases of rectangular waveguide dimensions.

For $a = 3$ cm, and assuming free space for the dielectric in the waveguide,

$$[f_c]_{\text{TE}_{1,0}} = \frac{1}{2a\sqrt{\mu\epsilon}} = \frac{3 \times 10^8}{2 \times 0.03} = 5000 \text{ MHz}$$

Hence, for a signal of frequency $f = 9000$ MHz, all the modes for which $f_c/[f_c]_{\text{TE}_{1,0}}$ is less than 1.8 propagate. From Figure 8.14, these are

$$\text{TE}_{1,0}, \text{TE}_{0,1}, \text{TM}_{1,1}, \text{TE}_{1,1} \qquad \text{for } b/a = 1$$
$$\text{TE}_{1,0} \qquad\qquad\qquad\qquad\quad \text{for } b/a = 1/2$$
$$\text{TE}_{1,0} \qquad\qquad\qquad\qquad\quad \text{for } b/a = 1/3$$

It can be seen from Figure 8.14 that for $b/a \leq 1/2$, the second lowest cutoff frequency that corresponds to that of the $\text{TE}_{2,0}$ mode is twice the cutoff frequency of the dominant mode $\text{TE}_{1,0}$. For this reason, the dimension b of a rectangular waveguide is generally chosen to be less than or equal to $a/2$ in order to achieve single-mode transmission over a complete octave (factor of two) range of frequencies.

Let us now consider guided waves of equal magnitude propagating in the positive z- and negative z-directions in a rectangular waveguide. This can be achieved by terminating the guide by a perfectly conducting sheet in a constant-z plane, that is, a transverse plane of the guide. Due to perfect reflection from the sheet, the fields will then be characterized by standing wave nature along the guide axis, that is, in the z-direction, in addition to the standing wave nature in the x- and y-directions. The standing wave pattern along the guide axis will have nulls of transverse electric field on the terminating sheet and in planes parallel to it at distances of integer multiples of $\lambda_g/2$ from that sheet. Placing of perfect conductors in these planes will not disturb the fields, since the boundary condition of zero tangential electric field is satisfied in those planes.

Conversely, if we place two perfectly conducting sheets in two constant-z planes separated by a distance d, then, in order for the boundary conditions to be satisfied, d must be equal to an integer multiple of $\lambda_g/2$. We then have a rectangular box of dimensions a, b, and d in the x-, y-, and z-directions, respectively, as shown in Figure 8.15. Such a structure is known as a *cavity resonator* and is the counterpart of the low-frequency lumped parameter resonant circuit at microwave frequencies since it supports

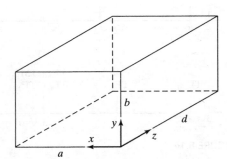

FIGURE 8.15

A rectangular cavity resonator.

oscillations at frequencies for which the above condition, that is,

$$d = l\frac{\lambda_g}{2}, \qquad l = 1, 2, 3, \ldots \tag{8.53}$$

is satisfied. Recalling that λ_g is simply the apparent wavelength of the obliquely bouncing uniform plane wave along the z-direction, we find that the wavelength corresponding to the mode of oscillation for which the fields have m number of one-half sinusoidal variations in the x-direction, n number of one-half sinusoidal variations in the y-direction, and l number of one-half sinusoidal variations in the z-direction is given by

$$\frac{1}{\lambda_{osc}^2} = \frac{1}{(2a/m)^2} + \frac{1}{(2b/n)^2} + \frac{1}{(2d/l)^2} \tag{8.54}$$

or

$$\lambda_{osc} = \frac{1}{\sqrt{(m/2a)^2 + (n/2b)^2 + (l/2d)^2}} \tag{8.55}$$

The expression for the frequency of oscillation is then given by

$$f_{osc} = \frac{v_p}{\lambda_{osc}} = \frac{1}{\sqrt{\mu\epsilon}} \sqrt{\left(\frac{m}{2a}\right)^2 + \left(\frac{n}{2b}\right)^2 + \left(\frac{l}{2d}\right)^2} \tag{8.56}$$

The modes are designated by three subscripts in the manner $TE_{m,n,l}$ and $TM_{m,n,l}$. Since m, n, and l can assume combinations of integer values, an infinite number of frequencies of oscillation are possible for a given set of dimensions for the cavity resonator. We shall now consider an example.

Example 8.4

The dimensions of a rectangular cavity resonator with air dielectric are $a = 4$ cm, $b = 2$ cm, and $d = 4$ cm. It is desired to determine the three lowest frequencies of oscillation and specify the mode(s) of oscillation, transverse with respect to the z-direction, for each frequency.

By substituting $\mu = \mu_0$, $\epsilon = \epsilon_0$, and the given dimensions for a, b, and d in (8.56), we obtain

$$f_{osc} = 3 \times 10^8 \sqrt{\left(\frac{m}{0.08}\right)^2 + \left(\frac{n}{0.04}\right)^2 + \left(\frac{l}{0.08}\right)^2}$$

$$= 3750\sqrt{m^2 + 4n^2 + l^2}\ \text{MHz}$$

By assigning combinations of integer values for m, n, and l and recalling that both m and n must be nonzero for TM modes, we obtain the three lowest frequencies of oscillation to be

$$3750 \times \sqrt{2} = 5303\ \text{MHz for } TE_{1,0,1} \text{ mode}$$

$$3750 \times \sqrt{5} = 8385\ \text{MHz for } TE_{0,1,1},\ TE_{2,0,1}, \text{ and } TE_{1,0,2} \text{ modes}$$

$$3750 \times \sqrt{6} = 9186\ \text{MHz for } TE_{1,1,1} \text{ and } TM_{1,1,1} \text{ modes}$$

8.5   REFLECTION AND REFRACTION OF PLANE WAVES

Let us now consider a uniform plane wave that is incident obliquely on a plane boundary between two different perfect dielectric media at an angle of incidence θ_i to the normal to the boundary, as shown in Figure 8.16. To satisfy the boundary conditions at the interface between the two media, a reflected wave and a transmitted wave will be set up. Let θ_r be the angle of reflection and θ_t be the angle of transmission. Then without writing the expressions for the fields, we can find the relationship among θ_i, θ_r, and θ_t by noting that for the incident, reflected, and transmitted waves to be in step at the boundary, their apparent phase velocities parallel to the boundary must be equal; that is,

$$\frac{v_{p1}}{\sin \theta_i} = \frac{v_{p1}}{\sin \theta_r} = \frac{v_{p2}}{\sin \theta_t} \tag{8.57}$$

where $v_{p1}(= 1/\sqrt{\mu_1 \epsilon_1})$ and $v_{p2}(= 1/\sqrt{\mu_2 \epsilon_2})$ are the phase velocities along the directions of propagation of the waves in medium 1 and medium 2, respectively.

From (8.57), we have

$$\sin \theta_r = \sin \theta_i \tag{8.58}$$

$$\sin \theta_t = \frac{v_{p2}}{v_{p1}} \sin \theta_i = \sqrt{\frac{\mu_1 \epsilon_1}{\mu_2 \epsilon_2}} \sin \theta_i \tag{8.59}$$

or

$$\theta_r = \theta_i \tag{8.60}$$

$$\theta_t = \sin^{-1}\left(\sqrt{\frac{\mu_1 \epsilon_1}{\mu_2 \epsilon_2}} \sin \theta_i \right) \tag{8.61}$$

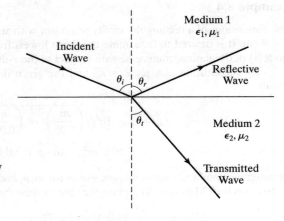

FIGURE 8.16

Reflection and transmission of an obliquely incident uniform plane wave on a plane boundary between two different perfect dielectric media.

Equation (8.60) is known as the *law of reflection* and (8.61) is known as the *law of refraction*, or *Snell's law*. Snell's law is commonly cast in terms of the refractive index, denoted by the symbol n and defined as the ratio of the velocity of light in free space to the phase velocity in the medium. Thus, if $n_1(= c/v_{p1})$ and $n_2(= c/v_{p2})$ are the (phase) refractive indices for media 1 and 2, respectively, then

$$\theta_t = \sin^{-1}\left(\frac{n_1}{n_2}\sin\theta_i\right) \qquad (8.62)$$

For two dielectrics having $\mu_1 = \mu_2 = \mu_0$, which is usually the case, (8.62) reduces to

$$\theta_t = \sin^{-1}\left(\sqrt{\frac{\epsilon_1}{\epsilon_2}}\sin\theta_i\right) \qquad (8.63)$$

We shall now consider the derivation of the expressions for the reflection and transmission coefficients at the boundary. To do this, we distinguish between two cases: (1) the electric field vector of the wave linearly polarized parallel to the interface and (2) the magnetic field vector of the wave linearly polarized parallel to the interface. The law of reflection and Snell's law hold for both cases, since they result from the fact that the apparent phase velocities of the incident, reflected, and transmitted waves parallel to the boundary must be equal.

The geometry pertinent to the case of the electric field vector parallel to the interface is shown in Figure 8.17, in which the interface is assumed to be in the $x = 0$ plane and the subscripts $i, r,$ and t associated with the field symbols denote incident, reflected, and transmitted waves, respectively. The plane of incidence, that is, the plane containing the normal to the interface and the propagation vectors, is assumed to be in the xz-plane, so that the electric field vectors are entirely in the y-direction. The corresponding magnetic field vectors are then as shown in the figure so as to be consistent

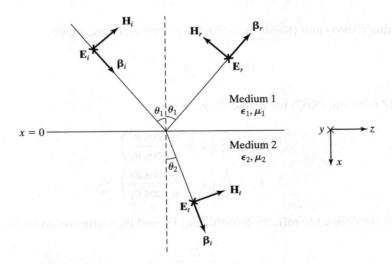

FIGURE 8.17

For obtaining the reflection and transmission coefficients for an obliquely incident uniform plane wave on a dielectric interface with its electric field perpendicular to the plane of incidence.

with the condition that **E**, **H**, and **β** form a right-handed mutually orthogonal set of vectors. Since the electric field vectors are perpendicular to the plane of incidence, this case is also said to correspond to perpendicular polarization. The angle of incidence is assumed to be θ_1. From the law of reflection (8.60), the angle of reflection is then also θ_1. The angle of transmission, assumed to be θ_2, is related to θ_1 by Snell's law, given by (8.61).

The boundary conditions to be satisfied at the interface $x = 0$ are that (1) the tangential component of the electric field intensity be continuous and (2) the tangential component of the magnetic field intensity be continuous. Thus, we have at the interface $x = 0$

$$E_{yi} + E_{yr} = E_{yt} \tag{8.64a}$$
$$H_{zi} + H_{zr} = H_{zt} \tag{8.64b}$$

Expressing the quantities in (8.64a) and (8.64b) in terms of the total fields, we obtain

$$E_i + E_r = E_t \tag{8.65a}$$
$$H_i \cos\theta_1 - H_r \cos\theta_1 = H_t \cos\theta_2 \tag{8.65b}$$

We also know from one of the properties of uniform plane waves that

$$\frac{E_i}{H_i} = \frac{E_r}{H_r} = \eta_1 = \sqrt{\frac{\mu_1}{\epsilon_1}} \tag{8.66a}$$
$$\frac{E_t}{H_t} = \eta_2 = \sqrt{\frac{\mu_2}{\epsilon_2}} \tag{8.66b}$$

Substituting (8.66a) and (8.66b) into (8.65b) and rearranging, we get

$$E_i - E_r = E_t \frac{\eta_1 \cos\theta_2}{\eta_2 \cos\theta_1} \tag{8.67}$$

Solving (8.65a) and (8.67) for E_i and E_r, we have

$$E_i = \frac{E_t}{2}\left(1 + \frac{\eta_1 \cos\theta_2}{\eta_2 \cos\theta_1}\right) \tag{8.68a}$$
$$E_r = \frac{E_t}{2}\left(1 - \frac{\eta_1 \cos\theta_2}{\eta_2 \cos\theta_1}\right) \tag{8.68b}$$

We now define the reflection coefficient $\Gamma_\perp$ and the transmission coefficient $\tau_\perp$ as

$$\Gamma_\perp = \frac{E_r}{E_i} = \frac{E_{yr}}{E_{yi}} \tag{8.69a}$$
$$\tau_\perp = \frac{E_t}{E_i} = \frac{E_{yt}}{E_{yi}} \tag{8.69b}$$

where the subscript $\perp$ refers to perpendicular polarization. From (8.68a) and (8.68b), we then obtain

$$\Gamma_\perp = \frac{\eta_2 \cos \theta_1 - \eta_1 \cos \theta_2}{\eta_2 \cos \theta_1 + \eta_1 \cos \theta_2} \tag{8.70a}$$

$$\tau_\perp = \frac{2\eta_2 \cos \theta_1}{\eta_2 \cos \theta_1 + \eta_1 \cos \theta_2} \tag{8.70b}$$

Equations (8.70a) and (8.70b) are known as the *Fresnel reflection and transmission coefficients* for perpendicular polarization.

Before we discuss the result given by (8.70a) and (8.70b), we shall derive the corresponding expressions for the case in which the magnetic field of the wave is parallel to the interface. The geometry pertinent to this case is shown in Figure 8.18. Here again the plane of incidence is chosen to be the xz-plane, so that the magnetic field vectors are entirely in the y-direction. The corresponding electric field vectors are then as shown in the figure so as to be consistent with the condition that $\mathbf{E}, \mathbf{H}$, and $\boldsymbol{\beta}$ form a right-handed mutually orthogonal set of vectors. Since the electric field vectors are parallel to the plane of incidence, this case is also said to correspond to parallel polarization.

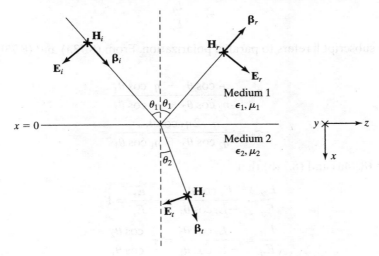

FIGURE 8.18

For obtaining the reflection and transmission coefficients for an obliquely incident uniform plane wave on a dielectric interface with its electric field parallel to the plane of incidence.

Once again the boundary conditions to be satisfied at the interface $x = 0$ are that (1) the tangential component of the electric field intensity be continuous and (2) the tangential component of the magnetic field intensity be continuous. Thus, we have at the interface $x = 0$,

$$E_{zi} + E_{zr} = E_{zt} \tag{8.71a}$$

$$H_{yi} + H_{yr} = H_{yt} \tag{8.71b}$$

Expressing the quantities in (8.71a) and (8.71b) in terms of the total fields and also using (8.66a) and (8.66b), we obtain

$$E_i - E_r = E_t \frac{\cos \theta_2}{\cos \theta_1} \qquad (8.72a)$$

$$E_i + E_r = E_t \frac{\eta_1}{\eta_2} \qquad (8.72b)$$

Solving (8.72a) and (8.72b) for E_i and E_r, we have

$$E_i = \frac{E_t}{2}\left(\frac{\eta_1}{\eta_2} + \frac{\cos \theta_2}{\cos \theta_1} \right) \qquad (8.73a)$$

$$E_r = \frac{E_t}{2}\left(\frac{\eta_1}{\eta_2} - \frac{\cos \theta_2}{\cos \theta_1} \right) \qquad (8.73b)$$

We now define the reflection coefficient $\Gamma_\parallel$ and the transmission coefficient $\tau_\parallel$ as

$$\Gamma_\parallel = -\frac{E_r}{E_i} \qquad (8.74a)$$

$$\tau_\parallel = \frac{E_t}{E_i} \qquad (8.74b)$$

where the subscript $\parallel$ refers to parallel polarization. From (8.73a) and (8.73b), we then obtain

$$\Gamma_\parallel = \frac{\eta_2 \cos \theta_2 - \eta_1 \cos \theta_1}{\eta_2 \cos \theta_2 + \eta_1 \cos \theta_1} \qquad (8.75a)$$

$$\tau_\parallel = \frac{2\eta_2 \cos \theta_1}{\eta_2 \cos \theta_2 + \eta_1 \cos \theta_1} \qquad (8.75b)$$

Note from (8.74a) and (8.74b) that

$$\frac{E_{zr}}{E_{zi}} = \frac{E_r \cos \theta_1}{-E_i \cos \theta_1} = -\frac{E_r}{E_i} = \Gamma_\parallel \qquad (8.76a)$$

$$\frac{E_{zt}}{E_{zi}} = \frac{-E_t \cos \theta_2}{-E_i \cos \theta_1} = \tau_\parallel \frac{\cos \theta_2}{\cos \theta_1} \qquad (8.76b)$$

Equations (8.75a) and (8.75b) are known as the *Fresnel reflection and transmission coefficients* for parallel polarization.

We shall now discuss the results given by (8.70a), (8.70b), (8.75a), and (8.75b) for the reflection and transmission coefficients for the two cases:

1. For $\theta_1 = 0$, that is, for the case of normal incidence of the uniform plane wave upon the interface, $\theta_2 = 0$ and

$$\Gamma_\perp = \frac{\eta_2 - \eta_1}{\eta_2 + \eta_1}, \qquad \Gamma_\parallel = \frac{\eta_2 - \eta_1}{\eta_2 + \eta_1}$$

$$\tau_\perp = \frac{2\eta_2}{\eta_2 + \eta_1}, \qquad \tau_\parallel = \frac{2\eta_2}{\eta_2 + \eta_1}$$

Thus, the reflection coefficients as well as the transmission coefficients for the two cases become equal as they should, since for normal incidence there is no difference between the two polarizations except for rotation by 90° parallel to the interface.

2. $\Gamma_\perp = 1$ and $\Gamma_\parallel = -1$ if $\cos \theta_2 = 0$; that is,

$$\sqrt{1 - \sin^2 \theta_2} = \sqrt{1 - \frac{\mu_1 \epsilon_1}{\mu_2 \epsilon_2} \sin^2 \theta_1} = 0$$

or

$$\sin \theta_1 = \sqrt{\frac{\mu_2 \epsilon_2}{\mu_1 \epsilon_1}} \tag{8.77}$$

where we have used Snell's law, given by (8.61), to express $\sin \theta_2$ in terms of $\sin \theta_1$. If we assume $\mu_2 = \mu_1 = \mu_0$, as is usually the case, (8.77) has real solutions for θ_1 for $\epsilon_2 < \epsilon_1$. Thus, for $\epsilon_2 < \epsilon_1$, that is, for transmission from a dielectric medium of higher permittivity into a dielectric medium of lower permittivity, there is a critical angle of incidence θ_c given by

$$\theta_c = \sin^{-1} \sqrt{\frac{\epsilon_2}{\epsilon_1}} \tag{8.78}$$

for which θ_2 is equal to 90° and $|\Gamma_\perp|$ and $|\Gamma_\parallel| = 1$. For $\theta_1 > \theta_c$, $\sin \theta_2$ becomes greater than 1, $\cos \theta_2$ becomes imaginary, and $\Gamma_\perp$ and $\Gamma_\parallel$ become complex, but with their magnitudes equal to unity, and *total internal reflection* occurs; that is, the time-average power of incident wave is entirely reflected, the boundary condition being satisfied by an evanescent field in medium 2. To explain the evanescent nature, we note with reference to the geometry of Figure 8.17 or Figure 8.18 that

$$\beta_{x2}^2 + \beta_{z2}^2 = \beta_t^2 = \omega^2 \mu_2 \epsilon_2$$

or

$$\beta_{x2}^2 = \omega^2 \mu_2 \epsilon_2 - \beta_{z2}^2$$

For $\theta_1 = \theta_c$, $\beta_{z2} = \beta_{z1} = \omega^2 \mu_1 \epsilon_1 \sin^2 \theta_c = \omega^2 \mu_2 \epsilon_2$, and $\beta_{x2}^2 = 0$. Therefore, for $\theta_1 > \theta_c$, $\beta_{z2} = \beta_{z1} = \omega^2 \mu_1 \epsilon_1 \sin^2 \theta_1 > \omega^2 \mu_2 \epsilon_2$, and $\beta_{x2}^2 < 0$. Thus, β_{x2} should be replaced by $-j\alpha_{x2}$, corresponding to exponential decay of the field in the x-direction without a propagating wave character. The phenomenon of total internal reflection is the fundamental principle of optical waveguides, since if we have a dielectric slab of permittivity ϵ_1 sandwiched between two dielectric media of permittivity $\epsilon_2 < \epsilon_1$, then by launching waves at an angle of incidence greater than the critical angle, it is possible to achieve guided wave propagation within the slab, as we shall learn in the next section.

3. $\Gamma_\perp = 0$ for $\eta_2 \cos \theta_1 = \eta_1 \cos \theta_2$; that is, for

$$\eta_2 \sqrt{1 - \sin^2 \theta_1} = \eta_1 \sqrt{1 - \frac{\mu_1 \epsilon_1}{\mu_2 \epsilon_2} \sin^2 \theta_1}$$

or

$$\sin^2\theta_1 = \frac{\eta_2^2 - \eta_1^2}{\eta_2^2 - \eta_1^2(\mu_1\epsilon_1/\mu_2\epsilon_2)} = \mu_2 \frac{\mu_2 - \mu_1(\epsilon_2/\epsilon_1)}{\mu_2^2 - \mu_1^2} \tag{8.79}$$

For the usual case of transmission between two dielectric materials, that is, for $\mu_2 = \mu_1$ and $\epsilon_2 \neq \epsilon_1$, this equation has no real solution for θ_1, and hence there is no angle of incidence for which the reflection coefficient is zero for the case of perpendicular polarization.

4. $\Gamma_{\parallel} = 0$ for $\eta_2 \cos\theta_2 = \eta_1 \cos\theta_1$; that is, for

$$\eta_2 \sqrt{1 - \frac{\mu_1\epsilon_1}{\mu_2\epsilon_2} \sin^2\theta_1} = \eta_1 \sqrt{1 - \sin^2\theta_1}$$

or

$$\sin^2\theta_1 = \frac{\eta_2^2 - \eta_1^2}{\eta_2^2(\mu_1\epsilon_1/\mu_2\epsilon_2) - \eta_1^2} = \epsilon_2 \frac{(\mu_2/\mu_1)\epsilon_1 - \epsilon_2}{\epsilon_1^2 - \epsilon_2^2} \tag{8.80}$$

If we assume $\mu_2 = \mu_1$, this equation reduces to

$$\sin^2\theta_1 = \frac{\epsilon_2}{\epsilon_1 + \epsilon_2}$$

which then gives

$$\cos^2\theta_1 = 1 - \sin^2\theta_1 = \frac{\epsilon_1}{\epsilon_1 + \epsilon_2}$$

and

$$\tan\theta_1 = \sqrt{\frac{\epsilon_2}{\epsilon_1}}$$

Thus, there exists a value of the angle of incidence θ_p, given by

$$\theta_p = \tan^{-1}\sqrt{\frac{\epsilon_2}{\epsilon_1}} \tag{8.81}$$

for which the reflection coefficient is zero, and hence there is complete transmission for the case of parallel polarization.

5. In view of cases 3 and 4, for an elliptically polarized wave incident on the interface at the angle θ_p, the reflected wave will be linearly polarized perpendicular to the plane of incidence. For this reason, the angle θ_p is known as the *polarizing angle*. It is also known as the *Brewster angle*. The phenomenon associated with the Brewster angle has several applications. An example is in gas lasers in which the discharge tube lying between the mirrors of a Fabry–Perot resonator is sealed

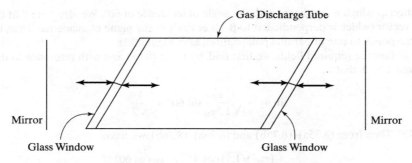

FIGURE 8.19

For illustrating the application of the Brewster angle effect in gas lasers.

by glass windows placed at the Brewster angle, as shown in Figure 8.19, to minimize reflections from the ends of the tube so that the laser behavior is governed by the mirrors external to the tube.

We shall now consider an example.

Example 8.5

A uniform plane wave having the electric field

$$\mathbf{E}_i = E_0\left(\frac{\sqrt{3}}{2}\mathbf{a}_x - \frac{1}{2}\mathbf{a}_z\right)\cos\left[6\pi \times 10^9 t - 10\pi(x + \sqrt{3}z)\right]$$

is incident on the interface between free space and a dielectric medium of $\epsilon = 1.5\epsilon_0$ and $\mu = \mu_0$, as shown in Figure 8.20. We wish to obtain the expressions for the electric fields of the reflected and transmitted waves.

First, we note from the given $\mathbf{E}_i$ that the propagation vector of the incident wave is given by

$$\boldsymbol{\beta}_i = 10\pi(\mathbf{a}_x + \sqrt{3}\mathbf{a}_z) = 20\pi\left(\frac{1}{2}\mathbf{a}_x + \frac{\sqrt{3}}{2}\mathbf{a}_z\right)$$

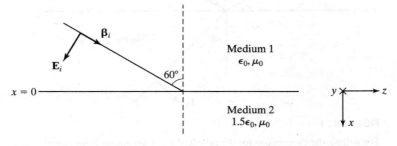

FIGURE 8.20

For Example 8.5.

the direction of which is consistent with the angle of incidence of 60°. We also note that the electric field vector (which is perpendicular to $\boldsymbol{\beta}_i$) is entirely in the plane of incidence. Thus, the situation corresponds to one of parallel polarization, as in Figure 8.18.

To obtain the required fields, we first find, by using (8.63) and with reference to the notation of Figure 8.18, that

$$\sin \theta_2 = \sqrt{\frac{\epsilon_0}{1.5\epsilon_0}} \sin 60° = \frac{1}{\sqrt{2}}$$

or $\theta_2 = 45°$. Then from (8.75a)–(8.75b) and (8.76a)–(8.76b), we have

$$\Gamma_\parallel = \frac{(\eta_0/\sqrt{1.5})\cos 45° - \eta_0 \cos 60°}{(\eta_0/\sqrt{1.5})\cos 45° + \eta_0 \cos 60°}$$

$$= \frac{2 - \sqrt{3}}{2 + \sqrt{3}} = 0.072$$

$$\tau_\parallel = \frac{2(\eta_0/\sqrt{1.5})\cos 60°}{(\eta_0/\sqrt{1.5})\cos 45° + \eta_0 \cos 60°}$$

$$= \frac{2\sqrt{2}}{2 + \sqrt{3}} = 0.758$$

$$\frac{E_r}{E_i} = -0.072$$

$$\frac{E_t}{E_i} = 0.758$$

Finally, noting with the aid of Figure 8.21 that

$$\boldsymbol{\beta}_r = 20\pi\left(-\frac{1}{2}\mathbf{a}_x + \frac{\sqrt{3}}{2}\mathbf{a}_z\right) = 10\pi(-\mathbf{a}_x + \sqrt{3}\mathbf{a}_z)$$

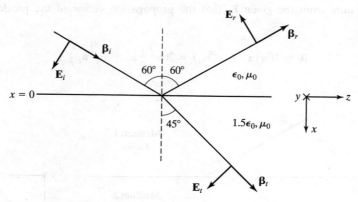

FIGURE 8.21

For writing the expressions for the reflected and transmitted wave electric fields for Example 8.5.

and

$$\boldsymbol{\beta}_t = 20\pi\sqrt{1.5}\left(\frac{1}{\sqrt{2}}\mathbf{a}_x + \frac{1}{\sqrt{2}}\mathbf{a}_z\right) = 10\sqrt{3}\pi(\mathbf{a}_x + \mathbf{a}_z)$$

we write the expressions for the reflected and transmitted wave fields to be

$$\mathbf{E}_r = -0.072E_0\left(\frac{\sqrt{3}}{2}\mathbf{a}_x + \frac{1}{2}\mathbf{a}_z\right)\cos\left[6\pi \times 10^9 t + 10\pi(x - \sqrt{3}z)\right]$$

and

$$\mathbf{E}_t = 0.758E_0\left(\frac{1}{\sqrt{2}}\mathbf{a}_x - \frac{1}{\sqrt{2}}\mathbf{a}_z\right)\cos\left[6\pi \times 10^9 t - 10\sqrt{3}\pi(x + z)\right]$$

Note that for $x = 0$, $E_{zi} + E_{zr} = E_{zt}$ and $E_{xi} + E_{xr} = 1.5E_{xt}$, so that the fields do indeed satisfy the boundary conditions.

8.6 DIELECTRIC SLAB GUIDE

In the preceding section, we learned that for a wave that is incident obliquely from a dielectric medium of permittivity ϵ_1 onto another dielectric medium of permittivity $\epsilon_2 < \epsilon_1$, total internal reflection occurs for angles of incidence θ_i exceeding the critical angle θ_c given by

$$\theta_c = \sin^{-1}\sqrt{\frac{\epsilon_2}{\epsilon_1}} \tag{8.82}$$

where it is assumed that $\mu = \mu_0$ everywhere. In this section, we shall consider the dielectric slab waveguide, which forms the basis for thin-film waveguides, used extensively in integrated optics.

The dielectric slab waveguide consists of a dielectric slab of permittivity ϵ_1, sandwiched between two dielectric media of permittivities less than ϵ_1. For simplicity, we shall consider the symmetric waveguide, that is, one for which the permittivities of the dielectrics on either side of the slab are the same and equal to ϵ_2, as shown in Figure 8.22. Then by launching waves at an angle of incidence $\theta_i > \theta_c$, where θ_c is given by (8.82), it is possible to achieve guided wave propagation within the slab, as shown in the figure. For a given thickness d of the slab and for a given frequency of the waves, there are only discrete values of θ_i for which the guiding can take place. In other words, guiding of a wave of a given frequency is not ensured simply because the condition for total internal reflection is met.

The allowed values of θ_i are dictated by the self-consistency condition, which can be explained with the aid of the construction in Figure 8.23, as follows. If we consider a point A on a given wavefront designated 1 and follow that wavefront as it moves to position $1'$ passing through point B, reflects at the interface $x = d/2$ giving rise to

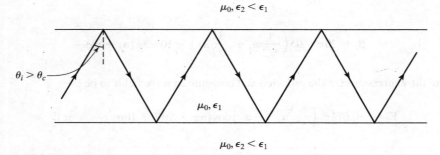

FIGURE 8.22

Total internal reflection in a dielectric slab waveguide.

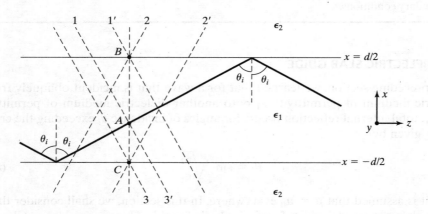

FIGURE 8.23

For explaining the self-consistency condition for waveguiding in a dielectric slab guide.

wavefront designated 2, then moves to position 2′ passing through point C, reflects at the interface $x = -d/2$ giving rise to wavefront designated 3, and finally moves to position 3′ passing through A, then we see that the total phase shift undergone must be equal to an integer multiple of 2π. If λ_0 is the wavelength in free space corresponding to the wave frequency, the self-consistency condition is given by

$$\frac{2\pi\sqrt{\epsilon_{r1}}}{\lambda_0}(AB\cos\theta_i) + \angle\bar{\Gamma}_B + \frac{2\pi\sqrt{\epsilon_{r1}}}{\lambda_0}(BC\cos\theta_i)$$

$$+ \angle\bar{\Gamma}_A + \frac{2\pi\sqrt{\epsilon_{r1}}}{\lambda_0}(CA\cos\theta_i) = 2m\pi, \qquad m = 0, 1, 2, \ldots \qquad (8.83)$$

where $\bar{\Gamma}_A$ and $\bar{\Gamma}_B$ are the reflection coefficients at the interfaces $x = -d/2$ and $x = d/2$, respectively, and $\epsilon_{r1} = \epsilon_1/\epsilon_0$. We recall that under conditions of total internal reflection, the reflection coefficients (8.70a) and (8.75a) become complex with their magnitudes equal to unity. For the symmetric waveguide, $\bar{\Gamma}_A = \bar{\Gamma}_B$. Thus, substituting $\bar{\Gamma}$ for $\bar{\Gamma}_A$ and $\bar{\Gamma}_B$ and $2d$ for $(AB + BC + CA)$, we write (8.83) as

$$\frac{4\pi d\sqrt{\epsilon_{r1}}}{\lambda_0} \cos\theta_i + 2\underline{/\bar{\Gamma}} = 2m\pi, \qquad m = 0, 1, 2, \ldots$$

or

$$\frac{2\pi d\sqrt{\epsilon_{r1}}}{\lambda_0} \cos\theta_i + \underline{/\bar{\Gamma}} = m\pi, \qquad m = 0, 1, 2, \ldots \tag{8.84}$$

To proceed further, we need to distinguish between the cases of perpendicular and parallel polarizations as defined in the preceding section, since the reflection coefficients for the two cases are different. We shall here consider only the case of perpendicular polarization. The situation then corresponds to TE modes, since the electric field has no longitudinal or z-component. Thus, substituting

$$\cos\theta_1 = \cos\theta_i$$

and

$$\cos\theta_2 = \sqrt{1 - \sin^2\theta_2}$$
$$= j\sqrt{\sin^2\theta_2 - 1}$$
$$= j\sqrt{\frac{\epsilon_1}{\epsilon_2}\sin^2\theta_i - 1}$$

in (8.70a), we obtain

$$\bar{\Gamma}_\perp = \frac{\eta_2\cos\theta_i - j\eta_1\sqrt{(\epsilon_1/\epsilon_2)\sin^2\theta_i - 1}}{\eta_2\cos\theta_i + j\eta_1\sqrt{(\epsilon_1/\epsilon_2)\sin^2\theta_i - 1}} \tag{8.85}$$

so that

$$\underline{/\bar{\Gamma}_\perp} = -2\tan^{-1}\frac{\eta_1\sqrt{(\epsilon_1/\epsilon_2)\sin^2\theta_i - 1}}{\eta_2\cos\theta_i}$$
$$= -2\tan^{-1}\frac{\sqrt{\sin^2\theta_i - (\epsilon_2/\epsilon_1)}}{\cos\theta_i} \tag{8.86}$$

Substituting (8.86) into (8.84), we then obtain

$$\frac{2\pi d\sqrt{\epsilon_{r1}}}{\lambda_0}\cos\theta_i - 2\tan^{-1}\frac{\sqrt{\sin^2\theta_i - (\epsilon_2/\epsilon_1)}}{\cos\theta_i} = m\pi, \qquad m = 0, 1, 2, \ldots$$

or

$$\tan\left(\frac{\pi d\sqrt{\epsilon_{r1}}}{\lambda_0}\cos\theta_i - \frac{m\pi}{2}\right) = \frac{\sqrt{\sin^2\theta_i - (\epsilon_2/\epsilon_1)}}{\cos\theta_i}, \qquad m = 0, 1, 2, \ldots$$

or

$$\tan\left[f(\theta_i)\right] = \begin{cases} g(\theta_i), & m = 0, 2, 4, \ldots \\ -\dfrac{1}{g(\theta_i)}, & m = 1, 3, 5, \ldots \end{cases} \qquad (8.87)$$

where

$$f(\theta_i) = \frac{\pi d\sqrt{\epsilon_{r1}}}{\lambda_0}\cos\theta_i \qquad (8.88a)$$

$$g(\theta_i) = \frac{\sqrt{\sin^2\theta_i - (\epsilon_2/\epsilon_1)}}{\cos\theta_i} \qquad (8.88b)$$

Equation (8.87) is the characteristic equation for the guiding of TE waves in the dielectric slab. For given values of ϵ_1, ϵ_2, d, and λ_0, the solutions for θ_i can be obtained by plotting the two sides of (8.87) versus θ_i and finding the points of intersection. The nature of this construction is shown in Figure 8.24. Each solution corresponds to one

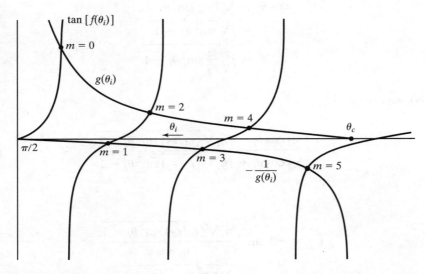

FIGURE 8.24

Graphical construction pertinent to the solution of equation (8.87).

mode. It can be seen from (8.88a) and Figure 8.24 that for a given set of values of ϵ_1 and ϵ_2, fewer solutions are obtained for θ_i as the ratio (d/λ_0) becomes smaller, since the number of branches of the plot of $\tan[f(\theta_i)]$ between $\theta_i = \pi/2$ and $\theta_i = \theta_c$ become fewer. It can also be seen that there is always one solution for a given d, even for arbitrarily low values of (d/λ_0)—that is, for large values of λ_0 or low frequencies.

Alternative to the graphical solution, we can use a computer to solve (8.87) for the allowed values of θ_i for specified values of $\epsilon_{r1}, \epsilon_{r2}, d$, and λ_0. Computed values of θ_i for values of $\epsilon_{r1} = 4, \epsilon_{r2} = 1, d = 10$ mm, and $\lambda_0 = 5$ mm are listed in Table 8.1.

TABLE 8.1   Allowed Values of θ_i for Dielectric Slab Guide Example

m	θ_i (deg)
0	83.42783
1	76.77756
2	69.96263
3	62.87805
4	55.38428
5	47.28283
6	38.30225

Returning now to Figure 8.24, we designate the modes associated with the solutions as TE_m modes, where $m = 0, 1, 2, \ldots$ correspond to the values of m on the plot. We note from the plot that the solution for a given TE_m mode for $m > 1$ does not exist if $f(\theta_c) < m\pi/2$. Therefore, the cutoff condition is given by

$$\frac{\pi d \sqrt{\epsilon_{r1}}}{\lambda_0} \cos \theta_c < \frac{m\pi}{2}$$

$$\frac{\pi d \sqrt{\epsilon_{r1}}}{\lambda_0} \sqrt{1 - \frac{\epsilon_2}{\epsilon_1}} < \frac{m\pi}{2}$$

$$\lambda_0 > \frac{2d \sqrt{\epsilon_{r1} - \epsilon_{r2}}}{m} \tag{8.89}$$

where we have used (8.82). The cutoff frequency is given by

$$f_c = \frac{c}{\lambda_0} = \frac{mc}{2d\sqrt{\epsilon_{r1} - \epsilon_{r2}}}$$

The fundamental mode, TE_0, has no cutoff frequency. Thus,

$$f_c = \frac{mc}{2d\sqrt{\epsilon_{r1} - \epsilon_{r2}}}, \qquad m = 0, 1, 2, \ldots \tag{8.90}$$

Example 8.6

For the symmetric dielectric slab waveguide of Figure 8.23, let $\epsilon_1 = 2.56\epsilon_0$, $\epsilon_2 = \epsilon_0$, and $d = 10\lambda_0$. We wish to find the number of TE modes that can propagate by guidance in the slab.

From (8.90),

$$f_c = \frac{mc}{20\lambda_0\sqrt{2.56-1}}$$

$$= \frac{mf}{24.98}, \qquad m = 0, 1, 2, \ldots$$

Thus, for $m > 24$, $f_c > f$ and the modes are cut off. Therefore, the number of propagating TE modes is 25, corresponding to $m = 0, 1, 2, \ldots, 24$.

The entire discussion for guided waves in the dielectric slab guide can be repeated for TM modes by using $\bar{\Gamma}_{\parallel}$ in the place of $\bar{\Gamma}_{\perp}$ in (8.84) to derive the characteristic equation for guidance. We shall include the derivation as Problem 8.32, and conclude this section with a brief description of an optical fiber, which is a common form of optical waveguide.

An optical fiber, so termed because of its filamentary appearance, consists typically of a core and a cladding, having cylindrical cross sections as shown in Figure 8.25(a). The core is made up of a material of permittivity greater than that of the cladding so that a critical angle exists for waves inside the core incident on the interface between the core and the cladding, and hence waveguiding is made possible in the core by total internal reflection. The phenomenon may be visualized by considering a longitudinal cross section of the fiber through its axis, shown in Figure 8.25(b), and comparing it with that of the slab waveguide, shown in Figure 8.22. Although the cladding is not essential for the purpose of waveguiding in the core, since the permittivity of the core material is greater than that of free space, it serves two useful purposes: (a) It avoids scattering and field distortion by the supporting structure of the fiber, since the field decays exponentially outside the core, and hence is negligible outside the cladding. (b) It allows single-mode propagation for a larger value of the radius of the core than permitted in the absence of the cladding.

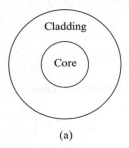

FIGURE 8.25

(a) Transverse and (b) longitudinal cross sections of an optical fiber.

SUMMARY

In this chapter, we studied the principles of waveguides. To introduce the waveguiding phenomenon, we first learned how to write the expressions for the electric and magnetic fields of a uniform plane wave propagating in an arbitrary direction with respect to the coordinate axes. These expressions are given by

$$\mathbf{E} = \mathbf{E}_0 \cos{(\omega t - \boldsymbol{\beta} \cdot \mathbf{r} + \phi_0)}$$
$$\mathbf{H} = \mathbf{H}_0 \cos{(\omega t - \boldsymbol{\beta} \cdot \mathbf{r} + \phi_0)}$$

where $\boldsymbol{\beta}$ and $\mathbf{r}$ are the propagation and position vectors given by

$$\boldsymbol{\beta} = \beta_x \mathbf{a}_x + \beta_y \mathbf{a}_y + \beta_z \mathbf{a}_z$$
$$\mathbf{r} = x\mathbf{a}_x + y\mathbf{a}_y + z\mathbf{a}_z$$

and ϕ_0 is the phase of the wave at the origin at $t = 0$. The magnitude of $\boldsymbol{\beta}$ is equal to $\omega\sqrt{\mu\epsilon}$, the phase constant along the direction of propagation of the wave. The direction of $\boldsymbol{\beta}$ is the direction of propagation of the wave. We learned that

$$\mathbf{E}_0 \cdot \boldsymbol{\beta} = 0$$
$$\mathbf{H}_0 \cdot \boldsymbol{\beta} = 0$$
$$\mathbf{E}_0 \cdot \mathbf{H}_0 = 0$$

that is, $\mathbf{E}_0$, $\mathbf{H}_0$, and $\boldsymbol{\beta}$ are mutually perpendicular, and that

$$\frac{|\mathbf{E}_0|}{|\mathbf{H}_0|} = \eta = \sqrt{\frac{\mu}{\epsilon}}$$

Also, since $\mathbf{E} \times \mathbf{H}$ should be directed along the propagation vector $\boldsymbol{\beta}$, it then follows that

$$\mathbf{H} = \frac{1}{\omega\mu}\boldsymbol{\beta} \times \mathbf{E}$$

The quantities β_x, β_y, and β_z are the phase constants along the x-, y-, and z-axes, respectively. The apparent wavelengths and the apparent phase velocities along the coordinate axes are given, respectively, by

$$\lambda_i = \frac{2\pi}{\beta_i}, \qquad i = x, y, z$$
$$v_{pi} = \frac{\omega}{\beta_i}, \qquad i = x, y, z$$

By considering the superposition of two uniform plane waves propagating at an angle to each other and placing two perfect conductors in appropriate planes such that the boundary condition of zero tangential electric field is satisfied, we introduced the parallel-plate waveguide. We learned that the composite wave is a transverse electric wave, or TE wave, since the electric field is entirely transverse to the direction of time-average power flow, that is, the guide axis, but the magnetic field is not. In terms of the uniform plane wave propagation, the phenomenon is one of waves bouncing obliquely between the conductors as they progress down the guide. For a fixed spacing a between the conductors of the guide, waves of different frequencies bounce obliquely at different angles such that the spacing a is equal to an integer, say, m number of one-half apparent wavelengths normal to the plates and hence the fields have m number of one-half-sinusoidal variations normal to the plates. These are said to correspond to $TE_{m,0}$ modes, where the subscript 0 implies no variations of the fields in the direction parallel to the plates and transverse to the guide axis. When the frequency is such that the spacing a is equal to m one-half wavelengths, the waves bounce normally to the plates without the feeling of being guided along the axis, thereby leading to the cutoff condition. Thus, the cutoff wavelengths corresponding to $TE_{m,0}$ modes are given by

$$\lambda_c = \frac{2a}{m}$$

and the cutoff frequencies are given by

$$f_c = \frac{v_p}{\lambda_c} = \frac{m}{2a\sqrt{\mu\epsilon}}$$

A given frequency signal can propagate in all modes for which $\lambda < \lambda_c$ or $f > f_c$. For the propagating range of frequencies, the wavelength along the guide axis, that is, the guide wavelength, and the phase velocity along the guide axis are given, respectively, by

$$\lambda_g = \frac{\lambda}{\sqrt{1 - (\lambda/\lambda_c)^2}} = \frac{\lambda}{\sqrt{1 - (f_c/f)^2}}$$

$$v_{pz} = \frac{v_p}{\sqrt{1 - (\lambda/\lambda_c)^2}} = \frac{v_p}{\sqrt{1 - (f_c/f)^2}}$$

We discussed the phenomenon of dispersion arising from the frequency dependence of the phase velocity along the guide axis, and we introduced the concept of group velocity. Group velocity is the velocity with which the envelope of a narrow-band modulated signal travels in the dispersive channel, and hence it is the velocity with which the information is transmitted. It is given by

$$v_g = \frac{d\omega}{d\beta_z} = v_p\sqrt{1 - \left(\frac{f_c}{f}\right)^2}$$

where β_z is the phase constant along the guide axis.

We extended the treatment of the parallel-plate waveguide to the rectangular waveguide, which is a metallic pipe of rectangular cross section. By considering a rectangular waveguide of cross-sectional dimensions a and b, we discussed transverse electric or TE modes as well as transverse magnetic or TM modes, and learned that while $TE_{m,n}$ modes can include values of m or n equal to zero, $TM_{m,n}$ modes require that both m and n be nonzero, where m and n refer to the number of one-half sinusoidal variations of the fields along the dimensions a and b, respectively. The cutoff wavelengths for the $TE_{m,n}$ or $TM_{m,n}$ modes are given by

$$\lambda_c = \frac{1}{\sqrt{(m/2a)^2 + (n/2b)^2}}$$

The mode that has the largest cutoff wavelength or the lowest cutoff frequency is the dominant mode, which here is the $TE_{1,0}$ mode. Waveguides are generally designed to transmit only the dominant mode.

By placing perfect conductors in two transverse planes of a rectangular waveguide separated by an integer multiple of one-half the guide wavelength, we introduced the cavity resonator, which is the microwave counterpart of the lumped parameter resonant circuit encountered in low-frequency circuit theory. For a rectangular cavity resonator having dimensions a, b, and d, the frequencies of oscillation for the $TE_{m,n,l}$ or $TM_{m,n,l}$ modes are given by

$$f_{osc} = \frac{1}{\sqrt{\mu\epsilon}} \sqrt{\left(\frac{m}{2a}\right)^2 + \left(\frac{n}{2b}\right)^2 + \left(\frac{l}{2d}\right)^2}$$

where l refers to the number of one-half sinusoidal variations of the fields along the dimension d.

We then considered oblique incidence of a uniform plane wave on the boundary between two perfect dielectric media. We derived the *laws of reflection and refraction*, given, respectively, by

$$\theta_r = \theta_i$$

$$\theta_t = \sin^{-1}\left(\sqrt{\frac{\mu_1\epsilon_1}{\mu_2\epsilon_2}} \sin \theta_i\right)$$

where θ_i, θ_r, and θ_t are the angles of incidence, reflection, and transmission, respectively, of a uniform plane wave incident from medium 1 (ϵ_1, μ_1) onto medium 2 (ϵ_2, μ_2). The law of refraction is also known as *Snell's law*. We then derived the expressions for the reflection and transmission coefficients for the cases of perpendicular and parallel polarizations. An examination of these expressions revealed the following, under the assumption of $\mu_1 = \mu_2$: (1) for incidence from a medium of higher permittivity onto one of lower permittivity, there is a critical angle of incidence given by

$$\theta_c = \sin^{-1}\sqrt{\frac{\epsilon_2}{\epsilon_1}}$$

beyond which total internal reflection occurs, and (2) for the case of parallel polarization, there is an angle of incidence, known as the *Brewster angle* and given by

$$\theta_p = \tan^{-1}\sqrt{\frac{\epsilon_2}{\epsilon_1}}$$

for which the reflection coefficient is zero.

Next, we introduced the dielectric slab waveguide, consisting of a dielectric slab of permittivity ϵ_1 sandwiched between two dielectric media of permittivities $\epsilon_2 < \epsilon_1$.

We learned that by launching waves at an angle of incidence θ_i greater than the critical angle for total internal reflection, it is possible to achieve guided wave propagation within the slab. For a given frequency, several modes are possible corresponding to values of θ_i that satisfy the self-consistency condition associated with the bouncing waves. We derived the characteristic equation for computing these values of θ_i for the TE case and discussed its solution. The modes are designated TE$_m$ modes and their cutoff frequencies are given by

$$f_c = \frac{mc}{2d\sqrt{\epsilon_{r1} - \epsilon_{r2}}}, \qquad m = 0, 1, 2, \ldots$$

where d is the thickness of the slab. The fundamental mode, TE$_0$, has no cutoff frequency. We concluded the discussion with a description of the optical fiber.

REVIEW QUESTIONS

8.1. What is the propagation vector? Interpret the significance of its magnitude and direction.

8.2. Discuss how the phase constants along the coordinate axes are less than the phase constant along the direction of propagation of a uniform plane wave propagating in an arbitrary direction.

8.3. Write the expressions for the electric and magnetic fields of a uniform plane wave propagating in an arbitrary direction and list all the conditions to be satisfied by the electric field, magnetic field, and propagation vectors.

8.4. What are apparent wavelengths? Why are they longer than the wavelength along the direction of propagation?

8.5. What are apparent phase velocities? Why are they greater than the phase velocity along the direction of propagation?

8.6. Discuss how the superposition of two uniform plane waves propagating at an angle to each other gives rise to a composite wave consisting of standing waves traveling bodily transverse to the standing waves.

8.7. What is a transverse electric wave? Discuss the reasoning behind the nomenclature TE$_{m,0}$ modes.

8.8. How would you characterize a transverse magnetic wave?

8.9. Compare the phenomenon of guiding of uniform plane waves in a parallel-plate waveguide with that in a parallel-plate transmission line.

8.10. Discuss how the cutoff condition arises in a waveguide.

8.11. Explain the relationship between the cutoff wavelength and the spacing between the plates of a parallel-plate waveguide based on the phenomenon at cutoff.

8.12. Is the cutoff wavelength dependent on the dielectric in the waveguide? Is the cutoff frequency dependent on the dielectric in the waveguide?

8.13. What is guide wavelength?

8.14. Provide a physical explanation for the frequency dependence of the phase velocity along the guide axis.

8.15. Discuss the phenomenon of dispersion.

8.16. Discuss the concept of group velocity with the aid of an example.

8.17. What is a dispersion diagram? Explain how the phase and group velocities can be determined from a dispersion diagram.

8.18. When is it meaningful to attribute a group velocity to a signal comprised of more than two frequencies? Why?

8.19. Discuss the propagation of a narrow-band amplitude modulated signal in a dispersive channel.

8.20. Discuss the nomenclature associated with the modes of propagation in a rectangular waveguide.

8.21. Explain the relationship between the cutoff wavelength and the dimensions of a rectangular waveguide based on the phenomenon at cutoff.

8.22. Why can there be no transverse magnetic modes having no variations for the fields along one of the dimensions of a rectangular waveguide?

8.23. What is meant by the dominant mode? Why are waveguides designed so that they propagate only the dominant mode?

8.24. Why is the dimension b of a rectangular waveguide generally chosen to be less than or equal to one-half the dimension a?

8.25. What is a cavity resonator?

8.26. How do the dimensions of a rectangular cavity resonator determine the frequencies of oscillation of the resonator?

8.27. Discuss the condition required to be satisfied by the incident, reflected, and transmitted waves at the interface between two dielectric media.

8.28. What is Snell's law?

8.29. What is meant by the plane of incidence? Distinguish between the two different linear polarizations pertinent to the derivation of the reflection and transmission coefficients for oblique incidence on a dielectric interface.

8.30. Briefly discuss the determination of the Fresnel reflection and transmission coefficients for an obliquely incident wave on a dielectric interface.

8.31. What is total internal reflection? Discuss the nature of the reflection coefficient and the manner in which the boundary condition is satisfied for an angle of incidence greater than the critical angle for total internal reflection.

8.32. What is the Brewster angle? What is the polarization of the reflected wave for an elliptically polarized wave incident on a dielectric interface at the Brewster angle? Discuss an application of the Brewster angle effect.

8.33. Discuss the principle of optical waveguides by considering the dielectric slab waveguide.

8.34. Explain the self-consistency condition for waveguiding in a dielectric slab waveguide.

8.35. Discuss the dependence of the number of propagating modes in a dielectric slab waveguide on the ratio of the thickness d of the dielectric slab to the wavelength λ_0.

8.36. Considering TE modes in a dielectric slab guide, specify the fundamental mode and discuss the associated cutoff condition.

8.37. Compare the phenomenon at cutoff in a metallic waveguide with that at cutoff in an optical waveguide.

8.38. Provide a brief description of an optical fiber.

PROBLEMS

8.1. Assuming the x- and y-axes to be directed eastward and northward, respectively, find the expression for the propagation vector of a uniform plane wave of frequency 15 MHz in free space propagating in the direction 30° north of east.

8.2. The propagation vector of a uniform plane wave in a perfect dielectric medium having $\epsilon = 4.5\epsilon_0$ and $\mu = \mu_0$ is given by

$$\boldsymbol{\beta} = 2\pi(3\mathbf{a}_x + 4\mathbf{a}_y + 5\mathbf{a}_z)$$

Find (a) the apparent wavelengths and (b) the apparent phase velocities, along the coordinate axes.

8.3. For a uniform plane wave propagating in free space, the apparent phase velocities along the x- and y-directions are found to be $6\sqrt{2} \times 10^8$ m/s and $2\sqrt{3} \times 10^8$ m/s, respectively. Find the direction of propagation of the wave.

8.4. The electric field vector of a uniform plane wave propagating in a perfect dielectric medium having $\epsilon = 9\epsilon_0$ and $\mu = \mu_0$ is given by

$$\mathbf{E} = 10(-\mathbf{a}_x - 2\sqrt{3}\mathbf{a}_y + \sqrt{3}\mathbf{a}_z) \cos [16\pi \times 10^6 t - 0.04\pi(\sqrt{3}x - 2y - 3z)]$$

Find (a) the frequency, (b) the direction of propagation, (c) the wavelength along the direction of propagation, (d) the apparent wavelengths along the x-, y-, and z-axes, and (e) the apparent phase velocities along the x-, y-, and z-axes.

8.5. Given

$$\mathbf{E} = 10\mathbf{a}_x \cos [6\pi \times 10^7 t - 0.1\pi(y + \sqrt{3}z)]$$

(a) Determine if the given $\mathbf{E}$ represents the electric field of a uniform plane wave propagating in free space. (b) If the answer to part (a) is *yes*, find the corresponding magnetic field vector $\mathbf{H}$.

8.6. Given

$$\mathbf{E} = (\mathbf{a}_x - 2\mathbf{a}_y - \sqrt{3}\mathbf{a}_z) \cos [15\pi \times 10^6 t - 0.05\pi(\sqrt{3}x + z)]$$

$$\mathbf{H} = \frac{1}{60\pi}(\mathbf{a}_x + 2\mathbf{a}_y - \sqrt{3}\mathbf{a}_z) \cos [15\pi \times 10^6 t - 0.05\pi(\sqrt{3}x + z)]$$

(a) Perform all the necessary tests and determine if these fields represent a uniform plane wave propagating in a perfect dielectric medium. (b) Find the permittivity and the permeability of the medium.

8.7. Two equal-amplitude uniform plane waves of frequency 25 MHz and having their electric fields along the y-direction propagate along the directions $\mathbf{a}_z$ and $\frac{1}{2}(\sqrt{3}\mathbf{a}_x + \mathbf{a}_z)$ in free space. (a) Find the direction of propagation of the composite wave. (b) Find the wavelength along the direction of propagation and the wavelength transverse to the direction of propagation of the composite wave.

8.8. Show that $\langle \sin^2 (\omega t - \beta z \sin \theta) \rangle$ and $\langle \sin 2(\omega t - \beta z \sin \theta) \rangle$ are equal to 1/2 and zero, respectively.

8.9. Find the spacing a for a parallel-plate waveguide having a dielectric of $\epsilon = 9\epsilon_0$ and $\mu = \mu_0$ such that 6000 MHz is 20 percent above the cutoff frequency of the dominant mode, that is, the mode with the lowest cutoff frequency.

8.10. The dimension a of a parallel-plate waveguide filled with a dielectric having $\epsilon = 4\epsilon_0$ and $\mu = \mu_0$ is 4 cm. Determine the propagating $\mathrm{TE}_{m,0}$ modes for a wave of frequency 6000 MHz. For each propagating mode, find f_c, θ, and λ_g.

8.11. The spacing a between the plates of a parallel-plate waveguide is equal to 5 cm. The dielectric between the plates is free space. If a generator of fundamental frequency

1800 MHz and rich in harmonics excites the waveguide, find all frequencies that propagate in $TE_{1,0}$ mode only.

8.12. The electric and magnetic fields of the composite wave resulting from the superposition of two uniform plane waves are given by

$$\mathbf{E} = E_{x0} \cos \beta_x x \cos (\omega t - \beta_z z) \mathbf{a}_x$$
$$+ E_{z0} \sin \beta_x x \sin (\omega t - \beta_z z) \mathbf{a}_z$$
$$\mathbf{H} = H_{y0} \cos \beta_x x \cos (\omega t - \beta_z z) \mathbf{a}_y$$

(a) Find the time-average Poynting vector. (b) Discuss the nature of the composite wave.

8.13. Transverse electric modes are excited in an air dielectric parallel-plate waveguide of dimension $a = 5$ cm by setting up at its mouth a field distribution having

$$\mathbf{E} = 10 \, (\sin 20\pi x + 0.5 \sin 60\pi x) \sin 10^{10}\pi t \, \mathbf{a}_y$$

Determine the propagating mode(s) and obtain the expression for the electric field of the propagating wave.

8.14. For the two-train example of Figure 8.8, find the group velocity if the speed of train numbered B is (a) 36 m/s and (b) 40 m/s, instead of 30 m/s. Discuss your results with the aid of sketches.

8.15. Find the velocity with which the group of two frequencies 2400 MHz and 2500 MHz travels in a parallel-plate waveguide of dimension $a = 2.5$ cm and having a perfect dielectric of $\epsilon = 9\epsilon_0$ and $\mu = \mu_0$.

8.16. For a narrow-band amplitude modulated signal having the carrier frequency 5000 MHz propagating in an air dielectric parallel-plate waveguide of dimension $a = 5$ cm, find the velocity with which the modulation envelope travels.

8.17. For an $\omega - \beta_z$ relationship given by

$$\omega = \omega_0 + k\beta_z^2$$

where ω_0 and k are positive constants, find the phase and group velocities for (a) $\omega = 1.5\omega_0$, (b) $\omega = 2\omega_0$, and (c) $\omega = 3\omega_0$.

8.18. By considering the parallel-plate waveguide, show that a point on the obliquely bouncing wavefront, traveling with the phase velocity along the oblique direction, progresses parallel to the guide axis with the group velocity.

8.19. For an air dielectric rectangular waveguide of dimensions $a = 3$ cm and $b = 1.5$ cm, find all propagating modes for $f = 12,000$ MHz.

8.20. For a rectangular waveguide of dimensions $a = 5$ cm and $b = 5/3$ cm, and having a dielectric of $\epsilon = 9\epsilon_0$ and $\mu = \mu_0$, find all propagating modes for $f = 2500$ MHz.

8.21. For $f = 3000$ MHz, find the dimensions a and b of an air dielectric rectangular waveguide such that $TE_{1,0}$ mode propagates with a 30 percent safety factor ($f = 1.30f_c$) but also such that the frequency is 30 percent below the cutoff frequency of the next higher order mode.

8.22. For an air dielectric rectangular cavity resonator having the dimensions $a = 2.5$ cm, $b = 2$ cm, and $d = 5$ cm, find the five lowest frequencies of oscillation. Identify the mode(s) for each frequency.

8.23. For a rectangular cavity resonator having the dimensions $a = b = d = 2$ cm, and filled with a dielectric of $\epsilon = 9\epsilon_0$ and $\mu = \mu_0$, find the three lowest frequencies of oscillation. Identify the mode(s) for each frequency.

8.24. In Figure 8.16, let $\epsilon_1 = 4\epsilon_0$, $\epsilon_2 = 9\epsilon_0$, and $\mu_1 = \mu_2 = \mu_0$. (a) For $\theta_i = 30°$, find θ_t. (b) Is there a critical angle of incidence for which $\theta_t = 90°$?

8.25. In Figure 8.16, let $\epsilon_1 = 4\epsilon_0$, $\epsilon_2 = 2.25\epsilon_0$, and $\mu_1 = \mu_2 = \mu_0$. (a) For $\theta_i = 30°$, find θ_t. (b) Find the value of the critical angle of incidence θ_c, for which $\theta_t = 90°$.

8.26. In Example 8.5, assume that

$$E_i = E_0(\mathbf{a}_x - \mathbf{a}_z) \cos [6\pi \times 10^8 t - \sqrt{2}\pi(x + z)]$$

and the angle of incidence is 45°. Obtain the expressions for the electric fields of the reflected and transmitted waves.

8.27. Repeat Problem 8.26 for

$$\mathbf{E}_i = E_0 \, \mathbf{a}_y \cos [6\pi \times 10^8 t - \sqrt{2}\pi(x + z)]$$

8.28. In Example 8.5, assume that the permittivity ϵ_2 of medium 2 is unknown and that

$$\mathbf{E}_i = E_0\left(\frac{\sqrt{3}}{2} \mathbf{a}_x - \frac{1}{2} \mathbf{a}_z\right) \cos [6\pi \times 10^9 t - 10\pi(x + \sqrt{3}z)]$$
$$+ E_0 \mathbf{a}_y \sin [6\pi \times 10^9 t - 10\pi(x + \sqrt{3}z)]$$

(a) Find the value of ϵ_2 for which the reflected wave is linearly polarized.
(b) For the value of ϵ_2 found in (a), find the expressions for the reflected and transmitted wave electric fields.

8.29. A thin-film waveguide employed in integrated optics consists of a substrate on which a thin film of refractive index (c/v_p) greater than that of the substrate is deposited. The medium above the film is air. For relative permittivities of the substrate and the film equal to 2.25 and 2.4, respectively, find the minimum bouncing angle of total internally reflected waves in the film. Assume $\mu = \mu_0$ for both substrate and film.

8.30. For a symmetric dielectric slab waveguide, $\epsilon_1 = 2.25\epsilon_0$ and $\epsilon_2 = \epsilon_0$. (a) Find the number of propagating TE modes for $d/\lambda_0 = 10$. (b) Find the maximum value of d/λ_0 for which the waveguide supports only one TE mode.

8.31. Design a symmetric dielectric slab waveguide, with $\epsilon_{r1} = 2.25$ and $\epsilon_{r2} = 2.13$, by finding the value of d/λ_0 such that the TE_1 mode operates at 20% above its cutoff frequency.

8.32. Consider the derivation of the characteristic equation for guiding of waves in the symmetric dielectric slab waveguide for the case of parallel polarization, which corresponds to TM modes. Noting that in Figure 8.18, $H_r/H_i = E_r/E_i = -\Gamma_\parallel$, where $\Gamma_\parallel$ is given by (8.75a), show that the characteristic equation is given by

$$\tan [f(\theta_i)] = \begin{cases} g(\theta_i), & m = 0, 2, 4, \ldots \\ -\dfrac{1}{g(\theta_i)}, & m = 1, 3, 5, \ldots \end{cases}$$

where

$$f(\theta_i) = \frac{\pi d \sqrt{\epsilon_{r1}}}{\lambda_0} \cos \theta_i$$

$$g(\theta_i) = \frac{\sqrt{\sin^2 \theta_i - (\epsilon_2/\epsilon_1)}}{(\epsilon_2/\epsilon_1) \cos \theta_i}$$

Antenna Basics

In the preceding chapters, we studied the principles of propagation and transmission of electromagnetic waves. The remaining important topic pertinent to electromagnetic wave phenomena is radiation of electromagnetic waves. We have, in fact, touched on the principle of radiation of electromagnetic waves in Chapter 4 when we derived the electromagnetic field due to the infinite plane sheet of sinusoidally time-varying, spatially uniform current density. We learned that the current sheet gives rise to uniform plane waves *radiating* away from the sheet to either side of it. We pointed out at that time that the infinite plane current sheet is, however, an idealized, hypothetical source. With the experience gained thus far in our study of the elements of engineering electromagnetics, we are now in a position to learn the principles of radiation from physical antennas, which is our goal in this chapter.

We shall begin the chapter with the derivation of the electromagnetic field due to an elemental wire antenna, known as the *Hertzian dipole*. After studying the radiation characteristics of the Hertzian dipole, we shall consider the example of a half-wave dipole to illustrate the use of superposition to represent an arbitrary wire antenna as a series of Hertzian dipoles in order to determine its radiation fields. We shall also discuss the principles of arrays of physical antennas and the concept of image antennas to take into account ground effects. Finally, we shall briefly consider the receiving properties of antennas and learn of their reciprocity with the radiating properties.

9.1 HERTZIAN DIPOLE

The Hertzian dipole is an elemental antenna consisting of an infinitesimally long piece of wire carrying an alternating current $I(t)$, as shown in Figure 9.1. To maintain the current flow in the wire, we postulate two point charges $Q_1(t)$ and $Q_2(t)$ terminating the wire at its two ends, so that the law of conservation of charge is satisfied. Thus, if

$$I(t) = I_0 \cos \omega t \tag{9.1}$$

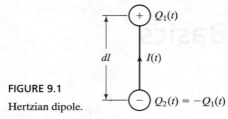

FIGURE 9.1

Hertzian dipole.

then

$$\frac{dQ_1}{dt} = I(t) = I_0 \cos \omega t \tag{9.2a}$$

$$\frac{dQ_2}{dt} = -I(t) = -I_0 \cos \omega t \tag{9.2b}$$

and

$$Q_1(t) = \frac{I_0}{\omega} \sin \omega t \tag{9.3a}$$

$$Q_2(t) = -\frac{I_0}{\omega} \sin \omega t = -Q_1(t) \tag{9.3b}$$

The time variations of I, Q_1, and Q_2, given by (9.1), (9.3a), and (9.3b), respectively, are illustrated by the curves and the series of sketches for the dipoles in Figure 9.2, corresponding to one complete period. The different sizes of the arrows associated with the dipoles denote the different strengths of the current, whereas the number of the plus or minus signs is indicative of the strength of the charges.

To determine the electromagnetic field due to the Hertzian dipole, we shall employ an intuitive approach based upon the knowledge gained in the previous chapters, as follows: From the application of what we have learned in Chapter 1, we can obtain the expressions for the electric and magnetic fields due to the point charges and the current element, respectively, associated with the Hertzian dipole, assuming that the fields follow exactly the time-variations of the charges and the current. These expressions do not, however, take into account the fact that time-varying electric and magnetic fields give rise to wave propagation. Hence, we shall extend them from considerations of our knowledge of wave propagation and then check if the resulting solutions satisfy Maxwell's equations. If they do not, we will then have to modify them so that they do satisfy Maxwell's equations and at the same time reduce to the originally derived expressions in the region where wave propagation effects are small, that is, at distances from the dipole that are small compared to a wavelength.

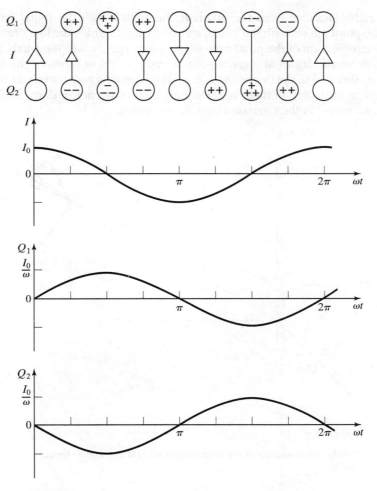

FIGURE 9.2

Time variations of charges and current associated with the Hertzian dipole.

To follow the approach outlined in the preceding paragraph, we locate the dipole at the origin with the current directed along the z-axis, as shown in Figure 9.3, and derive first the expressions for the fields by applying the simple laws learned in Sections 1.5 and 1.6. The symmetry associated with the problem is such that it is simpler to use a spherical coordinate system. Hence, if the reader is not already familiar with the spherical coordinate system, it is suggested that Appendix A be read at this stage. To review briefly, a point in the spherical coordinate system is defined by the intersection of a sphere centered at the origin, a cone having its apex at the origin and its surface symmetrical about the z-axis, and a plane containing the z-axis. Thus, the coordinates for a given point, say P, are r, the radial distance from the origin, θ, the angle which the radial line from the origin to the point makes with the z-axis, and ϕ, the angle which the line drawn from the origin to the projection of the point onto the xy-plane makes with the x-axis, as shown in Figure 9.3. A vector drawn at a given point is represented in terms of the unit vectors $\mathbf{a}_r$, $\mathbf{a}_\theta$, and $\mathbf{a}_\phi$ directed in the increasing r, θ, and ϕ directions, respectively, at that point. It is important to note that all three of these unit vectors are not uniform unlike the unit vectors $\mathbf{a}_x$, $\mathbf{a}_y$, and $\mathbf{a}_z$ in the Cartesian coordinate system.

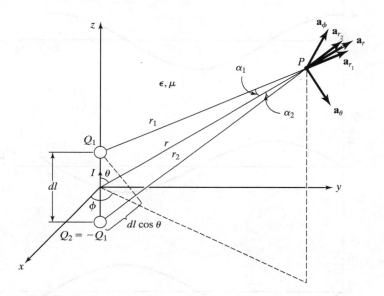

FIGURE 9.3

For the determination of the electromagnetic field due to the Hertzian dipole.

Now using the expression for the electric field due to a point charge given by (1.52), we can write the electric field at point P due to the arrangement of the two point charges Q_1 and $-Q_1$ in Figure 9.3 to be

$$\mathbf{E} = \frac{Q_1}{4\pi\epsilon r_1^2}\mathbf{a}_{r_1} - \frac{Q_1}{4\pi\epsilon r_2^2}\mathbf{a}_{r_2} \tag{9.4}$$

where r_1 and r_2 are the distances from Q_1 to P and $Q_2(=-Q_1)$ to P, respectively, and $\mathbf{a}_{r_1}$ and $\mathbf{a}_{r_2}$ are unit vectors directed along the lines from Q_1 to P and Q_2 to P, respectively, as shown in Figure 9.3. Noting that

$$\mathbf{a}_{r_1} = \cos\alpha_1\,\mathbf{a}_r + \sin\alpha_1\,\mathbf{a}_\theta \tag{9.5a}$$

$$\mathbf{a}_{r_2} = \cos\alpha_2\,\mathbf{a}_r - \sin\alpha_2\,\mathbf{a}_\theta \tag{9.5b}$$

we obtain the r and θ components of the electric field at P to be

$$E_r = \frac{Q_1}{4\pi\epsilon} \left(\frac{\cos\alpha_1}{r_1^2} - \frac{\cos\alpha_2}{r_2^2} \right) \tag{9.6a}$$

$$E_\theta = \frac{Q_1}{4\pi\epsilon} \left(\frac{\sin\alpha_1}{r_1^2} + \frac{\sin\alpha_2}{r_2^2} \right) \tag{9.6b}$$

For infinitesimal value of the length dl of the current element, that is, for $dl \ll r$,

$$\left(\frac{\cos\alpha_1}{r_1^2} - \frac{\cos\alpha_2}{r_2^2} \right) \approx \frac{1}{r_1^2} - \frac{1}{r_2^2}$$

$$= \frac{(r_2 - r_1)(r_2 + r_1)}{r_1^2 r_2^2} \approx \frac{(dl\cos\theta)\,2r}{r^4}$$

$$= \frac{2\,dl\cos\theta}{r^3} \tag{9.7a}$$

and

$$\left(\frac{\sin\alpha_1}{r_1^2} + \frac{\sin\alpha_2}{r_2^2} \right) \approx \frac{2\sin\alpha_1}{r^2}$$

$$\approx \frac{dl\sin\theta}{r^3} \tag{9.7b}$$

where we have also used the approximations that for $dl \ll r$, $(r_2 - r_1) \approx dl\cos\theta$ and $\sin\alpha_1 \approx [(dl/2)\sin\theta]/r$. These are, of course, exact in the limit that $dl \to 0$. Substituting (9.7a) and (9.7b) in (9.6a) and (9.6b), respectively, we obtain the electric field at point P due to the arrangement of the two point charges to be given by

$$\mathbf{E} = \frac{Q_1\,dl}{4\pi\epsilon r^3} (2\cos\theta\,\mathbf{a}_r + \sin\theta\,\mathbf{a}_\theta) \tag{9.8}$$

Note that $Q_1\,dl$ is the electric dipole moment associated with the Hertzian dipole.

Using the Biot–Savart law given by (1.68), we can write the magnetic field at point P due to the infinitesimal current element in Figure 9.3 to be

$$\mathbf{H} = \frac{\mathbf{B}}{\mu} = \frac{I\,dl\,\mathbf{a}_z \times \mathbf{a}_r}{4\pi r^2}$$

$$= \frac{I\,dl\sin\theta}{4\pi r^2}\,\mathbf{a}_\phi \tag{9.9}$$

To extend the expressions for $\mathbf{E}$ and $\mathbf{H}$ given by (9.8) and (9.9), respectively, we observe that when the charges and current vary with time, the fields also vary with time giving rise to wave propagation. The effect of a given time-variation of the source quantity is therefore felt at a point in space not instantaneously but only after a time lag. This time lag is equal to the time it takes for the wave to propagate from the source point to the observation point, that is, r/v_p, or $\beta r/\omega$, where $v_p(=1/\sqrt{\mu\epsilon})$ and $\beta(=\omega\sqrt{\mu\epsilon})$ are the phase velocity and the phase constant, respectively. Thus, for

$$Q_1 = \frac{I_0}{\omega} \sin \omega t \tag{9.10}$$

$$I = I_0 \cos \omega t \tag{9.11}$$

we would intuitively expect the fields at point P to be given by

$$\mathbf{E} = \frac{[(I_0/\omega) \sin \omega(t - \beta r/\omega)] \, dl}{4\pi\epsilon r^3}(2 \cos \theta \, \mathbf{a}_r + \sin \theta \, \mathbf{a}_\theta)$$

$$= \frac{I_0 \, dl \sin (\omega t - \beta r)}{4\pi\epsilon\omega r^3}(2 \cos \theta \, \mathbf{a}_r + \sin \theta \, \mathbf{a}_\theta) \tag{9.12a}$$

$$\mathbf{H} = \frac{[I_0 \cos \omega(t - \beta r/\omega)] \, dl}{4\pi r^2} \sin \theta \, \mathbf{a}_\phi$$

$$= \frac{I_0 \, dl \cos (\omega t - \beta r)}{4\pi r^2} \sin \theta \, \mathbf{a}_\phi \tag{9.12b}$$

There is, however, one thing wrong with our intuitive expectation of the fields due to the Hertzian dipole! The fields do not satisfy Maxwell's curl equations

$$\nabla \times \mathbf{E} = -\frac{\partial \mathbf{B}}{\partial t} = -\mu \frac{\partial \mathbf{H}}{\partial t} \tag{9.13a}$$

$$\nabla \times \mathbf{H} = \mathbf{J} + \frac{\partial \mathbf{D}}{\partial t} = \epsilon \frac{\partial \mathbf{E}}{\partial t} \tag{9.13b}$$

(where we have set $\mathbf{J} = 0$ in view of the perfect dielectric medium). For example, let us try the curl equation for $\mathbf{H}$. First, we note from Appendix B that the expansion for the curl of a vector in spherical coordinates is

$$\nabla \times \mathbf{A} = \frac{1}{r \sin \theta}\left[\frac{\partial}{\partial \theta}(A_\phi \sin \theta) - \frac{\partial A_\theta}{\partial \phi}\right]\mathbf{a}_r$$

$$+ \frac{1}{r}\left[\frac{1}{\sin \theta}\frac{\partial A_r}{\partial \phi} - \frac{\partial}{\partial r}(rA_\phi)\right]\mathbf{a}_\theta$$

$$+ \frac{1}{r}\left[\frac{\partial}{\partial r}(rA_\theta) - \frac{\partial A_r}{\partial \theta}\right]\mathbf{a}_\phi \tag{9.14}$$

Thus,

$$\nabla \times \mathbf{H} = \frac{1}{r \sin \theta} \frac{\partial}{\partial \theta} \left[\frac{I_0 \, dl \cos (\omega t - \beta r)}{4\pi r^2} \sin^2 \theta \right] \mathbf{a}_r$$

$$- \frac{1}{r} \frac{\partial}{\partial r} \left[\frac{I_0 \, dl \cos (\omega t - \beta r)}{4\pi r} \sin \theta \right] \mathbf{a}_\theta$$

$$= \frac{I_0 \, dl \cos (\omega t - \beta r)}{4\pi r^3} (2 \cos \theta \, \mathbf{a}_r + \sin \theta \, \mathbf{a}_\theta)$$

$$- \frac{\beta I_0 \, dl \sin (\omega t - \beta r)}{4\pi r^2} \sin \theta \, \mathbf{a}_\theta$$

$$= \epsilon \frac{\partial \mathbf{E}}{\partial t} - \frac{\beta I_0 \, dl \sin (\omega t - \beta r)}{4\pi r^2} \sin \theta \, \mathbf{a}_\theta$$

$$\neq \epsilon \frac{\partial \mathbf{E}}{\partial t} \tag{9.15}$$

The reason behind the discrepancy associated with the expressions for the fields due to the Hertzian dipole can be understood by recalling that in Section 4.6 we learned from considerations of the Poynting vector that the fields far from a physical antenna vary inversely with the radial distance away from the antenna. The expressions we have derived do not contain inverse distance dependent terms and hence they are not complete, thereby causing the discrepancy. The complete field expressions must contain terms involving $1/r$ in addition to those in (9.12a) and (9.12b). Since for small r, $1/r \ll 1/r^2 \ll 1/r^3$, the addition of terms involving $1/r$ and containing $\sin \theta$ to (9.12a) and (9.12b) would still maintain the fields in the region close to the dipole to be predominantly the same as those given by (9.12a) and (9.12b), while making the $1/r$ terms predominant for large r, since for large r, $1/r \gg 1/r^2 \gg 1/r^3$.

Thus, let us modify the expression for **H** given by (9.12b) by adding a second term containing $1/r$ in the following manner:

$$\mathbf{H} = \frac{I_0 \, dl \sin \theta}{4\pi} \left[\frac{\cos (\omega t - \beta r)}{r^2} + \frac{A \cos (\omega t - \beta r + \delta)}{r} \right] \mathbf{a}_\phi \tag{9.16}$$

where A and δ are constants to be determined. Then from Maxwell's curl equation for **H**, given by (9.13b), we have

$$\epsilon \frac{\partial \mathbf{E}}{\partial t} = \nabla \times \mathbf{H} = \frac{1}{r \sin \theta} \frac{\partial}{\partial \theta} (H_\phi \sin \theta) \mathbf{a}_r - \frac{1}{r} \frac{\partial}{\partial r} (r H_\phi) \mathbf{a}_\theta$$

$$= \frac{2 I_0 \, dl \cos \theta}{4\pi} \left[\frac{\cos (\omega t - \beta r)}{r^3} + \frac{A \cos (\omega t - \beta r + \delta)}{r^2} \right] \mathbf{a}_r$$

$$+ \frac{I_0 \, dl \sin \theta}{4\pi} \left[\frac{\cos (\omega t - \beta r)}{r^3} - \frac{\beta \sin (\omega t - \beta r)}{r^2} \right.$$

$$\left. - \frac{A\beta \sin (\omega t - \beta r + \delta)}{r} \right] \mathbf{a}_\theta \tag{9.17}$$

$$\mathbf{E} = \frac{2I_0 \, dl \cos \theta}{4\pi\epsilon\omega} \left[\frac{\sin (\omega t - \beta r)}{r^3} + \frac{A \sin (\omega t - \beta r + \delta)}{r^2} \right] \mathbf{a}_r$$

$$+ \frac{I_0 \, dl \sin \theta}{4\pi\epsilon\omega} \left[\frac{\sin (\omega t - \beta r)}{r^3} + \frac{\beta \cos (\omega t - \beta r)}{r^2} \right.$$

$$\left. + \frac{A\beta \cos (\omega t - \beta r + \delta)}{r} \right] \mathbf{a}_\theta \tag{9.18}$$

Now, from Maxwell's curl equation for $\mathbf{E}$ given by (9.13a), we have

$$\mu \frac{\partial \mathbf{H}}{\partial t} = -\nabla \times \mathbf{E} = -\frac{1}{r} \left[\frac{\partial}{\partial r} (rE_\theta) - \frac{\partial E_r}{\partial \theta} \right] \mathbf{a}_\phi$$

$$= \frac{I_0 dl \sin \theta}{4\pi\epsilon\omega} \left[\frac{2\beta \cos (\omega t - \beta r)}{r^3} - \frac{2A \sin (\omega t - \beta r + \delta)}{r^3} \right.$$

$$\left. - \frac{\beta^2 \sin (\omega t - \beta r)}{r^2} - \frac{A\beta^2 \sin (\omega t - \beta r + \delta)}{r} \right] \mathbf{a}_\phi \tag{9.19}$$

$$\mathbf{H} = \frac{I_0 \, dl \sin \theta}{4\pi} \left[\frac{2 \sin (\omega t - \beta r)}{\beta r^3} + \frac{2A \cos (\omega t - \beta r + \delta)}{\beta^2 r^3} \right.$$

$$\left. + \frac{\cos (\omega t - \beta r)}{r^2} + \frac{A \cos (\omega t - \beta r + \delta)}{r} \right] \mathbf{a}_\phi \tag{9.20}$$

We, however, have to rule out the $1/r^3$ terms in (9.20), since for small r these terms are more predominant than the $1/r^2$ dependence required by (9.12b). Equation (9.20) will then also be consistent with (9.16) from which we derived (9.18) and then (9.20). Thus, we set

$$\frac{2 \sin (\omega t - \beta r)}{\beta r^3} + \frac{2A \cos (\omega t - \beta r + \delta)}{\beta^2 r^3} = 0 \tag{9.21}$$

which gives us

$$\delta = \frac{\pi}{2} \tag{9.22a}$$

$$A = \beta \tag{9.22b}$$

Substituting (9.22a) and (9.22b) in (9.18) and (9.20), we then have

$$\mathbf{E} = \frac{2I_0 \, dl \cos \theta}{4\pi\epsilon\omega} \left[\frac{\sin (\omega t - \beta r)}{r^3} + \frac{\beta \cos (\omega t - \beta r)}{r^2} \right] \mathbf{a}_r$$

$$+ \frac{I_0 \, dl \sin \theta}{4\pi\epsilon\omega} \left[\frac{\sin (\omega t - \beta r)}{r^3} + \frac{\beta \cos (\omega t - \beta r)}{r^2} \right.$$

$$\left. - \frac{\beta^2 \sin (\omega t - \beta r)}{r} \right] \mathbf{a}_\theta \tag{9.23a}$$

$$\mathbf{H} = \frac{I_0 \, dl \sin \theta}{4\pi} \left[\frac{\cos (\omega t - \beta r)}{r^2} - \frac{\beta \sin (\omega t - \beta r)}{r} \right] \mathbf{a}_\phi \qquad (9.23b)$$

These expressions for **E** and **H** satisfy both of Maxwell's curl equations, reduce to (9.12a) and (9.12b), respectively, for small r ($\beta r \ll 1$), and they vary inversely with r for large r ($\beta r \gg 1$). They represent the complete electromagnetic field due to the Hertzian dipole.

9.2   RADIATION RESISTANCE AND DIRECTIVITY

In the previous section, we derived the expressions for the complete electromagnetic field due to the Hertzian dipole. These expressions look very complicated. Fortunately, it is seldom necessary to work with the complete field expressions because one is often interested in the field far from the dipole, which is governed predominantly by the terms involving $1/r$. We, however, had to derive the complete field in order to obtain the amplitude and phase of these $1/r$ terms relative to the amplitude and phase of the current in the Hertzian dipole, since these terms alone do not satisfy Maxwell's equations. Furthermore, by going through the exercise, we learned how to solve a difficult problem through intuitive extension and reasoning based on previously gained knowledge.

Thus from (9.23a) and (9.23b), we find that for a Hertzian dipole of length dl oriented along the z-axis and carrying current

$$I = I_0 \cos \omega t \qquad (9.24)$$

the electric and magnetic fields at values of r far from the dipole are given by

$$\mathbf{E} = -\frac{\beta^2 I_0 \, dl \sin \theta}{4\pi \epsilon \omega r} \sin (\omega t - \beta r) \, \mathbf{a}_\theta$$

$$= -\frac{\eta \beta I_0 \, dl \sin \theta}{4\pi r} \sin (\omega t - \beta r) \, \mathbf{a}_\theta \qquad (9.25a)$$

$$\mathbf{H} = -\frac{\beta I_0 \, dl \sin \theta}{4\pi r} \sin (\omega t - \beta r) \, \mathbf{a}_\phi \qquad (9.25b)$$

These fields are known as the *radiation fields*, since they are the components of the total fields that contribute to the time-average radiated power away from the dipole (see Problem 9.6). Before we discuss the nature of these fields, let us find out quantitatively what we mean by *far from the dipole*. To do this, we look at the expression for the complete magnetic field given by (9.23b) and note that the ratio of the amplitudes of the $1/r^2$ and $1/r$ terms is equal to $1/\beta r$. Hence for $\beta r \gg 1$, or $r \gg \lambda/2\pi$, the $1/r^2$ term is negligible compared to the $1/r$ term. Thus, even at a distance of a few wavelengths from the dipole, the fields are predominantly radiation fields.

Returning now to the expressions for the radiation fields given by (9.25a) and (9.25b), we note that at any given point, (a) the electric field (E_θ), the magnetic field (H_ϕ), and the direction of propagation (r) are mutually perpendicular, and (b) the ratio of E_θ to H_ϕ is equal to η, the intrinsic impedance of the medium, which are characteristic of uniform plane waves. The phase of the field, however, is uniform over the surfaces $r =$ constant, that is, spherical surfaces centered at the dipole, whereas the amplitude of the field is uniform over surfaces $(\sin\theta)/r = $ constant. Hence, the fields are only locally uniform plane waves, that is, over any infinitesimal area normal to the r-direction at a given point.

The Poynting vector due to the radiation fields is given by

$$
\begin{aligned}
\mathbf{P} &= \mathbf{E} \times \mathbf{H} \\
&= E_\theta \mathbf{a}_\theta \times H_\phi \mathbf{a}_\phi = E_\theta H_\phi \, \mathbf{a}_r \\
&= \frac{\eta \beta^2 I_0^2 \,(dl)^2 \sin^2\theta}{16\pi^2 r^2} \sin^2(\omega t - \beta r) \, \mathbf{a}_r
\end{aligned}
\tag{9.26}
$$

By evaluating the surface integral of the Poynting vector over any surface enclosing the dipole, we can find the power flow out of that surface, that is, the power *radiated* by the dipole. For convenience in evaluating the surface integral, we choose the spherical surface of radius r and centered at the dipole, as shown in Figure 9.4. Thus, noting that the differential surface area on the spherical surface is $(r\,d\theta)(r\sin\theta\,d\phi)\mathbf{a}_r$ or $r^2\sin\theta\,d\theta\,d\phi\,\mathbf{a}_r$, we obtain the instantaneous power radiated to be

$$
\begin{aligned}
P_{\text{rad}} &= \int_{\theta=0}^{\pi} \int_{\phi=0}^{2\pi} \mathbf{P} \cdot r^2 \sin\theta \, d\theta \, d\phi \, \mathbf{a}_r \\
&= \int_{\theta=0}^{\pi} \int_{\phi=0}^{2\pi} \frac{\eta \beta^2 I_0^2 \,(dl)^2 \sin^3\theta}{16\pi^2} \sin^2(\omega t - \beta r) \, d\theta \, d\phi \\
&= \frac{\eta \beta^2 I_0^2 \,(dl)^2}{8\pi} \sin^2(\omega t - \beta r) \int_{\theta=0}^{\pi} \sin^3\theta \, d\theta \\
&= \frac{\eta \beta^2 I_0^2 \,(dl)^2}{6\pi} \sin^2(\omega t - \beta r) \\
&= \frac{2\pi \eta I_0^2}{3} \left(\frac{dl}{\lambda}\right)^2 \sin^2(\omega t - \beta r)
\end{aligned}
\tag{9.27}
$$

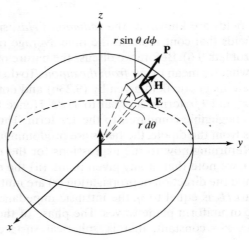

FIGURE 9.4

For computing the power radiated by the Hertzian dipole.

The time-average power radiated by the dipole, that is, the average of P_{rad} over one period of the current variation, is

$$\langle P_{rad} \rangle = \frac{2\pi\eta I_0^2}{3} \left(\frac{dl}{\lambda} \right)^2 \langle \sin^2(\omega t - \beta r) \rangle$$

$$= \frac{\pi\eta I_0^2}{3} \left(\frac{dl}{\lambda} \right)^2$$

$$= \frac{1}{2} I_0^2 \left[\frac{2\pi\eta}{3} \left(\frac{dl}{\lambda} \right)^2 \right] \tag{9.28}$$

We now define a quantity known as the *radiation resistance* of the antenna, denoted by the symbol R_{rad}, as the value of a fictitious resistor that dissipates the same amount of time-average power as that radiated by the antenna when a current of the same peak amplitude as that in the antenna is passed through it. Recalling that the average power dissipated in a resistor R when a current $I_0 \cos \omega t$ is passed through it is $\frac{1}{2} I_0^2 R$, we note from (9.28) that the radiation resistance of the Hertzian dipole is

$$R_{rad} = \frac{2\pi\eta}{3} \left(\frac{dl}{\lambda} \right)^2 \, \Omega \tag{9.29}$$

For free space, $\eta = \eta_0 = 120\pi \, \Omega$, and

$$R_{rad} = 80\pi^2 \left(\frac{dl}{\lambda} \right)^2 \, \Omega \tag{9.30}$$

As a numerical example, for (dl/λ) equal to 0.01, $R_{rad} = 80\pi^2(0.01)^2 = 0.08 \, \Omega$. Thus, for a current of peak amplitude 1 A, the time-average radiated power is equal to 0.04 W. This indicates that a Hertzian dipole of length 0.01λ is not a very effective radiator.

We note from (9.29) that the radiation resistance and hence the radiated power are proportional to the square of the electrical length, that is, the physical length expressed in terms of wavelength, of the dipole. The result given by (9.29) is, however, valid only for small values of dl/λ, since if dl/λ is not small, the amplitude of the current along the antenna can no longer be uniform and its variation must be taken into account in deriving the radiation fields and hence the radiation resistance. We shall do this in the following section for a half-wave dipole, that is, for a dipole of length equal to $\lambda/2$.

Let us now examine the directional characteristics of the radiation from the Hertzian dipole. We note from (9.25a) and (9.25b) that, for a constant r, the amplitude of the fields is proportional to $\sin \theta$. Similarly, we note from (9.26) that, for a constant r, the power density is proportional to $\sin^2 \theta$. Thus, an observer wandering on the surface of an imaginary sphere centered at the dipole views different amplitudes of the fields and of the power density at different points on the surface. The situation is illustrated in Figure 9.5(a) for the power density by attaching to different points on the spherical surface vectors having lengths proportional to the Poynting vectors at those points. It can be seen that the power density is largest for $\theta = \pi/2$, that is, in the plane normal to

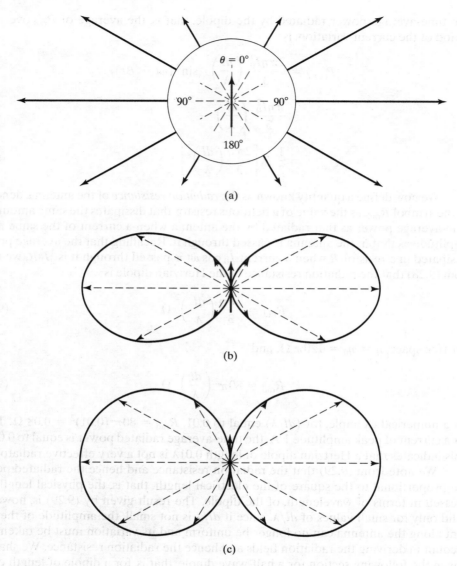

FIGURE 9.5

The directional characteristics of radiation from the Hertzian dipole.

the axis of the dipole, and decreases continuously toward the axis of the dipole, becoming zero along the axis.

It is customary to depict the radiation characteristic by means of a *radiation pattern*, as shown in Figure 9.5(b), which can be imagined to be obtained by shrinking the radius of the spherical surface in Figure 9.5(a) to zero with the Poynting vectors attached to it and then joining the tips of the Poynting vectors. Thus, the distance from the dipole point to a point on the radiation pattern is proportional to the power density in the direction of that point. Similarly, the radiation pattern for the fields can be drawn as shown in Figure 9.5(c), based upon the $\sin \theta$ dependence of the fields. In view of the independence of the fields from ϕ, the patterns of Figure 9.5(b)–(c) are valid for any plane containing the axis of the dipole. In fact, the three-dimensional radiation patterns can be imagined to be the figures obtained by revolving these patterns about the dipole axis. For a general case, the radiation may also depend on ϕ, and hence it will be necessary to draw a radiation pattern for the $\theta = \pi/2$ plane. Here, this pattern is merely a circle centered at the dipole.

We now define a parameter known as the *directivity* of the antenna, denoted by the symbol D, as the ratio of the maximum power density radiated by the antenna to the average power density. To elaborate on the definition of D, imagine that we take the power radiated by the antenna and distribute it equally in all directions by shortening some of the vectors in Figure 9.5(a) and lengthening the others so that they all have equal lengths. The pattern then becomes nondirectional and the power density, which is the same in all directions, will be less than the maximum power density of the original pattern. Obviously, the more directional the radiation pattern of an antenna is, the greater is the directivity.

From (9.26), we obtain the maximum power density radiated by the Hertzian dipole to be

$$[P_r]_{\text{max}} = \frac{\eta \beta^2 I_0^2 (dl)^2 [\sin^2 \theta]_{\text{max}}}{16\pi^2 r^2} \sin^2 (\omega t - \beta r)$$

$$= \frac{\eta \beta^2 I_0^2 (dl)^2}{16\pi^2 r^2} \sin^2 (\omega t - \beta r) \tag{9.31}$$

By dividing the radiated power given by (9.27) by the surface area $4\pi r^2$ of the sphere of radius r, we obtain the average power density to be

$$[P_r]_{\text{av}} = \frac{P_{\text{rad}}}{4\pi r^2} = \frac{\eta \beta^2 I_0^2 (dl)^2}{24\pi^2 r^2} \sin^2 (\omega t - \beta r) \tag{9.32}$$

Thus, the directivity of the Hertzian dipole is given by

$$D = \frac{[P_r]_{\text{max}}}{[P_r]_{\text{av}}} = 1.5 \tag{9.33}$$

To generalize the computation of directivity for an arbitrary radiation pattern, let us consider

$$P_r = \frac{P_0 \sin^2 (\omega t - \beta r)}{r^2} f(\theta, \phi) \tag{9.34}$$

where P_0 is a constant, and $f(\theta, \phi)$ is the power density pattern. Then

$$[P_r]_{\text{max}} = \frac{P_0 \sin^2 (\omega t - \beta r)}{r^2} [f(\theta, \phi)]_{\text{max}}$$

$$[P_r]_{\text{av}} = \frac{P_{\text{rad}}}{4\pi r^2}$$

$$= \frac{1}{4\pi r^2} \int_{\theta=0}^{2\pi} \int_{\phi=0}^{\pi} \frac{P_0 \sin^2 (\omega t - \beta r)}{r^2} f(\theta, \phi) \, \mathbf{a}_r \cdot r^2 \sin \theta \, d\theta \, d\phi \, \mathbf{a}_r$$

$$= \frac{P_0 \sin^2 (\omega t - \beta r)}{4\pi r^2} \int_{\theta=0}^{\pi} \int_{\phi=0}^{2\pi} f(\theta, \phi) \sin \theta \, d\theta \, d\phi$$

$$D = 4\pi \frac{[f(\theta, \phi)]_{\text{max}}}{\int_{\theta=0}^{\pi} \int_{\phi=0}^{2\pi} f(\theta, \phi) \sin \theta \, d\theta \, d\phi} \tag{9.35}$$

Example 9.1

Let us compute the directivity corresponding to the power density pattern function $f(\theta, \phi) = \sin^2 \theta \cos^2 \theta$.
From (9.35),

$$D = 4\pi \frac{[\sin^2 \theta \cos^2 \theta]_{\text{max}}}{\int_{\theta=0}^{\pi} \int_{\phi=0}^{2\pi} \sin^3 \theta \cos^2 \theta \, d\theta \, d\phi}$$

$$= 4\pi \frac{\left[\frac{1}{4} \sin^2 2\theta\right]_{\text{max}}}{2\pi \int_{\theta=0}^{\pi} (\sin^3 \theta - \sin^5 \theta) \, d\theta}$$

$$= \frac{1}{2} \frac{1}{(4/3) - (16/15)}$$

$$= 1\frac{7}{8}$$

The ratio of the power density radiated by the antenna as a function of direction to the average power density is given by $Df(\theta, \phi)$. This quantity is known as the *directive gain of the antenna.* Another useful parameter is the power gain of the antenna, which takes into account the ohmic power losses in the antenna. It is denoted by the symbol G and is proportional to the directive gain, the proportionality factor being the power efficiency of the antenna, which is the ratio of the power radiated by the antenna to the power supplied to it by the source of excitation.

9.3 HALF-WAVE DIPOLE

In the previous section, we found the radiation fields due to a Hertzian dipole, which is an elemental antenna of infinitesimal length. If we now have an antenna of any length having a specified current distribution, we can divide it into a series of Hertzian dipoles and by applying superposition can find the radiation fields for that antenna. We shall illustrate this procedure in this section by considering the half-wave dipole, which is a commonly used form of antenna.

The half-wave dipole is a center-fed, straight wire antenna of length L equal to $\lambda/2$ and having the current distribution

$$I(z) = I_0 \cos \frac{\pi z}{L} \cos \omega t \qquad \text{for} -\frac{L}{2} < z < \frac{L}{2} \qquad (9.36)$$

where the dipole is assumed to be oriented along the z-axis with its center at the origin, as shown in Figure 9.6(a). As can be seen from Figure 9.6(a), the amplitude of the current distribution varies cosinusoidally along the antenna with zeros at the ends and maximum at the center. To see how this distribution comes about, the half-wave dipole may be imagined to be the evolution of an open-circuited transmission line with the conductors folded perpendicularly to the line at points $\lambda/4$ from the end of the line. The current standing wave pattern for an open-circuited line is shown in Figure 9.6(b). It consists of zero current at the open circuit and maximum current at $\lambda/4$ from the open circuit, that is, at points a and a'. Hence, it can be seen that when the conductors are folded perpendicularly to the line at a and a', the half-wave dipole shown in Figure 9.6(a) results.

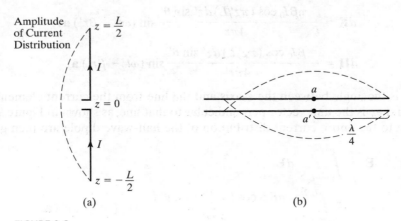

FIGURE 9.6

(a) Half-wave dipole. (b) Open-circuited transmission line for illustrating the evolution of the half-wave dipole.

Now, to find the radiation field due to the half-wave dipole, we divide it into a number of Hertzian dipoles, each of length dz', as shown in Figure 9.7. If we consider one of these dipoles situated at distance z' from the origin, then from (9.36) the current

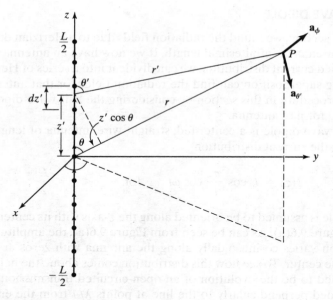

FIGURE 9.7

For the determination of the radiation field due to the half-wave dipole.

in this dipole is $I_0 \cos(\pi z'/L) \cos \omega t$. From (9.25a) and (9.25b), the radiation fields due to this dipole at point P situated at distance r' from it are given by

$$dE = -\frac{\eta \beta I_0 \cos(\pi z'/L)\, dz' \sin \theta'}{4\pi r'} \sin(\omega t - \beta r')\, \mathbf{a}_{\theta'} \qquad (9.37a)$$

$$d\mathbf{H} = -\frac{\beta I_0 \cos(\pi z'/L)\, dz' \sin \theta'}{4\pi r'} \sin(\omega t - \beta r')\, \mathbf{a}_{\phi} \qquad (9.37b)$$

where θ' is the angle between the z-axis and the line from the current element to the point P and $\mathbf{a}_{\theta'}$ is the unit vector perpendicular to that line, as shown in Figure 9.7. The fields due to the entire current distribution of the half-wave dipole are then given by

$$\mathbf{E} = \int_{z'=-L/2}^{L/2} d\mathbf{E}$$

$$= -\int_{z'=-L/2}^{L/2} \frac{\eta \beta I_0 \cos(\pi z'/L) \sin \theta'\, dz'}{4\pi r'} \sin(\omega t - \beta r')\, \mathbf{a}_{\theta'} \qquad (9.38a)$$

$$\mathbf{H} = \int_{z'=-L/2}^{L/2} d\mathbf{H}$$

$$= -\int_{z'=-L/2}^{L/2} \frac{\beta I_0 \cos(\pi z'/L) \sin \theta'\, dz'}{4\pi r'} \sin(\omega t - \beta r')\, \mathbf{a}_{\phi} \qquad (9.38b)$$

where r', θ', and $\mathbf{a}_{\theta'}$ are functions of z'.

For radiation fields, r' is at least equal to several wavelengths and hence $\gg L$. We can therefore set $\mathbf{a}_{\theta'} \approx \mathbf{a}_\theta$ and $\theta' \approx \theta$, since they do not vary significantly for $-L/2 < z' < L/2$. We can also set $r' \approx r$ in the amplitude factors for the same reason, but for r' in the phase factors we substitute $r - z' \cos \theta$, since $\sin (\omega t - \beta r') = \sin (\omega t - \pi r'/L)$ can vary appreciably over the range $-L/2 < z' < L/2$. Thus, we have

$$\mathbf{E} = E_\theta \mathbf{a}_\theta$$

where

$$E_\theta = -\int_{z'=-L/2}^{L/2} \frac{\eta \beta I_0 \cos (\pi z'/L) \sin \theta}{4\pi r} \sin (\omega t - \beta r + \beta z' \cos \theta)\, dz'$$

$$= -\frac{\eta (\pi/L) I_0 \sin \theta}{4\pi r} \int_{z'=-L/2}^{L/2} \cos \frac{\pi z'}{L} \sin \left(\omega t - \frac{\pi}{L}r + \frac{\pi}{L}z' \cos \theta \right) dz'$$

$$= -\frac{\eta I_0}{2\pi r} \frac{\cos [(\pi/2) \cos \theta]}{\sin \theta} \sin \left(\omega t - \frac{\pi}{L}r \right) \tag{9.39a}$$

Similarly,

$$\mathbf{H} = H_\phi \mathbf{a}_\phi$$

where

$$H_\phi = -\frac{I_0}{2\pi r} \frac{\cos [(\pi/2) \cos \theta]}{\sin \theta} \sin \left(\omega t - \frac{\pi}{L}r \right) \tag{9.39b}$$

The Poynting vector due to the radiation fields of the half-wave dipole is given by

$$\mathbf{P} = \mathbf{E} \times \mathbf{H} = E_\theta H_\phi \mathbf{a}_r$$

$$= \frac{\eta I_0^2}{4\pi^2 r^2} \frac{\cos^2 [(\pi/2) \cos \theta]}{\sin^2 \theta} \sin^2 \left(\omega t - \frac{\pi}{L}r \right) \mathbf{a}_r \tag{9.40}$$

The power radiated by the half-wave dipole is given by

$$P_{\text{rad}} = \int_{\theta=0}^{\pi} \int_{\phi=0}^{2\pi} \mathbf{P} \cdot r^2 \sin \theta \, d\theta \, d\phi \, \mathbf{a}_r$$

$$= \int_{\theta=0}^{\pi} \int_{\phi=0}^{2\pi} \frac{\eta I_0^2}{4\pi^2} \frac{\cos^2 [(\pi/2) \cos \theta]}{\sin \theta} \sin^2 \left(\omega t - \frac{\pi}{L}r \right) d\theta \, d\phi$$

$$= \frac{\eta I_0^2}{\pi} \sin^2 \left(\omega t - \frac{\pi}{L}r \right) \int_{\theta=0}^{\pi/2} \frac{\cos^2 [(\pi/2) \cos \theta]}{\sin \theta} \, d\theta$$

$$= \frac{0.609 \eta I_0^2}{\pi} \sin^2 \left(\omega t - \frac{\pi}{L}r \right) \tag{9.41}$$

The time-average radiated power is

$$\langle P_{rad} \rangle = \frac{0.609\eta I_0^2}{\pi} \left\langle \sin^2 \left(\omega t - \frac{\pi}{L} r \right) \right\rangle$$

$$= \frac{1}{2} I_0^2 \left(\frac{0.609\eta}{\pi} \right) \qquad (9.42)$$

Thus, the radiation resistance of the half-wave dipole is

$$R_{rad} = \frac{0.609\eta}{\pi} \ \Omega \qquad (9.43)$$

For free space, $\eta = \eta_0 = 120\pi \ \Omega$, and

$$R_{rad} = 0.609 \times 120 = 73 \ \Omega \qquad (9.44)$$

Turning our attention now to the directional characteristics of the half-wave dipole, we note from (9.39a) and (9.39b) that the radiation pattern for the fields is $\left[\cos \left(\frac{\pi}{2} \cos \theta \right) \right] \Big/ \sin \theta$ whereas for the power density, it is $\left[\cos^2 \left(\frac{\pi}{2} \cos \theta \right) \right] \Big/ \sin^2 \theta$. These patterns, which are sketched in Figure 9.8(a)–(b), are slightly more directional than the corresponding patterns for the Hertzian dipole. To find the directivity of the half-wave dipole, we note from (9.40) that the maximum power density is

$$[P_r]_{max} = \frac{\eta I_0^2}{4\pi^2 r^2} \left\{ \frac{\cos^2 \left[(\pi/2) \cos \theta \right]}{\sin^2 \theta} \right\}_{max} \sin^2 \left(\omega t - \frac{\pi}{L} r \right)$$

$$= \frac{\eta I_0^2}{4\pi^2 r^2} \sin^2 \left(\omega t - \frac{\pi}{L} r \right) \qquad (9.45)$$

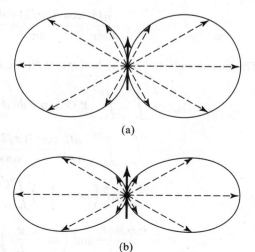

(a)

(b)

FIGURE 9.8

Radiation patterns for (a) the fields and (b) the power density due to the half-wave dipole.

The average power density obtained by dividing P_{rad} by $4\pi r^2$ is

$$[P_r]_{\text{av}} = \frac{0.609\eta I_0^2}{4\pi^2 r^2} \sin^2\left(\omega t - \frac{\pi}{L}r\right) \tag{9.46}$$

Thus, the directivity of the half-wave dipole is given by

$$D = \frac{[P_r]_{\text{max}}}{[P_r]_{\text{av}}} = \frac{1}{0.609} = 1.642 \tag{9.47}$$

9.4 ANTENNA ARRAYS

In Section 4.5, we illustrated the principle of an antenna array by considering an array of two parallel, infinite plane current sheets of uniform densities. We learned that by appropriately choosing the spacing between the current sheets and the amplitudes and phases of the current densities, a desired radiation characteristic can be obtained. The infinite plane current sheet is, however, a hypothetical antenna for which the fields are truly uniform plane waves propagating in the one dimension normal to the sheet. Now that we have gained some knowledge of physical antennas, in this section we shall consider arrays of such antennas.

The simplest array we can consider consists of two Hertzian dipoles, oriented parallel to the z-axis and situated at points on the x-axis on either side of and equidistant from the origin, as shown in Figure 9.9. We shall consider the amplitudes of the currents in the two dipoles to be equal, but we shall allow a phase difference α between them. Thus, if $I_1(t)$ and $I_2(t)$ are the currents in the dipoles situated at $(d/2, 0, 0)$ and $(-d/2, 0, 0)$, respectively, then

$$I_1 = I_0 \cos\left(\omega t + \frac{\alpha}{2}\right) \tag{9.48a}$$

$$I_2 = I_0 \cos\left(\omega t - \frac{\alpha}{2}\right) \tag{9.48b}$$

For simplicity, we shall consider a point P in the xz-plane and compute the field at that point due to the array of the two dipoles. To do this, we note from (9.25a) that the electric field intensities at the point P due to the individual dipoles are given by

$$\mathbf{E}_1 = -\frac{\eta\beta I_0 dl \sin\theta_1}{4\pi r_1} \sin\left(\omega t - \beta r_1 + \frac{\alpha}{2}\right)\mathbf{a}_{\theta_1} \tag{9.49a}$$

$$\mathbf{E}_2 = -\frac{\eta\beta I_0 dl \sin\theta_2}{4\pi r_2} \sin\left(\omega t - \beta r_2 - \frac{\alpha}{2}\right)\mathbf{a}_{\theta_2} \tag{9.49b}$$

where $\theta_1, \theta_2, r_1, r_2, \mathbf{a}_{\theta_1}$, and $\mathbf{a}_{\theta_2}$ are as shown in Figure 9.9.

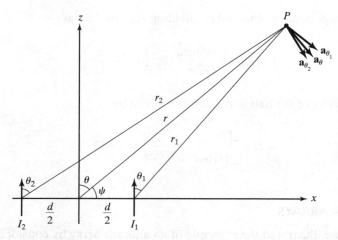

FIGURE 9.9

For computing the radiation field due to an array of two Hertzian dipoles.

For $r \gg d$, that is, for points far from the array, which is the region of interest, we can set $\theta_1 \approx \theta_2 \approx \theta$ and $\mathbf{a}_{\theta_1} \approx \mathbf{a}_{\theta_2} \approx \mathbf{a}_\theta$. Also, we can set $r_1 \approx r_2 \approx r$ in the amplitude factors, but for r_1 and r_2 in the phase factors, we substitute

$$r_1 \approx r - \frac{d}{2}\cos\psi \tag{9.50a}$$

$$r_2 \approx r + \frac{d}{2}\cos\psi \tag{9.50b}$$

where ψ is the angle made by the line form the origin to P with the axis of the array, that is, the x-axis, as shown in Figure 9.9. Thus we obtain the resultant field to be

$$
\begin{aligned}
\mathbf{E} &= \mathbf{E}_1 + \mathbf{E}_2 \\
&= -\frac{\eta\beta I_0\, dl\, \sin\theta}{4\pi r}\left[\sin\left(\omega t - \beta r + \frac{\beta d}{2}\cos\psi + \frac{\alpha}{2}\right)\right. \\
&\qquad \left. + \sin\left(\omega t - \beta r - \frac{\beta d}{2}\cos\psi - \frac{\alpha}{2}\right)\right]\mathbf{a}_\theta \\
&= -\frac{2\eta\beta I_0\, dl\, \sin\theta}{4\pi r}\cos\left(\frac{\beta d\cos\psi + \alpha}{2}\right)\sin(\omega t - \beta r)\mathbf{a}_\theta
\end{aligned}
\tag{9.51}
$$

Comparing (9.51) with the expression for the electric field at P due to a single dipole situated at the origin, we note that the resultant field of the array is simply equal to the single dipole field multiplied by the factor $2\cos\left(\dfrac{\beta d\cos\psi + \alpha}{2}\right)$, known as the *array factor*. Thus the radiation pattern of the resultant field is given by the product of $\sin\theta$, which is the radiation pattern of the single dipole field, and $\left|\cos\left(\dfrac{\beta d\cos\psi + \alpha}{2}\right)\right|$, which is the radiation pattern of the array if the antennas were isotropic. We shall call these three patterns the *resultant pattern*, the *unit pattern*, and the *group pattern*, respectively. It is apparent that the group pattern is independent of the nature of the individual antennas as long as they have the same spacing and carry currents having the same relative amplitudes and phase differences. It can also be seen that the group pattern is the same in any plane containing the axis of the array. In other words, the three-dimensional group pattern is simply the pattern obtained by revolving the group pattern in the xz-plane about the x-axis, that is, the axis of the array.

Example 9.2

For the array of two antennas carrying currents having equal amplitudes, let us consider several pairs of d and α and investigate the group patterns.

Case 1: $d = \lambda/2, \alpha = 0$. The group pattern is

$$\left| \cos\left(\frac{\beta\lambda}{4} \cos\psi \right) \right| = \cos\left(\frac{\pi}{2} \cos\psi \right)$$

This is shown sketched in Figure 9.10(a). It has maxima perpendicular to the axis of the array and nulls along the axis of the array. Such a pattern is known as a *broadside pattern*.

Case 2: $d = \lambda/2, \alpha = \pi$. The group pattern is

$$\left| \cos\left(\frac{\beta\lambda}{4} \cos\psi + \frac{\pi}{2} \right) \right| = \left| \sin\left(\frac{\pi}{2} \cos\psi \right) \right|$$

This is shown sketched in Figure 9.10(b). It has maxima along the axis of the array and nulls perpendicular to the axis of the array. Such a pattern is known as an *endfire pattern*.

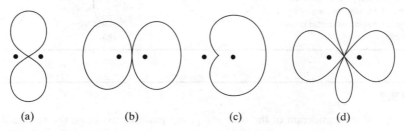

(a) (b) (c) (d)

FIGURE 9.10

Group patterns for an array of two antennas carrying currents of equal amplitude for (a) $d = \lambda/2, \alpha = 0$, (b) $d = \lambda/2, \alpha = \pi$, (c) $d = \lambda/4, \alpha = -\pi/2$, and (d) $d = \lambda, \alpha = 0$.

Case 3: $d = \lambda/4, \alpha = -\pi/2$. The group pattern is

$$\left| \cos\left(\frac{\beta\lambda}{8} \cos\psi - \frac{\pi}{4} \right) \right| = \cos\left(\frac{\pi}{4} \cos\psi - \frac{\pi}{4} \right)$$

This is shown sketched in Figure 9.10(c). It has a maximum along $\psi = 0$ and null along $\psi = \pi$. Again, this is an endfire pattern, but directed to one side. This case is the same as the one considered in Section 4.5.

Case 4: $d = \lambda, \alpha = 0$. The group pattern is

$$\left| \cos\left(\frac{\beta\lambda}{2} \cos\psi \right) \right| = \left| \cos\left(\pi \cos\psi \right) \right|$$

This is shown sketched in Figure 9.10(d). It has maxima along $\psi = 0°, 90°, 180°,$ and $270°$ and nulls along $\psi = 60°, 120°, 240°,$ and $300°$.

Proceeding further, we can obtain the resultant pattern for an array of two Hertzian dipoles by multiplying the unit pattern by the group pattern. Thus, recalling that the unit pattern for the Hertzian dipole is $\sin\theta$ in the plane of the dipole and considering values of $\lambda/2$ and 0 for d and α, respectively, for which the group pattern is given in Figure 9.10(a), we obtain the resultant pattern in the xz-plane, as shown in Figure 9.11(a). In the xy-plane, that is, the plane normal to the axis of the dipole, the unit pattern is a circle and hence the resultant pattern is the same as the group pattern, as illustrated in Figure 9.11(b).

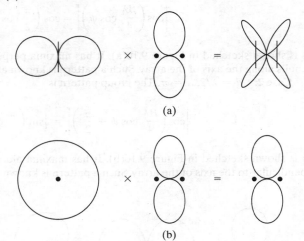

(a)

FIGURE 9.11

Determination of the resultant pattern of an antenna array by multiplication of unit and group patterns.

(b)

Example 9.3

The procedure of multiplication of the unit and group patterns to obtain the resultant pattern illustrated in Example 9.2 can be extended to an array containing any number of antennas. Let us, for example, consider a linear array of four isotropic antennas spaced $\lambda/2$ apart and fed in phase, as shown in Figure 9.12(a), and obtain the resultant pattern.

To obtain the resultant pattern of the four-element array, we replace it by a two-element array of spacing λ, as shown in Figure 9.12(b), in which each element forms a unit representing a two-element array of spacing $\lambda/2$. The unit pattern is then the pattern shown in Figure 9.10(a). The group pattern, which is the pattern of two isotropic radiators having $d = \lambda$ and $\alpha = 0$, is the pattern given in Figure 9.10(d). The resultant pattern of the four-element array is the product of these two patterns, as illustrated in Figure 9.12(c). If the individual elements of the four-element array are not isotropic, then this pattern becomes the group pattern for the determination of the new resultant pattern.

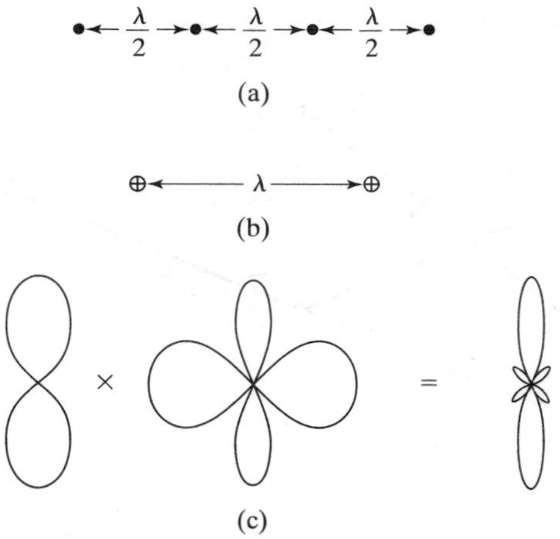

(a)

(b)

(c)

FIGURE 9.12

Determination of the resultant pattern for a linear array of four isotropic antennas.

9.5 IMAGE ANTENNAS

Thus far, we have considered the antennas to be situated in an unbounded medium so that the waves radiate in all directions from the antenna without giving rise to reflections from any obstacles. In practice, however, we have to consider the effect of the ground even if no other obstacles are present. To do this, it is reasonable to assume that the ground is a perfect conductor. Hence, in this section we shall consider antennas situated above an infinite plane, perfect-conductor surface and introduce the concept of image sources, a technique that is also useful in solving static field problems.

Thus, let us consider a Hertzian dipole oriented vertically and located at a height h above a plane, perfect-conductor surface, as shown in Figure 9.13(a). Since no waves can penetrate into the perfect conductor, as we learned in Section 5.5, the waves radiated from the dipole onto the conductor give rise to reflected waves, as shown in Figure 9.13(a) for two directions of incidence. For a given incident wave onto the conductor surface, the angle of reflection is equal to the angle of incidence, as can be seen intuitively from the following reasons: (a) The reflected wave must propagate away from the conductor surface, (b) the apparent wavelengths of the incident and reflected waves parallel to the conductor surface must be equal, and (c) the tangential component of the resultant electric field on the conductor surface must be zero, which also determines the polarity of the reflected wave electric field.

If we now produce the directions of propagation of the two reflected waves backward, they meet at a point which is directly beneath the dipole and at the same distance h below the conductor surface as the dipole is above it. Thus, the reflected waves appear to be originating from an antenna, which is the *image* of the actual antenna about the conductor surface. This image antenna must also be a vertical antenna since in order for the boundary condition of zero tangential electric field to be satisfied at all points on the conductor surface, the image antenna must have the same radiation pattern as that of the actual antenna, as shown in Figure 9.13(a). In particular, the current in the image antenna must be directed in the same sense as that in the actual antenna to be consistent with the polarity of the reflected wave electric field. It can therefore be

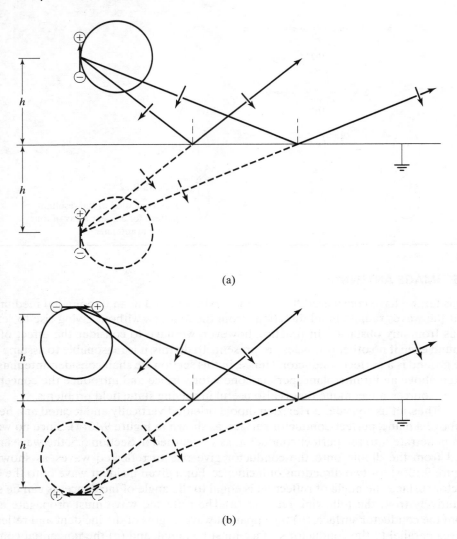

(a)

(b)

FIGURE 9.13

For illustrating the concept of image antennas. (a) Vertical Hertzian dipole and
(b) horizontal Hertzian dipole above a plane, perfect-conductor surface.

seen that the charges associated with the image dipole have signs opposite to those of
the corresponding charges associated with the actual dipole.

A similar reasoning can be applied to the case of a horizontal dipole above a perfect conductor surface, as shown in Figure 9.13(b). Here it can be seen that the current
in the image antenna is directed in the opposite sense to that in the actual antenna.
This again results in charges associated with the image dipole having signs opposite to
those of the corresponding charges associated with the actual dipole. In fact, this is
always the case.

From the foregoing discussion it can be seen that the field due to an antenna in the presence of the conductor is the same as the resultant field of the array formed by the actual antenna and the image antenna. There is, of course, no field inside the conductor. The image antenna is only a virtual antenna that seves to simplify the field determination outside the conductor. The simplicity arises from the fact that we can use the knowledge gained on antenna arrays in the previous section to determine the radiation pattern. Thus, for example, for a vertical Hertzian dipole at a height of $\lambda/2$ above the conductor surface, the radiation pattern in the vertical plane is the product of the unit pattern, which is the radiation pattern of the single dipole in the plane of its axis, and the group pattern corresponding to an array of two isotropic radiators spaced λ apart and fed in phase. This multiplication and the resultant pattern are illustrated in Figure 9.14. The radiation patterns for the case of the horizontal dipole can be obtained in a similar manner.

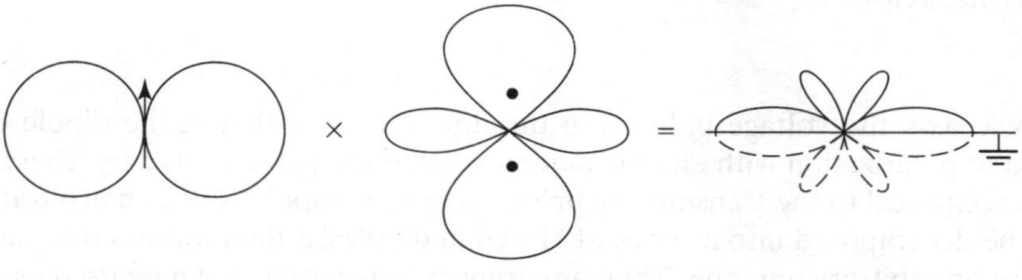

FIGURE 9.14

Determination of radiation pattern in the vertical plane for a vertical Hertzian dipole above a plane, perfect-conductor surface.

9.6   RECEIVING PROPERTIES

Thus far, we have considered the radiating, or transmitting, properties of antennas. Fortunately, it is not necessary to repeat all the derivations for the discussion of the receiving properties of antennas, since reciprocity dictates that the receiving pattern of an antenna be the same as its transmitting pattern. To illustrate this in simple terms without going through the general proof of reciprocity, let us consider a Hertzian dipole situated at the origin and directed along the z-axis, as shown in Figure 9.15. We know that the radiation pattern is then given by $\sin\theta$ and that the polarization of the radiated field is such that the electric field is in the plane of the dipole axis.

To investigate the receiving properties of the Hertzian dipole, we assume that it is situated in the radiation field of a second antenna so that the incoming waves are essentially uniform plane waves. Thus, let us consider a uniform plane wave with its electric field $\mathbf{E}$ in the plane of the dipole and incident on the dipole at an angle θ with its axis, as shown in Figure 9.15. Then the component of the incident electric field parallel to the dipole is $E\sin\theta$. Since the dipole is infinitesimal in length, the voltage induced in the dipole, which is the line integral of the electric field intensity along the length of the dipole, is simply equal to $(E\sin\theta)\,dl$ or to $E\,dl\sin\theta$. This indicates that for a given amplitude of the incident wave field, the induced voltage in the dipole is proportional to $\sin\theta$. Furthermore, for an incident uniform plane wave having its electric field normal to

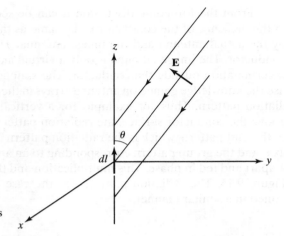

FIGURE 9.15

For investigating the receiving properties
of a Hertzian dipole.

the dipole axis, the voltage induced in the dipole is zero, that is, the dipole does not
respond to polarization with electric field normal to the plane of its axis. These properties are reciprocal to the transmitting properties of the dipole. Since an arbitrary antenna can be decomposed into a series of Hertzian dipoles, it then follows that reciprocity
holds for an arbitrary antenna. Thus, any transmitting antenna can be used as a receiving antenna, and vice versa.

We shall now briefly consider the loop antenna, a common type of receiving antenna. A simple form of loop antenna consists of a circular loop of wire with a pair of
terminals. We shall orient the circular loop antenna with its axis aligned with the z-axis,
as shown in Figure 9.16, and we shall assume that it is electrically short, that is, its
dimensions are small compared to the wavelength of the incident wave, so that the spatial variation of the field over the area of the loop is negligible. For a uniform plane
wave incident on the loop, we can find the voltage induced in the loop, that is, the line
integral of the electric field intensity around the loop, by using Faraday's law. Thus, if **H**

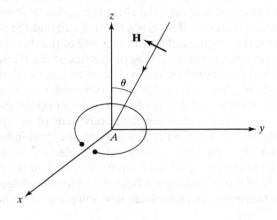

FIGURE 9.16

A circular loop antenna.

is the magnetic field intensity associated with the wave, the magnitude of the induced voltage is given by

$$|V| = \left| -\frac{d}{dt} \int_{\substack{\text{area of} \\ \text{the loop}}} \mathbf{B} \cdot d\mathbf{S} \right|$$

$$= \left| -\mu \frac{d}{dt} \int_{\substack{\text{area of} \\ \text{the loop}}} \mathbf{H} \cdot d\mathbf{S} \, \mathbf{a}_z \right|$$

$$= \mu A \left| \frac{\partial H_z}{\partial t} \right| \tag{9.52}$$

where A is the area of the loop. Hence the loop does not respond to a wave having its magnetic field entirely parallel to the plane of the loop, that is, normal to the axis of the loop.

For a wave having its magnetic field in the plane of the axis of the loop, and incident on the loop at an angle θ with its axis, as shown in Figure 9.16, $H_z = H \sin \theta$, and hence the induced voltage has a magnitude

$$|V| = \mu A \left| \frac{\partial H}{\partial t} \right| \sin \theta \tag{9.53}$$

Thus, the receiving pattern of the loop antenna is given by $\sin \theta$, same as that of a Hertzian dipole aligned with the axis of the loop antenna. The loop antenna, however, responds best to polarization with magnetic field in the plane of its axis, whereas the Hertzian dipole responds best to polarization with electric field in the plane of its axis.

Example 9.4

The directional properties of a receiving antenna can be used to locate the source of an incident signal. To illustrate the principle, let us consider two vertical loop antennas, numbered 1 and 2, situated on the x-axis at $x = 0$ m and $x = 200$ m, respectively. By rotating the loop antennas about the vertical (z-axis), it is found that no (or minimum) signal is induced in antenna 1 when it is in the xz-plane and in antenna 2 when it is in a plane making an angle of 5° with the axis, as shown by the top view in Figure 9.17. Let us find the location of the source of the signal.

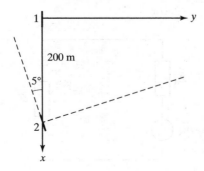

FIGURE 9.17

Top view of two loop antennas used to locate the source of an incident signal.

Since the receiving properties of a loop antenna are such that no signal is induced for a wave arriving along its axis, the source of the signal is located at the intersection of the axes of the two loops when they are oriented so as to receive no (or minimum) signal. From simple geometrical considerations, the source of the signal is therefore located on the y-axis at $y = 200/\tan 5°$ or 2.286 km.

A useful parameter associated with the receiving properties of an antenna is the effective area, denoted A_e and defined as the ratio of the time-average power delivered to a matched load connected to the antenna to the time-average power density of the appropriately polarized incident wave at the antenna. The matched condition is achieved when the load impedance is equal to the complex conjugate of the antenna impedance.

Let us consider the Hertzian dipole and derive the expression for its effective area. First, with reference to the equivalent circuit shown in Figure 9.18, where $\bar{V}_{oc}$ is the open-circuit voltage induced between the terminals of the antenna, $\bar{Z}_A = R_A + jX_A$ is the antenna impedance, and $\bar{Z}_L = \bar{Z}_A^*$ is the load impedance, we note that the time-average power delivered to the matched load is

$$P_R = \frac{1}{2}\left(\frac{|\bar{V}_{oc}|}{2R_A}\right)^2 R_A = \frac{|\bar{V}_{oc}|^2}{8R_A} \tag{9.54}$$

For a Hertzian dipole of length l, the open-circuit voltage is

$$\bar{V}_{oc} = \bar{E}l \tag{9.55}$$

where $\bar{E}$ is the electric field of an incident wave linearly polarized parallel to the dipole axis. Substituting (9.55) into (9.54), we get

$$P_R = \frac{|\bar{E}|^2 l^2}{8R_A} \tag{9.56}$$

For a lossless dipole, $R_A = R_{rad} = 80\pi^2(l/\lambda)^2$, so that

$$P_R = \frac{|\bar{E}|^2 \lambda^2}{640\pi^2} \tag{9.57}$$

The time-average power density at the antenna is

$$\frac{|\bar{E}|^2}{2\eta_0} = \frac{|\bar{E}|^2}{240\pi} \tag{9.58}$$

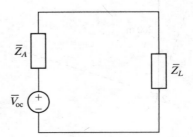

FIGURE 9.18

Equivalent circuit for a receiving antenna connected to a load.

Thus, the effective area is

$$A_e = \frac{|\bar{E}|^2 \lambda^2 / 640\pi^2}{|\bar{E}|^2 / 240\pi} = \frac{3\lambda^2}{8\pi} \tag{9.59}$$

or

$$A_e = 0.1194\lambda^2 \tag{9.60}$$

In practice, R_A is greater than R_{rad} due to losses in the antenna, and the effective area is less than that given by (9.60). Rewriting (9.59) as

$$A_e = 1.5 \times \frac{\lambda^2}{4\pi}$$

and recalling that the directivity of the Hertzian dipole is 1.5, we observe that

$$A_e = \frac{\lambda^2}{4\pi}D \tag{9.61}$$

Although we have obtained this result for a Hertzian dipole, it can be shown that it holds for any antenna.

We shall now derive the *Friis transmission formula*, an important equation in making communication link calculations. To do this, let us consider two antennas, one transmitting and the other receiving, separated by a distance d. Let us assume that the antennas are oriented and polarization matched so as to maximize the received signal. Then if P_T is the transmitter power radiated by the transmitting antenna, the power density at the receiving antenna is $(P_T/4\pi d^2)D_T$, where D_T is the directivity of the transmitting antenna. The power received by a matched load connected to the terminals of the receiving antenna is then given by

$$P_R = \frac{P_T D_T}{4\pi d^2} A_{eR} \tag{9.62}$$

where A_{eR} is the effective area of the receiving antenna. Thus, the ratio of P_R to P_T is given by

$$\frac{P_R}{P_T} = \frac{D_T A_{eR}}{4\pi d^2} \tag{9.63}$$

Denoting A_{eT} to be the effective area of the transmitting antenna if it were receiving, and using (9.61), we obtain

$$\frac{P_R}{P_T} = \frac{A_{eT} A_{eR}}{\lambda^2 d^2} \tag{9.64}$$

Equation (9.64) is the Friis transmission formula. It gives the maximum value of P_R/P_T for a given d and for a given pair of transmitting and receiving antennas. If the antennas are not oriented to receive the maximum signal, or if a polarization mismatch exists, or if the receiving antenna is not matched to its load, P_R/P_T would be less than that given by (9.64). Losses in the antennas would also decrease the value of P_R/P_T.

An alternative formula to (9.64) is obtained by substituting for A_{eR} in (9.63) in terms of the directivity D_R of the receiving antenna if it were used for transmitting. Thus, we obtain

$$\frac{P_R}{P_T} = \frac{D_T D_R \lambda^2}{16\pi^2 d^2} \tag{9.65}$$

SUMMARY

In this chapter we studied the principles of antennas. We first introduced the Hertzian dipole, which is an elemental wire antenna, and derived the complete electromagnetic field due to the Hertzian dipole by employing an intuitive approach based on the knowledge gained in the previous chapters. For a Hertzian dipole of length dl, oriented along the z-axis at the origin, and carrying current

$$I(t) = I_0 \cos \omega t$$

we found the complete electromagnetic field to be given by

$$\mathbf{E} = \frac{2I_0 \, dl \cos \theta}{4\pi\epsilon\omega} \left[\frac{\sin(\omega t - \beta r)}{r^3} + \frac{\beta \cos(\omega t - \beta r)}{r^2} \right] \mathbf{a}_r$$

$$+ \frac{I_0 \, dl \sin \theta}{4\pi\epsilon\omega} \left[\frac{\sin(\omega t - \beta r)}{r^3} + \frac{\beta \cos(\omega t - \beta r)}{r^2} - \frac{\beta^2 \sin(\omega t - \beta r)}{r} \right] \mathbf{a}_\theta$$

$$\mathbf{H} = \frac{I_0 \, dl \sin \theta}{4\pi} \left[\frac{\cos(\omega t - \beta r)}{r^2} - \frac{\beta \sin(\omega t - \beta r)}{r} \right] \mathbf{a}_\phi$$

where $\beta = \omega\sqrt{\mu\epsilon}$ is the phase constant.

For $\beta r \gg 1$ or for $r \gg \lambda/2\pi$, the only important terms in the complete field expressions are the $1/r$ terms, since the remaining terms are negligible compared to these terms. Thus for $r \gg \lambda/2\pi$, the Hertzian dipole fields are given by

$$\mathbf{E} = -\frac{\eta\beta I_0 \, dl \sin \theta}{4\pi r} \sin(\omega t - \beta r) \mathbf{a}_\theta$$

$$\mathbf{H} = -\frac{\beta I_0 \, dl \sin \theta}{4\pi r} \sin(\omega t - \beta r) \mathbf{a}_\phi$$

where $\eta = \sqrt{\mu/\epsilon}$ is the intrinsic impedance of the medium. These fields, known as the radiation fields, correspond to locally uniform plane waves radiating away from the dipole and, in fact, are the only components of the complete fields contributing to the time-average radiated power. We found the time-average power radiated by the Hertzian dipole to be given by

$$\langle P_{\text{rad}} \rangle = \frac{1}{2} I_0^2 \left[\frac{2\pi\eta}{3} \left(\frac{dl}{\lambda} \right)^2 \right]$$

and identified the quantity inside the brackets to be its radiation resistance. The radiation resistance, R_{rad}, of an antenna is the value of a fictitious resistor that will dissipate the same amount of time-average power as that radiated by the antenna when a current of the same peak amplitude as that in the antenna is passed through it. Thus, for the Hertzian dipole,

$$R_{\text{rad}} = \frac{2\pi\eta}{3} \left(\frac{dl}{\lambda} \right)^2$$

We then examined the directional characteristics of the radiation fields of the Hertzian dipole, as indicated by the factor $\sin\theta$ in the field expressions and hence by the factor $\sin^2\theta$ for the power density. We discussed the radiation patterns and introduced the concept of the directivity of an antenna. The directivity, D, of an antenna is defined as the ratio of the maximum power density radiated by the antenna to the average power density. For the Hertzian dipole,

$$D = 1.5$$

For the general case of a power density pattern $f(\theta, \phi)$, the directivity is given by

$$D = 4\pi \, \frac{[f(\theta, \phi)]_{\text{max}}}{\displaystyle\int_{\theta=0}^{\pi} \int_{\phi=0}^{2\pi} f(\theta, \phi) \sin\theta \, d\theta \, d\phi}$$

As an illustration of obtaining the radiation fields due to a wire antenna of arbitrary length and arbitrary current distribution by representing it as a series of Hertzian dipoles and using superposition, we considered the example of a half-wave dipole and derived its radiation fields. We found that for a center-fed half-wave dipole of length $L(=\lambda/2)$, oriented along the z-axis with its center at the origin, and having the current distribution given by

$$I(z) = I_0 \cos \frac{\pi z}{L} \cos \omega t \qquad \text{for } -\frac{L}{2} < z < \frac{L}{2}$$

the radiation fields are

$$\mathbf{E} = -\frac{\eta I_0}{2\pi r} \frac{\cos\left[(\pi/2)\cos\theta\right]}{\sin\theta} \sin\left(\omega t - \frac{\pi}{L} r\right) \mathbf{a}_\theta$$

$$\mathbf{H} = -\frac{I_0}{2\pi r} \frac{\cos\left[(\pi/2)\cos\theta\right]}{\sin\theta} \sin\left(\omega t - \frac{\pi}{L} r\right) \mathbf{a}_\phi$$

From these, we sketched the radiation patterns and computed the radiation resistance and the directivity of the half-wave dipole to be

$$R_{\text{rad}} = 73 \text{ ohms} \quad \text{for free space}$$
$$D = 1.642$$

We discussed antenna arrays and introduced the technique of obtaining the resultant radiation pattern of an array by multiplication of the unit and the group patterns. For an array of two antennas having the spacing d and fed with currents of equal amplitude but differing in phase by α, we found the group pattern for the fields to be $|\cos[(\beta d \cos \psi + \alpha)/2]|$, where ψ is the angle measured from the axis of the array, and we investigated the group patterns for several pairs of values of d and α. For example, for $d = \lambda/2$ and $\alpha = 0$, the pattern corresponds to maximum radiation broadside to the axis of the array, whereas for $d = \lambda/2$ and $\alpha = \pi$, the pattern corresponds to maximum radiation endfire to the axis of the array.

To take into account the effect of ground on antennas, we introduced the concept of an image antenna in a perfect conductor and discussed the application of the array techniques in conjunction with the actual and the image antennas to obtain the radiation pattern of the actual antenna in the presence of the ground.

Finally, we discussed receiving properties of antennas. In particular, (1) we discussed the reciprocity between the receiving and radiating properties of an antenna by considering the simple case of a Hertzian dipole, (2) we considered the loop antenna and illustrated the application of its directional properties for locating the source of a radio signal, and (3) we introduced the effective area concept and derived the Friis transmission formula.

REVIEW QUESTIONS

9.1. What is a Hertzian dipole?
9.2. Discuss the time-variations of the current and charges associated with the Hertzian dipole.
9.3. Briefly describe the spherical coordinate system.
9.4. Explain why it is simpler to use the spherical coordinate system to find the fields due to the Hertzian dipole.
9.5. Discuss the reasoning associated with the intuitive extension of the fields due to the time-varying current and charges of the Hertzian dipole based on time-varying electromagnetic phenomena.
9.6. Explain the reason for the inconsistency with Maxwell's equations of the intuitively derived fields due to the time-varying current and charges of the Hertzian dipole.
9.7. Briefly outline the reasoning used for the removal of the inconsistency with Maxwell's equations of the intuitively derived fields due to the Hertzian dipole.
9.8. Discuss the characteristics of the complete electromagnetic field due to the Hertzian dipole.
9.9. Consult an appropriate reference book and compare the procedure used for obtaining the electromagnetic field due to the Hertzian dipole with the procedure used here.
9.10. What are radiation fields? Why are they important?
9.11. Discuss the characteristics of the radiation fields.
9.12. Define the radiation resistance of an antenna.
9.13. Why is the expression for the radiation resistance of a Hertzian dipole not valid for a linear antenna of any length?
9.14. Explain why power lines are not effective radiators.
9.15. What is a radiation pattern?
9.16. Discuss the radiation pattern for the power density due to the Hertzian dipole.
9.17. Define the directivity of an antenna. What is the directivity of a Hertzian dipole?

9.18. What is the directivity of a fictitious antenna that radiates equally in all directions into one hemisphere?

9.19. How do you find the radiation fields due to an antenna of arbitrary length and arbitrary current distribution?

9.20. Discuss the evolution of the half-wave dipole from an open-circuited transmission line.

9.21. Justify the approximations involved in evaluating the integrals in the determination of the radiation fields due to the half-wave dipole.

9.22. What are the values of the radiation resistance and the directivity for a half-wave dipole?

9.23. What is an antenna array?

9.24. Justify the approximations involved in the determination of the resultant field of an array of two antennas.

9.25. Why is it that the distances r_1 and r_2 in the phase factors in equations (8.47a) and (8.47b) cannot be set equal to r, but the same quantities in the amplitude factors can be set equal to r?

9.26. What is an array factor? Provide a physical explanation for the array factor.

9.27. Discuss the concept of unit and group patterns and their multiplication to obtain the resultant pattern of an array.

9.28. Distinguish between broadside and endfire radiation patterns.

9.29. Discuss the concept of an image antenna to find the field of an antenna in the vicinity of a perfect conductor.

9.30. What determines the sense of the current flow in an image antenna relative to that in the actual antenna?

9.31. How does the concept of an image antenna simplify the determination of the radiation pattern of an antenna above a perfect-conductor surface?

9.32. Discuss the reciprocity associated with the transmitting and receiving properties of an antenna. Can you think of a situation in which reciprocity does not hold?

9.33. What is the receiving pattern of a loop antenna?

9.34. How should you orient a loop antenna to receive (a) a maximum signal and (b) a minimum signal?

9.35. Discuss the application of the directional receiving properties of a loop antenna in the location of the source of a radio signal.

9.36. How is the effective area of a receiving antenna defined?

9.37. Outline the derivation of the expression for the effective area of a Hertzian dipole.

9.38. Discuss the derivation of the Friis transmission formula.

PROBLEMS

9.1. The electric dipole moment associated with a Hertzian dipole of length 0.1 m is given by

$$\mathbf{p} = 10^{-9} \sin 2\pi \times 10^7 t \, \mathbf{a}_z \text{ C-m}$$

Find the current in the dipole.

9.2. Evaluate the curl of $\mathbf{E}$ given by equation (9.12a) and show that it is not equal to $-\mu \dfrac{\partial \mathbf{H}}{\partial t}$, where $\mathbf{H}$ is given by equation (9.12b).

9.3. Show that in the limit $\omega \to 0$, the complete field expressions given by equations (9.23a) and (9.23b) tend to equations (9.12a) and (9.12b), respectively.

9.4. Show that the radiation fields given by equations (9.25a) and (9.25b) do not by themselves satisfy both of Maxwell's curl equations.

9.5. Find the value of r at which the amplitude of the radiation field term in equation (9.23a) is equal to the resultant amplitude of the remaining two terms in the θ-component.

9.6. Obtain the Poynting vector corresponding to the complete electromagnetic field due to the Hertzian dipole and show that the $1/r^3$ and $1/r^2$ terms do not contribute to the time-average power flow from the dipole.

9.7. A straight wire of length 1 m situated in free space carries a uniform current $10 \cos 4\pi \times 10^6 t$ A. (a) Calculate the amplitude of the electric field intensity at a distance of 10 km in a direction at right angle to the wire. (b) Calculate the radiation resistance and the time-average power radiated by the wire.

9.8. Compute the radiation resistance per kilometer length of a straight power-line wire. Comment on the effectiveness of the power line as a radiator.

9.9. Find the time-average power required to be radiated by a Hertzian dipole in order to produce an electric field intensity of peak amplitude 0.01 V/m at a distance of 1 km broadside to the dipole.

9.10. A Hertzian dipole situated at the origin and oriented along the x-axis carries a current $I_1 = I_0 \cos \omega t$. A second Hertzian dipole, having the same length and also situated at the origin but oriented along the z-axis, carries a current $I_2 = I_0 \sin \omega t$. Find the polarization of the radiated electric field at (a) a point on the x-axis, (b) a point on the z-axis, (c) a point on the y-axis, and (d) a point on the line $x = y, z = 0$.

9.11. Find the ratio of the currents in two antennas having directivities D_1 and D_2 and radiation resistances $R_{\text{rad}\,1}$ and $R_{\text{rad}\,2}$ for which the maximum radiated power densities are equal.

9.12. The radiation pattern for the power density of an antenna located at the origin is dependent on θ in the manner $\sin^4 \theta$. Find the directivity of the antenna.

9.13. The radiation pattern for the power density of an antenna located at the origin is dependent on θ in the manner

$$f(\theta, \phi) = \begin{cases} \csc^2 \theta & \text{for } \pi/6 \le \theta \le \pi/2 \\ 0 & \text{otherwise} \end{cases}$$

Find the directivity of the antenna.

9.14. In Figure 9.7, let $L = 2$ m, and investigate the variations of r' and $\pi r'/L$ for $-L/2 < z' < L/2$ for (a) a point in the xy-plane at $r = 1$ km and (b) a point on the z-axis at $r = 1$ km.

9.15. By dividing the interval $0 < \theta < \pi/2$ into nine equal parts, numerically compute the value of

$$\int_{\theta=0}^{\pi/2} \frac{\cos^2 [(\pi/2) \cos \theta]}{\sin \theta} \, d\theta$$

9.16. Complete the missing steps in the evaluation of the integral in equation (9.39a).

9.17. Find the time-average power required to be radiated by a half-wave dipole in order to produce an electric field intensity of peak amplitude 0.01 V/m at a distance of 1 km broadside to the dipole.

9.18. Compare the correct value of the radiation resistance of the half-wave dipole with the incorrect value that would result from using the expression for the radiation resistance of the Hertzian dipole.

9.19. A short dipole is a center-fed straight wire antenna having a length that is small compared to a wavelength. The amplitude of the current distribution can then be approximated as decreasing linearly from a maximum at the center to zero at the ends. Thus, for a short dipole of length L lying along the z-axis between $z = -L/2$ and $z = L/2$, the current distribution is given by

$$I(z) = \begin{cases} I_0\left(1 + \dfrac{2z}{L}\right)\cos \omega t & \text{for } -\dfrac{L}{2} < z < 0 \\[2mm] I_0\left(1 - \dfrac{2z}{L}\right)\cos \omega t & \text{for } 0 < z < \dfrac{L}{2} \end{cases}$$

(a) Obtain the radiation fields of the short dipole. (b) Find the radiation resistance and the directivity of the short dipole.

9.20. For the array of two antennas of Example 9.2, find and sketch the group patterns for (a) $d = \lambda/4, \alpha = \pi/2$ and (b) $d = 2\lambda, \alpha = 0$.

9.21. For the array of two antennas of Example 9.2, having $d = \lambda/4$, find the value of α for which the maxima of the group pattern are directed along $\psi = \pm 60°$, and then sketch the group pattern.

9.22. Obtain the resultant pattern for a linear array of eight isotropic antennas, spaced $\lambda/2$ apart, carrying equal currents, and fed in phase.

9.23. Obtain the resultant pattern for a linear array of three isotropic antennas, spaced $\lambda/2$ apart, carrying unequal currents in the ratio $1 : 2 : 1$, and fed in phase.

9.24. For the array of two Hertzian dipoles of Figure 9.9, find and sketch the resultant pattern in the xz-plane for $d = \lambda/2$ and $\alpha = \pi$.

9.25. For the array of two Hertzian dipoles of Figure 9.9, find and sketch the resultant pattern in the xz-plane for $d = \lambda/4$ and $\alpha = -\pi/2$.

9.26. For a horizontal Hertzian dipole at a height $\lambda/4$ above a plane, perfect-conductor surface, find and sketch the radiation pattern in (a) the vertical plane perpendicular to the axis of the antenna and (b) the vertical plane containing the axis of the antenna.

9.27. For a vertical antenna of length $\lambda/4$ above a plane, perfect-conductor surface, find (a) the radiation pattern in the vertical plane and (b) the directivity.

9.28. A Hertzian dipole is situated parallel to a corner reflector, which is an arrangement of two plane, perfect conductors at right angles to each other, as shown by the cross-sectional view in Figure 9.19. (a) Locate the image antennas required to satisfy the boundary conditions on the corner reflector surface. (b) Find and sketch the radiation pattern in the cross-sectional plane.

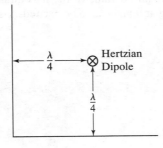

FIGURE 9.19

For Problem 9.28.

9.29. If the Hertzian dipole in Figure 9.19 is situated at a distance $\lambda/2$ from the corner and equidistant from the two planes, find the ratio of the radiation field at a point broadside to the dipole and away from the corner to the radiation field in the absence of the corner reflector.

9.30. An arrangement of two identical Hertzian dipoles situated at the origin and oriented along the x- and y-axes, known as the turnstile antenna, is used for receiving circularly polarized signals arriving along the z-axis. Determine how you would combine the voltages induced in the two dipoles so that the turnstile antenna is responsive to circular polarization rotating in the clockwise sense as viewed by the antenna but not to that of the counterclockwise sense of rotation.

9.31. A vertical loop antenna of area 1 m^2 is situated at a distance of 10 km from a vertical wire antenna of length $\lambda/4$ above a perfectly conducting ground (directivity $= 3.28$; see Problem 9.27) radiating at 2 MHz. The loop antenna is oriented so as to maximize the signal induced in it. For a time-average radiated power of 10 kW, find the amplitude of the voltage induced in the loop antenna.

9.32. An interferometer consists of an array of two identical antennas with spacing d. Show that for a uniform plane wave incident on the array at an angle ψ to the axis of the array, as shown in Figure 9.20, the phase difference $\Delta\phi$ between the voltage induced in antenna 1 and the voltage induced in antenna 2 is $(2\pi d/\lambda)\cos\psi$, where λ is the wavelength of the incident wave. For $d = 2\lambda$ and for $\Delta\phi = 30°$, find all possible values of ψ. Take into account the fact that the phase measurement is ambiguous by the amount $\pm 2n\pi$, where n is an integer.

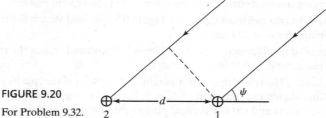

FIGURE 9.20

For Problem 9.32.

9.33. A communication link at a frequency of 30 MHz uses a half-wave dipole for the transmitting antenna and a small loop (directivity equal to 1.5) for the receiving antenna, involving a distance of 100 km. The antennas are oriented so as to receive maximum signal and the receiving antenna is matched to its load. If the received time-average power is to be $1 \mu\text{W}$, find the minimum required value of the maximum amplitude I_0 of the current with which the transmitting antenna has to be excited. Assume the antennas to be lossless.

Supplementary Topics

In Chapter 1, we learned the basic mathematical tools and physical concepts of vectors and fields. In Chapters 2 and 3, we learned the fundamental laws of electromagnetics, namely, Maxwell's equations, first in integral form and then in differential form. Then in Chapters 4 through 9, we extended our study to the fundamental electromagnetic concepts and phenomena, as relevant to electrical and computer engineering. These comprised the propagation, transmission, and radiation of electromagnetic waves, as well as the thread of statics-quasistatics-waves to bring out the frequency behavior of physical structures.

This final chapter is devoted to six independent topics, each one based on, and hence supplementary to, one or more of Chapters 4 through 9. The six topics can be studied independently following the respective chapters on which they are based. These supplementary topics, although independent of each other, have the common goal of extending the knowledge gained in the corresponding previous chapter(s) for the purpose of illustrating a concept, phenomenon, or application.

10.1 WAVE PROPAGATION IN IONIZED MEDIUM

In Chapter 4, we studied uniform plane wave propagation in free space. In this section, we shall extend the discussion to wave propagation in ionized medium. An example of ionized medium is the earth's ionosphere, which is a region of the upper atmosphere extending from approximately 50 km to more than 1000 km above the earth. In this region, the constituent gases are ionized, mostly because of ultraviolet radiation from the sun, thereby resulting in the production of positive ions and electrons that are free to move under the influence of the fields of a wave incident upon the medium. The positive ions are, however, heavy compared to electrons and hence they are relatively immobile. The electron motion produces a current that influences the wave propagation.

In fact, in Section 1.5 we considered the motion of a cloud of electrons of uniform density N under the influence of a time-varying electric field

$$\mathbf{E} = E_0 \cos \omega t \, \mathbf{a}_x \qquad (10.1)$$

and found that the resulting current density is given by

$$\mathbf{J} = \frac{Ne^2}{m\omega}E_0 \sin \omega t \, \mathbf{a}_x = \frac{Ne^2}{m}\int \mathbf{E} \, dt \tag{10.2}$$

where e and m are the electronic charge and mass, respectively. This result is based on the mechanism of continuous acceleration of the electrons by the force due to the applied electric field. In the case of the ionized medium, the electron motion is, however, impeded by the collisions of the electrons with the heavy particles and other electrons. We shall ignore these collisions as well as the negligible influence of the magnetic field associated with the wave.

Considering uniform plane wave propagation in the z-direction in an unbounded ionized medium, and with the electric field oriented in the x-direction, we then have

$$\frac{\partial E_x}{\partial z} = -\frac{\partial B_y}{\partial t} = -\mu_0 \frac{\partial H_y}{\partial t} \tag{10.3a}$$

$$\frac{\partial H_y}{\partial z} = -J_x - \frac{\partial D_x}{\partial t} = -\frac{Ne^2}{m}\int E_x \, dt - \epsilon_0 \frac{\partial E_x}{\partial t} \tag{10.3b}$$

Differentiating (10.3a) with respect to z and then substituting for $\partial H_y/\partial z$ from (10.3b), we obtain the wave equation

$$\frac{\partial^2 E_x}{\partial z^2} = -\mu_0 \frac{\partial}{\partial t}\left[-\frac{Ne^2}{m}\int E_x \, dt - \epsilon_0 \frac{\partial E_x}{\partial t}\right]$$

$$= \frac{\mu_0 Ne^2}{m}E_x + \mu_0\epsilon_0 \frac{\partial^2 E_x}{\partial t^2} \tag{10.4}$$

Substituting

$$E_x = E_0 \cos (\omega t - \beta z) \tag{10.5}$$

corresponding to the uniform plane wave solution into (10.4) and simplifying, we get

$$\beta^2 = \omega^2 \mu_0\epsilon_0 - \frac{\mu_0 Ne^2}{m}$$

$$= \omega^2 \mu_0\epsilon_0 \left(1 - \frac{Ne^2}{m\epsilon_0\omega^2}\right)$$

Thus, the phase constant for propagation in the ionized medium is given by

$$\beta = \omega\sqrt{\mu_0\epsilon_0 \left(1 - \frac{Ne^2}{m\epsilon_0\omega^2}\right)} \tag{10.6}$$

This result indicates that the ionized medium behaves as though the permittivity of free space is modified by the multiplying factor $[1 - (Ne^2/m\epsilon_0\omega^2)]$. We may therefore write

$$\beta = \omega\sqrt{\mu_0\epsilon_{\text{eff}}} \tag{10.7}$$

where

$$\epsilon_{\text{eff}} = \epsilon_0\left(1 - \frac{Ne^2}{m\epsilon_0\omega^2}\right) \tag{10.8}$$

is the *effective permittivity* of the ionized medium. We note that for $\omega \to \infty$, $\epsilon_{\text{eff}} \to \epsilon_0$ and the medium behaves just as free space. This is to be expected since (10.2) indicates that for $\omega \to \infty$, $\mathbf{J} \to 0$. As ω decreases from ∞, ϵ_{eff} becomes less and less until for ω equal to $\sqrt{Ne^2/m\epsilon_0}$, ϵ_{eff} becomes zero. Hence for $\omega > \sqrt{Ne^2/m\epsilon_0}$, ϵ_{eff} is positive, β is real, and the solution for the electric field remains to be that of a propagating wave. For $\omega < \sqrt{Ne^2/m\epsilon_0}$, ϵ_{eff} is negative, β becomes imaginary, and the solution for the electric field corresponds to no propagation.

Thus, waves of frequency $f > \sqrt{Ne^2/4\pi^2m\epsilon_0}$ propagate in the ionized medium and waves of frequency $f < \sqrt{Ne^2/4\pi^2m\epsilon_0}$ do not propagate. The quantity $\sqrt{Ne^2/4\pi^2m\epsilon_0}$ is known as the *plasma frequency* and is denoted by the symbol, f_N. Substituting values for e, m, and ϵ_0, we get

$$f_N = \sqrt{80.6N} \text{ Hz} \tag{10.9}$$

where N is in electrons per meter cubed. We can now write ϵ_{eff} as

$$\epsilon_{\text{eff}} = \epsilon_0\left(1 - \frac{f_N^2}{f^2}\right) \tag{10.10}$$

Proceeding further, we obtain the phase velocity for the propagating range of frequencies, that is, for $f > f_N$, to be

$$v_p = \frac{1}{\sqrt{\mu_0\epsilon_{\text{eff}}}} = \frac{1}{\sqrt{\mu_0\epsilon_0(1 - f_N^2/f^2)}}$$

$$= \frac{c}{\sqrt{1 - f_N^2/f^2}} \tag{10.11}$$

where $c = 1/\sqrt{\mu_0\epsilon_0}$ is the velocity of light in free space. From (10.11), we observe that $v_p > c$ and is a function of the wave frequency. The fact that $v_p > c$ is not a violation of the principle of relativity, since the dispersive nature of the medium resulting from the dependence of v_p upon f ensures that information always travels with a velocity less than c. The topic of *dispersion* is discussed in Section 8.3.

To apply what we have learned above concerning propagation in an ionized medium to the case of the earth's ionosphere, we first provide a brief description of the ionosphere. A typical distribution of the ionospheric electron density versus height above the earth is shown in Figure 10.1. The electron density exists in the form of several layers known as D, E, and F layers in which the ionization changes with the hour of the day, the

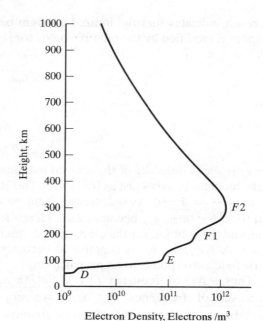

Height, km

Electron Density, Electrons /m^3

FIGURE 10.1

A typical distribution of ionospheric electron density versus height above the earth.

season, the sunspot cycle, and the geographic location. The nomenclature behind the designation of the letters for the layers is due to Appleton in England, who in 1925 and at about the same time as Breit and Tuve in the United States demonstrated experimentally the reflection of radio waves by the ionosphere. In his early work, Appleton was accustomed to writing E for the electric field of the wave reflected from the first layer he recognized. Later, when he recognized a second layer, at a greater height, he wrote F for the field of the wave reflected from it. Still later, he conjectured that there might be a third layer lower from either of the first two and thus he decided to name the possible lower layer D, thereby leaving earlier letters of the alphabet for other possible undiscovered, still lower layers. Electrons were indeed detected later in the D region.

The D region extends over the altitude range of about 50 km to about 90 km. Since collisions between electrons and heavy particles cannot be neglected in this region, it is mainly an absorbing region. The E region extends from about 90 km to about 150 km. Diurnal and seasonal variations of the E layer electron density are strongly correlated with the zenith angle of the sun. In the F region, the lower of the two strata is designated as the $F1$ layer and the higher, more intense ionized stratum is designated as the $F2$ layer. The $F1$ ledge is usually located between 160 km and 200 km. Above this region, the $F2$ layer electron density increases with altitude, reaching a peak at a height generally lying between 250 km and 400 km. Above this peak the electron density decreases monotonically with altitude. The $F1$ ledge is present only during the day. During the night, the $F1$ and $F2$ layers are identified as a single F layer. The $F2$ layer is the most important from the point of view of radio communication since it contains the greatest concentration of electrons. Paradoxically, it also exhibits several anomalies.

Wave propagation in the ionosphere is complicated by the presence of the earth's magnetic field. If we neglect the earth's magnetic field, then for a wave of frequency f incident vertically on the ionosphere from a transmitter on the ground, it is evident from the propagation condition $f > f_N$ that the wave propagates up to the height at which $f = f_N$, and since it cannot propagate beyond that height, it gets reflected at that height. Thus, waves of frequencies less than the maximum plasma frequency corresponding to the peak of the $F2$ layer cannot penetrate the ionosphere. Hence, for communication with satellites orbiting above the peak of the ionosphere, frequencies greater than this maximum plasma frequency, also known as the *critical frequency*, must be employed. While this critical frequency is a function of the time of day, the season, the sunspot cycle, and the geographic location, it is not greater than about 15 MHz and can be as low as a few megahertz. For a wave incident obliquely on the ionosphere, reflection is possible for frequencies greater than the critical frequency, up to about three times its value. Hence, for earth-to-satellite communication, frequencies generally exceeding about 40 MHz are employed. Lower frequencies permit long-distance, ground-to-ground communication via reflections from the ionospheric layers. This mode of propagation is familiarly known as the *sky wave mode* of propagation. For very low frequencies of the order of several kilohertz and less, the lower boundary of the ionosphere and the earth form a waveguide, thereby permitting waveguide mode of propagation.

In this section, we learned that in an ionized medium, wave propagation occurs only for frequencies exceeding the plasma frequency corresponding to the electron density. Applying this to the case of the earth's ionosphere, we found that this imposes a lower limit in frequency for communication with satellites.

REVIEW QUESTIONS

10.1. What is an ionized medium? What influences wave propagation in an ionized medium?

10.2. Provide physical explanation for the frequency dependence of the effective permittivity of an ionized medium.

10.3. Discuss the condition for propagation in an ionized medium.

10.4. What is plasma frequency? How is it related to the electron density?

10.5. Provide a brief description of the earth's ionosphere and discuss how it affects communication.

PROBLEMS

10.1. Show that the units of $\sqrt{Ne^2/m\epsilon_0}$ is (seconds)$^{-1}$ and that $e^2/4\pi^2 m\epsilon_0$ is equal to 80.6.

10.2. Assume the ionosphere to be represented by a parabolic distribution of electron density as given by

$$N(h) = \frac{10^{14}}{80.6}\left[1 - \left(\frac{h - 300}{100}\right)^2\right] \text{electrons/m}^3 \qquad \text{for } 200 < h < 400$$

where h is the height above the ground in kilometers. (a) Find the height at which a vertically incident wave of frequency 8 MHz is reflected. (b) Find the frequency of a vertically incident wave which gets reflected at a height of 220 km. (c) What is the lowest frequency below which communication is not possible across the peak of the layer?

10.3. For a uniform plane wave of frequency 10 MHz propagating normal to a slab of ionized medium of thickness 50 km and uniform plasma frequency 8 MHz, find (a) the phase velocity in the slab, (b) the wavelength in the slab, and (c) the number of wavelengths undergone by the wave in the slab.

10.2 WAVE PROPAGATION IN ANISOTROPIC MEDIUM

In Section 5.1, we learned that for certain dielectric materials known as *anisotropic dielectric materials*, **D** is not in general parallel to **E** and the relationship between **D** and **E** is expressed by means of a permittivity tensor consisting of a 3 × 3 matrix. Similarly, in Section 5.2 we learned of the anisotropic property of certain magnetic materials. There are several important applications based on wave propagation in anisotropic materials. A general treatment is, however, very involved. Hence, we shall consider two simple cases.

For the first example, we consider an anisotropic dielectric medium characterized by the **D** to **E** relationship given by

$$\begin{bmatrix} D_x \\ D_y \\ D_z \end{bmatrix} = \begin{bmatrix} \epsilon_{xx} & 0 & 0 \\ 0 & \epsilon_{yy} & 0 \\ 0 & 0 & \epsilon_{zz} \end{bmatrix} \begin{bmatrix} E_x \\ E_y \\ E_z \end{bmatrix} \tag{10.12}$$

and having the permeability μ_0. This simple form of permittivity tensor can be achieved in certain anisotropic liquids and crystals by an appropriate choice of the coordinate system. It is easy to see that the characteristic polarizations for this case are all linear, directed along the coordinate axes and having the effective permittivities ϵ_{xx}, ϵ_{yy}, and ϵ_{zz} for the x-, y-, and z-directed polarizations, respectively. Let us consider a uniform plane wave propagating in the z-direction. The wave will generally contain both x- and y-components of the fields. It can be decomposed into two waves, one having an x-directed electric field and the other having a y-directed electric field. These component waves travel individually in the anisotropic medium as though it is isotropic but with different phase velocities, since the effective permittivities are different. In view of this, the phase relationship between the two waves, and hence the polarization of the composite wave, changes with distance along the direction of propagation.

To illustrate the foregoing discussion quantitatively, let us consider the electric field of the wave to be linearly polarized at $z = 0$, as given by

$$\mathbf{E}(0) = (E_{x0}\,\mathbf{a}_x + E_{y0}\,\mathbf{a}_y)\cos\omega t \tag{10.13}$$

Then assuming $(+)$ wave only, the electric field at an arbitrary value of z is given by

$$\mathbf{E}(z) = E_{x0}\cos(\omega t - \beta_1 z)\,\mathbf{a}_x + E_{y0}\cos(\omega t - \beta_2 z)\,\mathbf{a}_y \tag{10.14}$$

where

$$\beta_1 = \omega\sqrt{\mu_0\epsilon_{xx}} \tag{10.15a}$$

$$\beta_2 = \omega\sqrt{\mu_0\epsilon_{yy}} \tag{10.15b}$$

are the phase constants corresponding to the x-polarized and y-polarized component waves, respectively. Thus, the phase difference between the x- and y-components of the field is given by

$$\Delta\phi = (\beta_2 - \beta_1)z \tag{10.16}$$

As the composite wave progresses along the z-direction, $\Delta\phi$ changes from zero at $z = 0$ to $\pi/2$ at $z = \pi/2(\beta_2 - \beta_1)$ to π at $z = \pi/(\beta_2 - \beta_1)$, and so on. The polarization of the composite wave thus changes from linear at $z = 0$ to elliptical for $z > 0$, becoming linear again at $z = \pi/(\beta_2 - \beta_1)$, but rotated by an angle of $2\tan^{-1}(E_{y0}/E_{x0})$, as shown in Figure 10.2. Thereafter, it becomes elliptical again, returning back to the original linear polarization at $z = 2\pi/(\beta_2 - \beta_1)$, and so on.

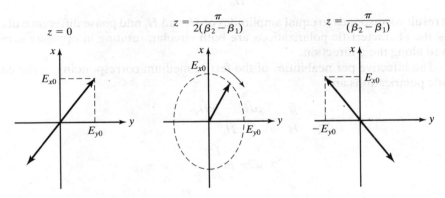

FIGURE 10.2

The change in polarization of the field of a wave propagating in the anisotropic dielectric medium characterized by equation (10.12).

For the second example, we consider propagation in a ferrite medium. Ferrites are a class of magnetic materials which, when subject to a d.c. magnetizing field, exhibit anisotropic magnetic properties. Since there are phase differences associated with the relationships between the components of **B** and the components of **H** due to this anisotropy, it is convenient to use the phasor notation and write the relationship in terms of the phasor components. For an applied d.c. magnetic field along the direction of propagation of the wave, which we assume to be the z-direction, this relationship is given by

$$\begin{bmatrix} \bar{B}_x \\ \bar{B}_y \\ \bar{B}_z \end{bmatrix} = \begin{bmatrix} \mu & -j\kappa & 0 \\ j\kappa & \mu & 0 \\ 0 & 0 & \mu_0 \end{bmatrix} \begin{bmatrix} \bar{H}_x \\ \bar{H}_y \\ \bar{H}_z \end{bmatrix} \tag{10.17}$$

where μ and κ depend upon the material, the strength of the d.c. magnetic field, and the wave frequency.

To find the characteristic polarizations, we first note from (10.17) that

$$\bar{B}_x = \mu \bar{H}_x - j\kappa \bar{H}_y \tag{10.18a}$$

$$\bar{B}_y = j\kappa \bar{H}_x + \mu \bar{H}_y \tag{10.18b}$$

Setting $\bar{B}_x/\bar{B}_y$ equal to $\bar{H}_x/\bar{H}_y$, we then have

$$\frac{\mu \bar{H}_x - j\kappa \bar{H}_y}{j\kappa \bar{H}_x + \mu \bar{H}_y} = \frac{\bar{H}_x}{\bar{H}_y}$$

which upon solution for $\bar{H}_x/\bar{H}_y$ gives

$$\frac{\bar{H}_x}{\bar{H}_y} = \pm j \tag{10.19}$$

This result corresponds to equal amplitudes of H_x and H_y and phase difference of $\pm 90°$. Thus, the characteristic polarizations are both circular, rotating in opposite senses as viewed along the z-direction.

The effective permeabilities of the ferrite medium corresponding to the characteristic polarizations are

$$\frac{\bar{B}_x}{\bar{H}_x} = \frac{\mu \bar{H}_x - j\kappa \bar{H}_y}{\bar{H}_x}$$

$$= \mu - j\kappa \frac{\bar{H}_y}{\bar{H}_x}$$

$$= \mu \mp \kappa \quad \text{for} \quad \frac{\bar{H}_x}{\bar{H}_y} = \pm j \tag{10.20}$$

The phase constants associated with the propagation of the characteristic waves are

$$\beta_{\pm} = \omega \sqrt{\epsilon(\mu \mp \kappa)} \tag{10.21}$$

where the subscripts $+$ and $-$ refer to $\bar{H}_x/\bar{H}_y = +j$ and $\bar{H}_x/\bar{H}_y = -j$, respectively. We note from (10.21) that β_+ can become imaginary if $(\mu - \kappa) < 0$. When this happens, wave propagation does not occur for that characteristic polarization. We shall hereafter assume that the wave frequency is such that both characteristic waves propagate.

Let us now consider the magnetic field of the wave to be linearly polarized in the x direction at $z = 0$, that is,

$$\mathbf{H}(0) = H_0 \cos \omega t \, \mathbf{a}_x \tag{10.22}$$

Then we can express (10.22) as the superposition of two circularly polarized fields having opposite senses of rotation in the xy-plane in the manner

$$\mathbf{H}(0) = \left(\frac{H_0}{2} \cos \omega t \, \mathbf{a}_x + \frac{H_0}{2} \sin \omega t \, \mathbf{a}_y \right)$$

$$+ \left(\frac{H_0}{2} \cos \omega t \, \mathbf{a}_x - \frac{H_0}{2} \sin \omega t \, \mathbf{a}_y \right) \tag{10.23}$$

The circularly polarized field inside the first pair of parentheses on the right side of (10.23) corresponds to

$$\frac{\bar{H}_x}{\bar{H}_y} = \frac{H_0/2}{-jH_0/2} = +j$$

whereas that inside the second pair of parentheses corresponds to

$$\frac{\bar{H}_x}{\bar{H}_y} = \frac{H_0/2}{jH_0/2} = -j$$

Assuming propagation in the positive z-direction, the field at an arbitrary value of z is then given by

$$
\begin{aligned}
\mathbf{H}(z) = &\left[\frac{H_0}{2} \cos(\omega t - \beta_+ z)\, \mathbf{a}_x + \frac{H_0}{2} \sin(\omega t - \beta_+ z)\, \mathbf{a}_y \right] \\
&+ \left[\frac{H_0}{2} \cos(\omega t - \beta_- z)\, \mathbf{a}_x - \frac{H_0}{2} \sin(\omega t - \beta_- z)\, \mathbf{a}_y \right] \\
= &\left[\frac{H_0}{2} \cos\left(\omega t - \frac{\beta_+ + \beta_-}{2} z - \frac{\beta_+ - \beta_-}{2} z \right) \mathbf{a}_x \right. \\
&+ \frac{H_0}{2} \sin\left(\omega t - \frac{\beta_+ + \beta_-}{2} z - \frac{\beta_+ - \beta_-}{2} z \right) \mathbf{a}_y \Bigg] \\
&+ \left[\frac{H_0}{2} \cos\left(\omega t - \frac{\beta_+ + \beta_-}{2} z + \frac{\beta_+ - \beta_-}{2} z \right) \mathbf{a}_x \right. \\
&- \frac{H_0}{2} \sin\left(\omega t - \frac{\beta_+ + \beta_-}{2} z + \frac{\beta_+ - \beta_-}{2} z \right) \mathbf{a}_y \Bigg] \\
= &\left[H_0 \cos\left(\frac{\beta_- - \beta_+}{2} z \right) \mathbf{a}_x + H_0 \sin\left(\frac{\beta_- - \beta_+}{2} z \right) \mathbf{a}_y \right] \\
&\cdot \cos\left(\omega t - \frac{\beta_+ + \beta_-}{2} z \right)
\end{aligned}
\tag{10.24}
$$

The result given by (10.24) indicates that the x- and y-components of the field are in phase at any given value of z. Hence, the field is linearly polarized for all values of z. The direction of polarization is, however, a function of z since

$$\frac{H_y}{H_x} = \frac{H_0 \sin[(\beta_- - \beta_+)/2]z}{H_0 \cos[(\beta_- - \beta_+)/2]z} = \tan\frac{\beta_- - \beta_+}{2} z \tag{10.25}$$

and hence the angle made by the field vector with the x-axis is $[(\beta_- - \beta_+)/2]z$. Thus, the direction of polarization rotates linearly with z at a rate of $(\beta_- - \beta_+)/2$. This phenomenon is known as *Faraday rotation* and is illustrated with the aid of the sketches in Figure 10.3. The sketches in any given column correspond to a fixed value of z, whereas the sketches in a given row correspond to a fixed value of t. At $z = 0$, the field

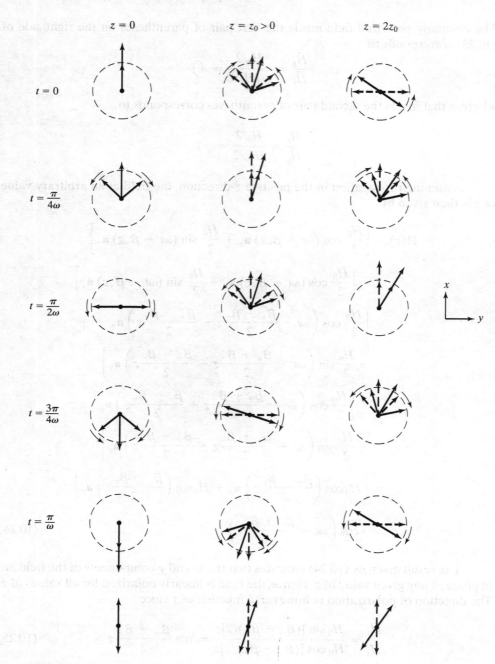

FIGURE 10.3

For illustrating the phenomenon of Faraday rotation.

is linearly polarized in the x-direction and is the superposition of two counter-rotating circularly polarized fields, as shown by the time series of sketches in the first column. If the medium is isotropic, the two counter-rotating circularly polarized fields undergo the same amount of phase lag with z and the field remains linearly polarized in the x-direction, as shown by the dashed lines in the second and third columns. For the case of the anisotropic medium, the two circularly polarized fields undergo different amounts of phase lag with z. Hence, their superposition results in a linear polarization making an angle with the x-direction and increasing linearly with z, as shown by the solid lines in the second and third columns.

The phenomenon of Faraday rotation in a ferrite medium that we have just discussed forms the basis for a number of devices in the microwave field. The phenomenon itself is not restricted to ferrites. For example, an ionized medium immersed in a d.c. magnetic field possesses anisotropic properties that give rise to Faraday rotation of a linearly polarized wave propagating along the d.c. magnetic field. A natural example of this is propagation along the earth's magnetic field in the ionosphere. A simple modern example of the application of Faraday rotation is, however, illustrated by the magneto-optical switch. In fact, Faraday rotation was originally discovered in the optics regime.

The magneto-optical switch is a device for modulating a laser beam by switching on and off an electric current. The electric current generates a magnetic field that rotates the magnetization vector in a magnetic iron-garnet film on a substrate of garnet, in the plane of the film through which a light wave passes. When it enters the film, the light wave field is linearly polarized normal to the plane of the film. If the current in the electric circuit is off, the magnetization vector is normal to the direction of propagation of the wave and the wave emerges out of the film without change of polarization, as shown in Figure 10.4(a). If the current in the electric circuit is on,

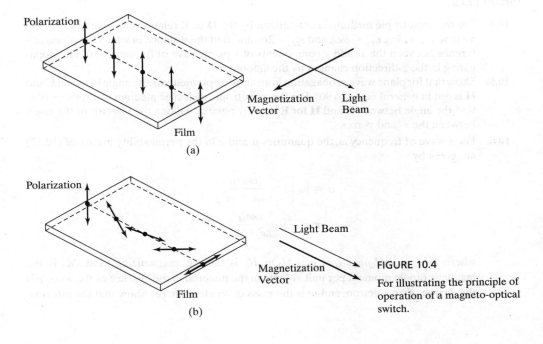

(a)

(b)

FIGURE 10.4

For illustrating the principle of operation of a magneto-optical switch.

the magnetization vector is parallel to the direction of propagation of the wave, the light wave undergoes Faraday rotation and emerges out of the film with its polarization rotated by 90°, as shown in Figure 10.4(b). After it emerges out of the film, the light beam is passed through a polarizer which has the property of absorbing light of the original polarization but passing through the light of the 90°-rotated polarization. Thus, the beam is made to turn on and off by the switching on and off of the current in the electric circuit. In this manner, any coded message can be made to be carried by the light beam.

In this section, we discussed wave propagation in an anisotropic medium. In particular, we learned that in a ferrite medium, a linearly polarized wave propagating along the direction of an applied d.c. magnetic field undergoes Faraday rotation. We then briefly mentioned other examples of media in which Faraday rotation takes place and finally discussed the operation of the magneto-optical switch, a device employing Faraday rotation for modulating a light beam.

REVIEW QUESTIONS

10.6. Discuss the principle behind wave propagation in an anisotropic medium based on the decomposition of the wave into characteristic waves.

10.7. When does a wave propagate in an anisotropic medium without change in polarization?

10.8. What is Faraday rotation? When does Faraday rotation take place in an anisotropic medium?

10.9. Consult appropriate reference books and list three applications of Faraday rotation.

10.10. What is a magneto-optical switch? Discuss its operation.

PROBLEMS

10.4. For the anisotropic medium characterized by the **D** to **E** relationship given by (10.12), assume $\epsilon_{xx} = 4\epsilon_0$, $\epsilon_{yy} = 9\epsilon_0$, and $\epsilon_{zz} = 2\epsilon_0$, and find the distance in which the phase difference between the x- and y-components of a plane wave of frequency 10^9 Hz propagating in the z-direction changes by the amount π.

10.5. Show that for plane wave propagation in an anisotropic medium, the angle between **E** and **H** is not in general equal to 90°. For the anisotropic dielectric medium of Problem 10.4, find the angle between **E** and **H** for **E** linearly polarized along the bisector of the angle between the x- and y-axes.

10.6. For a wave of frequency ω, the quantities μ and κ in the permeability matrix of (10.17) are given by

$$\mu = \mu_0 \left[1 + \frac{\omega_0 \omega_M}{\omega_0^2 - \omega^2} \right]$$

$$\kappa = -\mu_0 \frac{\omega \omega_M}{\omega_0^2 - \omega^2}$$

where $\omega_0 = \mu_0 |e| H_0 / m$, $\omega_M = \mu_0 |e| M_0 / m$, H_0 is the d.c. magnetizing field, M_0 is the magnetic dipole moment per unit volume in the material in the absence of the wave, e is the charge of an electron, and m is the mass of an electron. (a) Show that the effective permeabilities corresponding to the characteristic polarizations are $\mu_0\left[1 + \dfrac{\omega_M}{\omega_0 \mp \omega}\right]$ for $\bar{H}_x/\bar{H}_y = \pm j$. (b) Compute the Faraday rotation angle in degrees per centimeter along the z-direction for $\omega = 10^{10}$ rad/s, if $\omega_M = 5 \times 10^{10}$ rad/s, $\omega_0 = 1.5 \times 10^{10}$ rad/s, and $\epsilon = 9\epsilon_0$.

10.7. For the quantities defined in Problem 10.6 for the ferrite medium, show that for $\omega_0 \ll \omega$ and $\omega_M \ll \omega$, the Faraday rotation per unit distance along the z-direction is $\dfrac{\omega_M}{2}\sqrt{\mu_0\epsilon}$. Compute its value in degrees per centimeter if $\omega_M = 5 \times 10^{10}$ rad/s and $\epsilon = 9\epsilon_0$.

10.3   ELECTROMAGNETIC COMPATIBILITY AND SHIELDING

As stated in the preface of the book, electromagnetics is all around us. Every time we turn on a switch for electrical power or for electronic equipment, every time we press a key on our computer keyboard or on our cell phone, or every time we perform a similar action involving an everyday electrical device, electromagnetics comes into play. While these actions are performed for intentional purposes, the resulting electromagnetic energy may cause unintentional interference of a given system on another system or even one part of a given system on another part of the same system. For example, reception of an FM radio signal may be noisy when the radio is located near a computer, due to radiation from the digital circuits of the computer being received as noise by the radio antenna, thereby degrading the performance of the radio. The computer is said to be causing electromagnetic interference (EMI) in the radio. EMI demonstrates the need for designing systems which are compatible with their electromagnetic environment, which comprises the field of electromagnetic compatibility (EMC).

EMC is defined by IEC (International Electrotechnical Commission) as "the ability of a device, unit of equipment, or system to function satisfactorily in its electromagnetic environment without introducing intolerable electromagnetic disturbances to anything in that environment." An electromagnetic disturbance may be electromagnetic noise, an unwanted signal, or a change in the propagation medium itself. A system is said to be electromagnetically compatible if (1) it does not cause interference with other systems, (2) it is not susceptible to emission from other systems, and (3) it does not cause interference with itself.

In the analysis and design of systems for EMC work, quasistatic concepts are employed wherever they are applicable, because of simplicity compared to working with complete field solutions. We have learned in Chapter 6 that quasistatic approximations apply when the physical dimensions of the system are much smaller than the wavelength corresponding to the frequency of operation. Thus, three regimes come into play, as follows:

1. When the system is electrically small in all three of its dimensions, that is, when the physical size of the system is such that all three of its dimensions are smaller than the wavelength corresponding to the frequency of operation, then quasistatic approximations hold in all three dimensions and the system can be represented by a lumped circuit equivalent and circuit analysis techniques can be employed.

2. When the physical size of the system is smaller than the wavelength in two of its dimensions and comparable to or larger than the wavelength in the third dimension, then the system becomes a transmission line extending along the longer dimension.

3. When the physical size of the system is such that all three of its dimensions are comparable to or larger than the wavelength, then the analysis entails full field basis using the complete set of Maxwell's equations.

In general, a signal is composed of a spectrum of frequencies. The wavelength above is then the shortest significant wavelength, that is, the wavelength corresponding to the highest frequency of importance in the frequency spectrum of the signal.

As shown in Figure 10.5, all EMC problems can be divided into three parts: (a) source of emission or emitter, (b) receiver of emission or victim, and (c) coupling path or mechanism by means of which emission from the source is transferred to the receiver. In the example of noise in the FM radio due to computer, the source of noise is the computer, the victim is the radio, and the coupling mechanism is the medium between the digital circuits in the computer and the antenna of the radio.

FIGURE 10.5

The three parts of an EMC problem.

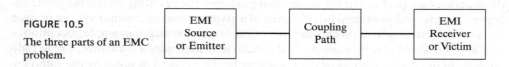

EMI Source or Emitter — Coupling Path — EMI Receiver or Victim

EMC problems can be solved by reducing or eliminating EMI, using one or more of the following three methods: (1) decreasing the emission from the source producing the EMI, (2) making the victim of EMI less susceptible, and (3) making the coupling path less efficient. Although often the only option available to solve an EMC problem is the third one, we shall first consider a simple example of the application of the first two methods.

Thus, let us consider a parallel-wire line consisting of a pair of long, parallel wires of spacing a and carrying currents $I(t)$ in opposite directions in the $z = 0$ plane, as shown in Figure 10.6. Let a small metallic loop of area A be located in the plane of the loop (the xy-plane) and such that the distance between the center of the wires to the center of the loop is $d \gg a$. The currents in the parallel wires produce a magnetic field, resulting in a time-varying magnetic flux enclosed by the loop and hence voltage induced in the loop, in accordance with Faraday's law, causing EMI in the loop. The EMC problem is to find ways to minimize the EMI in the loop.

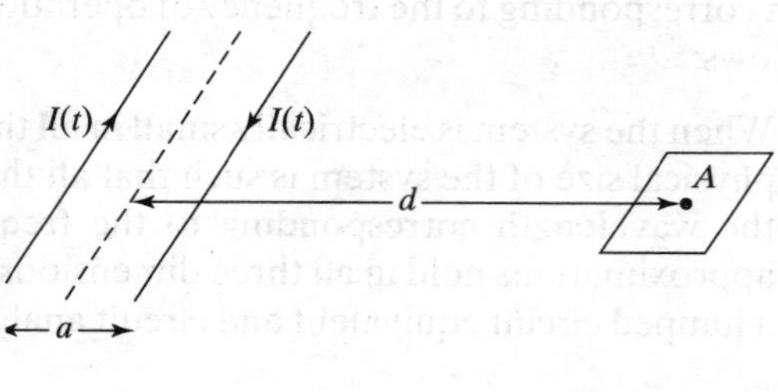

FIGURE 10.6

Arrangement of a metallic loop in the field of a parallel-wire line for illustrating the EMC problem.

$I(t)$ $I(t)$

d

A

$\leftarrow a \rightarrow$

The magnetic field due to the parallel wires can be computed from the fact that in the plane transverse to the wires, the fields have the same spatial character as for the static fields corresponding to the same geometry. Thus, applying the result for the magnetic field due to a long wire in Example 2.9 to the two wires and introducing the time variation, we can write the magnetic flux density due to the parallel wires at the center point of the loop to be

$$\mathbf{B} \text{ at center of loop} = \frac{\mu_0 I(t)}{2\pi} \left[\frac{1}{(d - a/2)} - \frac{1}{(d + a/2)} \right] \qquad (10.26\text{a})$$

and directed normal to the area of the loop. For $d \gg a$,

$$\mathbf{B} \text{ at center of loop} \approx \frac{\mu_0 a I(t)}{2\pi d^2} \qquad (10.26\text{b})$$

Since the area of the loop is very small compared to its distance from the line, we can assume that the magnetic field does not vary significantly within the area. Assuming also that the current in the wires and hence the magnetic field due to it does not vary significantly in the z-direction, we obtain the magnetic flux enclosed by the loop to be

$$\psi = \frac{\mu_0 a A I(t)}{2\pi d^2} \qquad (10.27)$$

The voltage induced in the loop is then given by

$$V = -\frac{d\psi}{dt} = -\frac{\mu_0 a A}{2\pi d^2} \frac{dI(t)}{dt} \qquad (10.28)$$

For $I(t) = I_0 \cos \omega t$,

$$V = \frac{\mu_0 a A I_0 \omega}{2\pi d^2} \sin \omega t \qquad (10.29)$$

It can be seen from (10.29) that for the induced voltage to be small, I_0, a, A, and ω must be as small as possible, and d should be as large as possible. In a practical situation, some of these parameters may be fixed and only the others may be varied. If the size of the loop cannot be varied, the effective area of the loop can be made smaller by rotating it to make an angle with the plane of the wires. When the angle is 90°, the magnetic field is parallel to the area of the loop and the induced voltage is zero, eliminating the EMC problem. If the spacing between the wires can be varied, another way to decrease EMI is to decrease the spacing, and if possible, for the wires to be twisted. An important observation from (10.29) is that the induced voltage in the loop and hence the EMI increases with frequency. This means that for a nonsinusoidal source, the interference from its frequency components is amplified proportional to the frequency.

As stated earlier, often the only option available to solve an EMC problem is to make the coupling path less efficient. Therefore, it is important to understand the coupling mechanisms. Depending on the separation distance between the source and the victim, different techniques of analysis are used. For small separation distances, circuit models can be used by representing the electric field coupling as *capacitive coupling* and magnetic field coupling as *inductive coupling*. An example of analysis involving capacitive and inductive couplings is considered in Section 10.4, devoted to *crosstalk* on transmission lines, which is interference due to a wave propagating along one transmission line inducing a wave on a neighboring second transmission line. When the source and victim share a common conductor, interference occurs through the common impedance of the conductor, and hence the coupling is termed *common impedance coupling*. The analysis is performed using circuit techniques. For large separation distances between the source and the victim, involving an intervening medium, field techniques are employed involving radiation from the source into the medium and the transfer of the radiated energy from the medium into the victim.

The techniques for the solution of EMC problems, that is, for decreasing the impact of EMI on the victim by making the coupling path less efficient, fall into four categories: (a) proper layout of components and cables, (b) system grounding and bonding, (c) surge suppression and filtering, and (d) shielding. The scope of each of these techniques is extensive by itself. We shall here consider only the topic of shielding by providing an example that makes use of the knowledge from Chapters 5 and 7. Specifically, we shall consider the problem of a plane metallic sheet as a shield for an incident plane wave from a distant source.

The geometry pertaining to the problem is shown in Figure 10.7, in which media 1 and 3 are free space, and medium 2 is a metallic sheet of thickness d. A uniform plane wave of radian frequency ω is incident normally on to the metallic sheet from medium 1. Thus, media 1 and 3 are characterized by the propagation parameters

$$\bar{\gamma}_1 = \bar{\gamma}_3 = j\beta_0 = j\omega/c \tag{10.30a}$$

$$\bar{\eta}_1 = \bar{\eta}_3 = \eta_0 = 120\pi \tag{10.30b}$$

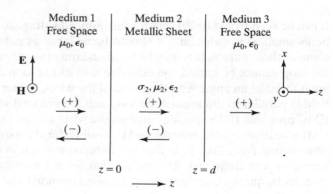

FIGURE 10.7

Geometry of the arrangement for the analysis of shielding by a metallic sheet.

and medium 2 is characterized by the propagation parameters

$$\bar{\gamma}_2 = \sqrt{j\omega\mu_2(\sigma_2 + j\omega\epsilon_2)} \tag{10.31a}$$

$$\bar{\eta}_2 = \sqrt{\frac{j\omega\mu_2}{\sigma_2 + j\omega\epsilon_2}} \tag{10.31b}$$

It is desired to analyze the system for the shielding effectiveness between medium 1 and medium 3. The *shielding effectiveness* or the *shielding factor*, denoted S, is defined to be the ratio of the amplitude of the incident electric field in medium 1 to the amplitude of the transmitted electric field in medium 3.

The incident plane wave sets up a reflected wave and a transmitted wave at the interface $z = 0$, with the reflected wave propagating back in the negative z-direction in medium 1 and the transmitted wave propagating in the positive z-direction in medium 2. When the transmitted wave in medium 2 reaches the interface $z = d$, it sets up a reflected wave which propagates back towards the interface $z = 0$, and a transmitted wave into medium 3. The reflected wave, when it reaches the interface $z = 0$, sets up its own reflection that adds up to the previous transmitted wave due to the incident wave from medium 1. It also sets up a transmitted wave into medium 1, which propagates in the negative z-direction. The transmitted waves into media 3 and 1 will not set up any reflections, because these media are assumed to extend to infinity in the positive z- and negative z-directions, respectively. But each wave in medium 2 sets up a reflected wave and a transmitted wave at the interface on which it is incident. In the steady state, all these transient waves add up and the situation is equivalent to a single $(+)$ wave and a single $(-)$ wave in medium 1, a single $(+)$ wave and a single $(-)$ wave in medium 2, and a single $(+)$ wave in medium 3. Therefore, the complex electric and magnetic field components of the waves in the three media can be written as follows:

Medium 1:

$$\bar{E}_{x1} = \bar{A}_1 e^{-j\beta_0 z} + \bar{B}_1 e^{j\beta_0 z} \tag{10.32a}$$

$$\bar{H}_{y1} = \frac{1}{\eta_0}\left(\bar{A}_1 e^{-j\beta_0 z} - \bar{B}_1 e^{j\beta_0 z}\right) \tag{10.32b}$$

Medium 2:

$$\bar{E}_{x2} = \bar{A}_2 e^{-\bar{\gamma}_2 z} + \bar{B}_2 e^{\bar{\gamma}_2 z} \tag{10.33a}$$

$$\bar{H}_{y2} = \frac{1}{\bar{\eta}_2}\left(\bar{A}_2 e^{-\bar{\gamma}_2 z} - \bar{B}_2 e^{\bar{\gamma}_2 z}\right) \tag{10.33b}$$

Medium 3:

$$\bar{E}_{x3} = \bar{A}_3 e^{-j\beta_0 z} \tag{10.34a}$$

$$\bar{H}_{y3} = \frac{1}{\eta_0} \bar{A}_3 e^{-j\beta_0 z} \tag{10.34b}$$

According to the definition, the shielding factor, S, is then equal to $|\bar{A}_1|/|\bar{A}_3|$. To find this quantity, we note that the constants $\bar{A}_1, \bar{A}_2, \bar{A}_3, \bar{B}_1,$ and $\bar{B}_2$ are related through the boundary conditions at the interfaces $z = 0$ and $z = d$. These are given by

$$\bar{E}_{x1} = \bar{E}_{x2} \quad \text{and} \quad \bar{H}_{y1} = \bar{H}_{y2} \quad \text{at } z = 0 \tag{10.35a}$$

$$\bar{E}_{x2} = \bar{E}_{x3} \quad \text{and} \quad \bar{H}_{y2} = \bar{H}_{y3} \quad \text{at } z = d \tag{10.35b}$$

Thus, we have

$$\bar{A}_1 + \bar{B}_1 = \bar{A}_2 + \bar{B}_2 \tag{10.36a}$$

$$\frac{1}{\eta_0}(\bar{A}_1 - \bar{B}_1) = \frac{1}{\eta_2}(\bar{A}_2 - \bar{B}_2) \tag{10.36b}$$

$$\bar{A}_2 e^{-\bar{\gamma}_2 d} + \bar{B}_2 e^{\bar{\gamma}_2 d} = \bar{A}_3 e^{-j\beta_0 d} \tag{10.36c}$$

$$\frac{1}{\eta_2}(\bar{A}_2 e^{-\bar{\gamma}_2 d} - \bar{B}_2 e^{\bar{\gamma}_2 d}) = \frac{1}{\eta_0} \bar{A}_3 e^{-j\beta_0 d} \tag{10.36d}$$

Solving (10.36c) and (10.36d) for $\bar{A}_2$ and $\bar{B}_2$ in terms of $\bar{A}_3$, we obtain

$$\bar{A}_2 = \bar{A}_3 \frac{1}{1 + \bar{\Gamma}_{23}} e^{-j\beta_0 d} e^{\bar{\gamma}_2 d} \tag{10.37a}$$

$$\bar{B}_2 = \bar{A}_2 \bar{\Gamma}_{23} e^{-2\bar{\gamma}_2 d} \tag{10.37b}$$

where

$$\bar{\Gamma}_{23} = \frac{\eta_0 - \eta_2}{\eta_0 + \eta_2} \tag{10.38}$$

is the electric field reflection coefficient, analogous to the voltage reflection coefficient in transmission-line analysis, for a single transient (+) wave incident from medium 2 onto the interface $z = d$. Substituting for $\bar{B}_2$ in (10.36a) and (10.36b) from (10.37b) and solving for $\bar{A}_1$ in terms of $\bar{A}_2$, we get

$$\bar{A}_1 = \bar{A}_2 \frac{1 + \bar{\Gamma}_{12}\bar{\Gamma}_{23} e^{-2\bar{\gamma}_2 d}}{1 + \bar{\Gamma}_{12}} \tag{10.39}$$

where

$$\bar{\Gamma}_{12} = \frac{\eta_2 - \eta_0}{\eta_2 + \eta_0} \tag{10.40}$$

is the electric field reflection coefficient for a single transient (+) wave incident from medium 1 onto the interface $z = 0$. Note that $\bar{\Gamma}_{12} = -\bar{\Gamma}_{23}$. From (10.39) and (10.37a), we then have

$$\frac{\bar{A}_1}{\bar{A}_3} = \frac{1 + \bar{\Gamma}_{12}\bar{\Gamma}_{23} e^{-2\bar{\gamma}_2 d}}{(1 + \bar{\Gamma}_{12})(1 + \bar{\Gamma}_{23})} e^{-j\beta_0 d} e^{\bar{\gamma}_2 d} \tag{10.41}$$

and the shielding factor is given by

$$S = \frac{|\bar{A}_1|}{|\bar{A}_3|} = \frac{\left|1 + \bar{\Gamma}_{12}\bar{\Gamma}_{23}e^{-2\gamma_2 d}\right|e^{\alpha_2 d}}{|1 + \bar{\Gamma}_{12}||1 + \bar{\Gamma}_{23}|} \tag{10.42}$$

It can be shown (see Problem 10.8) that this result is also obtainable by formulating the solution in terms of the individual transient waves resulting from bouncing back and forth between the interfaces $z = d$ and $z = 0$, writing field expressions for the individual transient waves and adding them up. Three contributions to the right side of (10.42) can then be identified as follows:

$$e^{\alpha_2 d} \text{ — contribution from attenuation in the metallic sheet } (A)$$

$$\left(\frac{1}{|1 + \bar{\Gamma}_{12}||1 + \bar{\Gamma}_{23}|}\right) \text{ — contribution due to transmission from free space to the metallic sheet and from the metallic sheet to free space } (T)$$

$$\left(1 + \bar{\Gamma}_{12}\bar{\Gamma}_{23}e^{-2\gamma_2 d}\right) \text{ — contribution from multiple reflections within the metallic sheet } (M)$$

The general formula for S given by (10.42) can be simplified for good conductor range of frequencies $(\sigma_2 \gg \omega\epsilon_2)$ for the metallic sheet, by recalling from Section 5.4 that for good conductors,

$$\bar{\eta}_2 \approx (1 + j)\sqrt{\pi f\, \mu_2/\sigma_2} \tag{10.43}$$

and $|\bar{\eta}_2| \ll \eta_0$, so that $\bar{\Gamma}_{12} \approx -1, \bar{\Gamma}_{23} \approx 1, (1 + \bar{\Gamma}_{12}) \approx 2\bar{\eta}_2/\eta_0$, and $(1 + \bar{\Gamma}_{23}) \approx 2$. Also, for good conductors, $\alpha_2 = \beta_2 \approx \sqrt{\pi f \mu_2 \sigma_2}$. Thus,

$$S \approx \frac{\eta_0}{4|\bar{\eta}_2|}\left|1 - e^{-2\alpha_2 d}e^{-j2\alpha_2 d}\right|e^{\alpha_2 d} \tag{10.44}$$

In terms of skin depth $\delta = 1/\alpha = 1/\sqrt{\pi f\mu_2\sigma_2}$, the distance in which the fields are attenuated in the good conductor by the factor e^{-1},

$$S \approx \frac{\eta_0}{4|\bar{\eta}_2|}\left|1 - e^{-2d/\delta}e^{-j2d/\delta}\right|e^{d/\delta} \tag{10.45}$$

In terms of decibels,

$$S \text{ in dB} = 20 \log_{10}\left(\eta_0/4|\bar{\eta}_2|\right)$$
$$+ 20 \log_{10}\left|1 - e^{-2d/\delta}e^{-j2d/\delta}\right|$$
$$+ 20 \log_{10} e^{d/\delta} \tag{10.46}$$

with the three terms on the right side identifying the three contributions, T, M, and A, respectively.

For a numerical example, for copper sheet, $\sigma_2 = 5.80 \times 10^7$ S/m, $\mu_2 = \mu_0$, $\epsilon_2 = \epsilon_0$, $\delta = 0.066/\sqrt{f}$ m, and $|\bar{\eta}_2| = 3.69 \times 10^{-7}\sqrt{f}$ Ω. For a given set of values of d and f, the quantities T, M, and A, and hence S can be computed. Table 10.1 shows these quantities for four pairs of values of d and f.

TABLE 10.1   Values of T, M, A, and S, for Several Pairs of Values of d and f for the Shielding Arrangement of Figure 10.7

d (mm)	f (MHz)	δ (mm)	$\|\bar{\eta}_2\|$ (Ω)	d/δ	T (db)	M (db)	A (db)	S (db)
1	1	0.066	3.69×10^{-4}	15.15	108.14	~0	131.59	239.73
1	0.1	0.209	1.167×10^{-4}	4.785	118.14	~0	41.56	159.70
0.001	1	0.066	3.69×10^{-4}	0.015	108.14	−27.59	0.13	80.68
0.001	0.1	0.209	1.167×10^{-4}	0.0048	118.14	−37.27	0.042	80.91

From Table 10.1, we can make the following observations:

1. For thick sheets, M is approximately zero and hence not important.
2. For thin sheets, A is negligible and M is important. Furthermore, since M is negative, meaning that the field is enhanced instead of getting attenuated, it acts counter to the shielding requirement.
3. For thin sheets, increase in T with decrease in frequency is countered by the increase in magnitude of M.
4. T is independent of thickness.

In this section, we introduced the topic of electromagnetic compatibility (EMC), having to do with the design of electrical systems which are compatible with the electromagnetic environment. We learned that the EMC problem can be divided into three parts, (a) source, (b) receiver, and (c) coupling path, and that it can be solved by three methods: (a) decreasing emission from the source, (b) making the receiver less susceptible, and (c) making the coupling path less efficient, which is often the only available option. We provided a simple example of the application of the first two methods. While there are several categories pertinent to the third method, we provided the example of electromagnetic shielding by considering the problem of a plane metallic sheet as a shield for an incident plane wave from a distant source.

REVIEW QUESTIONS

10.11. Describe EMI and EMC. What is IEEE's definition of EMC?

10.12. Outline the three regimes that come into play in the design of systems for EMC work.

10.13. Specify and discuss the three parts of an EMC problem.

10.14. Discuss the example of EMI in a metallic loop located in the field of a parallel-wire line and ways to minimize the EMI.

10.15. Outline the solution of the problem of a plane metallic sheet as a shield for an incident plane wave from a distant source.

10.16. What is *shielding factor*? Discuss the three contributions to the shielding factor for the plane metallic sheet arrangement.

PROBLEMS

10.8. For the arrangement of Figure 10.7, obtain the expression for the shielding factor by formulating the solution in terms of the individual transient waves bouncing back and forth between the interfaces $z = d$ and $z = 0$, writing the field expressions for the individual transient waves, and adding them up.

10.9. Compute the value of the shielding factor for a copper shield of thickness 0.01 mm at a frequency of 1 MHz.

10.10. Compute the value of the shielding factor for a steel shield of thickness 0.01 mm at a frequency of 10 MHz. Values of material parameters are as follows: $\sigma = 5.80 \times 10^6$ S/m, $\mu = 500\mu_0$, and $\epsilon = \epsilon_0$.

10.4 CROSSTALK ON TRANSMISSION LINES

When two or more transmission lines are in the vicinity of one another, a wave propagating along one line, which we shall call the primary line, can induce a wave on another line, the secondary line, due to capacitive (electric field) and inductive (magnetic field) coupling between the two lines, resulting in the undesirable phenomenon of crosstalk between the lines. An example is illustrated by the arrangement of Figure 10.8(a), which is a printed-circuit board (PCB) representation of two closely spaced transmission lines. Figure 10.8(b) represents the distributed circuit equivalent, where $\mathscr{C}_m$ and $\mathscr{L}_m$ are the coupling capacitance and coupling inductance, respectively, per unit length of the arrangement.

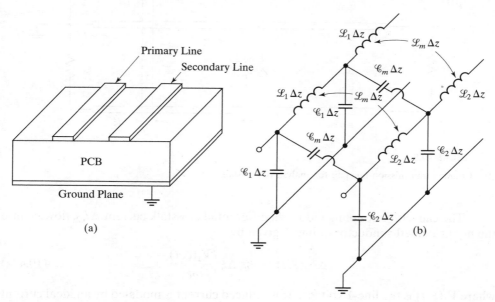

FIGURE 10.8

(a) PCB representation of two closely spaced transmission lines. (b) Distributed equivalent circuit for (a).

In this section, we shall analyze a pair of coupled transmission lines for the determination of induced waves on the secondary line for a given wave on the primary line. To keep the analysis simple, we shall consider both lines to be of the same characteristic impedance, velocity of propagation, and length, and terminated by their characteristic impedances, so that no reflections occur from the ends of either line. It is also convenient to assume the coupling to be weak, so that the effects on the primary line of waves induced in the secondary line can be neglected. Thus, we shall be concerned only with the crosstalk from the primary line to the secondary line and not vice versa. Briefly, as the (+) wave propagates on the primary line from source toward load, each infinitesimal length of that line induces voltage and current in the adjacent infinitesimal length of the secondary line, which set up (+) and (−) waves on that line. The contributions due to the infinitesimal lengths add up to give the induced voltage and current at a given location on the secondary line.

We shall represent the coupled-line pair, as shown in Figure 10.9, with the primary line as line 1 and the secondary line as line 2. Then, when the switch S is closed at $t = 0$, a (+) wave originates at $z = 0$ on line 1 and propagates toward the load. Let us consider a differential length $d\xi$ at the location $z = \xi$ of line 1 charged to the (+) wave voltage and current and obtain its contributions to the induced voltages and currents in line 2.

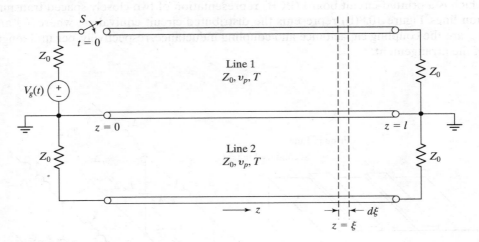

FIGURE 10.9

Coupled transmission-line pair for analysis of crosstalk.

The capacitive coupling induces a differential crosstalk current ΔI_{c2}, flowing into the nongrounded conductor of line 2, given by

$$\Delta I_{c2}(\xi, t) = \mathscr{C}_m \, \Delta\xi \, \frac{\partial V_1(\xi, t)}{\partial t} \tag{10.47a}$$

where $V_1(\xi, t)$ is the line-1 voltage. This induced current is modeled by an ideal current source, connected in parallel with line 2 at $z = \xi$ on that line, as shown in Figure 10.10(a). The current source views the characteristic impedance of the line to either side of $z = \xi$,

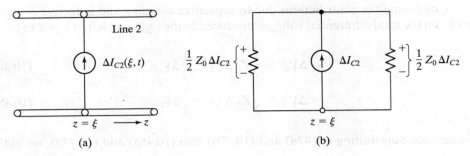

FIGURE 10.10

(a) Modeling for capacitive coupling in crosstalk analysis. (b) Equivalent circuit for (a).

so that the equivalent circuit is as shown in Figure 10.10(b). Thus, voltages of $\frac{1}{2}Z_0\,\Delta I_{c2}$ are produced to the right and left of $z = \xi$ and propagate as forward-crosstalk and backward-crosstalk voltages, respectively, on line 2.

The inductive coupling induces a differential crosstalk voltage, ΔV_{c2}, which is given by

$$\Delta V_{c2}(\xi, t) = \mathcal{L}_m\,\Delta\xi\,\frac{\partial I_1(\xi, t)}{\partial t} \tag{10.47b}$$

This induced voltage is modeled by an ideal voltage source in series with line 2 at $z = \xi$ on that line, as shown in Figure 10.11(a). The polarity of the voltage source is such that the current due to it in line 2 produces a magnetic flux, which opposes the change in the flux due to the current in line 1, in accordance with Lenz's law. The voltage source views the characteristic impedance of the line to either side of it, so that the equivalent circuit is as shown in Figure 10.11(b). Thus, voltages of $\frac{1}{2}\Delta V_{c2}$ and $-\frac{1}{2}\Delta V_{c2}$ are produced to the left and right of $z = \xi$, respectively, and propagate as backward-crosstalk and forward-crosstalk voltages, respectively, on line 2.

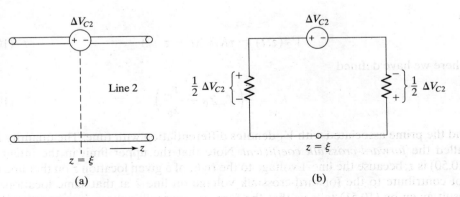

FIGURE 10.11

(a) Modeling for inductive coupling in crosstalk analysis. (b) Equivalent circuit for (a).

Combining the contributions due to capacitive coupling and inductive coupling, we obtain the total differential voltages produced to the right and left of $z = \xi$ to be

$$\Delta V_2^+ = \frac{1}{2} Z_0 \, \Delta I_{c2} - \frac{1}{2} \Delta V_{c2} \tag{10.48a}$$

$$\Delta V_2^- = \frac{1}{2} Z_0 \, \Delta I_{c2} + \frac{1}{2} \Delta V_{c2} \tag{10.48b}$$

respectively. Substituting (10.47a) and (10.47b) into (10.48a) and (10.48b), we obtain

$$\Delta V_2^+(\xi, t) = \left[\frac{1}{2} \mathscr{C}_m Z_0 \frac{\partial V_1(\xi, t)}{\partial t} - \frac{1}{2} \mathscr{L}_m \frac{\partial I_1(\xi, t)}{\partial t} \right] \Delta \xi$$

$$= \frac{1}{2} \left(\mathscr{C}_m Z_0 - \frac{\mathscr{L}_m}{Z_0} \right) \frac{\partial V_1(\xi, t)}{\partial t} \Delta \xi \tag{10.49a}$$

$$\Delta V_2^-(\xi, t) = \frac{1}{2} \left(\mathscr{C}_m Z_0 + \frac{\mathscr{L}_m}{Z_0} \right) \frac{\partial V_1(\xi, t)}{\partial t} \Delta \xi \tag{10.49b}$$

where we have substituted $I_1 = V_1/Z_0$, in accordance with the relationship between the voltage and current of a (+) wave.

We are now ready to apply (10.49a) and (10.49b) in conjunction with super-position to obtain the (+) and (−) wave voltages at any location on line 2, due to a (+) wave of voltage $V_1(t - z/v_p)$ on line 1. Thus, noting that the effect of V_1 at $z = \xi$ at a given time t is felt at a location $z > \xi$ on line 2 at time $t + (z - \xi)/v_p$, we can write

$$V_2^+(z, t) = \int_0^z \frac{1}{2} \left(\mathscr{C}_m Z_0 - \frac{\mathscr{L}_m}{Z_0} \right) \frac{\partial}{\partial t} \left[V_1\left(t - \frac{\xi}{v_p} - \frac{z - \xi}{v_p} \right) \right] d\xi$$

$$= \frac{1}{2} \left(\mathscr{C}_m Z_0 - \frac{\mathscr{L}_m}{Z_0} \right) \int_0^z \frac{\partial V_1(t - z/v_p)}{\partial t} d\xi \tag{10.50}$$

or

$$V_2^+(z, t) = z K_f V_1'(t - z/v_p) \tag{10.51}$$

where we have defined

$$K_f = \frac{1}{2} \left(\mathscr{C}_m Z_0 - \frac{\mathscr{L}_m}{Z_0} \right) \tag{10.52}$$

and the prime associated with V_1 denotes differentiation with time. The quantity K_f is called the *forward-crosstalk coefficient*. Note that the upper limit in the integral in (10.50) is z, because the line-1 voltage to the right of a given location z on that line does not contribute to the forward-crosstalk voltage on line 2 at that same location. The result given by (10.51) tells us that the forward-crosstalk voltage is proportional to z and the time derivative of the primary line voltage.

To obtain $V_2^-(z, t)$, we note that the effect of V_1 at $z = \xi$ at a given time t is felt at a location $z < \xi$ on line 2 at time $t + (\xi - z)/v_p$. Hence,

$$V_2^-(z, t) = \int_z^l \frac{1}{2}\left(\mathscr{C}_m Z_0 + \frac{\mathscr{L}_m}{Z_0}\right)\frac{\partial}{\partial t}\left[V_1\left(t - \frac{\xi}{v_p} - \frac{\xi - z}{v_p}\right)\right]d\xi$$

$$= \frac{1}{2}\left(\mathscr{C}_m Z_0 + \frac{\mathscr{L}_m}{Z_0}\right)\int_z^l \frac{\partial}{\partial t}\left[V_1\left(t + \frac{z}{v_p} - \frac{2\xi}{v_p}\right)\right]d\xi$$

$$= -\frac{1}{4}v_p\left(\mathscr{C}_m Z_0 + \frac{\mathscr{L}_m}{Z_0}\right)\int_z^l \frac{\partial}{\partial \xi}\left[V_1\left(t + \frac{z}{v_p} - \frac{2\xi}{v_p}\right)\right]d\xi$$

$$= -\frac{1}{4}v_p\left(\mathscr{C}_m Z_0 + \frac{\mathscr{L}_m}{Z_0}\right)\left[V_1\left(t + \frac{z}{v_p} - \frac{2\xi}{v_p}\right)\right]_{\xi=z}^l \quad (10.53)$$

or

$$V_2^-(z, t) = K_b\left[V_1\left(t - \frac{z}{v_p}\right) - V_1\left(t - \frac{2l}{v_p} + \frac{z}{v_p}\right)\right] \quad (10.54)$$

where we have defined the *backward-crosstalk coefficient*

$$K_b = \frac{1}{4}v_p\left(\mathscr{C}_m Z_0 + \frac{\mathscr{L}_m}{Z_0}\right) \quad (10.55)$$

Note that the lower limit in the integral in (10.53) is z, because the line-1 voltage to the left of a given location z on that line does not contribute to the backward-crosstalk voltage on line 2 at that same location.

For an example to illustrate the application of (10.51) and (10.54), let $V_g(t)$ in Figure 10.9 be the function shown in Figure 10.12, where $T_0 < T(= l/v_p)$. We wish to determine the (+) and (−) wave voltages on line 2.

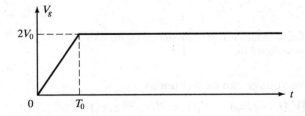

FIGURE 10.12

Source voltage for the system of Figure 10.9.

Noting that

$$V_1(t) = \frac{1}{2}V_g(t) = \begin{cases} (V_0/T_0)t & \text{for } 0 < t < T_0 \\ V_0 & \text{for } t > T_0 \end{cases}$$

and hence

$$V_1'(t) = \begin{cases} V_0/T_0 & \text{for } 0 < t < T_0 \\ 0 & \text{for } t > T_0 \end{cases}$$

and using (10.51), we can write the (+) wave voltage on line 2 as

$$V_2^+(z, t) = z K_f V_1'(t - z/v_p)$$

$$= \begin{cases} z K_f V_0/T_0 & \text{for } 0 < (t - z/v_p) < T_0 \\ 0 & \text{otherwise} \end{cases}$$

$$= \begin{cases} z K_f V_0/T_0 & \text{for } (z/v_p) < t < (z/v_p + T_0) \\ 0 & \text{otherwise} \end{cases}$$

$$= \begin{cases} z K_f V_0/T_0 & \text{for } (z/l)T < t < [(z/l)t + T_0] \\ 0 & \text{otherwise} \end{cases}$$

This is shown in the three-dimensional plot of Figure 10.13, in which the cross section in any constant-z plane is a pulse of voltage $z K_f V_0/T_0$ for $(z/l)T < t < (z/l)T + T_0$. Note that the pulse voltage is shown to be negative. This is because normally the effect of inductive coupling dominates that of the capacitive coupling, so that K_f is negative.

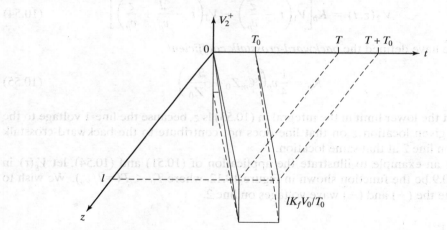

FIGURE 10.13

Three-dimensional depiction of forward-crosstalk voltage for the system of Figure 10.9, with $V_g(t)$ as in Figure 10.12.

Using (10.54), the (−) wave voltage can be written as

$$V_2^-(z, t) = K_b[V_1(t - z/v_p) - V_1(t - 2l/v_p + z/v_p)]$$

where

$$V_1\left(t - \frac{z}{v_p}\right) = \begin{cases} \dfrac{V_0}{T_0}\left(t - \dfrac{z}{v_p}\right) & \text{for } 0 < \left(t - \dfrac{z}{v_p}\right) < T_0 \\ V_0 & \text{for } \left(t - \dfrac{z}{v_p}\right) > T_0 \end{cases}$$

$$= \begin{cases} \dfrac{V_0}{T_0}\left(t - \dfrac{z}{l}T\right) & \text{for } \dfrac{z}{l}T < t < \left(\dfrac{z}{l}T + T_0\right) \\ V_0 & \text{for } t > \left(\dfrac{z}{l}T + T_0\right) \end{cases}$$

$$V_1\left(t - \frac{2l}{v_p} + \frac{z}{v_p}\right) = \begin{cases} \frac{V_0}{T_0}\left(t - \frac{2l}{v_p} + \frac{z}{v_p}\right) & \text{for } 0 < \left(t - \frac{2l}{v_p} + \frac{z}{v_p}\right) < T_0 \\ V_0 & \text{for } \left(t - \frac{2l}{v_p} + \frac{z}{v_p}\right) > T_0 \end{cases}$$

$$= \begin{cases} \frac{V_0}{T_0}\left(t - 2T + \frac{z}{l}T\right) & \text{for } \left(2T - \frac{z}{l}T\right) < t < \left(2T - \frac{z}{l}T + T_0\right) \\ V_0 & \text{for } t > \left(2T - \frac{z}{l}T + T_0\right) \end{cases}$$

These two voltages and the $(-)$ wave voltage for a value of z for which $(z/l)T + T_0 < 2T - (z/l)T$ are shown in Figure 10.14. Figure 10.15 shows the three-dimensional plot of $V_2^-(z, t)$, in which the cross section in any given constant-z plane gives the time variation of V_2^- for that value of z. Note that as z varies from zero to l, the shape of V_2^- changes from a trapezoidal pulse with a height of $K_b V_0$ at $z = 0$ to a triangular pulse of height $K_b V_0$ and width $2T_0$ at $z = (1 - T_0/2T)l$ and then changes to a trapezoidal pulse again but with a height continuously decreasing from $K_b V_0$ to zero at $z = l$.

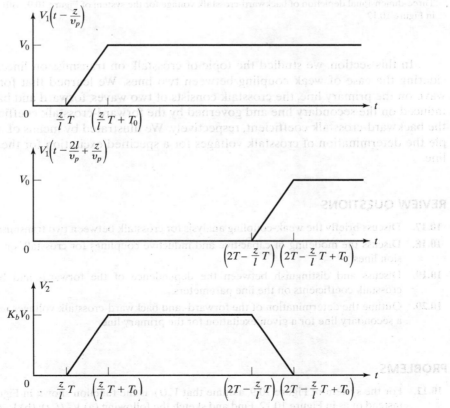

FIGURE 10.14

Determination of backward-crosstalk voltage for the system of Figure 10.9, with $V_g(t)$ as in Figure 10.12.

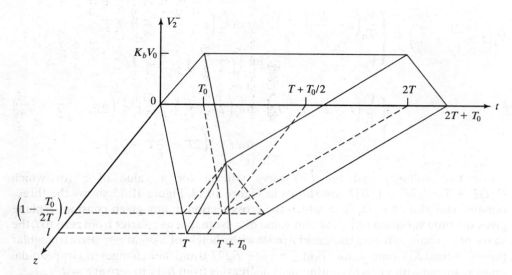

FIGURE 10.15

Three-dimensional depiction of backward-crosstalk voltage for the system of Figure 10.9, with $V_g(t)$ as in Figure 10.12.

In this section, we studied the topic of crosstalk on transmission lines, by considering the case of weak coupling between two lines. We learned that for a given wave on the primary line, the crosstalk consists of two waves, forward and backward, induced on the secondary line and governed by the forward-crosstalk coefficient and the backward-crosstalk coefficient, respectively. We illustrated by means of an example the determination of crosstalk voltages for a specified excitation for the primary line.

REVIEW QUESTIONS

10.17. Discuss briefly the weak-coupling analysis for crosstalk between two transmission lines.

10.18. Discuss the modeling of capacitive and inductive couplings for crosstalk on transmission lines.

10.19. Discuss and distinguish between the dependence of the forward- and backward-crosstalk coefficients on the line parameters.

10.20. Outline the determination of the forward- and backward-crosstalk voltages induced on a secondary line for a given excitation for the primary line.

PROBLEMS

10.11. For the system of Figure 10.9, assume that $V_g(t)$ is the function shown in Figure 10.16, instead of as in Figure 10.12. Find and sketch the following (a) $V_2^+(l, t)$; (b) $V_2^-(0, t)$; and (c) $V_2^-(0.8l, t)$.

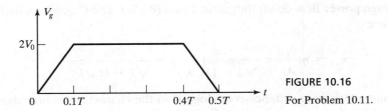

FIGURE 10.16
For Problem 10.11.

10.12. For the system of Figure 10.9, assume that

$$V_g(t) = \begin{cases} 2V_0 \sin^2 \pi t/T & \text{for } 0 < t < T \\ 0 & \text{otherwise} \end{cases}$$

Find and sketch the following: (a) $V_2^+(l, t)$; (b) $V_2^-(0, t)$; (c) $V_2^-(0.75l, t)$.

10.13. For the system of Figure 10.9, assume that $K_b/K_f = -25v_p$ and $T_0 = 0.2T$. For $V_g(t)$ given in Figure 10.12, find and sketch the following: (a) $V_2^+(z, 1.1T)$; (b) $V_2^-(z, 1.1T)$; and (c) $V_2(z, 1.1T)$.

10.5 PARALLEL-PLATE WAVEGUIDE DISCONTINUITY

In Section 8.2, we introduced $TE_{m,0}$ waves in a parallel-plate waveguide. Let us now consider reflection and transmission at a dielectric discontinuity in a parallel-plate guide, as shown in Figure 10.17. If a $TE_{m,0}$ wave is incident on the junction from section 1, then it will set up a reflected wave into section 1 and a transmitted wave into section 2, provided that mode propagates in that section. The fields corresponding to these incident, reflected, and transmitted waves must satisfy the boundary conditions at the dielectric discontinuity. These boundary conditions were derived in Section 5.5. Denoting the incident, reflected, and transmitted wave fields by the subscripts i, r, and t, respectively, we have from the continuity of the tangential component of **E** at a dielectric discontinuity,

$$E_{yi} + E_{yr} = E_{yt} \quad \text{at } z = 0 \tag{10.56}$$

and from the continuity of the tangential component of **H** at a dielectric discontinuity,

$$H_{xi} + H_{xr} = H_{xt} \quad \text{at } z = 0 \tag{10.57}$$

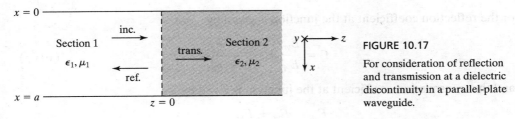

FIGURE 10.17

For consideration of reflection and transmission at a dielectric discontinuity in a parallel-plate waveguide.

We now define the guide characteristic impedance, η_{g1}, of section 1 as

$$\eta_{g1} = \frac{E_{yi}}{-H_{xi}} \tag{10.58}$$

Recognizing that $\mathbf{a}_y \times (-\mathbf{a}_x) = \mathbf{a}_z$, we note that η_{g1} is simply the ratio of the transverse components of the electric and magnetic fields of the $TE_{m,0}$ wave that give rise to time-average power flow down the guide. From (8.35a) and (8.35b) applied to section 1, we have

$$\eta_{g1} = \eta_1 \frac{\lambda_{g1}}{\lambda_1} = \frac{\eta_1}{\sqrt{1-(\lambda_1/\lambda_c)^2}} = \frac{\eta_1}{\sqrt{1-(f_{c1}/f)^2}} \tag{10.59}$$

The guide characteristic impedance is analogous to the characteristic impedance of a transmission line, if we recognize that E_{yi} and $-H_{xi}$ are analogous to V^+ and I^+, respectively. In terms of the reflected wave fields, it then follows that

$$\eta_{g1} = -\left(\frac{E_{yr}}{-H_{xr}}\right) = \frac{E_{yr}}{H_{xr}} \tag{10.60}$$

This result can also be seen from the fact that for the reflected wave, the power flow is in the negative z-direction and since $\mathbf{a}_y \times \mathbf{a}_x = -\mathbf{a}_z$, η_{g1} is equal to E_{yr}/H_{xr}. For the transmitted wave fields, we have

$$\frac{E_{yt}}{-H_{xt}} = \eta_{g2} \tag{10.61}$$

where

$$\eta_{g2} = \eta_2 \frac{\lambda_{g2}}{\lambda_2} = \frac{\eta_2}{\sqrt{1-(\lambda_2/\lambda_c)^2}} = \frac{\eta_2}{\sqrt{1-(f_{c2}/f)^2}} \tag{10.62}$$

is the guide characteristic impedance of section 2.

Using (10.58), (10.60), and (10.61), (10.57) can be written as

$$\frac{E_{yi}}{\eta_{g1}} - \frac{E_{yr}}{\eta_{g1}} = \frac{E_{yt}}{\eta_{g2}} \tag{10.63}$$

Solving (10.56) and (10.63), we get

$$E_{yi}\left(1 - \frac{\eta_{g2}}{\eta_{g1}}\right) + E_{yr}\left(1 + \frac{\eta_{g2}}{\eta_{g1}}\right) = 0 \tag{10.64}$$

or the reflection coefficient at the junction is given by

$$\Gamma = \frac{E_{yr}}{E_{yi}} = \frac{\eta_{g2} - \eta_{g1}}{\eta_{g2} + \eta_{g1}} \tag{10.65}$$

and the transmission coefficient at the junction is given by

$$\tau = \frac{E_{yt}}{E_{yi}} = \frac{E_{yi} + E_{yr}}{E_{yi}} = 1 + \Gamma \tag{10.66}$$

These expressions for Γ and τ are similar to those obtained in Section 7.2 for reflection and transmission at a transmission-line discontinuity. Hence, insofar as reflection and transmission at the junction are concerned, we can replace the waveguide sections by transmission lines having characteristic impedances equal to the guide characteristic impedances, as shown in Figure 10.18. It should be noted that unlike the characteristic impedance of a lossless line, which is a constant independent of frequency, the guide characteristic impedance of the lossless waveguide is a function of the frequency.

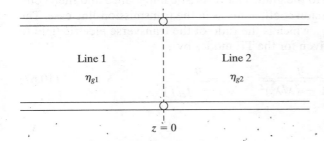

FIGURE 10.18

Transmission-line equivalent of parallel-plate waveguide discontinuity.

For a numerical example of computing Γ and τ, let us consider the parallel-plate waveguide discontinuity shown in Figure 10.19, and $TE_{1,0}$ waves of frequency $f = 5000$ MHz, incident on the junction from the free space side.

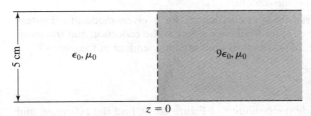

FIGURE 10.19

For illustrating the computation of reflection and transmission coefficients at a parallel-plate waveguide discontinuity.

For the $TE_{1,0}$ mode, $\lambda_c = 2a = 10$ cm, independent of the dielectric. For $f = 5000$ MHz,

$$\lambda_1 = \text{wavelength on the free space side} = \frac{3 \times 10^8}{5 \times 10^9} = 6 \text{ cm}$$

$$\lambda_2 = \text{wavelength on the dielectric side} = \frac{3 \times 10^8}{\sqrt{9} \times 5 \times 10^9} = \frac{6}{3} = 2 \text{ cm}$$

Since $\lambda < \lambda_c$ in both sections, $TE_{1,0}$ mode propagates in both sections. Thus,

$$\eta_{g1} = \frac{\eta_1}{\sqrt{1 - (\lambda_1/\lambda_c)^2}} = \frac{120\pi}{\sqrt{1 - (6/10)^2}} = 471.24 \ \Omega$$

$$\eta_{g2} = \frac{\eta_2}{\sqrt{1 - (\lambda_2/\lambda_c)^2}} = \frac{120\pi/\sqrt{9}}{\sqrt{1 - (2/10)^2}} = \frac{40\pi}{\sqrt{1 - 0.04}} = 128.25 \ \Omega$$

$$\Gamma = \frac{\eta_{g2} - \eta_{g1}}{\eta_{g2} + \eta_{g1}} = \frac{128.25 - 471.24}{128.25 + 471.24} = -0.572$$

$$\tau = 1 + \Gamma = 1 - 0.572 = 0.428$$

For $f = 4000$ MHz, we would obtain $\Gamma = -0.629$ and $\tau = 0.371$.

In this section, we discussed the solution of problems involving reflection and transmission at a discontinuity in a waveguide by using the transmission-line analogy. This consists of replacing each section of the waveguide by a transmission line whose characteristic impedance is equal to the guide characteristic impedance and then computing the reflection and transmission coefficients as in the transmission-line case. The guide characteristic impedance, η_g, which is the ratio of the transverse electric field to the transverse magnetic field, is given for the TE modes by

$$\eta_g = \frac{\eta}{\sqrt{1 - (\lambda/\lambda_c)^2}} = \frac{\eta}{\sqrt{1 - (f_c/f)^2}} \tag{10.67}$$

REVIEW QUESTIONS

10.21. Define guide characteristic impedance.

10.22. Provide a physical explanation for why the guide characteristic impedance is different from the intrinsic impedance of the medium in the guide.

10.23. Discuss the use of the transmission-line analogy for solving problems involving reflection and transmission at a waveguide discontinuity.

10.24. Why are the reflection and transmission coefficients for a given mode at a lossless waveguide discontinuity dependent on frequency whereas the reflection and transmission coefficients at the junction of two lossless lines are independent of frequency?

PROBLEMS

10.14. For the parallel-plate waveguide discontinuity of Figure 10.19, find the reflection and transmission coefficients for $f = 7500$ MHz propagating in (a) $TE_{1,0}$ mode and (b) $TE_{2,0}$ mode.

10.15. The left half of a parallel-plate waveguide of dimension $a = 4$ cm is filled with a dielectric of $\epsilon = 4\epsilon_0$ and $\mu = \mu_0$. The right half is filled with a dielectric of $\epsilon = 9\epsilon_0$ and $\mu = \mu_0$. For $TE_{1,0}$ waves of frequency 2500 MHz incident on the discontinuity from the left, find the reflection and transmission coefficients.

10.16. Assume that the permittivity of the dielectric to the right side of the parallel-plate waveguide discontinuity of Figure 10.19 is unknown. If the reflection coefficient for $TE_{1,0}$ waves of frequency 5000 MHz incident on the junction from the free space side is -0.2643, find the permittivity of the dielectric.

10.6   MAGNETIC VECTOR POTENTIAL AND THE LOOP ANTENNA

In Section 6.1, we learned that since

$$\nabla \times \mathbf{E} = 0$$

for the static electric field, $\mathbf{E}$ can be expressed as the gradient of a scalar potential in the manner

$$\mathbf{E} = -\nabla V$$

We then proceeded with the discussion of the electric scalar potential and its application for the computation of static electric fields. In this section, we shall introduce a similar tool for the magnetic field computation, namely, the magnetic vector potential. When extended to the time-varying case, the magnetic vector potential has useful application in the determination of fields due to antennas.

To introduce the magnetic vector potential concept, we recall that the divergence of the magnetic flux density vector, whether static or time-varying, is equal to zero, that is,

$$\nabla \cdot \mathbf{B} = 0 \tag{10.68}$$

If the divergence of a vector is zero, then that vector can be expressed as the curl of another vector, since the divergence of the curl of a vector is identically equal to zero, as can be seen by expansion in Cartesian coordinates:

$$\nabla \cdot \nabla \times \mathbf{A} = \left(\mathbf{a}_x \frac{\partial}{\partial x} + \mathbf{a}_y \frac{\partial}{\partial y} + \mathbf{a}_z \frac{\partial}{\partial z} \right) \cdot \begin{vmatrix} \mathbf{a}_x & \mathbf{a}_y & \mathbf{a}_z \\ \dfrac{\partial}{\partial x} & \dfrac{\partial}{\partial y} & \dfrac{\partial}{\partial z} \\ A_x & A_y & A_z \end{vmatrix}$$

$$= \begin{vmatrix} \dfrac{\partial}{\partial x} & \dfrac{\partial}{\partial y} & \dfrac{\partial}{\partial z} \\ \dfrac{\partial}{\partial x} & \dfrac{\partial}{\partial y} & \dfrac{\partial}{\partial z} \\ A_x & A_y & A_z \end{vmatrix} = 0$$

Thus, the magnetic field vector $\mathbf{B}$ can be expressed as the curl of a vector $\mathbf{A}$, that is,

$$\mathbf{B} = \nabla \times \mathbf{A} \tag{10.69}$$

The vector $\mathbf{A}$ is known as the magnetic vector potential in analogy with the electric scalar potential for V.

If we can now find $\mathbf{A}$ due to an infinitesimal current element, we can then find $\mathbf{A}$ for a given current distribution and determine $\mathbf{B}$ by using (10.69). Let us therefore consider an infinitesimal current element of length $d\mathbf{l}$ situated at the origin and oriented along the z-axis, as shown in Figure 10.20. Assuming first that the current is constant

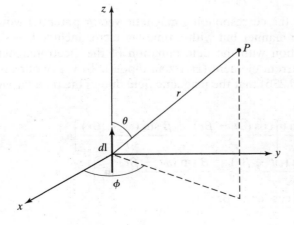

FIGURE 10.20

For finding the magnetic vector potential due to an infinitesimal current element.

and equal to I, we note from (1.68) that the magnetic field at a point P due to the current element is given by

$$\mathbf{B} = \frac{\mu}{4\pi} \frac{I d\mathbf{l} \times \mathbf{a}_r}{r^2} \tag{10.70}$$

where r is the distance from the current element to the point P and $\mathbf{a}_r$ is the unit vector directed from the element toward P. Expressing $\mathbf{B}$ as

$$\mathbf{B} = \frac{\mu}{4\pi} I d\mathbf{l} \times \left(-\nabla\frac{1}{r} \right) \tag{10.71}$$

and using the vector identity

$$\mathbf{A} \times \nabla V = V\nabla \times \mathbf{A} - \nabla \times (V\mathbf{A}) \tag{10.72}$$

we obtain

$$\mathbf{B} = -\frac{\mu I}{4\pi r}\nabla \times d\mathbf{l} + \nabla \times \left(\frac{\mu I d\mathbf{l}}{4\pi r} \right) \tag{10.73}$$

Since $d\mathbf{l}$ is a constant, $\nabla \times d\mathbf{l} = 0$, and (10.73) reduces to

$$\mathbf{B} = \nabla \times \left(\frac{\mu I d\mathbf{l}}{4\pi r} \right) \tag{10.74}$$

Comparing (10.74) with (10.69), we now see that the vector potential due to the current element situated at the origin is simply given by

$$\mathbf{A} = \frac{\mu I d\mathbf{l}}{4\pi r} \tag{10.75}$$

Thus, it has a magnitude inversely proportional to the radial distance from the element (similar to the inverse distance dependence of the scalar potential due to a point charge) and direction parallel to the element.

If the current in the element is now assumed to be time-varying in the manner

$$I = I_0 \cos \omega t$$

we would intuitively expect that the corresponding magnetic vector potential would also be time-varying in the same manner but with a time-lag factor included, as discussed in Section 9.1 in connection with the determination of the electromagnetic fields due to the time-varying current element (Hertzian dipole). To verify our intuitive expectation, we note from (9.23b) that the magnetic field due to the time-varying current element is given by

$$
\begin{aligned}
\mathbf{B} = \mu\mathbf{H} &= \frac{\mu I_0\, dl \sin\theta}{4\pi}\left[\frac{\cos(\omega t - \beta r)}{r^2} - \frac{\beta \sin(\omega t - \beta r)}{r} \right]\mathbf{a}_\phi \\
&= \frac{\mu I_0\, d\mathbf{l}}{4\pi} \times \left\{ \left[\frac{\cos(\omega t - \beta r)}{r^2} - \frac{\beta \sin(\omega t - \beta r)}{r} \right]\mathbf{a}_r \right\} \\
&= \frac{\mu I_0\, d\mathbf{l}}{4\pi} \times \left\{ -\nabla\left[\frac{\cos(\omega t - \beta r)}{r} \right] \right\}
\end{aligned}
$$

and proceed in the same manner as for the constant current case to obtain the vector potential to be

$$\mathbf{A} = \frac{\mu I_0 \, d\mathbf{l}}{4\pi r} \cos(\omega t - \beta r) \tag{10.76}$$

Comparing (10.76) with (10.75), we find that our intuitive expectation is indeed correct for the vector potential case, unlike the case of the fields in Section 9.1! The result given by (10.76) is familiarly known as the *retarded* vector potential in view of the phase-lag factor βr contained in it.

To illustrate an example of the application of (10.76), we now consider a circular loop antenna having circumference that is small compared to the wavelength so that it is an electrically small antenna. Under this condition, the current flowing in the loop can be assumed to be uniform around the loop. Recall that the electrically small loop antenna as a receiving antenna was introduced in Section 9.6. Let us assume the loop to be in the xy-plane with its center at the origin, as shown in Figure 10.21, and the loop current to be $I = I_0 \cos \omega t$ in the ϕ direction. In view of the circular symmetry around the z-axis, we can consider a point P in the xz-plane without loss of generality to find the vector potential. To do this, we divide the loop into a series of infinitesimal elements. Considering one such current element $d\mathbf{l} = a \, d\alpha \, (-\sin \alpha \, \mathbf{a}_x + \cos \alpha \, \mathbf{a}_y)$, as shown in Figure 10.21, and using (10.76), we obtain the vector potential at P due to that current element to be

$$d\mathbf{A} = \frac{\mu I_0 a \, d\alpha \, (-\sin \alpha \, \mathbf{a}_x + \cos \alpha \, \mathbf{a}_y)}{4\pi R} \cos(\omega t - \beta R) \tag{10.77}$$

where

$$R = [(r \sin \theta - a \cos \alpha)^2 + (a \sin \alpha)^2 + (r \cos \theta)^2]^{1/2}$$
$$= [r^2 + a^2 - 2ar \sin \theta \cos \alpha]^{1/2} \tag{10.78}$$

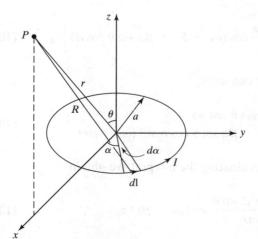

FIGURE 10.21

For finding the magnetic vector potential due to a small circular loop antenna.

The vector potential at point P due to the entire current loop is then given by

$$\mathbf{A} = \int_{\alpha=0}^{2\pi} d\mathbf{A}$$

$$= -\left[\int_{\alpha=0}^{2\pi} \frac{\mu I_0 a \sin \alpha \, d\alpha}{4\pi R} \cos(\omega t - \beta R)\right]\mathbf{a}_x$$

$$+ \left[\int_{\alpha=0}^{2\pi} \frac{\mu I_0 a \cos \alpha \, d\alpha}{4\pi R} \cos(\omega t - \beta R)\right]\mathbf{a}_y \qquad (10.79)$$

The first integral on the right side of (10.79) is, however, zero since the contributions to it due to elements situated symmetrically about the xz-plane cancel. Replacing $\mathbf{a}_y$ in the second term by $\mathbf{a}_\phi$ to generalize the result to an arbitrary point $P(r, \theta, \phi)$, we then obtain

$$\mathbf{A} = \left[\int_{\alpha=0}^{2\pi} \frac{\mu I_0 a \cos \alpha \, d\alpha}{4\pi R} \cos(\omega t - \beta R)\right]\mathbf{a}_\phi \qquad (10.80)$$

Although the evaluation of the integral in (10.80) is complicated, some approximations can be made for obtaining the *radiation fields*. For these fields, we can set the quantity R in the amplitude factor of the integrand equal to r. For R in the phase factor of the integrand, we write

$$R = r\left[1 + \frac{a^2}{r^2} - \frac{2a}{r}\sin\theta\cos\alpha\right]^{1/2}$$

$$\approx r\left[1 - \frac{a}{r}\sin\theta\cos\alpha\right] \qquad (10.81)$$

Thus, for the radiation fields,

$$\mathbf{A} = \left[\int_{\alpha=0}^{2\pi} \frac{\mu I_0 a \cos \alpha \, d\alpha}{4\pi r} \cos(\omega t - \beta r + \beta a \sin\theta\cos\alpha)\right]\mathbf{a}_\phi \qquad (10.82)$$

Now, since $2\pi a \ll \lambda$, or $\beta a \ll 1$, we can write

$$\cos(\omega t - \beta r + \beta a \sin\theta\cos\alpha)$$
$$\approx \cos(\omega t - \beta r) - \beta a \sin\theta\cos\alpha\sin(\omega t - \beta r) \qquad (10.83)$$

Substituting (10.83) into (10.82) and evaluating the integral, we obtain

$$\mathbf{A} = -\frac{\mu I_0 \pi a^2 \beta \sin\theta}{4\pi r}\sin(\omega t - \beta r)\,\mathbf{a}_\phi \qquad (10.84)$$

Having obtained the required magnetic vector potential, we can now determine the radiation fields. Thus from (10.69),

$$\mathbf{H} = \frac{\mathbf{B}}{\mu} = \frac{1}{\mu}\nabla \times \mathbf{A}$$

$$= -\frac{1}{\mu r}\frac{\partial}{\partial r}(rA_\phi)\,\mathbf{a}_\theta$$

$$= -\frac{I_0\pi a^2\beta^2\sin\theta}{4\pi r}\cos(\omega t - \beta r)\,\mathbf{a}_\theta \tag{10.85}$$

From $\nabla \times \mathbf{H} = \dfrac{\partial \mathbf{D}}{\partial t} = \epsilon\dfrac{\partial \mathbf{E}}{\partial t}$, we have

$$\frac{\partial \mathbf{E}}{\partial t} = \frac{1}{\epsilon}\nabla \times \mathbf{H} = \frac{1}{\epsilon r}\frac{\partial}{\partial r}(rH_\theta)\,\mathbf{a}_\phi$$

$$= -\frac{I_0\pi a^2\beta^3\sin\theta}{4\pi\epsilon r}\sin(\omega t - \beta r)\,\mathbf{a}_\phi$$

$$\mathbf{E} = \frac{I_0\pi a^2\beta^3\sin\theta}{4\pi\omega\epsilon r}\cos(\omega t - \beta r)\,\mathbf{a}_\phi$$

$$= \frac{\eta I_0\pi a^2\beta^2\sin\theta}{4\pi r}\cos(\omega t - \beta r)\,\mathbf{a}_\phi \tag{10.86}$$

Comparing (10.85) and (10.86) with (9.25a) and (9.25b), respectively, we note that a duality exists between the radiation fields of the small current loop and those of the infinitesimal current element aligned along the axis of the current loop.

Proceeding further, we can find the Poynting vector, the instantaneous radiated power and the time-average radiated power due to the loop antenna by following steps similar to those employed for the Hertzian dipole in Section 9.2. Thus,

$$\mathbf{P} = \mathbf{E} \times \mathbf{H} = E_\phi\mathbf{a}_\phi \times H_\theta\mathbf{a}_\theta = -E_\phi H_\theta\mathbf{a}_r$$

$$= \frac{\eta\beta^4 I_0^2\pi^2 a^4\sin^2\theta}{16\pi^2 r^2}\cos^2(\omega t - \beta r)\,\mathbf{a}_r$$

$$P_{\text{rad}} = \int_{\theta=0}^{\pi}\int_{\phi=0}^{2\pi}\mathbf{P}\cdot r^2\sin\theta\,d\theta\,d\phi\,\mathbf{a}_r$$

$$= \int_{\theta=0}^{\pi}\int_{\phi=0}^{2\pi}\frac{\eta\beta^4 I_0^2\pi^2 a^4\sin^3\theta}{16\pi^2}\cos^2(\omega t - \beta r)\,d\theta\,d\phi$$

$$= \frac{\eta\beta^4 I_0^2\pi^2 a^4}{6\pi}\cos^2(\omega t - \beta r)$$

$$\langle P_{\text{rad}}\rangle = \frac{\eta\beta^4 I_0^2\pi^2 a^4}{6\pi}\langle\cos^2(\omega t - \beta r)\rangle$$

$$= \frac{1}{2}I_0^2\left[\frac{8\pi^5\eta}{3}\left(\frac{a}{\lambda}\right)^4\right]$$

We now identify the radiation resistance of the small loop antenna to be

$$R_{rad} = \frac{8\pi^5 \eta}{3}\left(\frac{a}{\lambda}\right)^4 \tag{10.87}$$

For free space, $\eta = \eta_0 = 120\pi \ \Omega$, and

$$R_{rad} = 320\pi^6\left(\frac{a}{\lambda}\right)^4 = 20\pi^2\left(\frac{2\pi a}{\lambda}\right)^4 \tag{10.88}$$

Comparing this result with the radiation resistance of the Hertzian dipole given by (9.30), we note that the radiation resistance of the small loop antenna is proportional to the fourth power of its electrical size (circumference/wavelength) whereas that of the Hertzian dipole is proportional to the square of its electrical size (length/wavelength). The directivity of the small loop antenna is, however, the same as that of the Hertzian dipole, that is, 1.5, as given by (9.33), in view of the proportionality of the Poynting vectors to $\sin^2 \theta$ in both cases.

In this section, we introduced the magnetic vector potential as a tool for computing the magnetic fields due to current distributions. In particular, we derived the expression for the retarded magnetic vector potential for a Hertzian dipole and illustrated its application by considering the case of a small circular loop antenna. We derived the radiation fields for the loop antenna and compared its characteristics with those of the Hertzian dipole.

REVIEW QUESTIONS

10.25. Why can the magnetic flux density vector be expressed as the curl of another vector?

10.26. Discuss the analogy between the magnetic vector potential due to an infinitesimal current element and the electric scalar potential due to a point charge.

10.27. What does the word *retarded* in the terminology *retarded magnetic vector potential* refer to? Explain.

10.28. Discuss the application of the magnetic vector potential in the determination of the electromagnetic fields due to an antenna.

10.29. Discuss the duality between the radiation fields of a small circular loop antenna with those of a Hertzian dipole at the center of the loop and aligned with its axis.

10.30. Compare the radiation resistance and directivity of a small circular loop antenna with those of a Hertzian dipole.

PROBLEMS

10.17. By expansion in Cartesian coordinates, show that

$$\mathbf{A} \times \nabla V = V\nabla \times \mathbf{A} - \nabla \times (V\mathbf{A}).$$

10.18. For the half-wave dipole of Section 9.3, determine the magnetic vector potential for the radiation fields. Verify your result by finding the radiation fields and comparing with the results of Section 9.3.

10.19. A circular loop antenna of radius 1 m in free space carries a uniform current $10 \cos 4\pi \times 10^6 t$ A. (a) Calculate the amplitude of the electric field intensity at a distance of 10 km in the plane of the loop. (b) Calculate the radiation resistance and the time-average power radiated by the loop.

10.20. Find the length of a Hertzian dipole that would radiate the same time-average power as the loop antenna of Problem 10.19 for the same current and frequency as in Problem 10.19.

Cylindrical and Spherical Coordinate Systems

In Section 1.2, we learned that the Cartesian coordinate system is defined by a set of three mutually orthogonal surfaces, all of which are planes. The cylindrical and spherical coordinate systems also involve sets of three mutually orthogonal surfaces. For the cylindrical coordinate system, the three surfaces are a cylinder and two planes, as shown in Figure A.1(a). One of these planes is the same as the z = constant plane in the Cartesian coordinate system. The second plane contains the z-axis and makes an angle ϕ with a reference plane, conveniently chosen to be the xz-plane of the Cartesian coordinate system. This plane is therefore defined by ϕ = constant. The cylindrical surface has the z-axis as its axis. Since the radial distance r from the z-axis to points on the cylindrical surface is a constant, this surface is defined by r = constant. Thus, the three orthogonal surfaces defining the cylindrical coordinates of a point are r = constant, ϕ = constant, and z = constant. Only two of these coordinates (r and z) are distances; the third coordinate (ϕ) is an angle. We note that the entire space is spanned by varying r from 0 to ∞, ϕ from 0 to 2π, and z from $-\infty$ to ∞.

The origin is given by $r = 0$, $\phi = 0$, and $z = 0$. Any other point in space is given by the intersection of three mutually orthogonal surfaces obtained by incrementing the coordinates by appropriate amounts. For example, the intersection of the three surfaces $r = 2$, $\phi = \pi/4$, and $z = 3$ defines the point $A(2, \pi/4, 3)$, as shown in Figure A.1(a). These three orthogonal surfaces define three curves that are mutually perpendicular. Two of these are straight lines and the third is a circle. We draw unit vectors, $\mathbf{a}_r$, $\mathbf{a}_\phi$, and $\mathbf{a}_z$ tangential to these curves at the point A and directed toward increasing values of r, ϕ, and z, respectively. These three unit vectors form a set of mutually orthogonal unit vectors in terms of which vectors drawn at A can be described. In a similar manner, we can draw unit vectors at any other point in the cylindrical coordinate system, as shown, for example, for point $B(1, 3\pi/4, 5)$ in Figure A.1(a). It can now be seen that the unit vectors $\mathbf{a}_r$ and $\mathbf{a}_\phi$ at point B are not parallel to the corresponding unit vectors at point A. Thus, unlike in the Cartesian coordinate system, the unit vectors $\mathbf{a}_r$ and $\mathbf{a}_\phi$ in the cylindrical coordinate system do not have the same directions everywhere, that is, they are not uniform. Only the unit vector $\mathbf{a}_z$, which is the same as in the Cartesian coordinate system, is uniform. Finally, we note that for the choice of ϕ as in Figure A.1(a),

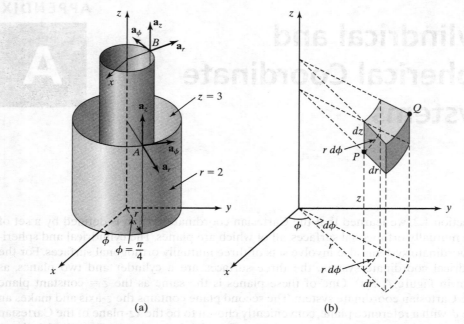

FIGURE A.1

Cylindrical coordinate system. (a) Orthogonal surfaces and unit vectors.
(b) Differential volume formed by incrementing the coordinates.

that is, increasing from the positive x-axis toward the positive y-axis, the coordinate system is right-handed, that is, $\mathbf{a}_r \times \mathbf{a}_\phi = \mathbf{a}_z$.

To obtain expressions for the differential lengths, surfaces, and volume in the cylindrical coordinate system, we now consider two points $P(r, \phi, z)$ and $Q(r + dr, \phi + d\phi, z + dz)$, where Q is obtained by incrementing infinitesimally each coordinate from its value at P, as shown in Figure A.1(b). The three orthogonal surfaces intersecting at P and the three orthogonal surfaces intersecting at Q define a box which can be considered to be rectangular, since dr, $d\phi$, and dz are infinitesimally small. The three differential length elements, forming the contiguous sides of this box are $dr\,\mathbf{a}_r$, $r\,d\phi\,\mathbf{a}_\phi$, and $dz\,\mathbf{a}_z$. The differential length vector $d\mathbf{l}$ from P to Q is thus given by

$$d\mathbf{l} = dr\,\mathbf{a}_r + r\,d\phi\,\mathbf{a}_\phi + dz\,\mathbf{a}_z \qquad (A.1)$$

The differential surfaces formed by pairs of the differential length elements are

$$\pm dS\,\mathbf{a}_z = \pm(dr)(r\,d\phi)\mathbf{a}_z = \pm dr\,\mathbf{a}_r \times r\,d\phi\,\mathbf{a}_\phi \qquad (A.2a)$$

$$\pm dS\,\mathbf{a}_r = \pm(r\,d\phi)(dz)\mathbf{a}_r = \pm r\,d\phi\,\mathbf{a}_\phi \times dz\,\mathbf{a}_z \qquad (A.2b)$$

$$\pm dS\,\mathbf{a}_\phi = \pm(dz)(dr)\mathbf{a}_\phi = \pm dz\,\mathbf{a}_z \times dr\,\mathbf{a}_r \qquad (A.2c)$$

Finally, the differential volume dv formed by the three differential lengths is simply the volume of the box, that is,

$$dv = (dr)(r\,d\phi)(dz) = r\,dr\,d\phi\,dz \qquad (A.3)$$

For the spherical coordinate system, the three mutually orthogonal surfaces are a sphere, a cone, and a plane, as shown in Figure A.2(a). The plane is the same as the $\phi =$ constant plane in the cylindrical coordinate system. The sphere has the origin as its center. Since the radial distance r from the origin to points on the spherical surface is a constant, this surface is defined by $r =$ constant. The spherical coordinate r should not be confused with the cylindrical coordinate r. When these two coordinates appear in the same expression, we shall use the subscripts c and s to distinguish between cylindrical and spherical. The cone has its vertex at the origin and its surface is symmetrical about the z-axis. Since the angle θ is the angle that the conical surface makes with the z-axis, this surface is defined by $\theta =$ constant. Thus, the three orthogonal surfaces defining the spherical coordinates of a point are $r =$ constant, $\theta =$ constant, and $\phi =$ constant. Only one of these coordinates (r) is distance; the other two coordinates (θ and ϕ) are angles. We note that the entire space is spanned by varying r from 0 to ∞, θ from 0 to π, and ϕ from 0 to 2π.

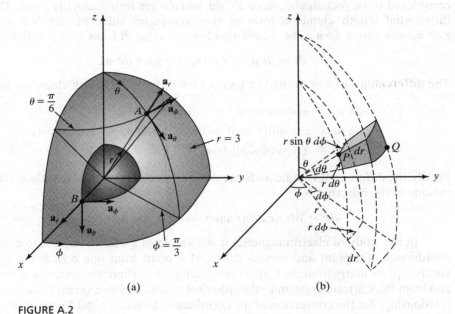

(a)   (b)

FIGURE A.2

Spherical coordinate system. (a) Orthogonal surfaces and unit vectors. (b) Differential volume formed by incrementing the coordinates.

The origin is given by $r = 0$, $\theta = 0$, and $\phi = 0$. Any other point in space is given by the intersection of three mutually orthogonal surfaces obtained by incrementing the coordinates by appropriate amounts. For example, the intersection of the three surfaces $r = 3$, $\theta = \pi/6$, and $\phi = \pi/3$ defines the point $A(3, \pi/6, \pi/3)$ as shown in Figure A.2(a). These three orthogonal surfaces define three curves that are mutually perpendicular. One of these is a straight line and the other two are circles. We draw unit vectors $\mathbf{a}_r$, $\mathbf{a}_\theta$, and $\mathbf{a}_\phi$ tangential to these curves at point A and directed toward increasing values of r, θ, and ϕ, respectively. These three unit vectors form a set of mutually orthogonal unit vectors in terms of which vectors drawn at A can be described. In a similar manner, we can draw unit vectors at any other point in the spherical coordinate system, as shown, for example, for point $B(1, \pi/2, 0)$ in Figure A.2(a). It can now be seen that these unit vectors at point B are not parallel to the corresponding unit vectors at point A. Thus, in the spherical coordinate system, all three unit vectors $\mathbf{a}_r$, $\mathbf{a}_\theta$, and $\mathbf{a}_\phi$ do not have the same directions everywhere, that is, they are not uniform. Finally, we note that for the choice of θ as in Figure A.2(a), that is, increasing from the positive z-axis toward the xy-plane, the coordinate system is right-handed, that is, $\mathbf{a}_r \times \mathbf{a}_\theta = \mathbf{a}_\phi$.

To obtain expressions for the differential lengths, surfaces, and volume in the spherical coordinate system, we now consider two points $P(r, \theta, \phi)$ and $Q(r + dr, \theta + d\theta, \phi + d\phi)$, where Q is obtained by incrementing infinitesimally each coordinate from its value at P, as shown in Figure A.2(b). The three orthogonal surfaces intersecting at P and the three orthogonal surfaces intersecting at Q define a box that can be considered to be rectangular, since dr, $d\theta$, and $d\phi$ are infinitesimally small. The three differential length elements forming the contiguous sides of this box are $dr\,\mathbf{a}_r$, $r\,d\theta\,\mathbf{a}_\theta$, and $r \sin \theta\, d\phi\, \mathbf{a}_\phi$. The differential length vector $d\mathbf{l}$ from P to Q is thus given by

$$d\mathbf{l} = dr\,\mathbf{a}_r + r\,d\theta\,\mathbf{a}_\theta + r \sin\theta\, d\phi\,\mathbf{a}_\phi \tag{A.4}$$

The differential surfaces formed by pairs of the differential length elements are

$$\pm dS\,\mathbf{a}_\phi = \pm(dr)(r\,d\theta)\mathbf{a}_\phi = \pm dr\,\mathbf{a}_r \times r\,d\theta\,\mathbf{a}_\theta \tag{A.5a}$$

$$\pm dS\,\mathbf{a}_r = \pm(r\,d\theta)(r \sin\theta\,d\phi)\mathbf{a}_r = \pm r\,d\theta\,\mathbf{a}_\theta \times r \sin\theta\,d\phi\,\mathbf{a}_\phi \tag{A.5b}$$

$$\pm dS\,\mathbf{a}_\theta = \pm(r \sin\theta\,d\phi)(dr)\mathbf{a}_\theta = \pm r \sin\theta\,d\phi\,\mathbf{a}_\phi \times dr\,\mathbf{a}_r \tag{A.5c}$$

Finally, the differential volume dv formed by the three differential lengths is simply the volume of the box, that is,

$$dv = (dr)(r\,d\theta)(r \sin\theta\,d\phi) = r^2 \sin\theta\, dr\, d\theta\, d\phi \tag{A.6}$$

In the study of electromagnetics it is sometimes useful to be able to convert the coordinates of a point and vectors drawn at a point from one coordinate system to another, particularly from the Cartesian system to the cylindrical system and vice versa, and from the Cartesian system to the spherical system and vice versa. To derive first the relationships for the conversion of the coordinates, let us consider Figure A.3(a), which illustrates the geometry pertinent to the coordinates of a point P in the three different coordinate systems. Thus, from simple geometrical considerations, we have

$$x = r_c \cos\phi \qquad y = r_c \sin\phi \qquad z = z \tag{A.7}$$

$$x = r_s \sin\theta \cos\phi \qquad y = r_s \sin\theta \sin\phi \qquad z = r_s \cos\theta \tag{A.8}$$

Conversely, we have

$$r_c = \sqrt{x^2 + y^2} \qquad \phi = \tan^{-1}\frac{y}{x} \qquad z = z \tag{A.9}$$

$$r_s = \sqrt{x^2 + y^2 + z^2} \qquad \theta = \tan^{-1}\frac{\sqrt{x^2 + y^2}}{z} \qquad \phi = \tan^{-1}\frac{y}{x} \tag{A.10}$$

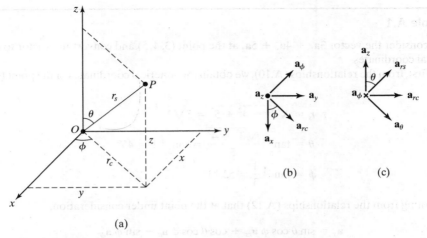

FIGURE A.3

(a) For conversion of coordinates of a point from one coordinate system to another.
(b) and (c) For expressing unit vectors in cylindrical and spherical coordinate systems,
respectively, in terms of unit vectors in the Cartesian coordinate system.

Relationships (A.7) and (A.9) correspond to conversion from cylindrical coordinates
to Cartesian coordinates and vice versa. Relationships (A.8) and (A.10) correspond to
conversion from spherical coordinates to Cartesian coordinates and vice versa.

Considering next the conversion of vectors from one coordinate system to another,
we note that in order to do this, we need to express each of the unit vectors of the first
coordinate system in terms of its components along the unit vectors in the second coordinate system. From the definition of the dot product of two vectors, the component of a
unit vector along another unit vector, that is, the cosine of the angle between the unit vectors, is simply the dot product of the two unit vectors. Thus, considering the sets of unit
vectors in the cylindrical and Cartesian coordinate systems, we have with the aid of
Figure A.3(b),

$$\mathbf{a}_{rc} \cdot \mathbf{a}_x = \cos \phi \qquad \mathbf{a}_{rc} \cdot \mathbf{a}_y = \sin \phi \qquad \mathbf{a}_{rc} \cdot \mathbf{a}_z = 0 \qquad \text{(A.11a)}$$
$$\mathbf{a}_\phi \cdot \mathbf{a}_x = -\sin \phi \qquad \mathbf{a}_\phi \cdot \mathbf{a}_y = \cos \phi \qquad \mathbf{a}_\phi \cdot \mathbf{a}_z = 0 \qquad \text{(A.11b)}$$
$$\mathbf{a}_z \cdot \mathbf{a}_x = 0 \qquad \mathbf{a}_z \cdot \mathbf{a}_y = 0 \qquad \mathbf{a}_z \cdot \mathbf{a}_z = 1 \qquad \text{(A.11c)}$$

Similarly, for the sets of unit vectors in the spherical and Cartesian coordinate systems,
we obtain with the aid of Figure A.3(c) and Figure A.3(b),

$$\mathbf{a}_{rs} \cdot \mathbf{a}_x = \sin \theta \cos \phi \qquad \mathbf{a}_{rs} \cdot \mathbf{a}_y = \sin \theta \sin \phi \qquad \mathbf{a}_{rs} \cdot \mathbf{a}_z = \cos \theta \qquad \text{(A.12a)}$$
$$\mathbf{a}_\theta \cdot \mathbf{a}_x = \cos \theta \cos \phi \qquad \mathbf{a}_\theta \cdot \mathbf{a}_y = \cos \theta \sin \phi \qquad \mathbf{a}_\theta \cdot \mathbf{a}_z = -\sin \theta \qquad \text{(A.12b)}$$
$$\mathbf{a}_\phi \cdot \mathbf{a}_x = -\sin \phi \qquad \mathbf{a}_\phi \cdot \mathbf{a}_y = \cos \phi \qquad \mathbf{a}_\phi \cdot \mathbf{a}_z = 0 \qquad \text{(A.12c)}$$

We shall now illustrate the use of these relationships by means of an example.

Example A.1

Let us consider the vector $3\mathbf{a}_x + 4\mathbf{a}_y + 5\mathbf{a}_z$ at the point $(3, 4, 5)$ and convert the vector to one in spherical coordinates.

First, from the relationships (A.10), we obtain the spherical coordinates of the point $(3, 4, 5)$ to be

$$r_s = \sqrt{3^2 + 4^2 + 5^2} = 5\sqrt{2}$$

$$\theta = \tan^{-1} \frac{\sqrt{3^2 + 4^2}}{5} = \tan^{-1} 1 = 45°$$

$$\phi = \tan^{-1} \frac{4}{3} = 53.13°$$

Then noting from the relationships (A.12) that at the point under consideration,

$$\mathbf{a}_x = \sin\theta \cos\phi \, \mathbf{a}_{rs} + \cos\theta \cos\phi \, \mathbf{a}_\theta - \sin\phi \, \mathbf{a}_\phi$$

$$= 0.3\sqrt{2}\mathbf{a}_{rs} + 0.3\sqrt{2}\mathbf{a}_\theta - 0.8\mathbf{a}_\phi$$

$$\mathbf{a}_y = \sin\theta \sin\phi \, \mathbf{a}_{rs} + \cos\theta \sin\phi \, \mathbf{a}_\theta + \cos\phi \, \mathbf{a}_\phi$$

$$= 0.4\sqrt{2}\mathbf{a}_{rs} + 0.4\sqrt{2}\mathbf{a}_\theta + 0.6\mathbf{a}_\phi$$

$$\mathbf{a}_z = \cos\theta \, \mathbf{a}_{rs} - \sin\theta \, \mathbf{a}_\theta = 0.5\sqrt{2}\mathbf{a}_{rs} - 0.5\sqrt{2}\mathbf{a}_\theta$$

we obtain

$$3\mathbf{a}_x + 4\mathbf{a}_y + 5\mathbf{a}_z = (0.9\sqrt{2} + 1.6\sqrt{2} + 2.5\sqrt{2})\mathbf{a}_{rs}$$

$$+ (0.9\sqrt{2} + 1.6\sqrt{2} - 2.5\sqrt{2})\mathbf{a}_\theta + (-2.4 + 2.4)\mathbf{a}_\phi = 5\sqrt{2}\mathbf{a}_{rs}$$

This result is to be expected since the given vector has components equal to the coordinates of the point at which it is specified. Hence, its magnitude is equal to the distance of the point from the origin, that is, the spherical coordinate r of the point, and its direction is along the line drawn from the origin to the point, that is, along the unit vector $\mathbf{a}_{rs}$ at that point. In fact, the given vector is a particular case of the vector $x\mathbf{a}_x + y\mathbf{a}_y + z\mathbf{a}_z = r_s\mathbf{a}_{rs}$, known as the *position vector*, since it is the same as the vector drawn from the origin to the point (x, y, z).

REVIEW QUESTIONS

A.1. What are the three orthogonal surfaces defining the cylindrical coordinate system?

A.2. What are the limits of variation of the cylindrical coordinates?

A.3. Which of the unit vectors in the cylindrical coordinate system are not uniform?

A.4. State whether the vector $3\mathbf{a}_r + 4\mathbf{a}_\phi + 5\mathbf{a}_z$ at the point $(1, 0, 2)$ and the vector $3\mathbf{a}_r + 4\mathbf{a}_\phi + 5\mathbf{a}_z$ at the point $(2, \pi/2, 3)$ are equal or not.

A.5. What are the differential length vectors in cylindrical coordinates?

A.6. What are the three orthogonal surfaces defining the spherical coordinate system?

A.7. What are the limits of variation of the spherical coordinates?

A.8. Which of the unit vectors in the spherical coordinate system are not uniform?

A.9. State if the vector $3\mathbf{a}_r + 4\mathbf{a}_\theta$ at the point $(1, \pi/2, 0)$ and the vector $3\mathbf{a}_r + 4\mathbf{a}_\theta$ at the point $(2, 0, \pi/2)$ are equal or not.

A.10. What are the differential length vectors in spherical coordinates?

A.11. Outline the procedure for converting a vector at a point from one coordinate system to another.

A.12. What is the expression for the position vector in the cylindrical coordinate system?

PROBLEMS

A.1. Express in terms of Cartesian coordinates the vector drawn from the point $P(2, \pi/3, 1)$ to the point $Q(4, 2\pi/3, 2)$ in cylindrical coordinates.

A.2. Express in terms of Cartesian coordinates the vector drawn from the point $P(1, \pi/3, \pi/4)$ to the point $Q(2, 2\pi/3, 3\pi/4)$ in spherical coordinates.

A.3. Determine if the vector $\mathbf{a}_r + \mathbf{a}_\phi + 2\mathbf{a}_z$ at the point $(1, \pi/4, 2)$ and the vector $\sqrt{2}\mathbf{a}_r + 2\mathbf{a}_z$ at the point $(2, \pi/2, 3)$ are equal or not.

A.4. Determine if the vector $3\mathbf{a}_r + \sqrt{3}\mathbf{a}_\theta - 2\mathbf{a}_\phi$ at the point $(2, \pi/3, \pi/6)$ and the vector $\mathbf{a}_r + \sqrt{3}\mathbf{a}_\theta - 2\sqrt{3}\mathbf{a}_\phi$ at the point $(1, \pi/6, \pi/3)$ are equal or not.

A.5. Find the dot and cross products of the unit vector $\mathbf{a}_r$ at the point $(1, 0, 0)$ and the unit vector $\mathbf{a}_\phi$ at the point $(2, \pi/4, 1)$ in cylindrical coordinates.

A.6. Find the dot and cross products of the unit vector $\mathbf{a}_r$ at the point $(1, \pi/4, 0)$ and the unit vector $\mathbf{a}_\theta$ at the point $(2, \pi/2, \pi/2)$ in spherical coordinates.

A.7. Convert the vector $5\mathbf{a}_x + 12\mathbf{a}_y + 6\mathbf{a}_z$ at the point $(5, 12, 4)$ to one in cylindrical coordinates.

A.8. Convert the vector $3\mathbf{a}_x + 4\mathbf{a}_y - 5\mathbf{a}_z$ at the point $(3, 4, 5)$ to one in spherical coordinates.

B Curl, Divergence, and Gradient in Cylindrical and Spherical Coordinate Systems

In Sections 3.1, 3.4, and 6.1, we introduced the curl, divergence, and gradient, respectively, and derived the expressions for them in the Cartesian coordinate system. In this appendix, we shall derive the corresponding expressions in the cylindrical and spherical coordinate systems. Considering first the cylindrical coordinate system, we recall from Appendix A that the infinitesimal box defined by the three orthogonal surfaces intersecting at point $P(r, \theta, \phi)$ and the three orthogonal surfaces intersecting at point $Q(r + dr, \phi + d\phi, z + dz)$ is as shown in Figure B.1.

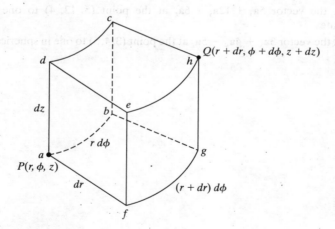

FIGURE B.1

Infinitesimal box formed by incrementing the coordinates in the cylindrical coordinate system.

From the basic definition of the curl of a vector introduced in Section 3.3 and given by

$$\nabla \times \mathbf{A} = \lim_{\Delta S \to 0} \left[\frac{\oint_C \mathbf{A} \cdot d\mathbf{l}}{\Delta S} \right]_{max} \mathbf{a}_n \tag{B.1}$$

we find the components of $\nabla \times \mathbf{A}$ as follows with the aid of Figure B.1:

$$
\begin{aligned}
(\nabla \times \mathbf{A})_r &= \lim_{\substack{d\phi \to 0 \\ dz \to 0}} \frac{\oint_{abcda} \mathbf{A} \cdot d\mathbf{l}}{\text{area } abcd} \\
&= \lim_{\substack{d\phi \to 0 \\ dz \to 0}} \frac{\left\{ \begin{array}{l} [A_\phi]_{(r,z)} r\, d\phi + [A_z]_{(r,\phi+d\phi)}\, dz \\ - [A_\phi]_{(r,z+dz)} r\, d\phi - [A_z]_{(r,\phi)}\, dz \end{array} \right\}}{r\, d\phi\, dz} \\
&= \lim_{d\phi \to 0} \frac{[A_z]_{(r,\phi+d\phi)} - [A_z]_{(r,\phi)}}{r\, d\phi} + \lim_{dz \to 0} \frac{[A_\phi]_{(r,z)} - [A_\phi]_{(r,z+dz)}}{dz} \\
&= \frac{1}{r} \frac{\partial A_z}{\partial \phi} - \frac{\partial A_\phi}{\partial z} \tag{B.2a}
\end{aligned}
$$

$$
\begin{aligned}
(\nabla \times \mathbf{A})_\phi &= \lim_{\substack{dz \to 0 \\ dr \to 0}} \frac{\oint_{adefa} \mathbf{A} \cdot d\mathbf{l}}{\text{area } adef} \\
&= \lim_{\substack{dz \to 0 \\ dr \to 0}} \frac{\left\{ \begin{array}{l} [A_z]_{(r,\phi)}\, dz + [A_r]_{(\phi,z+dz)}\, dr \\ - [A_z]_{(r+dr,\phi)}\, dz - [A_r]_{(\phi,z)}\, dr \end{array} \right\}}{dr\, dz} \\
&= \lim_{dz \to 0} \frac{[A_r]_{(\phi,z+dz)} - [A_r]_{(\phi,z)}}{dz} + \lim_{dr \to 0} \frac{[A_z]_{(r,\phi)} - [A_z]_{(r+dr,\phi)}}{dr} \\
&= \frac{\partial A_r}{\partial z} - \frac{\partial A_z}{\partial r} \tag{B.2b}
\end{aligned}
$$

$$
\begin{aligned}
(\nabla \times \mathbf{A})_z &= \lim_{\substack{dr \to 0 \\ d\phi \to 0}} \frac{\oint_{afgba} \mathbf{A} \cdot d\mathbf{l}}{\text{area } afgb} \\
&= \lim_{\substack{dr \to 0 \\ d\phi \to 0}} \frac{\left\{ \begin{array}{l} [A_r]_{(\phi,z)}\, dr + [A_\phi]_{(r+dr,z)} (r+dr)\, d\phi \\ - [A_r]_{(\phi+d\phi,z)}\, dr - [A_\phi]_{(r,z)} r\, d\phi \end{array} \right\}}{r\, dr\, d\phi} \\
&= \lim_{dr \to 0} \frac{[rA_\phi]_{(r+dr,z)} - [rA_\phi]_{(r,z)}}{r\, dr} + \lim_{d\phi \to 0} \frac{[A_r]_{(\phi,z)} - [A_r]_{(\phi+d\phi,z)}}{r\, d\phi} \\
&= \frac{1}{r} \frac{\partial}{\partial r} (rA_\phi) - \frac{1}{r} \frac{\partial A_r}{\partial \phi} \tag{B.2c}
\end{aligned}
$$

Combining (B.2a), (B.2b), and (B.2c), we obtain the expression for the curl of a vector in cylindrical coordinates as

$$
\nabla \times \mathbf{A} = \left[\frac{1}{r} \frac{\partial A_z}{\partial \phi} - \frac{\partial A_\phi}{\partial z} \right] \mathbf{a}_r + \left[\frac{\partial A_r}{\partial z} - \frac{\partial A_z}{\partial r} \right] \mathbf{a}_\phi + \frac{1}{r} \left[\frac{\partial}{\partial r}(rA_\phi) - \frac{\partial A_r}{\partial \phi} \right] \mathbf{a}_z
$$

$$
= \begin{vmatrix} \dfrac{\mathbf{a}_r}{r} & \mathbf{a}_\phi & \dfrac{\mathbf{a}_z}{r} \\ \dfrac{\partial}{\partial r} & \dfrac{\partial}{\partial \phi} & \dfrac{\partial}{\partial z} \\ A_r & rA_\phi & A_z \end{vmatrix} \tag{B.3}
$$

To find the expression for the divergence, we make use of the basic definition of the divergence of a vector, introduced in Section 3.6 and given by

$$
\nabla \cdot \mathbf{A} = \lim_{\Delta v \to 0} \frac{\oint_S \mathbf{A} \cdot d\mathbf{S}}{\Delta v} \tag{B.4}
$$

Evaluating the right side of (B.4) for the box of Figure B.1, we obtain

$$
\nabla \cdot \mathbf{A} = \lim_{\substack{dr \to 0 \\ d\phi \to 0 \\ dz \to 0}} \frac{\left\{ \begin{aligned} &[A_r]_{r+dr}(r+dr)\, d\phi\, dz - [A_r]_r r\, d\phi\, dz + [A_\phi]_{\phi+d\phi}\, dr\, dz \\ &- [A_\phi]_\phi\, dr\, dz + [A_z]_{z+dz} r\, dr\, d\phi - [A_z]_z r\, dr\, d\phi \end{aligned} \right\}}{r\, dr\, d\phi\, dz}
$$

$$
= \lim_{dr \to 0} \frac{[rA_r]_{r+dr} - [rA_r]_r}{r\, dr} + \lim_{d\phi \to 0} \frac{[A_\phi]_{\phi+d\phi} - [A_\phi]_\phi}{r\, d\phi}
$$

$$
+ \lim_{dz \to 0} \frac{[A_z]_{z+dz} - [A_z]_z}{dz}
$$

$$
= \frac{1}{r} \frac{\partial}{\partial r}(rA_r) + \frac{1}{r} \frac{\partial A_\phi}{\partial \phi} + \frac{\partial A_z}{\partial z} \tag{B.5}
$$

To obtain the expression for the gradient of a scalar, we recall from Appendix A that in cylindrical coordinates,

$$
d\mathbf{l} = dr\, \mathbf{a}_r + r\, d\phi\, \mathbf{a}_\phi + dz\, \mathbf{a}_z \tag{B.6}
$$

and hence

$$
d\Phi = \frac{\partial \Phi}{\partial r}\, dr + \frac{\partial \Phi}{\partial \phi}\, d\phi + \frac{\partial \Phi}{\partial z}\, dz
$$

$$
= \left(\frac{\partial \Phi}{\partial r} \mathbf{a}_r + \frac{1}{r} \frac{\partial \Phi}{\partial \phi} \mathbf{a}_\phi + \frac{\partial \Phi}{\partial z} \mathbf{a}_z \right) \cdot (dr\, \mathbf{a}_r + r\, d\phi\, \mathbf{a}_\phi + dz\, \mathbf{a}_z)
$$

$$
= \nabla \Phi \cdot d\mathbf{l} \tag{B.7}
$$

Thus,

$$\nabla\Phi = \frac{\partial\Phi}{\partial r}\,\mathbf{a}_r + \frac{1}{r}\frac{\partial\Phi}{\partial\phi}\,\mathbf{a}_\phi + \frac{\partial\Phi}{\partial z}\,\mathbf{a}_z \tag{B.8}$$

Turning now to the spherical coordinate system, we recall from Appendix A that the infinitesimal box defined by the three orthogonal surfaces intersecting at $P(r, \theta, \phi)$ and the three orthogonal surfaces intersecting at $Q(r + dr, \theta + d\theta, \phi + d\phi)$ is as shown in Figure B.2. From the basic definition of the curl of a vector given by (B.1), we then find the components of $\nabla \times \mathbf{A}$ as follows with the aid of Figure B.2:

$$(\nabla \times \mathbf{A})_r = \lim_{\substack{d\theta \to 0 \\ d\phi \to 0}} \frac{\oint_{abcda} \mathbf{A} \cdot d\mathbf{l}}{\text{area } abcd}$$

$$= \lim_{\substack{d\theta \to 0 \\ d\phi \to 0}} \frac{\left\{ \begin{array}{l} [A_\theta]_{(r, \phi)} r\, d\theta + [A_\phi]_{(r, \theta+d\theta)} r \sin(\theta + d\theta)\, d\phi \\ - [A_\theta]_{(r, \phi+d\phi)} r\, d\theta - [A_\phi]_{(r, \theta)} r \sin\theta\, d\phi \end{array} \right\}}{r^2 \sin\theta\, d\theta\, d\phi}$$

$$= \lim_{d\theta \to 0} \frac{[A_\phi \sin\theta]_{(r, \theta+d\theta)} - [A_\phi \sin\theta]_{(r, \theta)}}{r \sin\theta\, d\theta}$$

$$+ \lim_{d\phi \to 0} \frac{[A_\theta]_{(r, \phi)} - [A_\theta]_{(r, \phi+d\phi)}}{r \sin\theta\, d\phi}$$

$$= \frac{1}{r \sin\theta}\frac{\partial}{\partial\theta}(A_\phi \sin\theta) - \frac{1}{r \sin\theta}\frac{\partial A_\theta}{\partial\phi} \tag{B.9a}$$

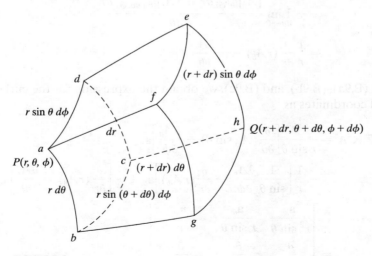

FIGURE B.2

Infinitesimal box formed by incrementing the coordinates in the spherical coordinate system.

$$(\nabla \times \mathbf{A})_\theta = \lim_{\substack{d\phi \to 0 \\ dr \to 0}} \frac{\oint_{adefa} \mathbf{A} \cdot d\mathbf{l}}{\text{area } adef}$$

$$= \lim_{\substack{d\phi \to 0 \\ dr \to 0}} \frac{\left\{ \begin{array}{l} [A_\phi]_{(r,\theta)} r \sin\theta \, d\phi + [A_r]_{(\theta,\phi+d\phi)} \, dr \\ - [A_\phi]_{(r+dr,\theta)}(r+dr)\sin\theta \, d\phi - [A_r]_{(\theta,\phi)} \, dr \end{array} \right\}}{r \sin\theta \, dr \, d\phi}$$

$$= \lim_{d\phi \to 0} \frac{[A_r]_{(\theta,\phi+d\phi)} - [A_r]_{(\theta,\phi)}}{r \sin\theta \, d\phi}$$

$$+ \lim_{dr \to 0} \frac{[rA_\phi]_{(r,\theta)} - [rA_\phi]_{(r+dr,\theta)}}{r \, dr}$$

$$= \frac{1}{r \sin\theta} \frac{\partial A_r}{\partial \phi} - \frac{1}{r} \frac{\partial}{\partial r}(rA_\phi) \tag{B.9b}$$

$$(\nabla \times \mathbf{A})_\phi = \lim_{\substack{dr \to 0 \\ d\theta \to 0}} \frac{\oint_{afgba} \mathbf{A} \cdot d\mathbf{l}}{\text{area } afgb}$$

$$= \lim_{\substack{dr \to 0 \\ d\theta \to 0}} \frac{\left\{ \begin{array}{l} [A_r]_{(\theta,\phi)} \, dr + [A_\theta]_{(r+dr,\phi)}(r+dr) \, d\theta \\ - [A_r]_{(\theta+d\theta,\phi)} \, dr - [A_\theta]_{(r,\phi)} r \, d\theta \end{array} \right\}}{r \, dr \, d\theta}$$

$$= \lim_{dr \to 0} \frac{[rA_\theta]_{(r+dr,\phi)} - [rA_\theta]_{(r,\phi)}}{r \, dr}$$

$$+ \lim_{d\theta \to 0} \frac{[A_r]_{(\theta,\phi)} \, dr - [A_r]_{(\theta+d\theta,\phi)} \, dr}{r \, d\theta}$$

$$= \frac{1}{r} \frac{\partial}{\partial r}(rA_\theta) - \frac{1}{r} \frac{\partial A_r}{\partial \theta} \tag{B.9c}$$

Combining (B.9a), (B.9b), and (B.9c), we obtain the expression for the curl of a vector in spherical coordinates as

$$\nabla \times \mathbf{A} = \frac{1}{r \sin\theta} \left[\frac{\partial}{\partial \theta}(A_\phi \sin\theta) - \frac{\partial A_\theta}{\partial \phi} \right] \mathbf{a}_r$$

$$+ \frac{1}{r} \left[\frac{1}{\sin\theta} \frac{\partial A_r}{\partial \phi} - \frac{\partial}{\partial r}(rA_\phi) \right] \mathbf{a}_\theta + \frac{1}{r} \left[\frac{\partial}{\partial r}(rA_\theta) - \frac{\partial A_r}{\partial \theta} \right] \mathbf{a}_\phi$$

$$= \begin{vmatrix} \dfrac{\mathbf{a}_r}{r^2 \sin\theta} & \dfrac{\mathbf{a}_\theta}{r \sin\theta} & \dfrac{\mathbf{a}_\phi}{r} \\[2ex] \dfrac{\partial}{\partial r} & \dfrac{\partial}{\partial \theta} & \dfrac{\partial}{\partial \phi} \\[2ex] A_r & rA_\theta & r \sin\theta \, A_\phi \end{vmatrix} \tag{B.10}$$

To find the expression for the divergence, we make use of the basic definition of the divergence of a vector given by (B.4) and by evaluating its right side for the box of Figure B.2, we obtain

$$\nabla \cdot \mathbf{A} = \underset{\substack{dr \to 0 \\ d\theta \to 0 \\ d\phi \to 0}}{\text{Lim}} \frac{\begin{cases} [A_r]_{r+dr}(r + dr)^2 \sin \theta \, d\theta \, d\phi - [A_r]_r r^2 \sin \theta \, d\theta \, d\phi \\ + [A_\theta]_{\theta+d\theta} r \sin (\theta + d\theta) \, dr \, d\phi - [A_\theta]_\theta r \sin \theta \, dr \, d\phi \\ + [A_\phi]_{\phi+d\phi} r \, dr \, d\theta - [A_\phi]_\phi r \, dr \, d\theta \end{cases}}{r^2 \sin \theta \, dr \, d\theta \, d\phi}$$

$$= \underset{dr \to 0}{\text{Lim}} \frac{[r^2 A_r]_{r+dr} - [r^2 A_r]_r}{r^2 \, dr} + \underset{d\theta \to 0}{\text{Lim}} \frac{[A_\theta \sin \theta]_{\theta+d\theta} - [A_\theta \sin \theta]_\theta}{r \sin \theta \, d\theta}$$

$$+ \underset{d\phi \to 0}{\text{Lim}} \frac{[A_\phi]_{\phi+d\phi} - [A_\phi]_\phi}{r \sin \theta \, d\phi}$$

$$= \frac{1}{r^2} \frac{\partial}{\partial r} (r^2 A_r) + \frac{1}{r \sin \theta} \frac{\partial}{\partial \theta} (A_\theta \sin \theta) + \frac{1}{r \sin \theta} \frac{\partial A_\phi}{\partial \phi} \qquad (B.11)$$

To obtain the expression for the gradient of a scalar, we recall from Appendix A that in spherical coordinates,

$$d\mathbf{l} = dr \, \mathbf{a}_r + r \, d\theta \, \mathbf{a}_\theta + r \sin \theta \, d\phi \, \mathbf{a}_\phi \qquad (B.12)$$

and hence

$$d\Phi = \frac{\partial \Phi}{\partial r} dr + \frac{\partial \Phi}{\partial \theta} d\theta + \frac{\partial \Phi}{\partial \phi} d\phi$$

$$= \left(\frac{\partial \Phi}{\partial r} \mathbf{a}_r + \frac{1}{r} \frac{\partial \Phi}{\partial \theta} \mathbf{a}_\theta + \frac{1}{r \sin \theta} \frac{\partial \Phi}{\partial \phi} \mathbf{a}_\phi \right) \cdot (dr \, \mathbf{a}_r + r \, d\theta \, \mathbf{a}_\theta + r \sin \theta \, d\phi \, \mathbf{a}_\phi)$$

$$= \nabla \Phi \cdot d\mathbf{l} \qquad (B.13)$$

Thus,

$$\nabla \Phi = \frac{\partial \Phi}{\partial r} \mathbf{a}_r + \frac{1}{r} \frac{\partial \Phi}{\partial \theta} \mathbf{a}_\theta + \frac{1}{r \sin \theta} \frac{\partial \Phi}{\partial \phi} \mathbf{a}_\phi \qquad (B.14)$$

REVIEW QUESTIONS

B.1. Briefly discuss the basic definition of the curl of a vector.

B.2. Justify the application of the basic definition of the curl of a vector to determine separately the individual components of the curl.

B.3. How would you generalize the interpretations for the components of the curl of a vector in terms of the lateral derivatives involving the components of the vector to hold in cylindrical and spherical coordinate systems?

B.4. Briefly discuss the basic definition of the divergence of a vector.

B.5. How would you generalize the interpretation for the divergence of a vector in terms of the longitudinal derivatives involving the components of the vector to hold in cylindrical and spherical coordinate systems?

B.6. Provide general interpretation for the components of the gradient of a scalar.

PROBLEMS

B.1. Find the curl and the divergence for each of the following vectors in cylindrical coordinates: (a) $r \cos \phi \, \mathbf{a}_r - r \sin \phi \, \mathbf{a}_\phi$; (b) $\dfrac{1}{r} \mathbf{a}_r$; (c) $\dfrac{1}{r} \mathbf{a}_\phi$.

B.2. Find the gradient for each of the following scalar functions in cylindrical coordinates: (a) $\dfrac{1}{r} \cos \phi$; (b) $r \sin \phi$.

B.3. Find the expansion for the Laplacian, that is, the divergence of the gradient, of a scalar in cylindrical coordinates.

B.4. Find the curl and the divergence for each of the following vectors in spherical coordinates: (a) $r^2 \mathbf{a}_r + r \sin \theta \, \mathbf{a}_\theta$; (b) $\dfrac{e^{-r}}{r} \mathbf{a}_\theta$; (c) $\dfrac{1}{r^2} \mathbf{a}_r$.

B.5. Find the gradient for each of the following scalar functions in spherical coordinates: (a) $\dfrac{\sin \theta}{r}$; (b) $r \cos \theta$.

B.6. Find the expansion for the Laplacian, that is, the divergence of the gradient, of a scalar in spherical coordinates.

Units and Dimensions

In 1960 the International System of Units was given official status at the Eleventh General Conference on weights and measures held in Paris, France. This system of units is an expanded version of the rationalized meter-kilogram-second-ampere (MKSA) system of units and is based on six fundamental or basic units. The six basic units are the units of length, mass, time, current, temperature, and luminous intensity.

The international unit of length is the meter. It is exactly 1,650,763.73 times the wavelength in vacuum of the radiation corresponding to the unperturbed transition between the levels $2p_{10}$ and $5d_5$ of the atom of krypton-86, the orange-red line. The international unit of mass is the kilogram. It is the mass of the International Prototype Kilogram which is a particular cylinder of platinum-iridium alloy preserved in a vault at Sevres, France, by the International Bureau of Weights and Measures. The international unit of time is the second. It is equal to 9,192,631,770 times the period corresponding to the frequency of the transition between the hyperfine levels $F = 4, M = 0$ and $F = 3, M = 0$ of the fundamental state $^2S_{1/2}$ of the cesium–133 atom unperturbed by external fields.

To present the definition for the international unit of current, we first define the newton, which is the unit of force, derived from the fundamental units meter, kilogram, and second in the following manner. Since velocity is rate of change of distance with time, its unit is meter per second. Since acceleration is rate of change of velocity with time, its unit is meter per second per second or meter per second squared. Since force is mass times acceleration, its unit is kilogram-meter per second squared, also known as the newton. Thus, the newton is that force which imparts an acceleration of 1 meter per second squared to a mass of 1 kilogram. The international unit of current, which is the ampere, can now be defined. It is the constant current that when maintained in two straight, infinitely long, parallel conductors of negligible cross section and placed 1 meter apart in vacuum produces a force of 2×10^{-7} newtons per meter length of the conductors.

The international unit of temperature is the Kelvin degree. It is based on the definition of the thermodynamic scale of temperature by designating the triple-point of water as a fixed fundamental point to which a temperature of exactly 273.16 degrees Kelvin is attributed. The international unit of luminous intensity is the candela. It is defined such that the luminance of a blackbody radiator at the freezing temperature of platinum is 60 candelas per square centimeter.

We have just defined the six basic units of the International System of Units. Two supplementary units are the radian and the steradian for plane angle and solid angle, respectively. All other units are derived units. For example, the unit of charge, the coulomb, is the amount of charge transported in 1 second by a current of 1 ampere; the unit of energy, the joule, is the work done when the point of application of a force of 1 newton is displaced a distance of 1 meter in the direction of the force; the unit of power, the watt, is the power that gives rise to the production of energy at the rate of 1 joule per second; the unit of electric potential difference, the volt, is the difference of electric potential between two points of a conducting wire carrying constant current of 1 ampere when the power dissipated between these points is equal to 1 watt; and so on. The units for the various quantities used in this book are listed in Table C.1, together with the symbols of the quantities and their dimensions.

TABLE C.1 Symbols, Units, and Dimensions of Various Quantities

Quantity	Symbol	Unit (symbol)	Dimensions
Admittance	$\overline{Y}$	siemens (S)	$M^{-1}L^{-2}TQ^2$
Area	A	square meter (m²)	L^2
Attenuation constant	α	neper/meter (Np/m)	L^{-1}
Capacitance	C	farad (F)	$M^{-1}L^{-2}T^2Q^2$
Capacitance per unit length	$\mathscr{C}$	farad/meter (F/m)	$M^{-1}L^{-3}T^2Q^2$
Cartesian coordinates	$\begin{cases} x \\ y \\ z \end{cases}$	meter (m) meter (m) meter (m)	L L L
Characteristic admittance	Y_0	siemens (S)	$M^{-1}L^{-2}TQ^2$
Characteristic impedance	Z_0	ohm (Ω)	$ML^2T^{-1}Q^{-2}$
Charge	Q, q	coulomb (C)	Q
Conductance	G	siemens (S)	$M^{-1}L^{-2}TQ^2$
Conductance per unit length	$\mathscr{G}$	siemens/meter (S/m)	$M^{-1}L^{-3}TQ^2$
Conduction current density	$\mathbf{J}_c$	ampere/square meter (A/m²)	$L^{-2}T^{-1}Q$
Conductivity	σ	siemens/meter (S/m)	$M^{-1}L^{-3}TQ^2$
Current	I	ampere (A)	$T^{-1}Q$
Cutoff frequency	f_c	hertz (Hz)	T^{-1}
Cutoff wavelength	λ_c	meter (m)	L
Cylindrical coordinates	$\begin{cases} r, r_c \\ \phi \\ z \end{cases}$	meter (m) radian (rad) meter (m)	L — L
Differential length element	$d\mathbf{l}$	meter (m)	L
Differential surface element	$d\mathbf{S}$	square meter (m²)	L^2
Differential volume element	dv	cubic meter (m³)	L^3
Directivity	D	—	—
Displacement flux density	$\mathbf{D}$	coulomb/square meter (C/m²)	$L^{-2}Q$
Electric dipole moment	$\mathbf{p}$	coulomb-meter (C-m)	LQ
Electric field intensity	$\mathbf{E}$	volt/meter (V/m)	$MLT^{-2}Q^{-1}$
Electric potential	V	volt (V)	$ML^2T^{-2}Q^{-1}$
Electric susceptibility	χ_e	—	—
Electron density	N	(meter)⁻³ (m⁻³)	L^{-3}
Electronic charge	e	coulomb (C)	Q
Energy	W	joule (J)	ML^2T^{-2}

TABLE C.1   Continued

Quantity	Symbol	Unit (symbol)	Dimensions
Energy density	w	joule/cubic meter (J/m^3)	$ML^{-1}T^{-2}$
Force	**F**	newton (N)	MLT^{-2}
Frequency	f	hertz (Hz)	T^{-1}
Group velocity	v_g	meter/second (m/s)	LT^{-1}
Guide characteristic impedance	η_g	ohm (Ω)	$ML^2T^{-1}Q^{-2}$
Guide wavelength	λ_g	meter (m)	L
Impedance	$\bar{Z}$	ohm (Ω)	$ML^2T^{-1}Q^{-2}$
Inductance	L	henry (H)	ML^2Q^{-2}
Inductance per unit length	$\mathscr{L}$	henry/meter (H/m)	MLQ^{-2}
Intrinsic impedance	η	ohm (Ω)	$ML^2T^{-1}Q^{-2}$
Length	l	meter (m)	L
Line charge density	ρ_L	coulomb/meter (C/m)	$L^{-1}Q$
Magnetic dipole moment	**m**	ampere-square meter (A-m^2)	$L^2T^{-1}Q$
Magnetic field intensity	**H**	ampere/meter (A/m)	$L^{-1}T^{-1}Q$
Magnetic flux	ψ	weber (Wb)	$ML^2T^{-1}Q^{-1}$
Magnetic flux density	**B**	tesla or weber/square meter (T or Wb/m^2)	$MT^{-1}Q^{-1}$
Magnetic susceptibility	χ_m	—	—
Magnetic vector potential	**A**	weber/meter (Wb/m)	$MLT^{-1}Q^{-1}$
Magnetization current density	**J**$_m$	ampere/square meter (A/m^2)	$L^{-2}T^{-1}Q$
Magnetization vector	**M**	ampere/meter (A/m)	$L^{-1}T^{-1}Q$
Mass	m	kilogram (kg)	M
Mobility	μ	square meter/volt-second (m^2/V-s)	$M^{-1}TQ$
Permeability	μ	henry/meter (H/m)	MLQ^{-2}
Permeability of free space	μ_0	henry/meter (H/m)	MLQ^{-2}
Permittivity	ϵ	farad/meter (F/m)	$M^{-1}L^{-3}T^2Q^2$
Permittivity of free space	ϵ_0	farad/meter (F/m)	$M^{-1}L^{-3}T^2Q^2$
Phase constant	β	radian/meter (rad/m)	L^{-1}
Phase velocity	v_p	meter/second (m/s)	LT^{-1}
Plasma frequency	f_N	hertz (Hz)	T^{-1}
Polarization current density	**J**$_p$	ampere/square meter (A/m^2)	$L^{-2}T^{-1}Q$
Polarization vector	**P**	coulomb/square meter (C/m^2)	$L^{-2}Q$
Power	P	watt (W)	ML^2T^{-3}
Power density	p	watt/square meter (W/m^2)	MT^{-3}
Poynting vector	**P**	watt/square meter (W/m^2)	MT^{-3}
Propagation constant	$\bar{\gamma}$	(meter)$^{-1}$(m^{-1})	L^{-1}
Propagation vector	$\boldsymbol{\beta}$	radian/meter (rad/m)	L^{-1}
Radian frequency	ω	radian/second (rad/s)	T^{-1}
Radiation resistance	R_{rad}	ohm (Ω)	$ML^2T^{-1}Q^{-2}$
Reactance	X	ohm (Ω)	$ML^2T^{-1}Q^{-2}$
Reflection coefficient	Γ	—	—
Refractive index	n	—	—
Relative permeability	μ_r	—	—
Relative permittivity	ϵ_r	—	—
Resistance	R	ohm (Ω)	$ML^2T^{-1}Q^{-2}$
Shielding factor	S	—	—
Skin depth	δ	meter (m)	L
Spherical coordinates	$\begin{cases} r, r_s \\ \theta \\ \phi \end{cases}$	meter (m) radian (rad) radian (rad)	L — —

(*continued*)

TABLE C.1   Continued

Quantity	Symbol	Unit (symbol)	Dimensions
Standing wave ratio	SWR	—	—
Surface charge density	ρ_S	coulomb/square meter (C/m^2)	$L^{-2}Q$
Surface current density	$\mathbf{J}_S$	ampere/meter (A/m)	$L^{-1}T^{-1}Q$
Susceptance	B	siemens (S)	$M^{-1}L^{-2}TQ^2$
Time	t	second (s)	T
Transmission coefficient	τ	—	—
Unit normal vector	$\mathbf{a}_n$	—	—
Velocity	v	meter/second (m/s)	LT^{-1}
Velocity of light in free space	c	meter/second (m/s)	LT^{-1}
Voltage	V	volt (V)	$ML^2T^{-2}Q^{-1}$
Volume	V	cubic meter (m^3)	L^3
Volume charge density	ρ	coulomb/cubic meter (C/m^3)	$L^{-3}Q$
Volume current density	$\mathbf{J}$	ampere/square meter (A/m^2)	$L^{-2}T^{-1}Q$
Wavelength	λ	meter (m)	L
Work	W	joule (J)	ML^2T^{-2}

Dimensions are a convenient means of checking the possible validity of a derived equation. The dimension of a given quantity can be expressed as some combination of a set of fundamental dimensions. These fundamental dimensions are mass (M), length (L), and time (T). In electromagnetics, it is the usual practice to consider the charge (Q), instead of the current, as the additional fundamental dimension. For the quantities listed in Table C.1, these four dimensions are sufficient. Thus, for example, the dimension of velocity is length (L) divided by time (T), that is, LT^{-1}; the dimension of acceleration is length (L) divided by time squared (T^2), that is, LT^{-2}; the dimension of force is mass (M) times acceleration (LT^{-2}), that is, MLT^{-2}; the dimension of ampere is charge (Q) divided by time (T), that is, QT^{-1}; and so on.

To illustrate the application of dimensions for checking the possible validity of a derived equation, let us consider the equation for the phase velocity of an electromagnetic wave in free space, given by

$$v_p = \frac{1}{\sqrt{\mu_0 \epsilon_0}}$$

We know that the dimension of v_p is LT^{-1}. Hence, we have to show that the dimension of $1/\sqrt{\mu_0 \epsilon_0}$ is also LT^{-1}. To do this, we note from Coulomb's law that

$$\epsilon_0 = \frac{Q_1 Q_2}{4\pi F R^2}$$

Hence, the dimension of ϵ_0 is $Q^2/[(MLT^{-2})(L^2)]$, or $M^{-1}L^{-3}T^2Q^2$. We note from Ampere's law of force applied to two infinitesimal current elements parallel to each other and normal to the line joining them that

$$\mu_0 = \frac{4\pi F R^2}{(I_1 \, dl_1)(I_2 \, dl_2)}$$

Hence, the dimension of μ_0 is $[(MLT^{-2})(L^2)]/(QT^{-1}L)^2$, or MLQ^{-2}. Now we obtain the dimension of $1/\sqrt{\mu_0\epsilon_0}$ as $1/\sqrt{(M^{-1}L^{-3}T^2Q^2)(MLQ^{-2})}$, or LT^{-1}, which is the same as the dimension of v_p. It should, however, be noted that the test for the equality of the dimensions of the two sides of a derived equation is not a sufficient test to establish the equality of the two sides, since any dimensionless constants associated with the equation may be in error.

It is not always necessary to refer to the table of dimensions for checking the possible validity of a derived equation. For example, let us assume that we have derived the expression for the characteristic impedance of a transmission line, that is, $\sqrt{\mathscr{L}/\mathscr{C}}$, and we wish to verify that $\sqrt{\mathscr{L}/\mathscr{C}}$ does indeed have the dimension of impedance. To do this, we write

$$\sqrt{\frac{\mathscr{L}}{\mathscr{C}}} = \sqrt{\frac{\omega\mathscr{L}l}{\omega\mathscr{C}l}} = \sqrt{\frac{\omega L}{\omega C}} = \sqrt{(\omega L)\left(\frac{1}{\omega C}\right)}$$

We now recognize from our knowledge of circuit theory that both ωL and $1/\omega C$, being the reactances of L and C, respectively, have the dimension of impedance. Hence, we conclude that $\sqrt{\mathscr{L}/\mathscr{C}}$ has the dimension of $\sqrt{(\text{impedance})^2}$, or impedance.

Suggested Collateral and Further Reading

Adler, R. B., L. J. Chu, and R. M. Fano, *Electromagnetic Energy Transmission and Radiation*, John Wiley & Sons, Inc., New York, 1960.

Bansal, R. (Ed.), *Handbook of Engineering Electromagnetics*, Marcel Dekker, Inc., New York, 2004.

Davidson, C. W., *Transmission Lines for Communications*, John Wiley & Sons, Inc., New York, 1978.

Fano, R. M., L. J. Chu, and R. B. Adler, *Electromagnetic Fields, Energy, and Forces*, John Wiley & Sons, Inc., New York, 1960.

Hayt, W. H., Jr., and J. A. Buck, *Engineering Electromagnetics*, 6th ed., McGraw-Hill Book Company, Inc., New York, 2001.

Jordan, E. C., and K. G. Balmain, *Electromagnetic Waves and Radiating Systems*, 2nd ed., Prentice-Hall, Inc., Englewood Cliffs, N. J., 1968.

Kraus, J. D., and D. A. Fleisch, *Electromagnetics with Applications*, 5th ed., McGraw-Hill Book Company, Inc., New York, 1999.

Matick, R. E., *Transmission Lines for Digital and Communication Networks*, IEEE Press, New York, 1995.

Ramo, S., J. R. Whinnery, and T. Van Duzer, *Fields and Waves in Communication Electronics*, 3rd ed., John Wiley & Sons, Inc., New York, 1994.

Rao, N. N., *Basic Electromagnetics with Applications*, Prentice-Hall, Inc., Englewood Cliffs, N. J., 1972.

Rao, N. N., *Elements of Engineering Electromagnetics*, 6th ed., Pearson Prentice-Hall, Upper Saddle River, N. J., 2004.

Rosenstark, S., *Transmission Lines in Computer Engineering*, McGraw-Hill Book Company, Inc., New York, 1994.

Answers to Odd-Numbered Problems

CHAPTER 1

1.1. (a) 2 m; (b) 0.8 m northward and 0.4 m eastward; (c) 0.8944 m

1.5. 21

1.7. $2\mathbf{a}_x + 2\mathbf{a}_y + \mathbf{a}_z$

1.9. $(4\mathbf{a}_x - 5\mathbf{a}_y + 3\mathbf{a}_z)/5\sqrt{2};\ 6\sqrt{2}$

1.11. $(4\mathbf{a}_x + 4\mathbf{a}_y + \mathbf{a}_z)\,dz$

1.13. $(4\mathbf{a}_x - \mathbf{a}_y)/\sqrt{17}$

1.15. $x + y + z = \text{constant}$

1.17. $\omega(-y\mathbf{a}_x + x\mathbf{a}_y)$

1.19. Traveling wave progressing in the negative z-direction

1.21. (a) Linear; (b) circular; (c) elliptical

1.23. Elliptical polarization

1.25. $5\cos(\omega t + 6.87°)$

1.27. $\sqrt{8\pi\epsilon_0 l^2 mg}$

1.29. $\dfrac{0.0555Q}{\epsilon_0}(\mathbf{a}_x + \mathbf{a}_y + \mathbf{a}_z)$ N/C

1.31. $\dfrac{10^{-7}}{\pi\epsilon_0}\sum\limits_{i=1}^{50}(2i - 1)[10^{-4}(2i - 1)^2 + 1]^{-3/2}\mathbf{a}_y$

1.33. $\dfrac{4 \times 10^{-7}}{\pi\epsilon_0}\sum\limits_{i=1}^{50}\sum\limits_{j=1}^{50}[10^{-4}(2i - 1)^2 + 10^{-4}(2j - 1)^2 + 1]^{-3/2}\mathbf{a}_z$

1.35. (a) $0.4485 \times 10^{-6}\sin 2\pi \times 10^7 t\ \mathbf{a}_x$ A/m²
(b) $0.4485 \times 10^{-8}\sin 2\pi \times 10^7 t$ A

1.37. $d\mathbf{F}_1 = 0;\ d\mathbf{F}_2 = \dfrac{\mu_0}{4\pi}I_1 I_2\,dx\,dy\,\mathbf{a}_x$

1.39. (a) $(5 \times 10^{-5}\mu_0/\pi)\mathbf{a}_z$; (b) $-(10^{-4}\mu_0/4\pi)\mathbf{a}_z$

1.41. $0.179\mu_0 I\mathbf{a}_z$

1.43. $-v_0 B_0(14\mathbf{a}_y + 7\mathbf{a}_z)$

CHAPTER 2

2.1. 0.855

2.3. 1

2.7. 1/6

2.9. $\dfrac{(4n^2 - 1)(1 - e^{-1})}{12n^3(1 - e^{-1/n})} e^{-1/2n}$; 0.20825, 0.21009, 0.21070, 0.21071

2.11. 16π

2.13. 30 A

2.15. $-B_0 b v_0 \left(\dfrac{1}{x_0 + a} - \dfrac{1}{x_0} \right)$

2.17. $B_0 b \omega \ln \dfrac{x_0 + a}{x_0} \sin \omega t - B_0 b v_0 \left(\dfrac{1}{x_0 + a} - \dfrac{1}{x_0} \right) \cos \omega t$

2.19. $2B_0 \omega \sin \omega t$

2.21. 0

2.23. (a) 0; (b) $I_1 - I_2$

2.25. $\dfrac{J_0 r}{2}$ for $r < a$ and $\dfrac{J_0 a^2}{2r}$ for $r > a$, where r is the radial distance from the axis; direction circular to the axis of the wire

2.27. (a) $I/4$; (b) $I/4$

2.29. 0.31606 C

2.31. $\rho_0 r / 3\epsilon_0$ for $r < a$ and $\rho_0 a^3 / 3\epsilon_0 r^2$ for $r > a$, where r is the radial distance from the center of the charge, and direction radially away from the center of the charge

2.33. -1 A

2.35. $\pi^2/2$

CHAPTER 3

3.1. $\omega B_0 \dfrac{z^2}{2} \sin \omega t \, \mathbf{a}_x$

3.3. (a) $z\mathbf{a}_x + x\mathbf{a}_y + y\mathbf{a}_z$; (b) 0

3.5. $\dfrac{1}{3} \times 10^{-7} \cos (6\pi \times 10^8 t - 2\pi z) \mathbf{a}_y$ Wb/m^2

3.7. $\mathbf{B} = -\omega\mu_0\epsilon_0 E_0 \dfrac{z^3}{3} \cos \omega t \, \mathbf{a}_y$

$\mathbf{E} = -\omega^2\mu_0\epsilon_0 E_0 \dfrac{z^4}{12} \sin \omega t \, \mathbf{a}_x$

3.9. $\mathbf{E} = 10 \cos (6\pi \times 10^8 t - 2\pi z) \mathbf{a}_x$

$\mathbf{B} = \dfrac{10^{-7}}{3} \cos (6\pi \times 10^8 t - 2\pi z) \mathbf{a}_y$

3.11. $J_0(a + z)\mathbf{a}_y$ for $-a < z < 0$, $J_0(a - z)\mathbf{a}_y$ for $0 < z < a$, 0 otherwise

3.13. Curl will have a component in the y-direction in addition to the x-component

3.15. Curl has only a z-component

3.17. $\oint_C \mathbf{A} \cdot d\mathbf{l} = 0$ for any C

3.19. (a) $3(x^2 + y^2 + z^2)$; (b) 0

3.21. (a) $-x\mathbf{a}_z, y$; (b) $-\mathbf{a}_z, 0$; (c) $0, 1$; (d) $0, 0$

3.23. $\dfrac{\rho_0}{2a\epsilon_0}(x^2 - a^2)\mathbf{a}_x$ for $-a < x < a$, 0 otherwise

3.25. (a) and (c)

3.27. $\nabla \cdot \mathbf{r} = 3$

3.29. $\oint_S \mathbf{A} \cdot d\mathbf{S} = 2\pi, \nabla \cdot \mathbf{A} = 3$

3.31. 0

CHAPTER 4

4.1. (a) 0.2 A; (b) 0; (c) 0.2 A

4.3. (a) $0.2 \cos \omega t$ A; (b) $0.2 \sin \omega t$ A; (c) $0.2828 \sin(\omega t + 45°)$ A

4.5. (a) $\pm 0.0368 \cos \omega t \, \mathbf{a}_y$ for $z = 0\pm$; (b) $\pm 0.0135 \cos \omega t \, \mathbf{a}_y$ for $z = 0\pm$

4.7. $J_0 \dfrac{a}{2} \mathbf{a}_y$ for $z < -a$, $-J_0\left(z + \dfrac{z^2}{2a}\right)\mathbf{a}_y$ for $-a < z < 0$, $-J_0\left(z - \dfrac{z^2}{2a}\right)\mathbf{a}_y$ for

$0 < z < a$, $-J_0\dfrac{a}{2}\mathbf{a}_y$ for $z > a$

4.9. $-(\rho_0 a/\epsilon_0)\mathbf{a}_x$ for $x < -a$, $(\rho_0 x/\epsilon_0)\mathbf{a}_x$ for $-a < x < a$, $(\rho_0 a/\epsilon_0)\mathbf{a}_x$ for $x > a$

4.15. $(t - z\sqrt{\mu_0\epsilon_0})^2$ corresponds to a $(+)$ wave; $(t + z\sqrt{\mu_0\epsilon_0})^2$ corresponds to a $(-)$ wave

4.17. $C = \dfrac{\eta_0 J_{S0}}{2}$

For Problem 4.13, $E_x = \dfrac{\eta_0 J_{S0}}{2}(t \mp z\sqrt{\mu_0\epsilon_0})^2$ for $z \gtrless 0$ and

$H_y = \pm\dfrac{E_x}{\eta_0}$ for $z \gtrless 0$

4.19. $\mathbf{E} = [0.1\eta_0 \cos(6\pi \times 10^8 t \mp 2\pi z) + 0.05\eta_0 \cos(12\pi \times 10^8 t \mp 4\pi z)]\mathbf{a}_x$

for $z \gtrless 0$

$\mathbf{H} = \pm\dfrac{E_x}{\eta_0}\mathbf{a}_y$ for $z \gtrless 0$

4.21. Spacing $= \lambda/4$; amplitudes $= J_{S0}, \dfrac{1}{3}J_{S0}$; phase difference $= \pi/2$

4.23. (a) right circular; (b) left circular

4.25. $\dfrac{E_0}{2}[\cos(\omega t + \beta z)\mathbf{a}_x - \sin(\omega t + \beta z)\mathbf{a}_y]$

$+ \dfrac{E_0}{2}[\cos(\omega t + \beta z)\mathbf{a}_x + \sin(\omega t + \beta z)\mathbf{a}_y]$

4.27. $1.25 E_0\left[\cos\left(2\pi \times 10^8 t - \dfrac{2\pi}{3}z + 0.2048\pi\right)\mathbf{a}_x\right.$

$\left. + \sin\left(2\pi \times 10^8 t - \dfrac{2\pi}{3}z + 0.2048\pi\right)\mathbf{a}_y\right]$

4.29. (a) Same as in Figure 4.17, except displaced to the left by $1/3$ μs;

(b) 75.4 V/m for $300(n - 1/3) < |z| < 300n$ and -37.7 V/m for $300(n - 1) < |z| < 300(n - 1/3)$, $n = 1, 2, 3, \ldots$;

(c) $0.2z/|z|$ A/m for $300(n - 1) < |z| < 300(n - 2/3)$ and $-0.1z/|z|$ A/m for $300(n - 2/3) < |z| < 300n$, $n = 1, 2, 3, \ldots$

4.31. 30 mV/m

CHAPTER 5

5.1. (a) 0.1724×10^{-4} V/m, 0.1724×10^{-6} V, 0.1724×10^{-5} Ω;

(b) 0.2857×10^{-4} V/m, 0.2857×10^{-6} V, 0.2857×10^{-5} Ω;

(c) 250 V/m, 2.5 V, 25 Ω

5.3. 1.5245×10^{-19} s

5.5. (a) $-8.667 \times 10^{-7} \sin 2\pi \times 10^9 t$ A; (b) $-2.778 \times 10^{-6} \sin 2\pi \times 10^9 t$ A; (c) $-4.444 \times 10^{-5} \sin 2\pi \times 10^9 t$ A

5.7. (a) $\epsilon_0 E_0(4\mathbf{a}_x + 2\mathbf{a}_y + 2\mathbf{a}_z)$; (b) $8\epsilon_0 E_0(\mathbf{a}_x + \mathbf{a}_y + \mathbf{a}_z)$; (c) $0.5E_0(3\mathbf{a}_x - \mathbf{a}_y - \mathbf{a}_z)$

5.9. $|e|^2 B_0 a^2 / 2m$, 0.7035×10^{-18} A-m^2

5.11. $0.5 \times 10^{-6} \sin 2\pi z$ A

5.13. $\dfrac{\partial^2 \bar{H}_y}{\partial z^2} = \bar{\gamma}^2 \bar{H}_y$

5.15. 0.00083 Np/m, 4.7562×10^{-3} rad/m, 1.32105×10^8 m/s, 1321.05 m, $(161.102 + j28.115)$ Ω

5.17. $\mathbf{E} = 3.736 e^{\mp 0.0404z} \cos\left(2\pi \times 10^6 t \mp 0.0976z + \dfrac{\pi}{8}\right) \mathbf{a}_x$ for $z \gtrless 0$

$\mathbf{H} = \pm 0.05 e^{\mp 0.0404z} \cos(2\pi \times 10^6 t \mp 0.0976z) \mathbf{a}_y$ for $z \gtrless 0$

5.19. 16.09 m, 1.917:1, 90° out of phase

5.21. (a) 30 MHz; (b) 5 m; (c) 1.5×10^8 m/s; (d) $4\epsilon_0$;

(e) $\dfrac{1}{6\pi} \cos(6\pi \times 10^7 t - 0.4\pi z)\, \mathbf{a}_y$ A/m

5.23. (a) Same as in Figure 4.17;

(b) 75.4 V/m for $100(n - 1/3) < |z| < 100n$ and -37.7 V/m for $100(n - 1) < |z| < 100(n - 1/3)$, $n = 1, 2, 3, \ldots$;

(c) $0.6z/|z|$ A/m for $100(n - 1) < |z| < 100(n - 2/3)$ and $-0.3z/|z|$ A/m for $100(n - 2/3) < |z| < 100n$, for $n = 1, 2, 3, \ldots$

5.25. (a) 0.0211 Np/m, 18.73 rad/m, 0.3354×10^8 m/s, 0.3354 m, 42.15 Ω;

(b) $2\pi \times 10^{-3}$ Np/m, $2\pi \times 10^{-3}$ rad/m, 10^7 m/s, 1000 m, $2\pi(1 + j)$ Ω

5.27. $E_0(4\mathbf{a}_x + 2\mathbf{a}_y - 6\mathbf{a}_z)$, $H_0(4\mathbf{a}_x - 3\mathbf{a}_y)$

5.29. $2\epsilon_0$

5.31. Yes

5.33. $4|D_0|$

5.35. (b) $\dfrac{E_0}{120\pi} \cos 10\pi x \sin 3\pi \times 10^9 t \, \mathbf{a}_y$;

(c) $\dfrac{E_0}{120\pi} \sin 3\pi \times 10^9 t \, \mathbf{a}_z$ on both sheets

5.37. (a) 4; (b) 16; (c) 4/9

5.39. $\mathbf{E}_r = -E_0 \cos(\omega t + \beta z) \mathbf{a}_x, \ \mathbf{H}_r = \dfrac{E_0}{\eta} \cos(\omega t + \beta z) \mathbf{a}_y$

$\mathbf{E} = 2E_0 \sin \omega t \sin \beta z \, \mathbf{a}_x, \ \mathbf{H} = \dfrac{2E_0}{\eta} \cos \omega t \cos \beta z \, \mathbf{a}_y$

$[\mathbf{J}_S]_{z=0} = \dfrac{2E_0}{\eta} \cos \omega t \, \mathbf{a}_x$

CHAPTER 6

6.1. (a) $\dfrac{x\mathbf{a}_x + y\mathbf{a}_y + z\mathbf{a}_z}{\sqrt{x^2 + y^2 + z^2}}$; (b) $yz\mathbf{a}_x + zx\mathbf{a}_y + xy\mathbf{a}_z$

6.3. $\dfrac{1}{3\sqrt{5}}(5\mathbf{a}_x + 2\mathbf{a}_y + 4\mathbf{a}_z)$

6.5. 2.121

6.7. $Q/30\pi\epsilon_0$

6.9. $V = \dfrac{10^{-5}}{\pi\epsilon_0} \sum\limits_{i=1}^{50} [10^{-4}(2i - 1)^2 + y^2]^{-1/2}$

$\mathbf{E} = \dfrac{10^{-5}}{\pi\epsilon_0} \sum\limits_{i=1}^{50} [10^{-4}(2i - 1)^2 + 1]^{-3/2} \mathbf{a}_y$

6.11. (a) $-\dfrac{4\epsilon_0 V_0}{9d^2}\left(\dfrac{x}{d}\right)^{-2/3}$; (b) $[\rho_s]_{x=0} = 0, \ [\rho_s]_{x=d} = \dfrac{4\epsilon_0 V_0}{3d}$

6.13. $V = -\dfrac{kx^3}{6\epsilon} + \dfrac{kd^2 x}{8\epsilon}$ for $-\dfrac{d}{2} < x < \dfrac{d}{2}$

6.15. (a) $\dfrac{\epsilon_1(d - t) + \epsilon_2(t - x)}{\epsilon_1(d - t) + \epsilon_2 t} V_0$ for $0 < x < t$,

$\dfrac{\epsilon_1(d - x)}{\epsilon_1(d - t) + \epsilon_2 t} V_0$ for $t < x < d$

(b) $\dfrac{\epsilon_1(d - t)}{\epsilon_1(d - t) + \epsilon_2 t} V_0$; (c) $\dfrac{\epsilon_1\epsilon_2 wl}{\epsilon_1(d - t) + \epsilon_2 t}$

6.17. $\dfrac{\mu_1\mu_2}{\mu_1 + \mu_2}\left(\dfrac{2dl}{w}\right)$

6.19. $\bar{Y}_{in} = j\omega \dfrac{\epsilon wl}{d}\left(1 + \dfrac{\omega^2\mu\epsilon l^2}{3}\right)$; equivalent circuit is a series combination of

$C = \dfrac{\epsilon wl}{d}$ and $\dfrac{1}{3} L$, where $L = \dfrac{\mu dl}{w}$

6.21. (b) $\bar{Z}_{in} = \dfrac{j\omega\mu dl}{w}\left(1 - \dfrac{j\omega\mu\sigma l^2}{3}\right)$; equivalent circuit consists of an inductor L in parallel with a resistor $3R$ where $L = \mu dl/w$ and $R = d/\sigma lw$

6.23. (a) $2\pi \cos(2\pi \times 10^6 t - 0.02\pi z)$ V; (b) $0.25 \cos(2\pi \times 10^6 t - 0.02\pi z)$ A; (c) $0.5\pi \cos^2(2\pi \times 10^6 t - 0.02\pi z)$ W

6.29. $\mathcal{L} = 0.429\mu$, $\mathcal{C} = 2.333\epsilon$, $\mathcal{G} = 2.333\sigma$; exact values are 0.4192μ, 2.3855ϵ, and 2.3855σ, respectively

6.31. $\dfrac{1}{9}\mu$, 9ϵ, 9σ

CHAPTER 7

7.1. (a) $\mathcal{L} = 0.278 \times 10^{-6}$ H/m, $\mathcal{G} = 4.524 \times 10^{-16}$ S/m; (b) $(52.73 + j0)\ \Omega$

7.3. $\bar{V}(z) = 2\bar{A}\cos\beta z$, $\bar{I}(z) = -j\dfrac{2\bar{A}}{Z_0}\sin\beta z$; $\bar{Z}_{in} = -jZ_0\cot\beta l$

7.5. (a) 0.60286 m; (b) 1.35286 m

7.7. (a) $50\ \Omega$; (b) 8.09 V, 0.1176 A

7.9. (a) 10^8 m/s; (b) 10^6 m, 2×10^{-6}, $j0.00079\ \Omega$; 1m, 2, 0; 8m, 0.25, ∞

7.11. (a) $\dfrac{1}{16}P_i$; (b) $\dfrac{9}{16}P_i$; (c) $\dfrac{3}{8}P_i$

7.13. $150\ \Omega$

7.17. -0.00533 S, 1.667

7.19. 0.14λ, 0.192λ

7.21. (a) 40 V for $-40 < z < 60$ m, 0 otherwise;

(b) $\dfrac{2}{3}$ A for $0 < z < 120$ m, $-\dfrac{1}{3}$ A for -80 m $< z < 0$, 0 otherwise;

(c) 0 for $0 < t < 0.1\ \mu$s, 40 V for $t > 0.1\ \mu$s;

(d) 0 for $0 < t < 0.2\ \mu$s, $-\dfrac{1}{3}$ A for $t > 0.2\ \mu$s

7.23. (a) 60 V, 1 A; (b) 67.5 V, 0.9 A; (c) -7.5 V, 0.1 A

7.25.

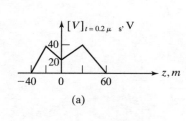

(a)

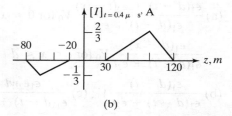

(b)

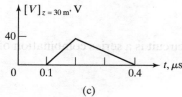

(c)

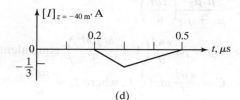

(d)

7.27.

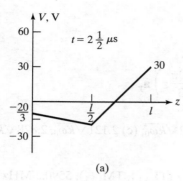

(a)

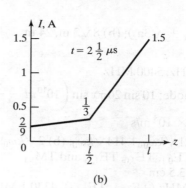

(b)

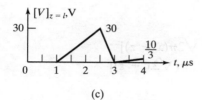

(c)

7.29. 1.46 V

7.31. (a) 40 V for $-l < z < l$; $-\dfrac{2}{3}$ A for $-l < z < 0$, $\dfrac{1}{3}$ A for $0 < z < l$;

(b) $\dfrac{130}{3} \times 10^{-6}$ J;

(c) Voltage across 60 Ω = 40 V for $0 < t < 1$ μs, 10 V for $1 < t < 3$ μs, 0 otherwise; voltage across 120 Ω = 40 V for $0 < t < 1$ μs, 0 otherwise;

(d) $\dfrac{130}{3} \times 10^{-6}$ J

7.33.

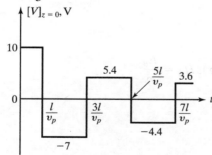

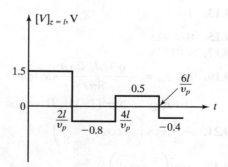

7.35. (a) 38.4 Ω; (b) 48.4 Ω

CHAPTER 8

8.1. $0.05\pi(\sqrt{3}\mathbf{a}_x + \mathbf{a}_y)$

8.3. $\dfrac{1}{2\sqrt{2}}\mathbf{a}_x + \dfrac{\sqrt{3}}{2}\mathbf{a}_y + \dfrac{1}{2\sqrt{2}}\mathbf{a}_z$

8.5. (a) Yes; (b) $\dfrac{1}{24\pi}(\sqrt{3}\mathbf{a}_y - \mathbf{a}_z)\cos[6\pi \times 10^7 t - 0.1\pi(y + \sqrt{3}z)]$

8.7. (a) $\frac{1}{2}(\mathbf{a}_x + \sqrt{3}\mathbf{a}_z)$; (b) $8\sqrt{3}$ m, 24 m

8.9. 1 cm

8.11. 3600 MHz, 5400 MHz

8.13. $TE_{1,0}$ mode; $10 \sin 20\pi x \sin\left(10^{10}\pi t - \frac{80\pi}{3}z\right)\mathbf{a}_y$

8.15. 0.5769×10^8 m/s

8.17. (a) $2.121\sqrt{k\omega_0}$, $1.414\sqrt{k\omega_0}$; (b) $2\sqrt{k\omega_0}$, $2\sqrt{k\omega_0}$; (c) $2.121\sqrt{k\omega_0}$, $2.828\sqrt{k\omega_0}$

8.19. $TE_{1,0}$, $TE_{0,1}$, $TE_{2,0}$, $TE_{1,1}$, and $TM_{1,1}$

8.21. 6.5 cm, 3.5 cm

8.23. 3535.5 MHz ($TE_{1,0,1}$, $TE_{0,1,1}$), 4330.1 MHz ($TE_{1,1,1}$, $TM_{1,1,1}$), 5590.2 MHz

($TE_{2,0,1}$, $TE_{0,2,1}$, $TE_{1,0,2}$, $TE_{0,1,2}$)

8.25. (a) $41.81°$; (b) $48.6°$

8.27. $\mathbf{E}_r = -0.1716E_0\mathbf{a}_y \cos[6\pi \times 10^8 t + \sqrt{2}\pi(x - z)]$

$\mathbf{E}_t = 0.8284E_0\mathbf{a}_y \cos[6\pi \times 10^8 t - \sqrt{2}\pi(\sqrt{2}x + z)]$

8.29. $75.52°$

8.31. $\sqrt{3}$

CHAPTER 9

9.1. $0.2\pi \cos 2\pi \times 10^7 t$ A

9.5. 0.2λ

9.7. (a) 1.257×10^{-3} V/m; (b) $R_{rad} = 0.0351\ \Omega$, $\langle P_{rad} \rangle = 1.7546$ W

9.9. 1.111 W

9.11. $\sqrt{(D_2 R_{rad2})/(D_1 R_{rad1})}$

9.13. $1\frac{7}{8}$

9.15. 0.60943

9.17. 1.015 W

9.19. (a) $E_\theta = -\dfrac{\eta\beta L I_0 \sin\theta}{8\pi r}\sin(\omega t - \beta r)$, $H_\phi = \dfrac{E_\theta}{\eta}$;

(b) $R_{rad} = 20\pi^2(L/\lambda)^2$, $D = 1.5$

9.21. $-\dfrac{\pi}{4}$, $\cos\left(\dfrac{\pi}{4}\cos\psi - \dfrac{\pi}{8}\right)$

9.23. $\cos^2\left(\dfrac{\pi}{2}\cos\psi\right)$

9.25. $\left|\cos\psi \cos\left(\dfrac{\pi}{4}\cos\psi - \dfrac{\pi}{4}\right)\right|$

9.27. $\left[\cos\left(\dfrac{\pi}{2}\cos\theta\right)\right]\bigg/\sin\theta$, where θ is the angle from the vertical, $D = 3.284$

9.29. 4

9.31. 0.00587 V

9.33. 13.262 A

CHAPTER 10

10.3. (a) 5×10^8 m/s; (b) 50 m; (c) 1000

10.5. 101.31°

10.7. 143.24

10.9. 121.71 db

10.11. (a) $10lK_fV_0/T$ for $T < t < 1.1T, -10lK_fV_0/T$ for $1.4T < t < 1.5T$, 0 otherwise

(b)

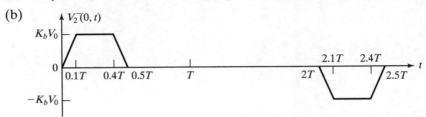

(c)

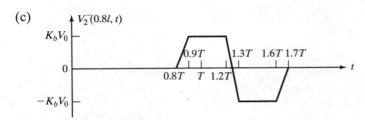

10.13. (a) $-0.2K_bV_0(z/l)$ for $0.9l < z < l$;

(b) K_bV_0 for $0 < z < 0.9l$, $10K_bV_0(1 - z/l)$ for $0.9l < z < l$;

(c) K_bV_0 for $0 < z < 0.9l$, $K_bV_0(10 - 10.2z/l)$ for $0.9l < z < l$

10.15. $\Gamma = -0.3252, \tau = 0.6748$

10.19. (a) 0.1654×10^{-3} V/m; (b) $R_{\text{rad}} = 0.6077 \times 10^{-3}\ \Omega, \langle P_{\text{rad}} \rangle = 0.0304$ W

APPENDIX A

A.1. $-3\mathbf{a}_x + \sqrt{3}\mathbf{a}_y + \mathbf{a}_z$

A.3. Equal

A.5. $-\dfrac{1}{\sqrt{2}}, \dfrac{1}{\sqrt{2}}\mathbf{a}_z$

A.7. $13\mathbf{a}_r + 6\mathbf{a}_z$

APPENDIX B

B.1. (a) $-\sin\phi\ \mathbf{a}_z, \cos\phi$; (b) 0, 0 except at $r = 0$; (c) 0 except at $r = 0, 0$

B.3. $\dfrac{1}{r}\dfrac{\partial}{\partial r}\left(r\dfrac{\partial \Phi}{\partial r}\right) + \dfrac{1}{r^2}\dfrac{\partial^2 \Phi}{\partial \phi^2} + \dfrac{\partial^2 \Phi}{\partial z^2}$

B.5. (a) $-\dfrac{1}{r^2}(\sin\theta\ \mathbf{a}_r - \cos\theta\ \mathbf{a}_\theta)$; (b) $\cos\theta\ \mathbf{a}_r - \sin\theta\ \mathbf{a}_\theta$

Index

443